8th Brazilian Congress on Metrology (Metrologia 2015)

Journal of Physics: Conference Series Volume 733

Bento Goncalves, Brazil
29 November – 2 December 2015

Editor:

Rodrigo P B Costa-Felix

ISBN: 978-1-5108-2817-9
ISSN: 1742-6588

Printed from e-media with permission by:

Curran Associates, Inc.
57 Morehouse Lane
Red Hook, NY 12571

CURRAN ASSOCIATES INC. proceedings.com

Some format issues inherent in the e-media version may also appear in this print version.

Printed by Curran Associates, Inc. (2016)

For permission requests, please contact the Institute of Physics
at the address below.

Institute of Physics
Dirac House, Temple Back
Bristol BS1 6BE UK

Phone: 44 1 17 929 7481
Fax: 44 1 17 920 0979

techtracking@iop.org

Additional copies of this publication are available from:

Curran Associates, Inc.
57 Morehouse Lane
Red Hook, NY 12571 USA
Phone: 845-758-0400
Fax: 845-758-2633
Email: curran@proceedings.com
Web: www.proceedings.com

TABLE OF CONTENTS

THE EIGHTH BRAZILIAN CONGRESS ON METROLOGY (METROLOGIA 2015)

The United Nations celebrated 2015 as the International Year of Light. By a curious coincidence, many notable events in science and technology completed a multiple of 50 or 100 years in 2015. From the pioneering work of the wise Ibn Al-Haytham in 1015, through Fresnel, Maxwell, Einstein, the discovery of the cosmic microwave background, to the use of optical fibres in communications in 1965.

Electromagnetic radiation is present in our daily lives in countless applications. It is remarkable that there is no way to think about these applications without thinking of measurements. From entangled photons to more prosaic public illumination of our daily life, we are intrinsically connected all the time with the luminous phenomena. Among other things, the light allows global communication on a large scale. It strengthens the internationalization of production processes, which brings considerable changes in relations, processes and economic structures, as well as it orients the social, political and cultural behaviour of any country. These conditions of this internationalization require interchangeability of parts of complex systems, translated into strict adherence to the standards and specifications that use increasingly accurate measurement techniques, as well as the growing demand from consumer markets for products and higher quality services. They also require innovation and improvements in domestic production to boost the competitiveness of industries in domestic and foreign markets. Thus, if the Science of Measurements is taken as a serious concern, countries are better prepared to evolve towards economic and social development.

In this 8th edition of the Brazilian Congress on Metrology (METROLOGIA 2015), in addition to the thematic sessions in various areas of Metrology and Conformity Assessment, we hold several satellite events. They are already traditional events or highlight important current issues such as those relating to the International Year of Light. The events are meetings on Electricity (XI International Seminar on Electrical Metrology – SEMETRO), Ionizing Radiation (II Brazilian Congress on Ionizing Radiation – CBMRI) and the Meeting of the Brazilian Chemical Metrology Network (REMEQ).

As in previous events, we hosted the Expo Metrology 2015 trade show and an exhibition of measuring equipment. During the conference, short courses in various areas of interest in Metrology and Conformity Assessment were given, organized by the Basic Industrial Technology National School – ENTIB, an initiative from the Brazilian Society of Metrology (SBM).

The 8th Brazilian Congress on Metrology consisted of Keynote Speeches (6) and regular papers (292). Among the regular papers, the 100 most outstanding ones, comprising a high quality content on Metrology, were selected to be published in this issue of the *Journal of Physics: Conference Series*.

With tremendous satisfaction and gratefulness, the organizers of METROLOGIA 2015 would like to present this volume of IOP *Journal of Physics: Conference Series* (JPCS) to the scientific community to promote further research in Metrology. We believe that this volume will be both an excellent source of scientific material in the fast evolving fields that were enclosed by METROLOGIA 2015.

PRESIDENT OF THE CONGRESS
Americo Tristao Bernardes
Federal University of Ouro Preto (UFOP)
atb@iceb.ufop.br

VICE-PRESIDENT OF THE CONGRESS
Adauto de Oliveira
Brazilian National Institute of Metrology, Quality and Technology
aoneto@inmetro.gov.br

EDITOR-IN-CHIEF
Rodrigo P B Costa-Félix
Brazilian National Institute of Metrology, Quality and Technology
rpfelix@inmetro.gov.br

CHAIR OF THE BRAZILIAN CONGRESS ON OPTICS METROLOGY (CBMO)
Iakyra Borrakuens Couceiro
Brazilian National Institute of Metrology, Quality and Technology
ibcouceiro@inmetro.gov.br

CHAIR OF THE XI INTERNATIONAL SEMINAR ON ELECTRICAL METROLOGY (SEMETRO)
Gelson Martins da Rocha
Brazilian National Institute of Metrology, Quality and Technology
gmrocha@inmetro.gov.br

CHAIR OF THE II BRAZILIAN CONGRESS ON IONIZING RADIATION (CBMRI)
José Guilherme Pereira Peixoto
Brazilian Institute of Radiation Protection and Dosimetry
guilherm@ird.gov.br

CHAIR OF THE MEETING OF THE BRAZILIAN CHEMICAL METROLOGY NETWORK (REMEQ-I)
Vanderléa de Souza
Brazilian National Institute of Metrology, Quality and Technology
vsouza@inmetro.gov.br

Technical and Scientific Committee for METROLOGIA 2015

Alexandre Mendes
Álvaro José Damião
Ana Maria Ribeiro Franco
Ana Paula Dornelles de Alvarenga

Antonio Serra
Bernardo Maranhão Dantas
Camila Salata
Carla Thereza Coelho
Carlos Alberto Achete
Carlos Alberto Aragão de Carvalho Filho
Carlos Eduardo de Almeida
Carlos Henrique Simões de Souza
Carlos José da Silva
Claudia Naves David Amorim
Claudio Lenz Cesar
Daniel da Silva Quaresma
Daniel Slomovitz
Domingos de Oliveira
Etiene Benini Mendes
Evaldo Simões da Fonseca
Felipe Albano
Flavio Alexandre da Silveira
Flávio Henrique Vasconcelos
Flávio José da Silva Dantas
Giovane de Jesus Teixeira
Gregory Amaral Kyriazis
Gunter Drexler
Hans Peter Henrik Grieneisen
Hector Laiz
Humberto Siqueira Brandi
Jailton Carreteiro Damasceno
José Guilherme Pereira Peixoto
Jose Mauro Granjeiro
José Ubiratan Delgado
Karla Cristina Souza Patrão
Kelly da Silva Barroco Barbosa
Leonardo Augusto Abreu de Souza
Lucas Di Lillo
Luis Alexandre G. Magalhães
Luiz Fernando Rust da Costa Carmo
Maria da Penha Potiens
Maria Helena Farias
Miguel Neira
Naira Maria Balzaretti
Noara Foiatto
Omer Pohlmann Filho
Patrícia Cals de Oliveira Vitorio
Pedro M. Ramos

Pedro Pacheco Queiroz Filho
Raphael Machado
Regis Pinheiro Landim
Renata Teixeira De Barros e Vasconcellos
Renato Di Prinzio
Ricardo Amorim Barbosa
Ricardo Iuzzolino
Ricardo José de Carvalho
Rodolfo Sabóia Lima de Souza
Rodrigo Pereira Barretto da Costa Félix
Romeu José Daroda
Rosane Moreira Debatin
Sebastian Yuri
Selma Junqueira
Silvio Francisco dos Santos
Teógenes Augusto da Silva
Thais Medeiros Rodrigues
Thiago Ferreira da Silva
Thiago Menogotto
Valnei Smarcaro da Cunha
Vanderlea de Souza
Vanderléa de Souza
Vladimir Rodrigues de Lima
Waldemar Guilherme Kurten Ihlenfeld
Walsan Wagner Pereira

Peer review statement

All papers published in this volume of *Journal of Physics: Conference Series* have been peer reviewed through processes administered by the proceedings Editors. Reviews were conducted by expert referees to the professional and scientific standards expected of a proceedings journal published by IOP Publishing.

Effects of screens set characteristics on the flow field in a wind tunnel

A M Santos, D B Souza, F O Costa, M H Farias, P de L Massari, S Araújo and Y B Zanirath

Fluid Dynamics Metrology Division, Scientific and Industrial Metrology, National Institute of Metrology, Quality and Technology (Inmetro), Rio de Janeiro, Brazil

E-mail: mhfarias@inmetro.gov.br

Abstract. Wind tunnels have broad range of applications, and although there are common elements among the different types of tunnels, the layout and configuration of each facility will depend on its particular purpose. The flow conditioners section is a common component for all tunnels, and frequently contains flow straighteners and screens [1]. The role of screens is to minimize non uniformities or turbulence level on the flow field. In this work, the development and characteristics of the vertical velocity profile along an atmospheric wind tunnel length were evaluated. By using such data as initial reference, the effects, on the flow, when screens set is inserted on it were analyzed, in order to identify which changes could be done in the wind tunnel configuration to improve the characteristics of the flow field. Such atmospheric wind tunnel is a facility which belongs to the Brazilian National Metrology Institute INMETRO.

1. Introduction

A wind tunnel is a facility designed to produce a regular and controlled stream of air in order to determine experimentally the flow conditions and to measure stress distribution in model testing. Atmospheric wind tunnels are used as important tool to evaluate various studies in the engineering field, such as small-scale testing in aircraft and cars models, study of boundary layers, pollutant dispersion investigations, simulation of atmospheric layers and other applications [2,3].

In order to perform the experiments aforementioned, the flow quality in the wind tunnel is necessary, as for example, a uniform and steady velocity profile at the wind tunnel entrance, the turbulence intensity reduction in test section etc. The last one is made by using screens and honeycombs. The screens reduces axial turbulence more than lateral turbulence, while honeycombs reduce lateral turbulence more than axial turbulence.

Wind tunnels screens are composed by thin wires interwoven to form square or rectangular meshes. Mehta [4] studied the effect of screen material in the flow quality and discovered significant differences between plastic and metal screens due to the fact that the weaving properties and elastic modulus of the fibers are distinct. Karnik and Tavoularis [5] discussed the influence of mesh size and spacing screens on the reduction of flow non-uniformities, considering a same wire solidity and shape, for a uniformly sheared flow.

In metrology area, several studies are concerned to characterization of the flow for velocity measurement instruments calibration and fluid dynamics investigations [6-8]. The National Institute of Metrology, Quality and Technology (Inmetro) from Brazil designed and constructed an atmospheric wind tunnel for research and technological applications. Based on previous results from this wind tunnel characterization, the need of flow quality improvement inside the tunnel was identified. Thus, the main goal of the present work is to investigate the effect in non-uniformity of the flow field by using screens in the flow. The measurement techniques employed to measure flow parameters such as turbulent intensity and average velocity were hot-wire anemometry and ultrasonic anemometry.

2. Experimental apparatus and procedures

2.1. Wind tunnel description

The atmospheric wind tunnel from Inmetro (Figure 1) is a low speed and high turbulence intensity level facility. It is an open-circuit tunnel with $10\,m$ long, divided in 5 modules of $2\,m$ lenght each one, $(1.0 \times 1.0\,m^2)$ closed test section.

The flow speed is controlled by a frequency inverter connected to a fan $(9\,kW)$ and the maximum mean velocity is about $10\,m/s$. Air cross a set of screens, a diffuser (ratio 1:2) and a honeycomb before entering into the settling section.

Figure 1. The atmospheric wind tunnel of Inmetro.

2.2. Hot-wire anemometry

The hot-wire anemometry is an intrusive technique used to measure flow parameters such as turbulent intensity and averaged velocity by sensing the changes in heat transfer from a small, electrically heated sensor exposed to the fluid motion [9]. There are two categories for hot-wire anemometers, i.e., single and x-probe. Single wire probes, where the wire is oriented normal to the mean flow direction, are used to measure the velocity component in the mean flow direction. The crossed wire probes or x-probes, consisting of two wires arranged in an x-configuration, are employed to measure both stream-wise and cross-flow velocity components simultaneously [10].

In this wind tunnel characterization, the hot-wire anemometer used was a single wire probe (55P11) with the software Streamware^{TM}, both from Dantec Dynamics. A Constant Temperature Anemometry (CTA) bridge was used to acquire the data, which were processed by Excel^{TM}. The Dantec Dynamics Hot Wire Calibrator was used before runing measurements in each position in the tunnel. The number of samples for each point measured was 65536, under the sampling rate of $10\,kHz$.

2.3. Ultrassonic anemometry

The anemometer used to measure the mean velocity was a 3D ultrasonic anemometer (Kaijo Sonic, model DA-600), with velocity range from $0\,m/s$ to $20\,m/s$ and resolution of $0.005\,m/s$, calibrated and traceable to the National Metrology Institute of Japan (NMIJ).

2.4. Experimental Procedure

In the previous work [6], the wind tunnel characterization consisted of velocity profiles determination in 5 test sections (see Figure 2) along the wind tunnel length, by using Pitot Tube. Six different levels of mean velocities for the wind were tested and all the measurements were made on vertical lines which crossed the symmetry axis of the tunnel. After measuring such velocity profiles it was observed high non-uniformity of it. Then, seeking out the uniformity of the flow field, changes of the ceiling inclination were made. The ceiling was tilted from the first module entrance up to the end of the third module. This resulted in 1.9 degrees of inclination along 6 meters of the tunnel length.

Based on the aforementioned results, in the present work hot wire anemometer technique was employed to measure instantaneous velocities. So, the turbulence intensity was also calculated in a region where the velocity profile was uniform (or almost).

Firstly, the measurements were done without the screens set in the end of the section III (Figure 2a) and in the end of the section V (Figure 2b), both in the symmetry axis. Seven vertical positions were analyzed and, for each position, six different levels of mean velocities for the wind were settled, as done in the tests aforementioned [6].

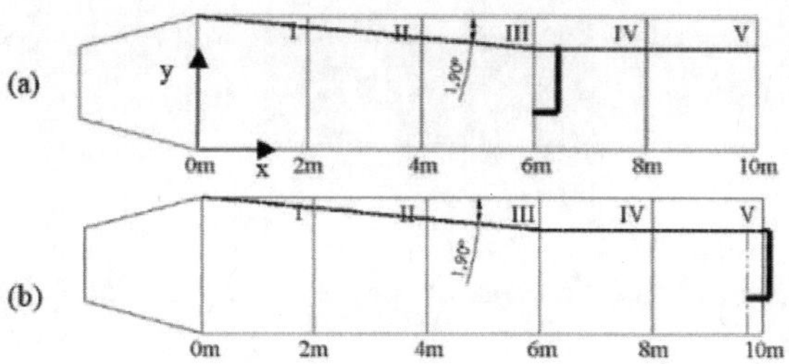

Figure 2. Position of the measuring instrument in wind tunnel (a) in the end of the section III, (b) in the end of the section V.

The second step included the measurements in the middle of the section V (Figure 3), in the symmetry axis of the cross section, by using hot-wire anemometer with and without screens set. Due to the screens size, five vertical positions were evaluated in both experimental conditions aforesaid. The screens set is composed by two stainless steel screens as described in Table 1.

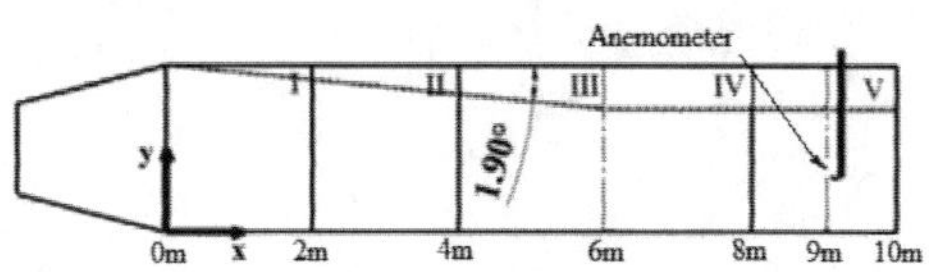

Figure 3. Position of the measuring instrument in wind tunnel with and without screens set.

The screens set were mounted with an interspace equal to 100 mm. The space between the anemometer and the screens set was also 100 mm, related to ten times the mesh size, related to screen II. The position of the anemometer was the same presented in the Figure 3. The Figure 4 shows the screens assembly inside the wind tunnel.

The last step consisted of measurements in the wind tunnel with the screens set by using ultrasonic anemometer.

The most representative results are presented in the next topic.

Table 1. Details of screens samples

Screen	Wire diameter (mm)	Mesh size (mm)
I	1.65	8.51
II	2.10	10.59

Figure 4. Screens set inside the wind tunnel.

3. Results and Discussion

The results obtained to the velocity profile in the symmetry axis to the sections III and V and its respective turbulence intensities are showed in the Figures 5 and 6.

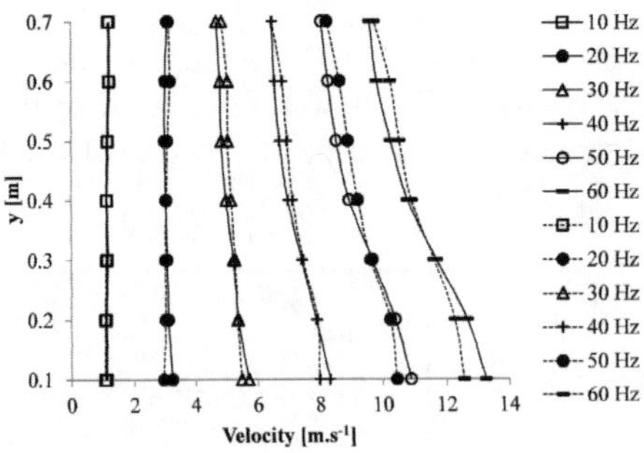

Figure 5. Velocity profile using hot wire anemometer positioned in the section III, as shown in Figure 2a (solid line) and in the section V, as shown in Figure 2b (dashed line).

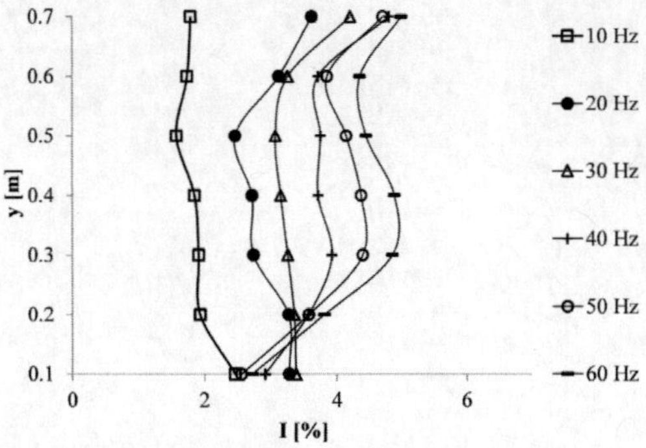

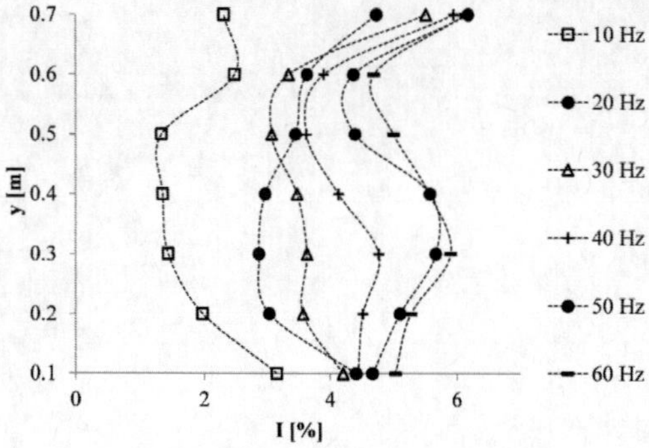

Figure 6. Turbulence intensity for the hot wire anemometer positioned in the section III (left) and the section V (right).

Analyzing the results above-mentioned, the velocity profiles on the vertical plane of symmetry along both tunnel sections were very similar. Nevertheless, the turbulence intensity showed distinct behaviors. In the section III, the turbulence intensity varied from 1.5% to 5% and in the section V, this range was from 1.5% to 6%. Besides, for both sections evaluated, there is a range of mean velocity of the wind tunnel (1 to 3.5 m/s) on which is possible to introduce appropriate designed devices for atmospheric boundary layer behavior reproduction, without diminishing considerable length of the test section, since in such velocity range the velocity profile is almost uniform. However, for higher mean velocities range than that, limitations can derive due to the short length of test section available for flow development inside the tunnel.

Based on the similarity between the velocity profiles obtained in the tunnel sections analyzed previously, the middle of the last section was chosen to evaluate the effect of the screens set as presented in Figure 7. The Figure 8 shows the effects on turbulence intensities due to screens.

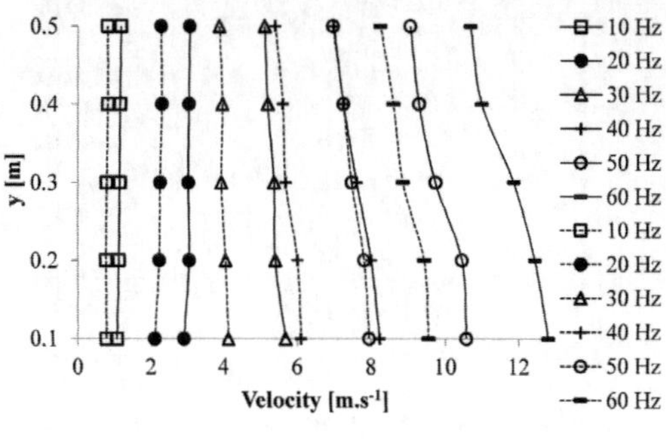

Figure 7. Velocity profile using hot wire anemometer positioned in the section V without screens set (solid line) and with screens set (dashed line).

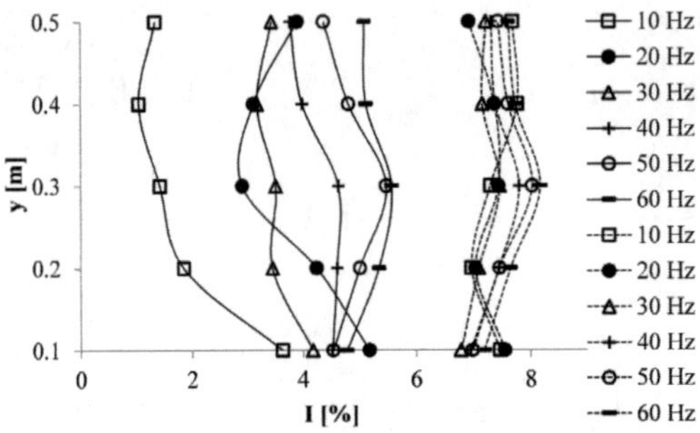

Figure 8. Turbulence intensity for the hot wire anemometer positioned in the section V without screens set (solid line) and with screens set (dashed line).

Analyzing the results above-mentioned, the velocity profiles on the vertical plane of symmetry were improved significantly with the addition of screens set. It means that screens set were effective not only in reducing the mean velocity, in the region of the screens set, but also in smoothing the velocity non-uniformity. The turbulence intensity (with and without screens) showed very distinct behaviors. The turbulence intensities became higher by using screens set, varying from 6.8% to 8.2%, while without screens this range was from 1.0% to 5.6%. However, turbulence level dispersion was diminished due to screens presence.

To validate the experiments which were carried out using hot wire anemometer, ultrasonic anemometer was employed, because it is traceable to the national measurement standards from NMIJ, as aforementioned. The Figure 9 presents the results obtained in the wind tunnel with screens set by employing both anemometers.

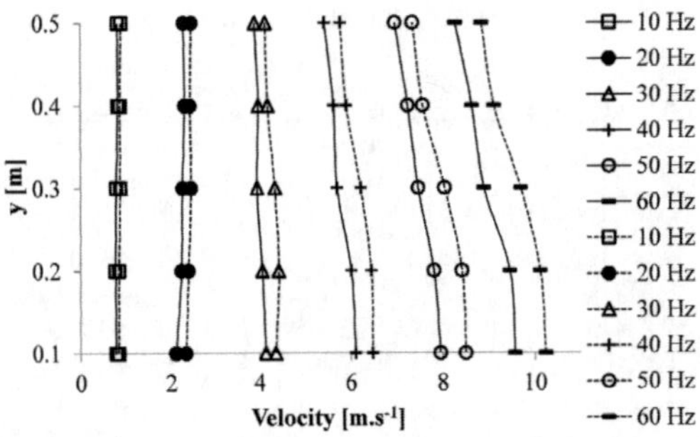

Figure 9. Velocity profile with screens set using hot wire anemometer (solid line) and ultrasonic anemometer (dashed line).

According the Figure 9, the mean velocity obtained by ultrasonic anemometer showed a slight increase when compared with the results from hot wire anemometer. In addition, the velocity profiles were very similar for both anemometers including non-uniformity for high velocities.

In relation to measurements using hot-wire anemometer, it was identified some uncertainties sources due to the hot-wire anemometer probe positioning, calibration equipment and method, frequency response, electrical noise, data acquisition and climatic conditions (temperature, pressure and humidity) [11-12]. Thus, it is necessary to investigate more deeply the contribution of each uncertainty source in the results obtained.

Journal of Physics: Conference Series **733** (2016) 012001 doi:10.1088/1742-6596/733/1/012001

3.1. Some considerations about uncertainty of hot-wire anemometers

This topic will present a brief discussion about uncertainties associated with temperature variations from calibration to experiment, which is one of the major uncertainty source [11-13].

Firstly, it is necessary to correct the voltages E_a for temperature variations during calibration as shown in (1):

$$E_{corr} = (\frac{T_w - T_0}{T_w - T_a})^{0.5}.E_a \tag{1}$$

where E_a = acquired voltage; T_w = sensor hot temperature; T_0 = ambient reference temperature related to the last overheat set-up before calibration; T_a= ambient temperature during acquisition.

Secondly, a 4th order polynomial is created (2), which relates the velocity to the corrected voltage read by the acquisition system:

$$v = C_0 + C_1.E_{corr} + C_2.E_{corr}^2 + C_3.E_{corr}^3 + C_4.E_{corr}^4 \tag{2}$$

where C_0 to C_4 are the calibration constants and v is the velocity. Thus, the corrected velocity, v_{corr}, is a function such that (3):

$$v_{corr} = f(v, T_w, T_0, T_a) \tag{3}$$

Since T_0 and T_w are constants in the partial derivative $\partial v_{corr}/\partial T_a$,$v_{corr}$ can be expressed as (4):

$$v_{corr} = f(v, T_a) \tag{4}$$

To determine v_{corr} as function of v, the suitable polynomial root from (2) is found and then applied in (1), such that (5):

$$E_{corr} = f(v) \tag{5}$$

Then, replacing (5) in (1), the corrected velocity can be found as function of the uncorrected velocity (6):

$$v_{corr} = f(v) \tag{6}$$

With (6), one can calculate the influence of temperature to the velocity uncertainty, which is presented in (7):

$$u(v_{Ta}) = \frac{dv_{corr}}{dT_a}.u(T_a) \tag{7}$$

Figure 10 shows the velocities uncertainties due to the temperature uncertainty, considering:
$T_0 - T_a = 5\,°C$
$u(Ta) = 0.05\,°C$

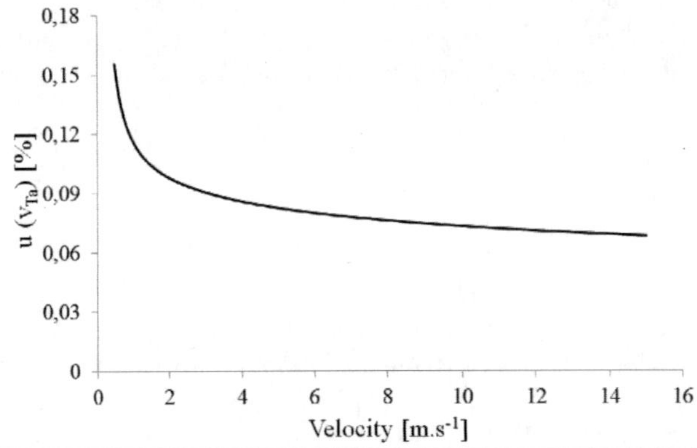

Figure 10. Velocities uncertainties due to the temperature uncertainties versus velocity.

As can be seen in the Figure 10, the relative velocity uncertainty decreases as velocity increases, as expected. It is important to notice that this is only one of the uncertainty sources; the full model for calculating the uncertainty may have up to 9 terms, including the calibrator certificate, which have a typical value of uncertainty of 1%.

4. Conclusion

In this work, the velocity profile and the turbulence intensity of the tunnel were investigated in the vertical symmetry axis of the cross section. The results obtained without the screens set indicated that for low average speed of the flow in the wind tunnel, the velocity profile on the vertical plane tends to be uniform. Nevertheless, for high velocities, non-uniformities in the velocity profile were noted. The addition of screens set reduced the mean velocity and also diminished the non-uniformities.

Although the turbulence intensity was higher with the screens set addition, it became more uniform. So, it is possible to conclude that the screens set employed in this work contributed significantly to improve not only the velocity profile but also the turbulence intensity dispersion. It is suggested that others configurations of screens must be assessed.

In this work, since the atmospheric wind tunnel from Inmetro is mounted and operational, the examination of its flow characteristics deals with two important purposes related to the Scientific Metrology activities in fluid dynamics area, which are: the effort for refinement of experimental procedures in order to minimize measurement uncertainties and the endeavor to add relevant and reliable experimental data on complex fluid flow to the literature, in order to contribute to knowledge advances.

Acknowledgments

The authors are grateful to CNPq, FAPERJ, FINEP, PETROBRAS and INMETRO for their support and sponsorship which have become possible the development of this research.

References

[1] Barlow J B, Jr. Rae W H and Pope A 1999 *Low speed wind tunnel.* **3th** Ed. (New York: John Wiley and Sons, Inc)
[2] Stull, R B 1988 *An introduction to boundary layer meteorology* (The Netherlands: Kluwer Academic Publishers)
[3] Kulkarni V, Sahoo N and Chavan S D 2011 *J. Wind. Eng. Ind. Aerodyn.* **99** **(1)** 37

[4] Mehta R D 1985 *AIAA Journal* **23 (9)**1335
[5] Karnik U and Tavoularis S 1987 *Exp. Fluids* **5 (5)** 247
[6] Farias M H, Santos A M, Souza D B, Ferreira, L L R, Massari P de L, Massari P L, Garcia D A and Costa F O 2014 *Anais Proc. - 3rd Cimmec* **art.** 127620
[7] Nader G, Santos C, Jabardo P J S, Cardoso M, Taira N M and Pereira M T 2006 *XVIII IMEKO World Congress*
[8] Piccato A, Spazzini P G and Malvano R 2009 *7th ISFFM*
[9] Erm L P 2000 *Calibrations of the flow in the test section of the low-speed wind tunnel at AMRL* (DSTO Aeronautical and Maritime Research Laboratory)
[10] Manshadi M D 2011 *The importance of turbulence in assessment of wind tunnel flow quality, wind tunnels and experimental fluid dynamic research* (Available from: www. intechopen.com/book/wind-tunnels-and experimental-fluid-dynamics-research-the importance-of-turbulence-reduction-in-assessment-of-wind-tunnelflow-quality)
[11] Bruun H H 2002 *Hot-wire anemometry principles and signal analysis.* **1st** Ed. (New York:Oxford University Press)
[12] Jørgensen F E 2002 to measure turbulence with hot-wire anemometers a practical guide. (Denmark: Dantec Dynamics)
[13] Yavuzkurt S 1984 *J.Fluids Eng.* **106** 181

Experimental studies toward the characterization of Inmetro's circulating water channel

A M Santos [1], A T P Alho [2], D A Garcia [1], M H Farias [1], P L Massari[3] and V V S Silva[1]

[1] National Institute of Metrology, Quality and Technology-INMETRO, Fluid Dynamics Metrology Division, Av. Nsa. Sra. das Graças, 50-Prédio 06-Xerém, CEP 22250-020-Duque de Caxias, RJ, Brazil;
[2] Federal University of Rio de Janeiro-UFRJ, Department of Naval and Oceanic Engineering, Cidade Universitária, Ilha do Fundão, CEP 21945-970-Rio de Janeiro, RJ, Brazil;
[3] Pontifical Catholic University of Rio de Janeiro-PUC-Rio, Mechanical Engineering Department, Rua Marquês de São Vicente, 225-Gávea, CEP 22451-900-Rio de Janeiro, RJ, Brazil

E-mail: mhfarias@inmetro.gov.br

Abstract. Circulating water channels are facilities which can be used for conducting environmental, metrological and engineering studies. The Brazilian National Institute of Metrology-INMETRO has a water channel of innovative design, and the present work deals with the prior experimental investigation of its hydrodynamics performance. By using the optical technique PIV – Particle Image Velocimetry, under certain conditions, the velocity profile behavior in a region inside the channel was analyzed in order to evaluate the scope of applicability of such bench.

1. Introduction

In research laboratories, water channels or flumes are benches used for fluid flow investigations or submerged bodies behavior analyses, for a variety of applications, e.g. scientific, environmental and technological areas [1]. In these facilities, by controlling operation conditions it is possible to reproduce effects of flow stratification (temperature, concentration), turbulence and boundary layer, free surface (waves generation or dissipation), among other aspects of a flow field. The performance and applicability limits of a water channel will depends on its design and internal components characteristics (flow straighteners, screens (mesh and distribution), working section details etc.) [2]. The more specific the apparatus application is, the more exigencies for accuracy on its design, building and ancillary measurement instruments to be employed.

The Fluid Dynamics Metrology Division (Dinam) of the Scientific and Industrial Metrology area of the Brazilian National Metrology Institute (Inmetro) has a Circulating Water Channel. The channel was constructed to be used in instrument calibrations and fluid flow researches. This bench was designed by the own Inmetro/Dinam researchers team [3] and it was built under rigorous specifications. Recently it was commissioned. The earliest steps towards its characterization process are object of the present work. Some aspects of the flow generated in such bench when it is free of a model testing inside are shown.

2. Facility overview

The water channel (figure 1) is an innovative design [3]. It has long visualization side walls (test section has 12m in length) which are composed by 05 monolithic glass plates interconnected by high strength silicone. Except along the upper and lower edges of the wall, there is no metallic structures interconnecting the glass plates which compose the channel. The channel floor is also made of glass, for flow visualization and measurements through the bottom. So, it is possible to observe and measure the flow along a wide window and at different views.

The cross section of the channel is rectangular, 0.6m length x 0.7m height, and at the channel entrance an accommodation section contains straighteners (figure 2) and a space for screens insertion (when required). At the channel exit, two types of customized wave suppressors were installed (figure 3 shows the triangular type). The position of such devices can be changed according to the water column and flow rate, in order to vanish generated waves.

Figure 1. The Inmetro's Circulating Water Channel – wide glass wall.

Figure 2. Aspect of the flow straighteners.

Figure 3. Set of triangular wave suppressors.

A centrifugal pump (30 kW) drives the flow, and the maximum mean flow velocity inside the channel is 0.5 m/s (max. flow rate is 648 m^3/h). An electromagnetic flowmeter (diameter 0.254 m; nominal capacity 53.6 m^3/h to 1770 m^3/h; calibrated in 10 points of the nominal range) is used for monitoring the mean flow rate in the channel. Adjacent to the channel was constructed a platform, on which a unity will be driven under controlled velocity (max. 0.5 m/s) for transporting instrumentation or a model to be studied in the channel.

3. Experimental procedure

The main aspects of the fluid flow, as well as the water channel operation capability determination were initiated during the bench commissioning phase. After that, a flow field mapping in the central plane of a region in the test section (figure 4), under certain conditions, was performed.

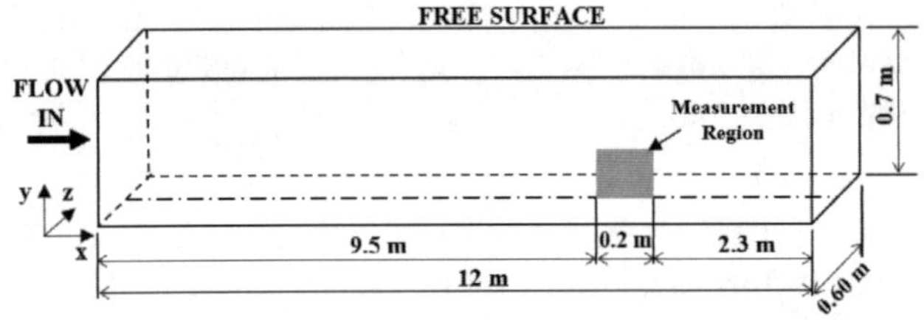

Figure 4. Details of the channel dimensions and the measurement region.

3.1. *Brief comment on the water channel commissioning process*
The commissioning phase consisted of evaluating the facility under several conditions, on which we can detach: **i)** a progressive channel filling with water, in four steps of a water column of 150 mm, up to reach the maximum water column intended for working (600 mm). During all water channel filling process, the deflections of its horizontal supporting structures were monitored along the all length (12 m), and the maximum deflection was less than 1 mm (at the middle length); **ii**) the free surface behavior after the flow left the straighteners and reached the wave suppressors; **iii**) the hydraulic pump (30 kW) performance influences: operation stability (under 25% of the maximum pump motor rotation and also with partial opening of a valve positioned at the pump exit).

3.2.Brief comment on the wave suppressors acting in the flow
Before introducing the wave suppressors in the channel, a fast increase of the wave amplitude was provoked due to a pumping sudden stop or due to the water flow rate increase. So, in the characterization process, a limitation of the pump rotation was imposed, in order to avoid stationary wave generation along the channel. Then, after the wave suppressors were installed, stationary waves vanished, and so, the pump could work at rotation level close to its maximum capacity (with the highest water column intended to work in the water channel, 600 mm).

3.3.Flow characterization using the particle image velocimetry technique-PIV
The PIV technique[4] was used in this work. It consists of a dual cavity laser source which pulses synchronized with high speed cameras (CCD or CMOS) aperture. Micro particles of reflexive material are seeded in the flow (tracer particles). When the laser sheet is projected into the flow, such particles reflect the incident light. So, sequential images of the illuminated region are captured by the cameras (01 camera for two-dimensional measurements and two cameras for 3D stereo). By comparing the position of the particles in small regions (called interrogation window) of two sequential images captured by the cameras, and relating them to the acquisition frequency of the images (time interval), through mathematical correlations (cross-correlation or auto correlation) is possible to estimate the flow velocity (considered equal to the tracer particles velocity). A velocity field and associated information on the flow (ex. turbulence characteristics) can be mapped in this way.

The PIV system from Dantec Dynamics™ used in the present work was in 2D configuration. Basically, this system is **i)** a double cavity Nd:YAG laser (Dual Power 200-15 Laser 2x200 mJ, 15 Hz, 532 nm) and **ii)** one camera CCD Flow Sense Mk2 4M – with resolution 2048 × 2048 pixels (pixel size 7.4 μm), with 60 mm lens and glass filter high pass (cut-off at 532 nm).

3.3.1. On the measurement uncertainties estimate when using the PIV technique. The PIV system is one of the most important state of art tools which has been employed in fluid dynamics experimental investigations. The scientific community considers this technique non-intrusive, however, under metrological sense, it is an intrusive technique, because itis required the presence of tracers particles in the flow, and most of time the tracer particles are not originate of thework fluid, i.e., it is necessary add tracer particles to the fluid. Due to the operation principle of this technique, there are some challenger gaps to be solved by the scientific and metrological communities, related to find the best refined methodology for measurement uncertainties quantification (UQ) when using such system. The community concerning about this subject is related on what should be considered as sources of uncertainties. The uncertainty level when using PIV is influenced by the flow dynamics, becausethe setting of several system parameters and adjustments which are need to be comprised before running each experiment depends strongly on the flow behavior. In this context, the main challenge faced nowadays is to find the best way for evaluating each uncertainty component[5], regarding the intricacy of sources. Such sources come from cameras and laser alignment, laser intensity and time between laser pulses, image distortions, physical to image scales (scaling magnification), tracer particles (displacement, concentration and density), mapping function, algorithm, image processing, and others. Then, as not only experiment conduction way and technological resources are influence factors, but also computational and mathematical tools have impact on the uncertainty ranges[6], it is clear that PIV UQ and related uncertainty reduction is a matter that requires multidisciplinary treatment.

In the last years the attention for PIV UQ has been increased [5]. There is a lot of subjects to be discussed and investigated. As interested part, last year Dinam/Inmetro has introduced in its activities the research line in such topic, including both computational and experimental focuses. For starting point of the experimental side, Dinam has been working intensively on PIV experimental procedures improvement. Nevertheless, these refined procedures have not been already implemented when the present work was developed.

In this preliminary evaluation of the resulting flow inside the Inmetro's water channel, the level of standard deviation of the local mean velocity (time averaged) was chosen as a parameter for this bench hydrodynamics performance approbation, since it permits infer some about stability and the development state of the flow in the measured region. Standard deviation is a representative criterion, since it carries influences of the system alignment, software, timing, etc, besides, of course, the flow behavior. The consistency of the hydrodynamic profile was also appreciated under the basis of the physical process and the theory of fluid flow.

3.4.Measurement conditions

Table 1 summarizes the measurement conditions. In this table, the flow rate uncertainty is different for each flow rate level because the flowmeter was calibrated in several points within its nominal range. The water column was 0.6 m.

Table 2 shows the PIV system setup, which was controlled by the Dynamics Studio™ (software for acquisition and processing).

Table 1. Measurement conditions.

Mean flow rate range (m^3/h)	Uncertainty (%)	Water[a] column (m)
147	0.1943	0.6
207	0.1943	0.6
265	0.1943	0.6
325	0.1214	0.6

[a] Seeded tracer particles: silver coated hollow glass spheres; size distribution diameter: 8 to 3 μm; density: 900 kg/m^3

Table 2. PIV system setup.

Number of captured images at each flow rate	Time between laser pulses (μs)	Trigger rate	Interrogation window size (pixel)
2000 (double frame)	5000 μs (at 147 m^3/h and 207 m^3/h) 4000 μs (at 265 m^3/h and 325 m^3/h)	7.4 Hz	Initial: 128x128 Final: 64x64

Journal of Physics: Conference Series **733** (2016) 012002 doi:10.1088/1742-6596/733/1/012002

4. Results and discussion

The water channel flow velocities profiles are shown in figures 5 to 8. The figures 5 and 6 show the velocity field for the minimum and maximum tested flow rates. U is the horizontal velocity component.

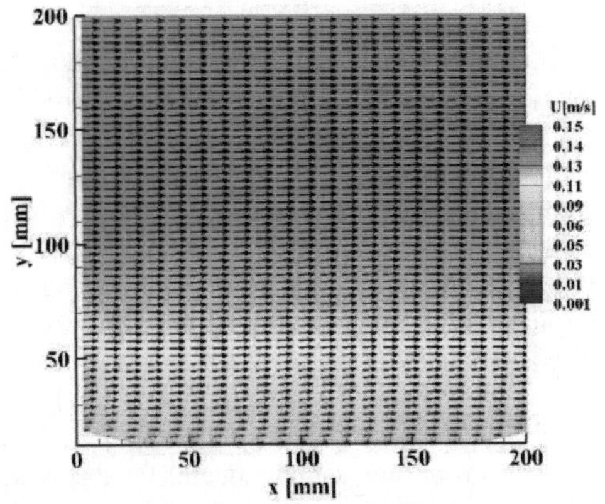

Figure 5. Mean velocities field for flow rate 147 m^3/h; (mean of 2000 captured images).

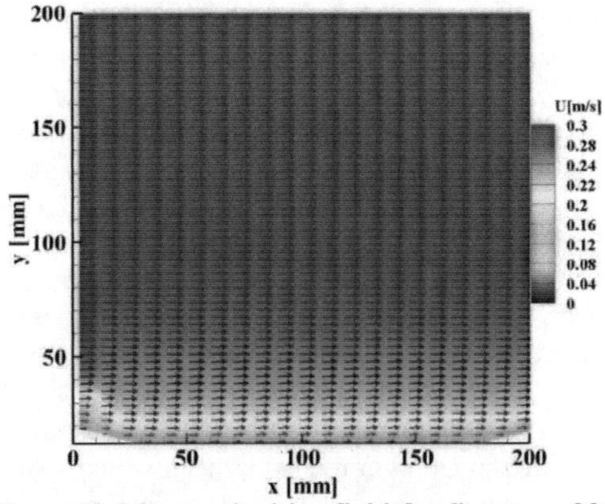

Figure 6. Mean velocities field for flow rate 325 m^3/h; (mean of 2000 captured images).

It can be observed that the vertical positions (y) in which the horizontal velocity becomes uniform are around 120 mm and 70 mm approximately, respectively, for minimum and maximum tested flow rates. In these flow fields, the maximum standard deviation value of the local horizontal velocity in the uniform region was 0.02 m/s.

The figure 7 shows the vertical position from which the horizontal velocity component does not change along the main flow direction, and the figure 8 shows the boundary layer thickness.

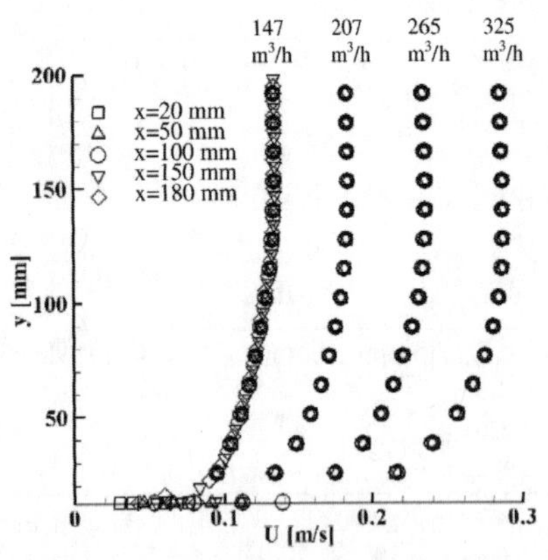

Figure 7. Horizontal velocity (U) profile along the channel axis for different flow rates.

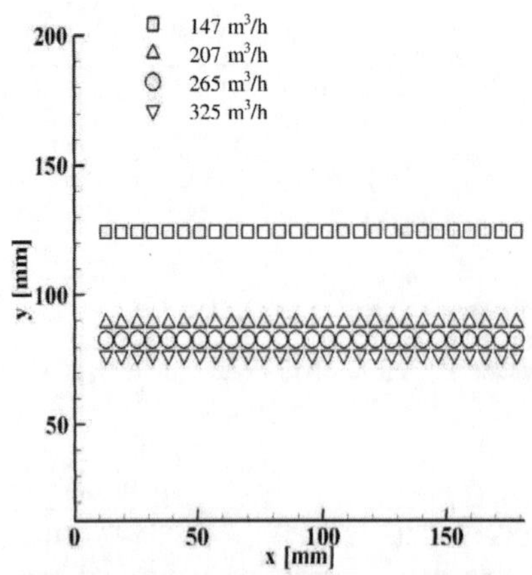

Figure 8. Boundary layer thickness at different flow rates and longitudinal positions.

According to the boundary layer theory [7], the maximum thickness of this flow region can be estimated as being the distance to the wall where the established local fluid velocity (U_i) is approximately 99% of the free stream velocity (U_∞), i.e.,

$$U_i = 0.99U_\infty \tag{1}$$

Based on the color scale legend on figures 5 and 6, applying equation (1) to the data, Ui gives 0.139m/s and 0.297m/s, respectively. Viewing, on the graphs, the lower vertical position (y) where these velocity levels are reached, it would correspond to boundary layers thickness around 125mm and 70mm, respectively. Otherwise, by plotting the evolution of the horizontal velocity component for different flow rates along the channel longitudinal axis, as shown in figure 8, it can be concluded that the boundary layer thickness is almost constant for different longitudinal positions of the flow measurement region. Thus, the experimental data shown in figures 7 and 8 confirm the ranges of estimated values to the boundary layer thickness, for the minimum and maximum flow rates, by using equation (1).

5. Conclusion

This work shows the earliest evaluation of the Inmetro's water channel. The experimental data show a good efficacy of the wave suppressors and the flow straighteners of the bench, since the standard deviation levels for local velocities are within acceptable ranges for this kind of flow and the pattern velocity profile is according to the expected. Besides, under the tested conditions, the flow field shows typical characteristics of developed flow and there is no perceptible influence of the free surface in that flow position. It is a satisfactory finding.

A complete characterization of the water channel is an experimental work that will demand long term to be finished, due to the need of investigating the flow field under a variety of conditions, such as different water column height, different vertical planes and cross section along the channel length etc. So, this research is being continued. Notwithstanding, the results of this preliminary analysis of the Inmetro's Water Channel have already indicated that this bench is appropriated for several fluid dynamics studies, as well as for devices performance or model testing investigations, in agreement to the aims for which such bench was carefully designed and constructed.

Acknowledgements

The authors are grateful to CNPq (including Proc. 478636/2012-1 and Prometro), FAPERJ, FINEP, PETROBRAS, PFRH-PETROBRAS and INMETRO for their support and sponsorship for the development of this research.

References

[1] Nezu I and Sanjou M 2011, *J Hydr. Envir. Res.* **5** 215-230
[2] Tavoularis S 2005 *Measurement in Fluid Mechanics* (Camb. Univ. Press, USA) p.157
[3] Alho A T P , Farias M H , Neto J L S 2010 *On the Design of a Circulating Water Channel for the Brazilian National Institute of Metrology*-INMETRO - 15th FLOMEKO, Taipei-Taiwan
[4] Raffel M, Willert C, Wereley S and Kompenhans J, *Particle Image Velocimetry- a Practical Guide*, 2007, Springer, New York
[5] Christensen K T and Scarano F, 2015,*Meas. Sci. Technol. 26 (2015) 070201*
[6] Massullo A and Theunissen R, 2016, *Adaptive vector validation in image velocimetry to minimise the influence of outlier clusters*,Experiments in Fluids(2016) Vol 57 (33)
[7] Schlichting H and Gersten K, *Boundary Layer Theory*, 2000, 8th ed, Springer Verlag, Berlim.

Modification of equipments for compliance in radiated emissions for certification according to CISPR normative

Alexandre Rodrigues da Silva [1], Marcelo Sanches Dias [2] , João Carlos Martins Coelho [3]

[1]alesilva@ipt.br, [2]mdias@ipt.br, [3]jcmc@ipt.br

[1] IPT- *Instituto de Pesquisas Tecnológicas do Estado de São Paulo*, [2] Institute, Brazil

Abstract: This is a study of radiated emissions generated by electronic equipment. Two different equipments were tested and their respective results presented, where it was observed that their emissions' values were worse than desired. After some modifications, and a second batch of tests, better results were achieved.

1. Introduction

Electromagnetic Interference (EMI) is a disturbance that affects an electrical circuit due to electromagnetic conduction or electromagnetic radiation emitted from an external source, and that source could be an electric or electronic equipment. Because of that reason, products that need electric power to operate could have the influence of EMI from another equipment while it could be itself producing it, creating problems to other units close to it. It is necessary to verify if products have emissions above the appropriate limits and if they are effectively immune to it.

Experimentally it can be verified that this interference between equipment could occur in two different ways, by conducted induction and radiated induction. This study is only for the radiated interference. In this case, it can also be in two different ways, differential mode and common mode. Some of the energy will propagate as the differential mode (forth on one cable, back on another), and the rest as the common mode (along all of the cables simultaneously and back via an unknown "ground" return). [1]
Cables are a source of EME that cause significant problems in complying with electromagnetic compatibility (EMC) regulations, and even can jeopardize signal integrity. Ferrite cores or flexible absorbing materials on cables are used to suppress common mode radiated from cables. [1]

Filter circuits are used in a wide variety of applications. This filters can be active or passive. They normally are electric circuits with capacitors, inductors and resistors. Low pass filters are those that allow low frequency signal to pass and block the ones with higher frequency. Generally, a major part of electronic equipment employ anti-aliasing low pass filters in their signal conditioning stages. In a higher frequency band, mostly passive components are used. [2]

Cable shielding was also developed to solve problems with EMI radiated interference. The type of shielding that is used for this type of interference is the braided shielding. The braided wire works because the surface area of the braid is significantly greater than solid wire or stranded wire. Since the skin effect applies to high frequency interference, the greater surface area of the braid can facilitate in reducing the effect of high frequency EMI. Braids are effective for controlling both immunity and emissions. When the interference is less than 100 MHZ within d the shield the braid will represent a large inductive path preventing the coupling in an unintended circuit. [5]

With those technics described, employed together or apart, can control susceptibility and radiation of EMI of electronic equipment. If the tested unit has its sensitivity to EMI more of desired, or its emissions are above the regulation values, those methods could decrease the emissions. This will assure that the equipment is suitable operate with other different equipment nearby without interfering with their operation.

This paper is about radiated emissions. Its main objective is to compare radiated emissions before and after changes to the equipments, adding ferrite cores, changing low pass filters by adding resistors in series, and applying braided shielding to cables. The results are positive, proving that the described methods are effective for all equipments that could present this kind of problems.

2. Testing method and changes made to the equipment

The electromagnetic compatibility tests were done in semi-anechoic chamber in IPT (Instituto de Pesquisas Tecnológicas do Estado de São Paulo). They follow the specifications of CISPR 11[3] normative e CISPR 22[4] normative.

The first test for both equipment was before of any change from of the shelf units. For both equipment, this first test result had results worse than desired. For the first equipment, the results were above desired in several frequencies although it was above regulatory values only for about 50 MHz and for values between 200 and 700 MHz.

Some changes were made to the cables. A ferrite core was added close the power supply and it was shortened considerably. The aim was to add a magnetic resistance to the circuit, in order to decrease the effect of the inducted current for the frequency were the larger emissions were, and for that the chosen core was the one designed for maximum attenuation at the frequency desired.

Low pass filters were employed. In some cases complete filters, and in other only resistors in series, changing the cut off frequency. This filters let lower frequency pass and block those larger than 50MHz, which were the ones with results higher than desired in the test.

Cables were also replaced by different ones, with braided shielding. Even with all the changes described, the equipment still had emissions above desired. With the shielding in the internal cables, finally the values were under the limit, making possible the operation in the presence of other equipment.

Following the same procedure, the second equipment was also tested without change, and presented results that have higher emissions than allowed.

The first step was to increase the resistance in series with the circuit, changing the frequency for the low pass filter. Installing a ferrite core complete the changes and brought the results to acceptable values.

3. **Results**

For the first equipments, before the changes, the result is on Figure 1. Many of the frequencies are above the limits prescribed by CISPR11. Particularly the frequencies between 40 MHz and 50 MHz, and above 1 MHz

After the before mentioned changes, the result is on Figure 2. The emissions results are a lot better, and there was no change on the equipment functionality. The results agree well with CISPR 11 imposed limits.

For the second equipments, the radiated results are presented on Figure 3. Figure 4 shows better results, after the changes. On that case only one frequency was above the limit, at 200 MHz, and that got to acceptable values.

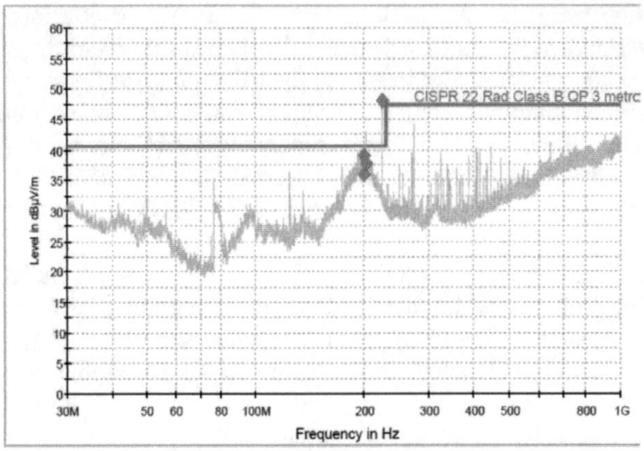

Figure 1: Radiated emissions, first equipment1, before modifications.

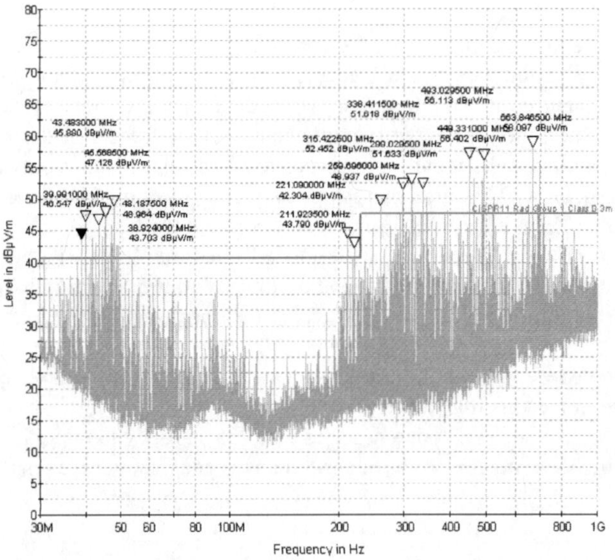

Figure 2: Radiated emissions, first equipment 1, after modifications.

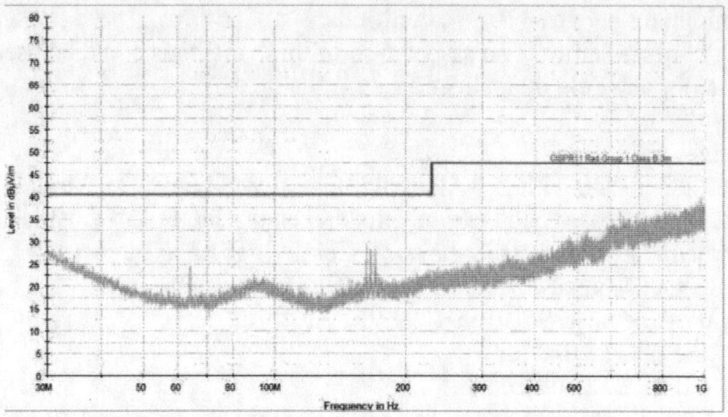

Figure 3: Radiated emissions, equipment 2, before modifications.

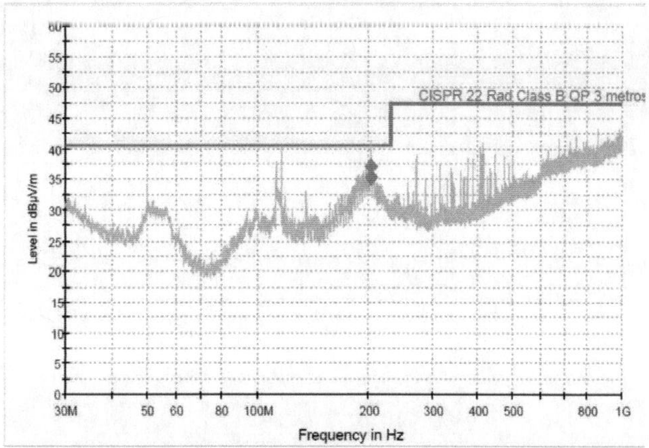

Figure 4: Radiated emissions, equipment 2, after modifications.

4. Discussion

Both equipments designed well for their functionalities, but were not suited to CISPR11 and CISPR22 normative, for electromagnetic emissions. The initial tests show that. After the modifications made and described, adding resistors, shielding and ferrites got the results to what is prescribed.

These results are seen on the presented graphs, and it can be observed that the results are below the limits prescribed.
Any type of electronic equipments could be modified following the same rules, using the same method described.

5. Conclusion

There are several ways to correct the electromagnetic emissions that are above the limits stablished by the normative.

It is demonstrated on the examples described on this article, the ways were successful, making possible for the equipments here tested to be certified.
If applied the described way, an association of these methods could be used for any electronic equipments, for any field of application.

6. References

[1] Jing Li, Yao-Jiang Zhang, Aleksandr Gafarov, Soumya De, Marina Y. Koledintseva, Joel Marchand, David Hess, Todd Durant, Eric Nickerson, James L. Drewniak, Jun Fan, *EMI Reduction Evaluation with flexible absorbing Ferrite Cores Applied on Cables*, IEEE, 2012
[2] Kugelstadt, Thomas. 2008. *Active Filter Design Techniques.* s.l. : Texas Instruments, 2008.Calgary Alberta, 1995, 0165-1684/95.
[3] IEC CISPR 11 2010 ED 5.1
[4] IEC CISPR 22 2008 Ed6.0
[5] IEE Power and Energy Society, *IEEE guide on shielding practice for low voltage cables, IEEE Std. 1143™-2012*, (Revision of IEEE Std. 1143-1994)

Measurement of creepage distance and air clearance: differences between different professionals

Aline de Oliveira Silva, Henrique Takachi Moriya, Tiago Cortez, José Carlos T. B. Moraes

Divisão de Ensaios e Calibração, Laboratório de Engenharia Biomédica, Escola Politécnica, Universidade de São Paulo, São Paulo, SP, Brazil

E-mail: aline@leb.usp.br

Abstract. The standard IEC/ISO 60601-1:2005 specifies general requirements for measuring creepage distance and air clearance for medical electrical equipment. Four experienced professionals were asked to measure creepage distances and air clearance in three different segments of an acrylic body of proof. The results were compared and the found differences were discussed in order to discover the misinterpretations of the standard requirements. After a final consensus between the professionals, the distances were measured again to obtain the final results.

1. Introduction

The IEC/ISO 60601 series specifies requirements for basic safety and essential performance applicable to medical electrical equipment. The item 8.9 prescribes measurement tests of creepage distance and air clearance. The main purpose of this test is to verify the insulation between electrical parts of the equipment as a mean of operator and patient protection.

The standard IEC/ISO 60601-1:2005 [3]defines creepage distance as 'the shortest distance along the surface of the insulating material between two conductive parts' and air clearance as 'the shortest path in air between two conductive parts'.

2. Objective

The main objective of this article is to show the misinterpretations of the standard requirements during the measurement of creepage distances and air clearance comparing the results obtained by four different experienced professionals [1].

3. Materials and methods

The measurements were performed on an acrylic body of proof provided by an independent organization (Qualabor, Brazil) [2]. Its diagram is shown in figure 1.

Journal of Physics: Conference Series **733** (2016) 012004 doi:10.1088/1742-6596/733/1/012004

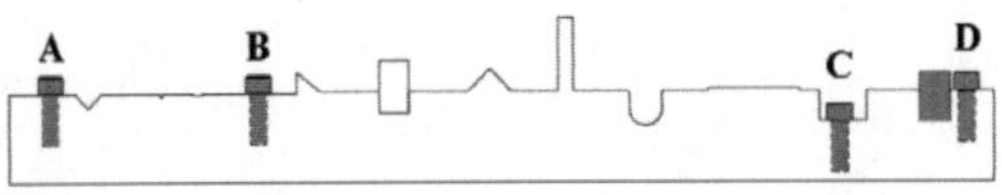

Figure 1. Body of proof. Gray parts are metal parts.

This bar has several grooves and some metal parts. There are four screws fixed in the bar determining the points A, B, C and D.

The measurements were made between points A-B (figure 2), B-C (figure 3) and C-D (figure 4). Every distance has a particular characteristic and format.

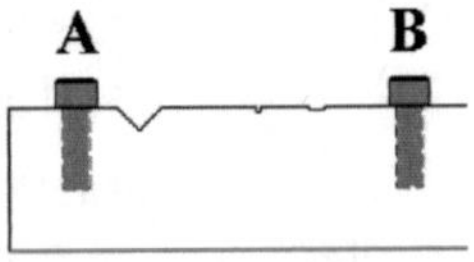

Figure 2. Diagram of the segment between points A and B.

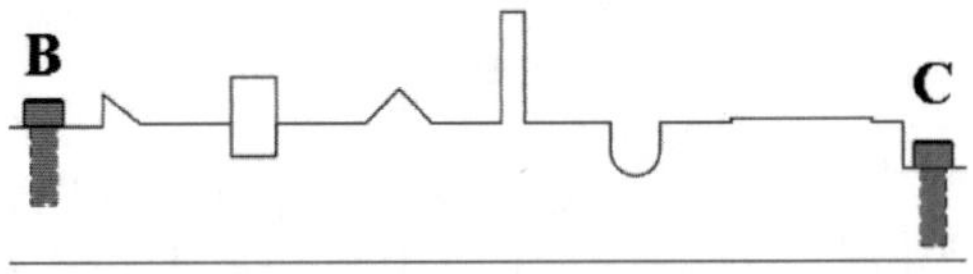

Figure 3. Diagram of the segment between points B and C.

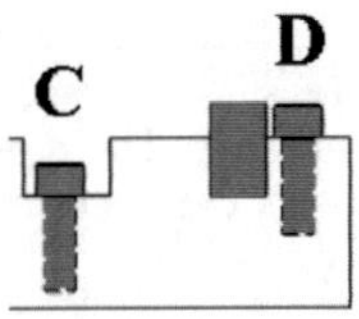

Figure 4. Diagram of the segment between points C and D.

The distances were measured with the same caliper rule (330-114, Mitutoyo Corporation, Japan) (Calibration certificate traceable by International System valid until 12/3/2016).

Four experienced professional (I, II, II, and IV) were asked to perform measurements.

During the tests, all professionals consulted the standard IEC/ISO 60601-1:2005 to remind the concepts.

After all measurements have been made, the results were compared and the problems were identified in a consensual way by the professionals. After a final consensus, the distances were measured again to obtain the final results.

3.1. Some applied standardization concepts

The method used to measure distances is based on IEC/ISO 60601-1:2005 illustrative figures as showed in figures 5, 6, 7 and 8, where the dashed line represents air clearance and the shaded bar representes creepage distance.

Illustrative figures can lead sometimes to misinterpretations depending on the reader. Therefore, it is very helpful to understand the reasons of some misinterpretations in order to prevent the occurrence of this kind of mistakes.

Figure 5 shows that both distances have to be measured directly across the groove if its distance is lower than 1 mm.

Figure 6 shows that if the groove is longer than 1 mm, creepage distance path follow the contour of the groove.

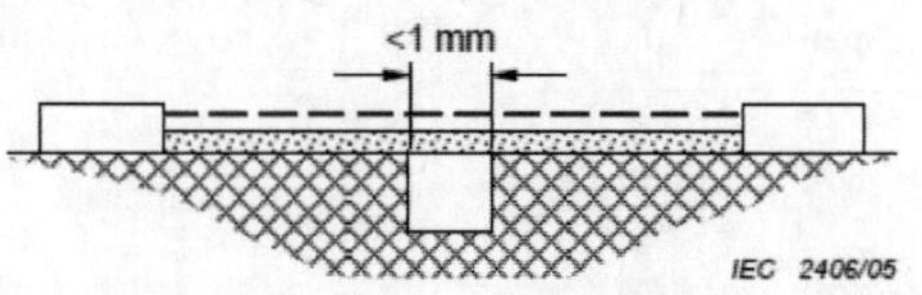

Figure 5. IEC/ISO 60601-1:2005 (Figure 23) [3].

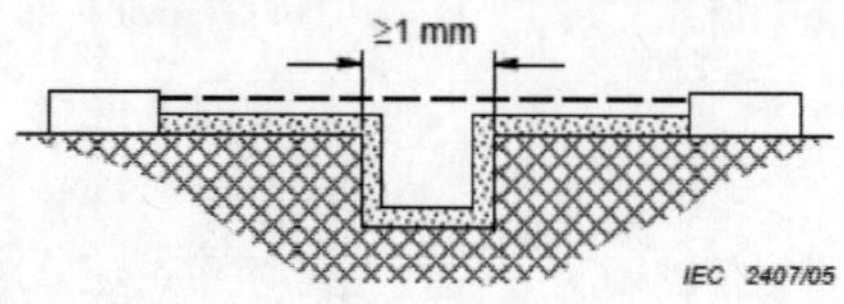

Figure 6. IEC/ISO 60601-1:2005 (Figure 24) [3].

Figure 7 shows that creepage distance path follows the contour of the rib and air clearance is the shortest direct air path over the top.

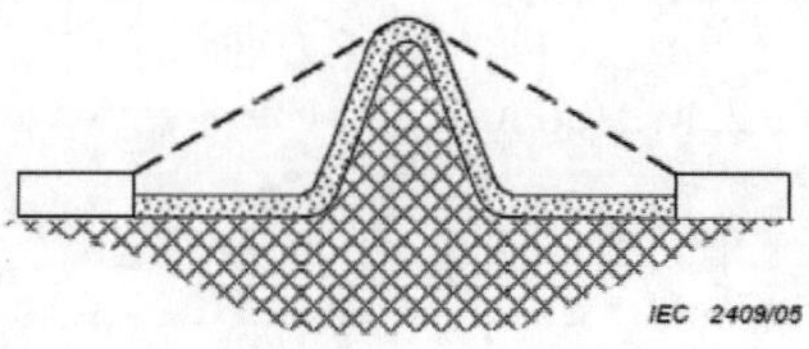

Figure 7. IEC/ISO 60601-1:2005 (Figure 26) [3].

Figure 8 shows that when the gap between the head of the screw and the wall of recess is wide enough to be taken into account so the air clearance is the shortest distance to any point on the head of the screw and creepage distance path follows the surface [1].

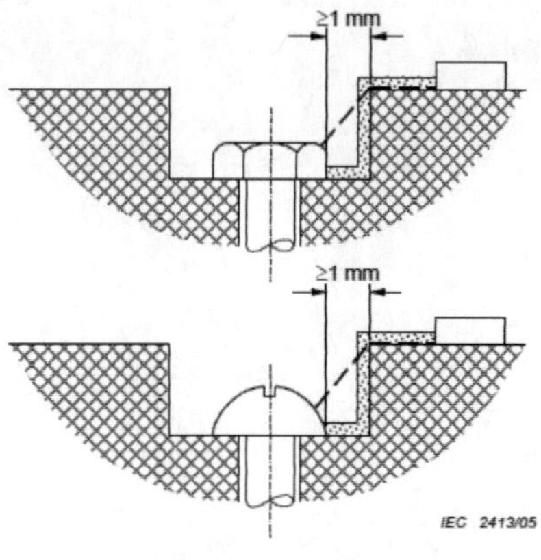

Figure 8. IEC/ISO 60601-1:2005 (Figure 30) [3].

4. Results

The measurements of all professionals were compared for each segment.

The table 1 presents the measurements of air clearance.

The table 2 presents the measurements of creepage distance.

Table 1. Measurements for air clearance.

Air Clearance	**A-B (mm)**	**B-C (mm)**	**C-D (mm)**
I	54.1	200.6	29.3
II	53.9	200.9	27.4
III	54.0	201.5	27.5
IV	54.1	200.5	29.1
Average	54.0	200.9	28.3
Standard deviation	0.1	0.5	1.0
Maximum error	0.1	0.6	1.0
Maximum Error/Average (%)	0.2	0.3	3.5

Table 2. Measurements for creepage distance.

Creepage Distance	**A-B (mm)**	**B-C (mm)**	**C-D (mm)**
I	58.3	302.8	36.8
II	60.8	306.4	36.8
III	58.8	302.3	37.2
IV	58.3	302.8	36.8
Average	59.1	303.6	36.9
Standard deviation	1.2	1.9	0.2
Maximum error	1.8	2.8	0.3
Maximum Error/Average (%)	3.0	0.9	0.8

The table 3 presents final results.

Table 3. Final results (mean ± standard deviation).

Final results	**A-B (mm)**	**B-C (mm)**	**C-D (mm)**
Air Clearance	54.2±0.1	200.4±0.4	29.1±0.4
Creepage Distance	58.7±1.3	302.5±1.6	36.8±1.9

5. Discussions

After analysis of table 1 (air clearance distance) the group could conclude that for distance A-B, the measurements were very close. The standard deviation between these points was small (0.1 mm) and the maximum error relating to average was 0.2%. For the distance B-C, the standard deviation was a little higher than the others (0.5 mm) and the maximum error relating to average was 0.3%. These first two results were considered acceptable by the group. For distance C-D, the standard deviation was 1.0 mm and the maximum error relating to average was considered high (3.5%). Professionals II and III had a misunderstanding of the concept. They used the distance of the base of chamfer instead of the method presented in figure 8.

After analysis of table 2 (creepage distance) the group could conclude that for distance A-B, the standard deviation was 1.2 mm and the maximum error relating to average was considered high (3.0%). Professional II measured the entire pathway, having a misunderstanding of the concept presented in figure 6. There was a groove smaller than 1 mm in this interval of the bar and all its surfaces were measured by professional II.

For distance B-C, the standard deviation was 1.9 mm and the maximum error relating to average was 0.9%. The measurement between these points was the most difficult one, due to the sample format. The number of parts to measure was higher, about 25. The maximum error was 2.8 mm, considered high (Professional II). The most complicated measurement was the one presented in figure 9. Basically, each professional measured in a different way once the standard IEC/ISO 60601-1 does not exemplify this case. Professionals I and IV considered the perimeter of the semi circle and professionals II and III imaginatively drew a square to make the measurements. After analysis, the professionals considered right the measurement of the semi circle.

Figure 9. Semi circle.

For distance C-D, the standard deviation was 0.2 mm and the maximum error relating to average was 0.8%, considered acceptable by the group.

Measurement of creepage distance is more complex than air clearance as presented by results. The first one has a higher number of measurements and some of them are difficult due to sample format.

The final measurements were performed after the comparisons and discussions as showed in table 3.

The standard IEC/ISO 60601-1:2005 already has an amendment 1:2012 [4] but it was not published in Brazil yet. This amendment changes the value presented in figures 5, 6, 7 and 8. Now this value is according to the micro-environment pollution degree.

6. Conclusion

Although the measurements of creepage distance and air clearance look straightforward, some rules and concepts have to be known by the professionals because misunderstandings of the standard directions based on illustrative figures can easily happen.

References

[1] ABNT NBR ISO/IEC 17025:2005 - Requisitos gerais para a competência de laboratórios de ensaio e calibração.
[2] NIT-DICLA-026 rev08 – Requisitos sobre a participação dos laboratórios de ensaio e de calibração em atividades de ensaio de proficiência.
[3] IEC/ISO 60601-1:2005 – Medical electrical equipment. Part 1: general requirements for basic safety and essential performance.

Journal of Physics: Conference Series **733** (2016) 012004 doi:10.1088/1742-6596/733/1/012004

[4] IEC/ISO 60601-1:2005 + Amendment 1:2012 – Medical electrical equipment. Part 1: general requirements for basic safety and essential performance.

Journal of Physics: Conference Series **733** (2016) 012005 doi:10.1088/1742-6596/733/1/012005

Embrapa's experience in the production and development of agriculture reference materials

A R A Nogueira[1], G B Souza[1], C M Bossu[1,2], S R Bianchi[1,3], T R Verhalen[1,2], P T Silva[1,2], A A J Peixoto[1] and C S Silva[1,2]
[1]Embrapa Pecuária Sudeste, 13560-970, S Carlos-SP, Brazil
[2]Dep. Química, Universidade Federal de São Carlos, S Carlos-SP, Brazil
[3]Embrapa Solos, Rio de Janeiro RJ, Brazil

E-mail: ana.nogueira@embrapa.br

Abstract. The main challenge of Embrapa is to develop a model of genuine Brazilian tropical agriculture and livestock. To get this task, the quality of laboratories results is mandatory, increasing the demand for reference materials. Projects were proposed to produce reference materials to support the national agriculture laboratories and consolidate a network able to perform reliable and reproducible analytical testing laboratory within the internationally standards required. Reference materials were produced and available to interested laboratories and collaborative tests were conducted to obtain consensus values. The results and statistical evaluations were performed with the use of software developed by Embrapa Pecuária Sudeste.

1. Introduction

Appropriate compositionally-similar reference materials, a way of transferring accuracy from precise methods of analysis to the laboratory are demanding task. It is observed especially for agricultural products produced in a tropical climate with different characteristics of those produced in temperate areas, the usual suppliers of reference materials. The COMAR database has registered more than 11000 RMs for various applications produced by 25 countries, including Brazil, which is represented by INMETRO. Otherwise, less than 150 of these samples are related to agriculture and livestock materials, none of them a Brazilian material [1]. Since the assurance of the traceability of measurements became indispensable to acquire accurate analytical information, get conclusions and make decisions, mainly in research and regulatory areas, the developed reference materials are required.

The variety of raw and processed agricultural products, foods, and feedstuffs normally constitute a broad range of chemical composition, and inconsistent values are normal [2]. As an effort to obtain most appropriate control of analytical results, Embrapa starts developed agriculture reference materials with a close correspondence, on analyte and matrix, suitable for data quality control for inorganic constituents.

This report describes the preparation of five candidate agriculture reference materials: mineral mixing (to feed bovine milk cattle) [3]; tropical forage [4]; tropical soil [5]; bovine liver; and phosphate rock [6]. The sample preparation and proficiency tests were performed at Embrapa Pecuária Sudeste with auxiliary input by outside expert laboratories. Volunteers cooperating laboratories was essential to obtain the materials.

2. Experimental

2.1. Sources of material

Mineral mixing and phosphate rock were commercial products. Forage, soil, and bovine liver were prepared in-house. The tropical forage (RM-Agro E1001a) (*Brachiaria brizantha* Staf. cv. Marandu) was obtained with the harvest in 400 pots containing 25 kg of a previously artificially contaminated oxisol. Arsenic, Cd, Cr, and Pb (25 mmol/pot) solutions prepared from nitrate salts were added in the corrected oxisol.

The first crop of the forage has been made approximately 60 days after planting and the second one after 120 days. The material was oven-dried at 45°C for 72 h and ground in a cutting mill with 500 µm sieve. The second ground was performed in a high-speed centrifuge mill with 250 µm sieve. It was obtained approximately 20 kg of dry and ground material, only with the aerial part of the forage plant, with particle sizes lower than 280 µm. The material was homogenized in a Y type homogenizer coated with polypropylene, and divided into 100 mL amber glass bottles, resulting 360 bottles each one with 50 g of the sample.

The tropical soil (RM-Agro E2002a) was the same oxisol that was previously prepared to forage. Next the end of forage experiment, the samples from the 400 pots were mixed, dried at ambient temperature and sieved, to avoid the presence of harmful materials, such as forage roots. After 12 months, the soil was ground in a mortar grind and homogenized in a concrete mixer coated with epoxy resin. It was obtained approximately 65 kg of dry sieved soil, with particle sizes lower than 390 µm, which was divided into 645 amber glass bottles, each one with 100 g of soil material.

The bovine liver (RM-Agro E3001a) was obtained from 57 bovines. Processing including bulk tissue separation, freeze-drying, milling, sieving, and packing resulted in 330 amber glass bottles with 100 g of the bovine liver with particle sizes lower than 145 µm.

The mineral mixing (RM-Agro E2001a) was a commercial product acquired by Embrapa Pecuária Sudeste to feed milk cattle. After ground, sieve and homogenizing process, it was obtained 300 polypropylene vials with 50 g of product with particle sizes lower than 80 µm.

The phosphate rock (RM-Agro E2003a) was also a commercial product, a phosphate rock from Morocco. After ground, sieve and homogenizing process, it was obtained 100 amber glass bottles with 100 g of product with particle sizes lower than 150 µm. The particle sizes measurements were performed at the Nuclear Center Agriculture Energy (CENA, Piracicaba, SP, Brazil), in an Analysette 22 (MicroTec Plus, Fritsch, Germany).

All the prepared materials were bulk sterilized by cobalt-60 irradiation to 20-25 kG (IPEN, São Paulo, SP, Brazil) to enhance long-term stability, and were additionally individually sealed in aluminium-nylon pouches.

2.2. Chemical Characterization

A series of preliminary in-house experiments was undertaken to evaluate the adequate sample preparation and quantification methodologies, to define the materials homogeneity and stability.

The materials were produced following the internationally accepted ISO Guides 30-35 norms [7-9] considering the preparation and packaging, material characterization, homogeneity tests (inter- and intra-bottle analysis of variance (ANOVA), short and long-term stability, and four collaborative campaigns.

Chemical characterization to establish reference elemental concentrations were performed by microwave assisted digestion. Arsenic, Cd, Cr, and Pb were determined by inductively coupled plasma mass spectrometry (ICP-MS). Calcium, Cu, Fe, K, Mg, Mn, Na, P, and Zn were determined by inductively coupled plasma optical emission spectrometry (ICP OES). For quality control purposes, certified reference materials (SRM – NIST) were included in the analytical determination and were based on reference and validated methodologies.

For the intra-bottle homogeneities evaluation, 10 subsamples were collected from one bottle of each prepared material. For inter-bottle homogeneity assessments, 10 bottles of each one of material were randomly taken and analyzed. The variation represents the random variation of the prepared materials batches. Statistical evaluation of results was obtained using the single factor analysis of variance (ANOVA). The homogeneity was evaluated by the Fisher test (F-test).

Short stabilities were assessed by submitting the samples (three bottles were randomly taken from each one of prepared material) to extremes humidity and temperature conditions. The determination of the elements was performed in triplicate before and at the end of the experiment, after 30 days. Long term stabilities were also carried out with the use of three bottles randomly selected, and the quantification of the analytes has been monthly performed. Each one of the collaborative campaigns was established with the chosen and an invitation to laboratories specialized that performed the selected analysis. A bottle of prepared material and a form with the instructions to carry out the determinations and returned the results were sent to laboratories that accepted the invitation to participate in the campaign. An attempt was consciously made to get the several analytical procedures, with different sample preparation steps, including no decomposition as in instrumental neutron activation analysis (INAA), as well as various detection techniques. The results were statistically evaluated to define the reference values by a software previously developed [10]. The uncertainties related to each step of the prepared materials were calculated to obtain the expanded uncertainty for each one of the analytes and to the final drafting of the control chart containing the results of the reference materials.

3. Results

The list of agricultural reference materials arising from this research is presented in Table 1. The five collaborative chemical characterization campaigns, together with contribution from initiating laboratories (Embrapa, CENA, and IPEN) produced a great number of submitted individual concentration results. The bovine liver is in the final data treatment and statistical evaluation.

Table 1. Agricultural reference materials for inorganic chemical composition.

Tropical forage	RM-Agro E1001a
Mineral mixing	RM-Agro E2001a
Tropical soil	RM-Agro E2002a
Phosphate Rock	RM-Agro E2003a
Bovine liver*	RM-Agro E3001a

*In the final stage of development.

Table 2 presents a summary list of actual mass fractions with associated uncertainties in the four firstly prepared reference materials, tropical forage, mineral mixing, tropical soil, and phosphate rock.

Table 2. Mass fractions with associated uncertainties of constituent elements in four agricultural reference materials.

Element	E1001a	E2001a	E2002a	E2003a
Al (g/kg)	-	-	-	2.84 ± 0.14
As (mg/kg)	1.69±0.70	0.48±0.20	59.3±7.2	15.53 ± 7.25
Ca (g/kg)	4.37±0.58	187.8±36.3	0.25±0.07	359.9 ± 56.4
Cd (mg/kg)	19.9±5.1	0.43±0.08	94.0±11.4	21.26 ± 3.02
Cr (mg/kg)	3.30±1.66	-	120±30	289 ± 19
Cu (mg/kg)	4.0±0.7	804±87	8.8±4.0	36.2 ± 12.6
Fe (mg/kg)	0.09±0.01	2.73±0.21	32.9±12.7	1709 ± 62
K (g/kg)	12.0±2.4	35.6± 4.3	0.24±0.07	568 ± 52
Mg (g/kg)	2.95±0.44	10.8±2.7	0.33±0.06	2.69 ± 0.13
Mn (mg/kg)	76.0±18.5	1567±141	130±20	16.29 ± 6.90
Na (g/kg)	0.19±0.10	80±15	0.08±0.02	5.35 ± 0.27
Pb (mg/kg)	4.0±1.8	18.5±9.6	173.8±18.8	-
P (g/kg)	-	84.5±17.3	0.21± 0.04	143 ± 39
V (mg/kg)	-	-	-	158 ± 9
Zn (mg/kg)	9.9±1.6	-	-	278 ± 13

Values based on a dry material basis obtained from independent collaborative campaigns. Uncertainties associated with reference values are estimates expressed typically like 95% confidence intervals for a single future determination based on a sample weight of at least 200 mg and include de expanded uncertainties related to homogeneities, characterizations, and long-term stabilities.

Acknowledgments: For many workers assisting with various aspects of the development, including the cooperating analysts and laboratories taking part in the characterization exercises. We also thank the financial support from CNPq, INCTAA, and CAPES.

References

[1] Comar - *International database of certified reference materials.* http://www.comar.bam.de/en/ Accessed on August 2015

[2] Ihnat M 2000 *J. Radional. Nuclear Chem.* **245** 65

[3] Silva P H T 2011 *Avaliação de parâmetros e produção de materiais de referência de mistura mineral para nutrientes e contaminantes inorgânicos*. PPGQ/UFSCar, São Carlos SP, Brazil. MSc Thesis

[4] Bossu C M 2013 *Produção e caracterização de material de referência de forrageira para nutrientes e contaminantes inorgânicos*. PPGQ/UFSCar, São Carlos SP, Brazil. PhD Thesis

[5] Verhalen T R 2014 *Preparo e caracterização de material de referência de solo para nutrientes e contaminantes inorgânicos.* PPGQ/UFSCar, São Carlos SP, Brazil. MSc Thesis

[6] Silva C S 2016 *Produção e avaliação de materiais de referência para nutrientes e contaminantes inorgânicos em amostras agronômicas*. PPGQ/UFSCar, São Carlos SP, Brazil. PhD Thesis

[7] ISO Guide 30:2015 *Reference materials - Selected terms and definitions*

Journal of Physics: Conference Series **733** (2016) 012005 doi:10.1088/1742-6596/733/1/012005

[8] ISO Guide 31:2015 *Reference materials - Contents of certificates, labels and accompanying documentation*

[9] ISO Guide 35:2006 *Reference Materials - General and statistical principles for certification*

[10] Souza G B *et al* 2009 *Accred Qual Assur* **14** 455

Journal of Physics: Conference Series **733** (2016) 012006 doi:10.1088/1742-6596/733/1/012006

Production and characterization of a bovine liver candidate reference material

S R Bianchi[1,2], A M J Peixoto[1], G B Souza[1], R R Tullio[1], A R A Nogueira[1]
[1]Embrapa Pecuaria Sudeste, Rodovia Washington Luiz Km 234, 13560-970, São Carlos-SP, Brazil; [2]Embrapa Solos, Rio de Janeiro-RJ, Brazil.

E-mail: ana.nogueira@embrapa.br

Abstract. The preparation of a bovine liver candidate reference material and the steps are taken to confirm its homogeneity, long and short term stabilities, and consensus values are described. Details of the sample preparation and the final collaborative exercise are presented. The material elemental composition was characterized by 17 elements (As, Ca, Cd, Co, Cu, Fe, K, Mg, Mo, Mn, Na, P, Pb, Se, Sr, V, and Zn) of nutritional and toxicological significance.

1. Introduction

Brazil is one of the largest exporters of beef cattle in the world. According to the Ministry of Agriculture, in 2020, it is expected that domestic production of meat will supply 44.5% of the global marketplace [1].

Increasing demands are being placed on the analytical scientific community to provide information on a growing range of inorganic and organic constituents in a vast array of commodities, including meat and meat products. Because of the rather wide variety and origins of such products and the wide range of elemental contents that are encountered, meats are subject to extensive routine monitoring for both toxic and essential elements. One of the ways to certify the quality and safety of meat for national consumer and exportation is through the determination of nutrients and toxic elements. These analyze call for the need for development and validation of analytical methods able to determine such elements with appropriate precision and accuracy.

To evaluate some figures of merit for method validation, control the quality of analyzes and consequently the quality of the product, certified reference materials (CRM) are needed, preferentially with properties similar to the matrix to be analyzed [2]. The production of CRMs and reference materials (RM) in Brazil is still incipient [3,4].

With the view to attempting to fill some of the gaps, there is a need to contribute to the technological development of the country, increasing the availability of new reference material in various areas have become targets of researchers and institutions [5,6]. In this context, the preparation and characterization of a bovine liver candidate reference material were performed, according to with ISO GUIDE series 30-35 [2,7-10].

2. Experimental

2.1. Preparation of bovine liver

The liver used in this study was harvested from 57 Nelore breed (*Bos indicus*) cattle provided from breeding experiments developed at Embrapa Southeast Livestock (São Carlos SP, Brazil). The bovine livers were collected according to the Embrapa's "Animals Ethic

Commission", and were derived from clinically healthy animals, produced under sanitary conditions.

The cattle were fed on pasture and in the last 90 days in the feedlot. The animals were slaughtered in a commercial environment slaughtered, and livers were collected 24 h after slaughter. Approximately 320 kg fresh weight of liver was collected, washed with ultrapure water and cutting in small pieces with ceramic knives to minimize contaminations. The livers were placed in polyethylene bags and frozen at -20°C until further processing. The frozen pieces of liver were then ground in a stainless steel comminuting machine (Skymsen, Brusque SC, Brazil), placed on polyethylene-lined trays, frozen, and lyophilized in a commercial food industry (Liofoods, Araras SP, Brazil). About 34 kg of lyophilized material was then ground in high-speed centrifuge mill (ZM 200, Retsch, Germany), by using 500 μm in the first step and 250 μm in a second one to obtain smaller particles sizes.

Afterward, the sample was divided into three buckets and homogenized using a Y homogenizer covered with inert resin (Model MA 201/5MO Marconi, Piracicaba SP, Brazil). About 3 kg of material was mixing during each run and stirred for 15 min. The homogenizing procedure was repeated three times to ensure homogeneity.

After homogenization, the sample was bottled in previously demineralized amber glass bottles. A total of 330 vials, containing 100 g of material was bulk sterilized with 25-30 kGy of cobalt 60 y-radiation by IPEN, Institute of Energy Nuclear Research, (São Paulo SP, Brazil).

A bottle was subjected to particle size distribution test performed at the Nuclear Center Agriculture Energy (CENA, Piracicaba–SP, Brazil), using equipment Analysette 22 MicroTec Plus (Fritsch, Germany).

2.2. Certification Project

A series of preliminary experiments was undertaken to delineate the method of determination procedure. The sample was microwave assisted digested, and inductively coupled plasma optical emission spectroscopy (ICP OES) and inductively plasma coupled mass spectroscopy (ICP-MS) were using as measurement techniques. Certified reference material of bovine liver (SRM® 1577b) from National Institute of Standards and Technology (NIST, USA) was used to check the accuracy of As, Ca, Cd, Co, Cu, Fe, K, Mg, Mo, Mn, Na, P, Pb, Se, Sr, V, and Zn mass fraction determinations.

2.3. Homogeneity Studies

The between-bottle homogeneity test aims to check for variability among bottles. Ten bottles of the candidate material were randomly selected, and to each flask were carried out three replicates of 250 mg, the elements were determined and subjected to analyzes of variance (ANOVA).

For achievement of within-bottle homogeneity testing, a flask was also randomly chosen (7 replicates) and the elements were evaluated in triplicates.

2.4. Test of Minimum Mass

The minimum mass study was conducted according to the standards recommended by the ISO GUIDE 34. This study was carried out using three different masses of beef liver candidate material: 100, 200, and 250 mg ± 0.001 mg. A total of 7 digestions for each mass were performed. The results were evaluated based on the standard deviation values, and it was concluded that the minimum sample intake should be 250 mg, that present lower values of

RSD, defined as the minimum mass. Otherwise, INAA analyzes were performed using test portions in the range of 150-180 mg and were observed homogeneity for distribution of Co, Cu, Fe, K, Mo, Mn, Na, Se, and Zn within the uncertainty of the method.

2.5. Determination of stability

Short term stability was assessed over a total of 30 days by using ICP OES determination method. Tree bottles were randomly selected and subjected to a storage regime, including two-time points (0 and 30 days) in the controlled condition of high humidity 95±5 % and temperature (40°C). Luminance was not evaluated, considering the samples were bottled in amber glasses and individually sealed in aluminum-nylon pouches.

The determination of the elements was performed in triplicate in the time 0, at the beginning of the experiment, and after 30 days.

Long-term stability study was also performed using ICP OES method. Three vials of the reference material candidate were also randomly selected, storage in a cold chamber with controlled temperature and humidity (10°C and 25%). The determinations of elements were made every 8 weeks for 12 months and then the results were available by ANOVA single factor.

2.6. Collaborative study

The collaborative study beginning with the chosen and an invitation to laboratories specialized that performed biological analysis. Each laboratory received a bottle containing 100 g of the candidate material.

The laboratories were invited to apply the methods they considered most reliable in their hands for the elements they are used to do in bovine liver. The contents of the bottles were required to be mixed before each analytical sub-sample was taken. The measurements for the elements were made in six (6) steps, and all the independent results were reported, expressed on a dry mass basis. The dry mass correction factor was determined on a separate portion of the sample, according to a specified procedure (oven drying at 105°C).

3. Results

3.1. Uncertainty evaluation

Ten separate vials were statistically evaluated with a single factor to determine the inner- and inter-bottle analysis of variance (ANOVA) to determine the uncertainty components of the available material.

3.2. Short and long-term stability

The single factor ANOVA results with a significance level of 5% ($\alpha = 0.05$) to check the short-term stability. All evaluated elements proved to be stable, with no statistically significant differences comparing the values recorded on the first day and after 30 days period.

The long-term stability data were evaluated during 12 months, and the single factor ANOVA with a significance level of 5% ($\alpha = 0.05$) and no significant differences were found in the samples (Fcritic > F anal). A regression analysis of the plotted data was also performed, indicating slope (*b*) close to zero. Thus, the samples were stable even with open vials and stored at 10°C and 25% humidity. The obtained values were considered in the evaluation of expanded uncertainty.

3.3. Assignment of property values and uncertainties obtained

The uncertainty implies confidence in the validity of the measurement result. Value with its uncertainty allows us to evaluate the precision or accuracy of an analytical method.

It was calculated the uncertainty associated with the characterization pattern (u*car*), homogeneity (u*bb*), the long-term stability (u*lts*) and the expanded uncertainty of the candidate reference material (UMR).

3.4. Characterization

The characterization was based on results from either a primary analytical technique carried out by instrumental nuclear atomic atomization (INAA), or the combined results from two or more chemically independent analytical techniques obtained at Embrapa and collaborating expert laboratories, according proposed by ISO GUIDE 35 [10].

4. Conclusions

The produced Brazilian bovine liver reference material is available to the laboratories and is particularly useful in the investigation of contaminants and nutrients in food analysis.

Acknowledgements: The authors warmly acknowledge the financial support from CNPq, INCTAA, and CAPES, and the collaboration of the following laboratories: LANAGRO/PE, LANAGRO/RS, LANAGRO/SP, CEA/UNESP Rio Claro/SP, Embrapa Solos/RJ, EXATA, Jataí/GO, DQ/UFSM, CENA/USP, IPEN, and IQ/USP.

References

[1] Ministério da Agricultura. www.agricultura.gov.br/animal/exportacao. Accessed on Augst 2015
[2] ISO Guide 33:2015 *Reference materials- Good practice in using reference materials*
[3] Cardoso M H W M et al 2010 *Ci Tecn Alimentos* **30** 429
[4] Park K et al 2011 *J. Food Comp Anal* **24** 1064
[5] Valente A et al 2014 *Food Chem* **154** 71
[6] Santos A M P et al 2015 *Food Chem* **178** 287
[7] ISO Guide 30:2015 *Reference materials - Selected terms and definitions*
[8] ISO Guide 31:2015 *Reference materials - Contents of certificates, labels and accompanying documentation*
[9] ISO Guide 34:2009 *General requirements for the competence of reference material producers*
[10] ISO Guide 35:2006 *Reference Materials - General and statistical principles for certification*

Capacity assessment of a system for metrological traceability on liquid micro flow rate measurement

B L Dias, M H Aquino, M H Farias, J Koiller, P C Gabriel, V Y Aibe

Fluid Dynamics Metrology Division; Institute of Metrology, Quality and Technology – INMETRO, Xerém, Duque de Caxias, Brazil

mhaquino@inmetro.gov.br

Abstract. The necessity of standardizing and traceability to SI in micro scale flow has been subject of intense discuss in the metrology community. The Bureau International of Poids et Mesures (BIPM) has encouraged National Metrology Institutes to be prepared to participate, in 2024, of the first Key Comparison event planned in this area. With this aim, the present work describes a standard system based on gravimetric method for fluid delivery and respective quantification, which was developed by the National Institute of Metrology, Quality and Technology (INMETRO) in order to provide traceability for microflow rate measurement. The system performance and capability were evaluated. An important goal of the work is contribute to the establishment and advancement of researches on microfluidic issues in the Institute.

.Keywords: micro flow rate; microfluidics; metrological traceability; gravimetric method

1. Introduction

Microfluidics can be applied to several fields of the technology. This science and their applications can optimize and improve processes as those found in areas such as biomedical, pharmaceutical, chemical, environmental, aerospace and others. However, to ensure reliability on quantifying fluid flow rate in micro scale is a challenging task, due to the need of controlling variables that influence in the measurement process. Nowadays, there is a gap in the metrological traceability chain for very low flow rate. Due to this gap, the Bureau International des Poids et Mesures – BIPM [1] has encouraged National Metrology Institutes to offer systems for providing metrological traceability on micro flow rate measurement up to 2024, when will run the first international key comparison event in microfluidic.

The metrological community has discussed about the microscale flow rate measurement. Several techniques and methods have been suggested by researchers, whom has characterized and calibrated microfluidic systems by using such techniques and methods [2-11].

The present work describes the new standard system for metrological traceability on liquid micro flow rate measurement which was developed in the Fluid Dynamics Division of the National Metrology Institute of Quality and Technology (INMETRO).The system capability is evaluated under different ranges of low flow rates. Since with this system implementation will be possible to study flow in micro channels with traceability to national standards, the results can be used as reference to fluid velocity measurement by applying the microPIV (micro Particle Image Velocimetry) technique.

2. Methods and Instrumentation

The measurement technique of the Inmetro´s standard system is based on gravimetric method because the good accuracy in conversion of measured mass to volume. This method consists on measuring the mass which is contained or delivered to a recipient, and the conversion from mass to volume of liquid is performed according to the mathematical model (Equation1) based on the equation described in ISO 4787 [13].

2.1. Setting up the system

The system assembling and setting is very relevant, since are aspects that strongly influence the measurement results. There are several types of gravimetric setups, syringes and infusion pumps. In this work the system was composed by:

- Syringe: 1 ml with resolution 1µl
- Balance: 610g with resolution 0.1mg
- Micro Pump with accuracy ± 0.5% and Resolution 0.00001 ml/min
- Glassware to receive water from syringe

In the system capacity evaluation were considered important effects such as evaporation, thermal and environmental influences. The evaporation can be a critical influence in gravimetric systems and the difficulty for controlling it is high. In these experiments, a plastic film was placed on the upper edge of the glassware for reducing water evaporation. Before receiving the liquid, the plastic film was crossed by the syringe needle (see "2"on figure 1). On the thermal influence, the water temperature variation can cause thermal expansion of the syringe material and then, change its volume. Therefore, it is advisable to create and to maintain a system with temperature, pressure and humidity monitored by calibrated instruments.

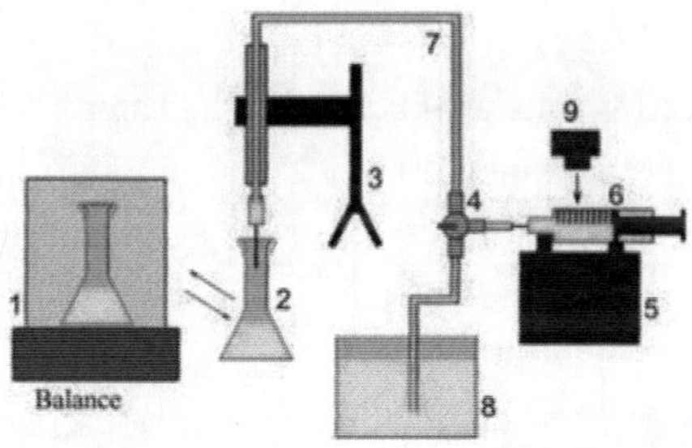

Figure 1. System for microflow rate measurement.

2.2. Techniques and measurement procedures

The apparatus schematic is shown in figure 1. The syringe (6) was attached to the infusion pump (5) and connected to a hose (7). With a three way valve, through hoses the syringe pump was connected to a glassware receiver (2) and a water reservoir (8). A camera (9) was used to read the volume of fluid (according to the syringe piston position) on the syringe scale. A flow rate was set in the pump, and before infusion, the initial water temperature was measured. The reference temperature was 20°C. While transferring water from the syringe (6) to the glassware (2), the air temperature, humidity and pressure were monitored. The mass of the glassware (before and after collecting water) was measured by the balance (1) at each run. There were performed six measurements for each flow rate, in order to assess the system repeatability. Also, at each run it were registered the volume of fluid read on the syringe scale by using the camera (figure 1). After that, the conversion from mass to volume at reference temperature 20°C, as described at the mathematical model (equation 1) was made. So, based on the calculated volume by gravimetric method and on the study of the setting up of the gravimetric system, including associated uncertainty components analysis, it was possible to ensure the traceability of micro flow rate to SI (International System of Units).

Considering the previously quoted variables, the flowrate capacity of the system under different syringe volumes were:

Table 1. The system capacity

Syringe Volume	100 µl	1 ml	10 ml
Highest Flow rate	266.447µl/min	2.47528 ml/min	25.5609ml/min
Lowest Flow rate	0.33µl/min	3.3µl/min	0.0323ml/min

After identifying the flowrate ranges, it was chosen the syringe of 1ml in capacity for running tests and evaluating the uncertainties by using the gravimetric method. The syringe capacity was 1 ml, but in each run the totalized volume was 0.8 ml in order to visualize appropriately the delivered volume of water, and avoid barrel end effects. In the experiments, to improve the visualization of the syringe scale and avoid parallax error in reading it, a camera was installed as shown on figure 1.

During the system assembly additional care was taken, for instance, environmental influence was controlled (causing a thermal equilibrium), the system was insulated of vibration (for stability and equilibrium during the measurements), air entrance in the system was prevented and air bubble trapping was minimized, etc. The syringe and hose were filled with deionized water very carefully before starting the first run. The basic sequence for filling the hydraulic system with water and delivering this fluid was: after delivering liquid to the glassware, the syringe was filled again by collecting water from the reservoir. The water density was determined using an equation recommended by BIPM, the Tanaka Equation [13].

3. Mathematical model

The following equation was used for volume of liquid determination:

$$\Delta Ve = \frac{Mc - Mv + M_E}{\left(\rho_L[T_L] + \delta\rho_L[T_L] - \rho_{ar}\right)} \cdot \left(1 - \frac{\rho_{ab}}{\rho_b}\right) \cdot \left(1 + \alpha_v \cdot (T_r - T_d)\right) + \delta R \quad (1)$$

Where:

- ΔV_e- is the volume of liquid at the reference temperature ($T_r = 20°C$)
- M_c- is the apparent mass (glassware), after receiving the liquid
- M_V–is the apparent mass (glassware), before receiving the liquid
- M_E – is the evaporated mass
- ρ_L- is the liquid density at temperature T_L
- T_L- is the liquid temperature
- δT_L - error due to the liquid temperature variation in space and time
- ρ_A- is the air density during the apparent water mass measurement
- ρ_{ab}– is the air density during balance calibration
- ρ_b – is the density of the standard weight used in balance calibration
- α_V - is the volumetric thermal expansion coefficient

- T_d - is the syringe(barrel) temperature
- δR - is the error due to repeatability of volume measurement

4. Analysis of results

Table 2 shows the results for different flow rate levels (mean of six measurement at each level). The reading correction is the difference between the calculated volume by equation 1 and the indicated volume on the syringe scale. The relative and expanded uncertainties were calculated considering all factors of the mathematical model [14, 15].

Table 2. Results

Calculated Volume (ml)	0.7988	0.7971	0.7924	0.7958
Nominal Flow rate	0.013 ml/min	6.6 µl/min	4.4 µl/min	3.3µl/min
Reading Correction (%)	0.0312	0.0859	0. 0382	0.0330
Expanded Uncertainty(ml)	0.00173	0.00185	0.00174	0.00173
Relative Uncertainty (%)	0.22	0.23	0.22	0.22

The relative contributions to the uncertainty are presented in table 3.

Table 3. Contribution to Uncertainty

Contribution to uncertainty	Relative Contribution To 0,013 ml/min	Relative Contribution To 6.6 µl/min	Relative Contribution To 4.4 µl/min	Relative Contribution To 3.3 µl/min
Read Volume with camera	32.206	28.627	31.680	32,060
Evaporation Mass	20.108	17.874	19.780	20.016
Mass before transferred	20.108	17.874	19.780	20.016
Mass delivered	20.108	17.874	19.780	20.016
Repetition	7.101	17.394	8.556	7.484

In table 3 can be seen that the read volume on the barrel scale represents the major contribution for uncertainties at all tested flow rates. While at the flow rate 6.6 µl/min the contribution due to the repetition is in the same range of that due to each mass evaluated, for other experimented flow rate levels there is not this tendency. More investigation has to be done in order to explore in details this aspect.

5. Conclusion

In this work, a system developed for providing metrological traceability in microflowrate measurement was evaluated.

The main sources for uncertainty of the bench were identified. It was found that the system performance could be improved if a syringe and a balance with better resolution were used, and also the way of reading the syringe scale was refined. However, as first studies of the bench, the results were considered satisfactory, and this system could be used as reference in experiments using the microPIV(micro Particle Image Velocity) technique, for example. The system development and validation for the bench improvement are in progress, and in this way, some points will be explored in more details, as e.g.:

- Evaporation effects
- Uncertainty components
- Aspects of the gravimetric configuration

The National Institute of Metrology, Quality and Technology(Inmetro) aims to improve this system and, in the future, participate of intercomparison.

6. Acknowledgment

The authors are grateful to CNPq, FAPERJ, FINEP, PETROBRAS, PFRH-PETROBRAS and INMETRO for their support and sponsorship which have become possible the development of this research.

7. References

[1] Bureau International des PoidsetMesures–BIPM.; strategy 2014-2024, *consultative committee for mass and related quantities (ccm).*

[2] Hannu a. Sairanen, Marti, O. Heinonem.; 2014. *Hygrometer-based calibration method for syringe pumps providing ultra-low liquid flows.Microfluidicnanofluidic*, **16:**187-193.

[3] Haswell, S.J., 1997. *Development and operating characteristics of micro flow injection analysis systems based on eletroosmotic flow, Analyst.*;**122 (1R-10R).**

[4] Marinozzi, F., Bini, F. andCappa, P., 2005. *Calibrator for micro flow delivery systems, Rev.Sci.* Instr., 76.

[5] Nath, P., Roy, S., Conlisk, T. and Fleischman, A.J., 2005. *A system for micro/nano fluidic flow diagnostics, Biomedical Devices*, 7.

[6] Nuno Miguel Matos Pires, Tao Dong, UlrikHanke and Nils Hoivik.; 2014.*Recent Developments in Optical Detection - Technologies in Lab-on-a-Chip Devices for Biosensing Applications, Sensors,* **14, pp15458-15479**

[7] Reitz, S., Kummrow, A., Kammel, M. And Neukammer, J., 2010. *Determination of micro-litre volumes with high accuracy for flow cytometric blood cell counting, Meas. Sci. Technol.,* **21**

[8] Wolf, H., 2010. *Microflow meets metrology, meas. Sci. Technol.,***21.**

[9] Bruss, Henrik., 2008. *Theoretical Microfluidics.***1**

[10] Yi-Jui Liu , Juin-Yi Yang , Yung-Mau Nie, Chun-Hung Lu, Eric Dowkon Huang, Chow-Shing Shin, Patrice Baldeck, Chih-Lang Lin.; 2015. *A simple and direct reading flow meter fabricated by Two-photonpolymerization for microfluidic channel* , *MicrofluidNanofluid* **18:**427–431

[11] König, J., Voigt, A., Büttner, L. and Czarske, J., 2010. *Precise micro flow rate measurements by a laser Doppler velocity profile sensor with time division multiplexing, Meas. Sci. Technol.,* **21**.

[12] ISO 4787:2010 – *laboratory glassware –volumetric instruments methods for testing of capacity and for use.*

[13] M. Tanaka, G. Giard, R. Davis, A. Peuto and N. Begnell, 2001.*Recommended table for the density of water between 0ºC and 40ºC based on recent experimental reports.*

[14] Evaluation of measurement data,; 2008. *Guide to the expression of uncertainty in measurement – GUM.*

[15] International Vocabulary of Metrology.2008. *Basic and General concepts and associated terms (VIM)* **3rd edition**.

Journal of Physics: Conference Series **733** (2016) 012008 doi:10.1088/1742-6596/733/1/012008

Information systems as a tool to improve legal metrology activities

B A Rodrigues Filho[1,2], A N R Soratto[1] and R F Gonçalves[2]

[1] National Institute of Metrology, Quality and Technology, Nossa Senhora das Graças Av., 50, Duque de Caxias, RJ, Brazil
[2] Paulista University, Dr. Bacelar St., 1212, 4th floor, Vila Clementino, São Paulo, SP Brazil

Email: bafilho@inmetro.gov.br

Abstract. This study explores the importance of information systems applied to legal metrology as a tool to improve the control of measuring instruments used in trade. The information system implanted in Brazil has also helped to understand and appraise the control of the measurements due to the behavior of the errors and deviations of instruments used in trade, allowing the allocation of resources wisely, leading to a more effective planning and control on the legal metrology field. A study case analyzing the fuel sector is carried out in order to show the conformity of fuel dispersers according to maximum permissible errors. The statistics of measurement errors of 167,310 fuel dispensers of gasoline, ethanol and diesel used in the field were analyzed demonstrating the accordance of the fuel market in Brazil to the legal requirements.

1. Introduction

Legal metrology is responsible for apply regulations to measuring instruments that impact both economy and society, establishing requirement, as maximum permissible error, in order to provide the accuracy of measurements, maintaining measurement deviations to acceptable levels, arising the confidence to the market [1].

Also, the impacts of measuring instruments can be significant to a country, affecting a significant part of its Gross National Product (GNP) [2], since a 0.1% error in measurements would represent 0.05% of the Gross Domestic Product (GDP) [1]. Moreover, deviations in the commerce can be related to fair competition and consumers´ protections issues.

The asymmetry between buyers and sellers in the market due to a measurement error, causing a monetary impact, is described as an economic distortion, or asymmetry of information, and it is higher as the instrument is less accurate. This asymmetry can also be used as an indicator to measure the economic impact of legal metrology [3].

The legal metrological control is the tool used to maintain under control this asymmetry, using type approval, verifications and surveillance activities to keep these errors according to established in the regulations regarding measuring instruments used in trade.

Information systems (IS) were introduced in organizations mainly to support managers [4]. The concept of an information system involves not only technical aspects of a system as the software or hardware, but also all the information flow, i.e. data and human resources [5].

The aspects of developing an information system, a technology information tool, are wider than developing software or a platform as it also incorporates human, administrative and organization

Journal of Physics: Conference Series **733** (2016) 012008 doi:10.1088/1742-6596/733/1/012008

aspects [6]. In addition, the IS when properly lined up to the strategic requirements of the organization contributes to innovation [7].

Recently, information systems have been introduced in legal metrology successfully in order to organize and improve the control, as the system proposed to assist the implementation of the Measuring Instrument Directive in Europe, simulating the behaviour of measuring instruments in the market [8]. Also in Europe a system was developed to organize and manage the information among notified bodies in legal metrology [9].

In prepackaged, a system was also proposed to control of the mark on the products in order to comply the regulation in the European Union [10].

The present study aims to explore how information systems applied to the legal metrology activities have helped Inmetro, the National Metrology Institute in Brazil, to improve the level of control over the market arising confidence in measurements. The fuel market is used as a study case to relate the enhancements achieved in the sector.

2. Legal Metrology

The legal metrological control comprises the activities that aim to provide confidence to measurements used in trade and economy. These activities are divided in three distinct levels of control [11]: type approval and initial verification; subsequent verification; surveillance.

Type approval represents the compliance of the devices to constructive characteristics before their production, in order to guarantee that the instruments manufactured or imported are according to the national regulation and initial verification checks the instruments accordance to the approved features. Afterwards, the subsequent verification represents the level of control of the instruments after they are put in the market, testing periodically or after a repair if the device fulfils the requirements. Finally, the surveillance is the level of control focused mainly in identifying metrological frauds and misuses of the instruments in the market. These activities may be conducted by public or private bodies according to each country necessity.

In Brazil, the activities of legal metrological control are aligned to International Organization of Legal Metrology recommendations. The National Metrology Institute (NMI), Inmetro is responsible for legal metrology in Brazil, aggregating the activities of regulation, accreditation, conformity assessment and traceability of standards to the International System (IS).

Specifically in the legal metrology field, while type approval is conducted by the Inmetro, both initial and subsequent verifications (periodic and after repair) as well as surveillance are conducted by the 26 delegated public bodies over the country [12].

Despite after repair verification being conducted by the delegate bodies, the repairs are executed by 3,281 notified bodies that are responsible for maintenance and putting the instrument in use in the market [13]. The delegate bodies conduct the after repair verification based on the repair reported informed by the repairer.

3. Information System

3.1. Acquiring data of subsequent verification

The Brazilian NMI has implanted in 2011 an Integrated Management System (IMS) in order to aggregate the information provided from legal control of measuring instruments carried out by the delegated bodies, unifying the results of metrological tests in the market, integrating the results, allowing Inmetro to plan and act more efficiently.

The metrologists in the field transmit the information regarding the metrological tests of subsequent verifications and inspections, according to OIML recommendations, using a data collector.

Afterwards, the information inserted during the tests are synchronized periodically to a database server, integrating all the results allowing to the system to produce reports for a more effective control of the legal metrology activities.

Figure 1 displays the data collector screen for a fuel dispenser verification test for high volume flow test (45 ℓ/min), where the instrument displays a volume of 19.8 ℓ and the standard 20.0 ℓ.

Figure 1. Data collector connected to IMS database used to perform metrological tests.

3.2. Harmonization and standardization

One of the main aspects of an integrated information system regards about harmonization of procedures, once metrological tests according to OIML recommendations are already programed in the data collector. The harmonization of procedures also contributes to the standardization of instruments used in trade as they are subjected to harmonized procedures of verification.

The consistence of data due to harmonized and standard information allows obtaining macroeconomic indicators regarding the economic impact of measuring errors in the market.

3.3. Output data reports and data mining

Once measuring errors of the instruments in the market are collected and storage in a single data base, it is possible to obtain reports informing metrological characteristics and geographical location of measuring instruments in the market; repairs executed by notified bodies; workforce necessary to meet the demand; measuring errors; and others.

The errors of the instruments used in the market are an important output of the system as they allow computing the economic distortion indicator, representing the impact of the legal metrology activities on the economy.

4. Methodology

The results represent the tests carried out by the delegated bodies in both periodic and after repair verification in fuel dispensers used in the market from June to December 2014, covering 20 of 26 states in Brazil. The subsequent verifications covered 78,804 dispersers of gasoline, 37,719 of ethanol and 50,787 of diesel during the mentioned period.

The metrological tests used for the present study are conducted considering the volume accuracy at maximum flow rate, according to Brazilian Regulation [14] that is aligned to OIML Recommendation [15]. The measure procedure for the volume tests consists in a measurement standard of nominal standard value of 20 ℓ and tests conducted at the maximum and minimum flow rate current regulation. The maximum permissible error for the tests is $\pm$ 0.5% what represents a volume of $\pm$ 100 ml for 20 ℓ dispensed.

Furthermore, the measurement error bias is also an important aspect to be considered in the analyses of the results. A measurement error is described as a measured quantity value minus a reference quantity value. Thus, in the commerce involving buyers and sellers, positive values would represent consumers′ losses and negative values sellers′ losses.

5. Results and discussion

The distribution of the measurement errors of fuel dispensers trading gasoline used in the market is showed in figure 2. Similar distributions are obtained to both ethanol and diesel dispensers.

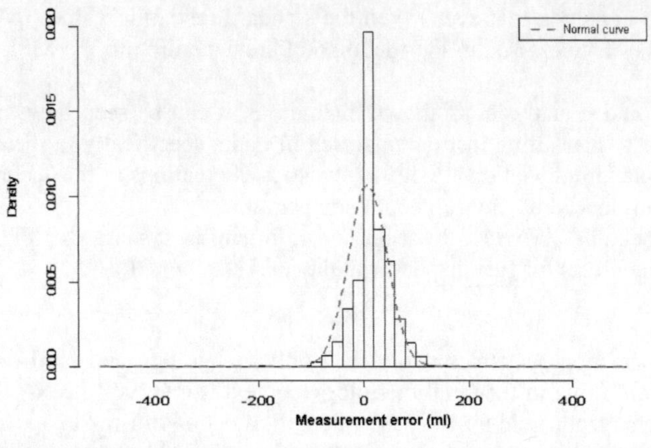

Figure 2. Measurement error distribution for gasoline fuel dispenser.

Table 1 displays the average value x of the measuring errors as well as the standard deviation σ for the dispensers according to the fuel traded, and figure 3 shows the average and standard deviation for gasoline, ethanol and diesel dispensers.

Table 1. Average and standard deviation of fuel dispensers used in trade.

Fuel dispenser	x	σ
Gasoline	7.99	45.99
Ethanol	4.22	40.83
Diesel	6.68	49.52

Average values of x =7.99 ml to gasoline, 4.22 ml to ethanol and 6.68 to diesel dispensers represent a bias toward consumer´s losses, however these values are inferior to the MPE, demonstrating the accordance of to the current regulation.

Despite average values closer to zero, the standard deviations, according to table 1 shows a big dispersion around the average. However, since the measurement errors of volume of the fuel dispensers varies from negative to positive values, and this measure is regulated, a standard deviation bigger than the average is expected.

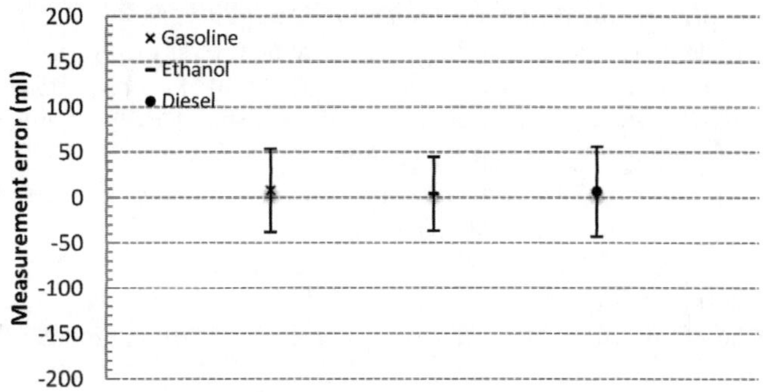

Figure 3. Average and standard deviation of fuel dispensers according to fuel delivered.

It is also possible to conclude that even when the standard deviation values (σ) 45.99, 40.83 and 49.52 are considered, the devices used in the fuel market are according to the MPE established in the national regulation.

Then, due to the x and σ analysis, as shown in figure 3, it can be seen how the legal metrology provides the confidence to measuring instruments used in trade, specifically in the fuel market in this study. High values of standard deviation are due to statistics fluctuations of the accuracy test, once fuel dispensers are greatly influenced by environmental temperature.

The harmonized procedures provided by a unified information system give the reliability for the distribution of measuring errors for fuel dispenser in the market.

6. Conclusion

The concept of an information system is wider than only technology and hardware, involving also human resources and data flow. In the legal metrology context it involves the NMI (Inmetro) and the 26 delegate public bodies responsible for the subsequent verification in Brazil.

The Integrated Management System used to connect the stakeholder has showed an adequate tool to harmonize procedures regarding verifications and metrological tests leading the whole system to an adequate level of standardization.

The metrological tests conducted in verifications show that the control of fuel dispensers is according to regulation, detailing both average and standard deviation for the instruments errors.

Both information (x and σ) allows the NMI to make decisions regarding technology used in the sector, regulations and procedures review, in order to decrease the impact of measures in trade due to either accuracy or precision control.

Both average and deviation results measuring errors are relevant information in order to obtain macroeconomic indicators and the economic distortions in order to measure the impact of legal metrology to the economy.

References

[1] Birch J. Economic and Social Benefits of Legal Metrology. *OIML Bull.* 2004 Apr; XLV 10–3.
[2] Carstens S. AFRIMETS (Intra–Africa Metrology System). *OIML Bull.* 2010 Jan; LI 5–10.
[3] Rodrigues Filho BA, Gonçalves RF. Legal metrology, the economy and society: A systematic literature review. *Measurement.* 2015 Jun;69:155–63.
[4] Gupta YP. Management information systems planning: analysis and techniques. *Technovation.* 1989 May; 9 63–81.
[5] Alter S. *Information Systems: a management perspective*. Addison-Wesley Publishing Co. Massachusetts; 1992.
[6] Keen PGW. Information Technology And The Management Theory: The Fusion Map. *IBM Syst J.* 1993; 32 17–38.
[7] Yang M-L, Wang AM-L, Cheng K-C. The impact of quality of IS information and budget slack on innovation performance. *Technovation.* 2009 Aug; 29 527–36.
[8] Crenna F, Rossi GB, Bovio L. Probabilistic measurement evaluation for the implementation of the Measuring Instrument Directive. *Measurement.* 2009 Dec; 42 1522–31.
[9] Oosterman C, Dixon P. NoBoMet: A European platform for Notified Bodies working in Legal Metrology. *OIML Bull.* 2012 Jan; LIII 11–3.
[10] Nitescu D. Design a successful architecture for measurement control system in legal metrology. *UPB Sci Bull Ser C Electr Eng.* 2008; 70 145–56.
[11] Klenovsky P. Legal metrological control over measuring instruments in use: current situation. *OIML Bull.* 2006 Jul; XLVII 22–33.
[12] Silva PPA, Réche MME, Silva ME. The legal metrology Brazilian model. XVIII IMEKO Word Congress Proceedings. Rio de Janeiro, Brazil; 2006. p. 18.
[13] Rodrigues Filho BA, Silva ME. The PSIE platform as a tool to improve metrological control in Brazil. 4th CAFMET Proceedings. Pretoria, South Africa; 2014. p. 3.

Journal of Physics: Conference Series **733** (2016) 012008 doi:10.1088/1742-6596/733/1/012008

[14] Instruções relativas às condições a que devem satisfazer as bombas medidoras utilizadas em medições de volume de combustíveis líquidos. *Portaria Inmetro n° 023* Feb 25, 1985 (in portuguese).
[15] Testing procedures and test report format for pattern evaluation of fuel dispensers for motor vehicles. 1995. Available in: <www.oiml.org/en/files/pdf_r/r118-e95.pdf>.

Acknowledgement
The authors acknowledge the support of Inmetro, which enabled us to carry out this research.

Quality assurance of absorbed energy in Charpy impact test

C L F Rocha[1], D A K Fabricio[1], V M Costa[1] and A Reguly[1]

[1] Physical Metallurgy Laboratory (LAMEF) / Federal University of Rio Grande do Sul, 99 Osvaldo Aranha Avenue, Porto Alegre, Rio Grande do Sul, Brazil.

E-mail: cfanezi@demet.ufrgs.br

Abstract. In order to ensure the quality assurance and comply with standard requirements, an intralaboratory study has been performed for impact Charpy tests, involving two operators. The results based on ANOVA (Analysis of Variance) and Normalized Error statistical techniques pointed out that the execution of the tests is appropriate, because the implementation of quality assurance methods showed acceptable results.

1. Introduction

According to ASTM E23, Charpy impact test determines the amount of energy absorbed by a standard sample during fracture [1]. ISO/IEC 17025 standard stablishes the general requirements for technical competence in test and calibration laboratories. One of the requirements is the quality assurance of results based on pre-stablished acceptance criteria [2-3].

A statistical methodology for the evaluation of the significance of different factors and their interactions is the Analysis of Variance (ANOVA). This technique allows the comparison of variances between average values of a variable at different levels of a factor, resulting in a statistical value called p-value. This value represents the probability of significance of the analyzed factor. When *p-value* $< P$ (significance level), the analyzed factor is considered significant in relation to the residual error.

In monitoring the quality of test results, another statistical method often used is Normalized Error (E_n). This method tests the compatibility of measured values with a reference value. The value of E_n can be calculated according to equation (1) [5]. The result is considered acceptable when $E_n \leq 1$.

$$E_n = |X_1 - X_2|\left[(u_1^2 + u_2^2)^{1/2}\right]^{-1} \quad (1)$$

Equation (1) represents a ratio between the difference of the average values of two analyzed parameters ($X_1 - X_2$) and the square roots of the square sum of the expanded measurement uncertainty of both parameters (u_1 and u_2).

Thus, this paper intends to carry out a comparative study of the absorbed energy in impact testing, through the statistical methods of Normalized Error and ANOVA.

2. Material and Methods

Impact tests have been carried out with standardized test specimens, according to figure 1, type 'A' notch was prepared, following standard orientations for dimensional tolerance.

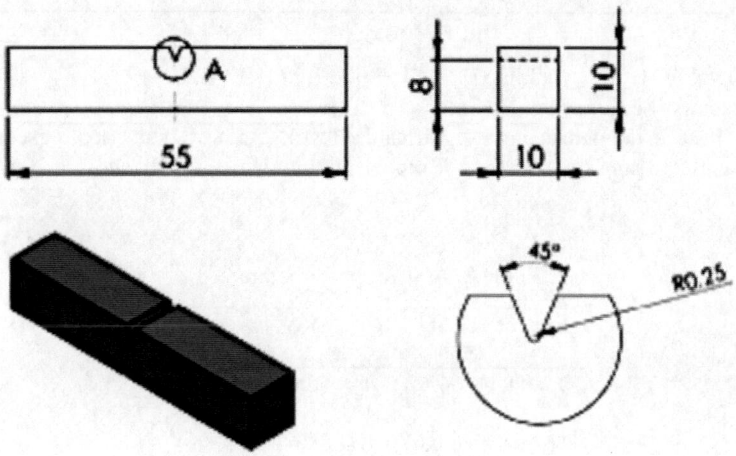

Figure 1. Test specimens, dimensions in millimeters.

For statistical analysis, two operators (A and B) have tested three test specimens each (i.e. 6 specimens under the same conditions). After measuring the samples, data has been analyzed through ANOVA and Normalized Error methods, considering operator as analyzed factor.

The acceptance criteria for normalized error has been $E_n \leq 1$ and for ANOVA has been *p-value* < *P*, as indicated in literature.

Impact tests have been carried out using an Instron analogic impact test machine, with 400 J of capacity and 2 J of resolution.

3. Results and Discussion

Results of absorbed energy (KV_8) in Charpy testing and the uncertainty associated to KV_8 ($u(KV_8)$) are shown in table 1. The estimation of the uncertainty was according to ISO/GUM [6], considering the resolution, standard deviation and calibration for KV_8.

Table 1. Charpy impact test results.

Specimen	KV_8	Mean KV_8	$u(KV_8)$	Operator
A.1	154 J			
A.2	152 J	156.67 J	17.04 J	A
A.3	164 J			
B.1	138 J			
B.2	152 J	150.00 J	29.24 J	B
B.3	160 J			

According to table 1, $E_n = |(156.67) - (150.00)|/[(17.04)^2+(29.24)^2]^{(1/2)} = 0.197$. Results are considered satisfactory, because $E_n \leq 1$.

ANOVA has been estimated using a significance level of 5%. Results are presented in table 2, where *SQ* represents the sum of squares, *DF* represents the degrees of freedom and *MS* the mean of squares.

Table 2. ANOVA for operators.

Source	*SQ*	*DF*	*MS*	*p-value*
Operator	66.67	1	66.67	0.42
Error	330.67	4	82.67	
Total	397.33	5		

According to the results in table 2, it was not possible to prove the effect of the operator at a significance value (P) of 5%, since *p-value* = 0.42 > P = 0.05. This result is satisfactory, since an effect of the operator in the result is undesired.

Thus, by both techniques (ANOVA and E_n), it is possible to conclude the operator effect does not affect the result. That is, Charpy impact test is considered compatible for this application.

4. Conclusion

The statistical analysis proposed in this paper, comply with the ISO/IEC 17025 quality assurance requirement for the Charpy impact test with in an easy and reliable method. Although the participation in interlaboratory tests is necessary.

The statistical test Normalized Error between operators has been resulted satisfactory, considering the value of E_n was smaller than 1.

The technique based on ANOVA has resulted in a conclusion equivalent to the normalized error, since *p-value* was larger than the specified significance level P = 5 %.

It is possible to conclude that these two statistical methods can be used for quality control of Charpy impact tests. The execution of the tests is appropriate, because the implementation of quality assurance methods showed acceptable results.

5. References

[1] Garcia A, Spim J and Santos C A 2012 Ensaio dos Materiais. Rio de Janeiro, LTC, vol. 2

[2] ASTM International 2007 ASTM E23: Standard Test Methods for Notched Bar Impact Testing of Metallic Materials. Pennsylvania

[3] Associação Brasileira de Normas Técnicas 2005 NBR ISO/IEC 17025: Requisitos gerais para a competência de laboratórios de ensaio e calibração. Rio de Janeiro

[4] Ribeiro J L D and Caten C S T 2012 Cartas de Controle para Variáveis, Cartas de Controle para Atributos, Função de Perda Quadrática, Análise de Sistemas de Medição. FEENG/UFRGS, Porto Alegre

[5] INMETRO 2010 DOQ-CGCRE-008: Orientação sobre Validação de Métodos Analíticos. Coordenação Geral de Acreditação, 3rd revision

[6] JCGM 2008 Evaluation of measurement data: Guide to the expression of uncertainty in measurement. Geneva

Intralaboratory comparison for the anisotropy test of sheet metals

D A K Fabricio[1], L Trevisan[2], A Reguly[3] and T R Strohaecker[3]

[1] Instituto Federal de Educação, Ciência e Tecnologia de Santa Catarina – Câmpus Chapecó. Rua Nereu Ramos, 3450 D, Chapecó, SC, Brasil
[2] Instituto Federal de Educação, Ciência e Tecnologia do Rio Grande do Sul – Câmpus Farroupilha. Av. São Vicente, 785, Farroupilha, RS, Brasil
[3] Laboratório de Metalurgia Física – Universidade Federal do Rio Grande do Sul. Avenida Bento Gonçalves, 9500 Prédio 43820, Porto Alegre, RS, Brasil

E-mail: danielkapper@gmail.com

Abstract. The plastic strain ratio is an important parameter for the evaluation of the anisotropy of sheet metals in deep-drawing processes. This paper presents the results of an intralaboratory comparison for the anisotropy test, using normalized error as statistic method of analysis. The study was carried out through the comparison of the results obtained from different operators for the same sheet metal. The results of the study are considered satisfactory.

1. Introduction

According to ASTM E517:2010 [1], the plastic strain ratio (r) is a parameter that indicates the ability of a sheet metal to resist thinning or thickening when subjected to either tensile or compressive forces in the plane of the sheet. It is a measure of plastic anisotropy and is related to the preferred crystallographic orientations within a polycrystalline metal. The r value, therefore, is considered a measure of sheet metal drawability.

The value of r can be calculated according to equation (1), where w_o and l_o represent the initial width and length of test specimens, and w_f and l_f represent the final width and length, respectively.

$$r = \frac{\ln\left(\frac{w_0}{w_f}\right)}{\ln\left(\frac{l_f w_f}{l_0 w_0}\right)} \tag{1}$$

Another parameter of the anisotropy test is the r_m value, which represents the weighted average of r values obtained in three directions: parallel (r_0), diagonal (r_{45}), and transverse (r_{90}) to the rolling direction [1], according to equation (2).

$$r_m = \frac{r_0 + 2r_{45} + r_{90}}{4} \tag{2}$$

ISO/IEC 17025:2005 standard stablishes management and technical requirements for the competence of testing and calibration laboratories. Among the technical requirements, there is the quality assurance of test results: the standard demands the laboratory to apply periodically procedures to monitor the quality of the measurements, using statistical techniques to analyze the results [2].

The Normalized Error (E_n) method tests the compatibility of measured values (X_1 and its uncertainty, u_1) to a reference value (X_2, with its uncertainty, u_2). E_n is calculated according to equation (3), and the result is considered compatible when $E_n \leq 1$. This method is widely used in quality assurance programs (intralaboratory and interlaboratory comparisons) according to the literature [3–7].

$$E_n = \frac{|X_1 - X_2|}{\sqrt{u_1^2 + u_2^2}} \tag{3}$$

Several cases of intralaboratory comparison are found in literature. For example, Cools *et al.* [8] have studied forest soil analysis, evaluating interlaboratory and intralaboratory variation. Lee and Hill [9] have evaluated intralaboratory repeatability of residual stress determined by the slitting method and, finally, Smagunova *et al.* [10] have discussed on-line control of the intralaboratory precision in chemical analysis.

Thus, the main purpose of this paper is to present the results of an intralaboratory comparison for the anisotropy (plastic strain ratio) test for sheet metals, using normalized error.

2. Material and methods

Test specimens have been taken from a sheet metal with dimensions 500 x 300 mm (figure 1) in three directions: 0° (parallel), 45° (diagonal), and 90° (transverse) to the rolling direction, in order to calculate r_m.

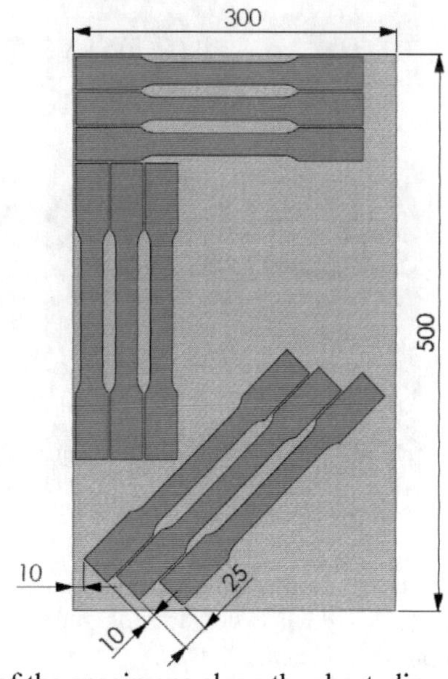

Figure 1. Position of the specimens along the sheet, dimensions in millimeters.

The geometry of each test specimen, accordingly to ASTM E517:2010 standard, is represented in figure 2.

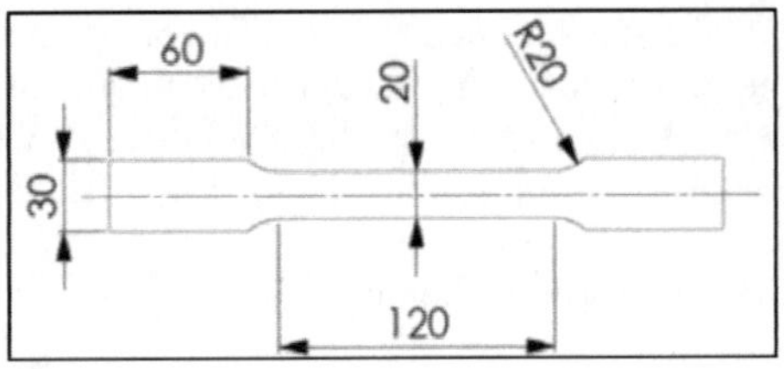

Figure 2. Test specimens, dimensions in millimeters.

Two operators ('A' and 'B') have received three test specimens, one from each direction (0°, 45° and 90°). They were individually instructed to carry out the test using a caliper rule for dimensional measurement, and a universal testing machine for the elongation of test specimens.

The values of l_o and w_o were measured before executing the test. Then, test specimens were inserted in the testing machine. The specified elongation was 10%, i. e., test specimens should be deformed until the relation $(l_f - l_o)/l_o$ was approximately equal to 0.1. After the deformation, the values of l_f and w_f were measured and used in the calculation of r_0, r_{45}, r_{90} and r_m.

Finally, measurement uncertainty of the results was estimated according to ISO/GUM [11].

3. Results and discussion

Tests were carried out using a universal tensile testing machine as illustrated in figure 3. Longitudinal strains were measured using a strain gauge extensometer while the strain in the width was measured with a digital caliper.

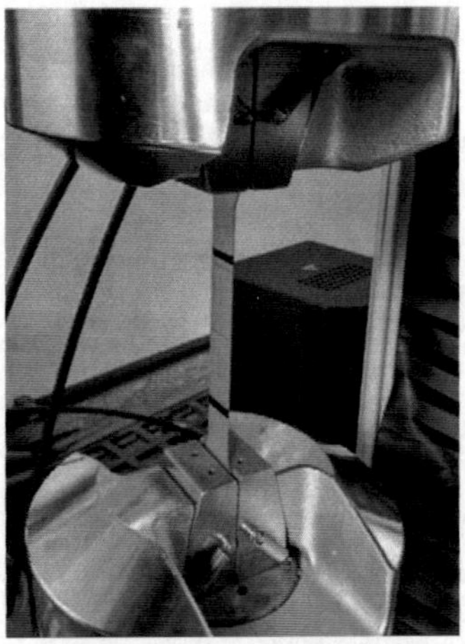

Figure 3. Test execution.

Results of r, r_m and the uncertainty (u) associated to r_m ($u(r_m)$) are shown in table 1. The estimation of the uncertainty was according to ISO/GUM, considering the resolution, standard deviation and calibration data for each dimensional variable (w_o, l_o, w_f, l_f).

Table 1. Anisotropy test results.

Specimen	r	r_m	$u(r_m)$	Operator
A-0º	1.32			
A-45º	0.80	0.99	0.20	A
A-90º	1.05			
B-0º	1.27			
B-45º	0.83	0,98	0.16	B
B-90º	0.99			

The value of the normalized error associated to r_m was calculated according to equation (4).

$$E_n = \frac{|0.99-0.98|}{\sqrt{(0.20)^2+(0.16)^2}} = 0.039 \tag{4}$$

Since $E_n = 0.039 \leq 1$, the result of the intralaboratory program is considered satisfactory. The differences between test executors (operators 'A' and 'B') were smaller than the measurement uncertainty.

The use of normalized error in intralaboratory comparisons and test validation is consistent with literature. For example, Alves *et al.* [12] have used this method in atomic absorption spectroscopy, obtaining $E_n \leq 1$ and, consequently, validating the results. Similar studies have been carried out by other authors [3–7], in the fields of hardness, dimensional and hygrometry measurement, using the same acceptance criteria.

However, the uncertainty values in this work have been large, approximately equal to 20% of the calculated values of r_m. Measurement uncertainty of r_m was estimated considering the following uncertainty sources for each dimensional variable: data from equipment calibration, equipment resolution and standard deviation of repeatability. The deviation of repeatability was responsible by approximately 75% of combined uncertainty, while data from calibration certificate was responsible by 16% of contribution and resolution by 9%.

The large deviation of repeatability is probably caused by the type of measurement equipment used (digital caliper). Perhaps different equipment could improve uncertainty values. Thus, future works could focus on further evaluation and improvement of the uncertainty of this test.

4. Conclusions

This paper has presented the implementation of a procedure to evaluate quality assurance of the anisotropy test of sheet metals. An intralaboratory comparison through the normalized error has been carried out, and the results were satisfactory.

Nevertheless, the uncertainty of the test was large, about 20% of the measured values of rm. Future work is necessary to analyze and reduce measurement uncertainty caused by deviation of repeatability in dimensional measurement for this test.

References

[1] ASTM E517-00 2010 Standard test method for plastic strain ratio r for sheet metal (Pennsylvania)

[2] ISO/IEC 17025 2005 General requirements for the competence of testing and calibration laboratories (Geneva)

[3] Costa P B, Muniz B C O, Machado R R, Koch C A and Barros W S 2008 Comparação entre métodos para medição de impressões de dureza Brinell e Vickers *1º Congresso Internacional de Metrologia Mecânica* (Rio de Janeiro)

[4] Marques A, Alves J A P, Costa P B and Barros W S 2008 Comparação interlaboratorial de métodos de verificação de máquinas de medir por coordenadas (MMC) *1º Congresso Internacional de Metrologia Mecânica* (Rio de Janeiro)

[5] Couto P R G, Oliveira L H P, Oliveira J S, Silva W S and Soares R S 2009 Comparação intralaboratorial do Lapre-Inmetro na faixa de medição de 1,4 kPa a 7 MPa *5° Congresso Brasileiro de Metrologia* (Salvador)

[6] Brionizio J D, Lima S G and Mainier F B 2005 Comparação intralaboratorial em Higrometria *5° Encontro para a Qualidade de Laboratórios* (São Paulo)

[7] Haag J, Fabricio D A K and Reguly A 2014 Análise estatística dos resultados do ensaio de impacto Charpy *3° Congresso Internacional de Metrologia Mecânica* (Gramado)

[8] Cools J, Delanote V, Scheldeman X, Quataert P, De Vos B and Roskams P 2004 Quality assurance and quality control in forest soil analyses: a comparison between European soil laboratories *Accreditation and Quality Assurance* 9 pp 688–694

[9] Lee M J and Hill M R 2007 Intralaboratory Repeatability of Residual Stress Determined by the Slitting Method *Experimental Mechanics* 47 pp 745–752

[10] Smagunova A N, Pan'kov S D and Kozlov V A 2006 Management of Quality Control in Analytical Laboratories *Journal of Analytical Chemistry* 61 (4) pp 312–319

[11] JCGM 2008 Evaluation of measurement data: guide to the expression of uncertainty in measurement (GUM) 1st ed (Geneva)

[12] Alves S, Santos M M C and Trancoso M A 2009 Evaluation of measurement uncertainties for the determination of total metal content in soils by atomic absorption spectrometry *Accreditation and Quality Assurance* 14 pp 87–93

Technology solutions to support supervisory activities and also to provide information access to the society

D Paladini [1] and A B Mello [2]

[1,2] National Institute of Metrology, Quality and Technology, Nossa Senhora das Graças Av., 50, Duque de Caxias, RJ, Brazil

[1] davi.paladini@gmail.com
[2] abeatriz@inmetro.gov.br

Abstract. Inmetro's data about the conformity of certificated products, process and services are, usually, displayed at fragmented databases of difficult access for several reasons, for instance, the lack of computational solutions which allow this kind of access to its users. A discussion about some of the technological solutions to support supervisory activities by the appropriate regulatory bodies and also to provide information access to society in general is herein presented, along with a theoretical explanation of the pros and cons of such technologies to the conclusion that a mobile platform seems to be the best tool for the requirements of Inmetro.

1. Introduction

Inmetro, as part of their duties, makes use of the delegation of enforcement activities of the conformity of products, as a tool to ensure the safety of society and the environment. Despite the operational decentralization, this centralization of strategy refers to a responsibility, especially regarding the quality of management of compliance information. This management remains with the institution, which today is facing difficulties in providing this information to its partner control agencies, who work directly in the inspection process. Therefore, it is necessary to study information management strategies, with emphasis on supporting fiscalization processes and dissemination of information to society.

With the advent of relevant legislation [1] which regulated the constitutional right of access to public information, the need to act more expressively, through quick response strategies and accurate provision of data and information which facilitate the monitoring of work, became clear to Inmetro; in line also with the principles of transparency and publicity which guide the delegation of its activities.

2. Actual Situation

Initiatives to provide information to regulatory agencies and society become problematic either in the absence of complete automated tools capable of integrating the databases necessary to consultations at fiscalization points and control and those requested by the citizen himself, or when tools that do not fully meet the initiative's needs are provided. In the present scenario, governments must mobilize in order to provide not only the supervisory bodies, but also citizens, greater transparency and possible accessibility to information, creating objective goals to expand their reach and using, to this end, all the tools available in the marketplace. The creation of information distribution platforms and online services, also known as e-Government, has been widely discussed, especially as their social influence and reliability [2].

It is necessary to understand whether portal which centralizes services and information would facilitate access for its stakeholders and also facilitate the dissemination of compliance information by Inmetro, and what would be the optimal way of making this portal available, without losing sight of the larger current demand for this type of resource.

The Democratic State requires that, "In addition to the principles of equality and freedom, the democratic ideal consists of action, participation, responsibility and interaction between different subjects" [3]. The extension of democracy demands an increasing participation of civil society in discussions and decision-making processes related to public policy [4]. The National Constitution, as established in 1988, considers the civil participation as essential, exercising democracy in a direct and participatory manner. The various levels of government should then provide sufficient mechanisms to secure the right progress of these rules.

Following the constitutional requirements for the exercise of citizenship, it is essential that the State shall ensure to all Brazilians the means to exercise their rights, therefore understand citizenship is also to understand the individual's relationship forms with the State [5]. It has been pointed out by several authors that "The democratic possibilities can be influenced by new technologies since they have the power to change the way people relate and communicate" [6], meanwhile, the constant and fast advances in Information and Communication Technology, mainly internet and connectivity, have been flaunted as the next opportunity to transform the relationship between governments and citizens, particularly in their role as service costumer [7].

The concept of Mobile Government, related to the Mobile Democracy, present to the citizens a great opportunity for two essential activities. First, the opportunity to express opinions directly to governments, exercising their representation in a clear and structured way. This allows governments to have access to an important set of information about authentic opinions directly acquired from the population, without the intervention of consultants and advisers. Second, governments can accomplish the dissemination of information to encourage citizen participation and question them directly [8].

3. Methodology

In order to better understand the mechanisms for providing information to existing stakeholders, the different features and the Government's needs in terms of availability of information on compliance have been outlined through documentary research.

It can be seen that through the development of new computing platforms, access to different forms and functions in software has expanded greatly. New lines of research point to the Internet of Things [9] and Pervasive Computing [10] as a reality ever closer.

These new computing paradigms require an equally innovative approach in the use of infrastructural resources. The continuous improvement of distributed systems [11] has culminated in the current availability of cloud systems, known as Cloud Computing [12]. It is a trend that supports the development of complex computer projects using various simpler devices working in a coordinated manner.

For software development within these paradigms, one of the most common structures is the Service-Oriented Architecture - SOA [13], which enables tight control over features available remotely. These controls are important for the security of the information disclosed, and for satisfying the requirements of the various available profiles [14].

The advent of mobile devices enables the distribution of computational functionality to individuals in the field, such as enforcement agents or citizens seeking information at any time. However, such devices have limited resources such as battery life, information storage capacity and processing performance. Such limitations would be minimized by sending information to more powerful servers, which would then be responsible for processing and sending the results to the mobile device [15].

Importantly, this type of solution has much more elaborate security demands, establishing new requirements for information security policies [16].

Based on the documentary research undertaken, the next step of this study was to enumerate the government's IT parameters which must be adopted for this work. Based on these parameters and documentary research conducted previously it was possible to identify the pros and cons of different computational solutions for the provision of compliance information to agents of control and inspection and citizen.

The continuation of this work, still under development, is a quantitative research, taking advantage of the collection and objective analysis of data in order to identify the services that may be incorporated into this structure, as well as new services which may be required in the future. It is a questionnaire directed to its target audience, enabling the identification of potential new services and ongoing projects which can be added to the current one. At this stage also it will be possible to identify other factors which may be for or against such computational solutions. The SurveyMonkey® software will be used as a tool for survey [17]. It is an available free software that allows the development of questionnaires and the automatic processing of the responses.

4. Conclusion

Following these concepts, it is concluded that creating an integrated framework for the provision of a mobile services platform is a possible solution to the problem presented. The proposal for a portal developed into a mobile platform, allowing access to information when it is needed seems to be the best alternative for the fiscalization process, according to the analysis of pros and cons of each computer solution studied.

A portal for mobile devices allows the distribution of computational functionality to individuals in the field, such as enforcement agents or citizens seeking information at any time. All regulatory agencies, the Brazilian Network of Metrology and Quality agents, the appraisers of the General Coordination of Accreditation, all of which use compliance information in diffuse Inmetro bases, are potential direct beneficiaries of this solution. In addition, the range of mobile devices has a wide prospect for growth [18] as well as being a necessary tool for the positioning of Inmetro in a context of modernity.

It is important to heed the computational requirements involved in aspects of information processing, because such devices have limited resources, a problem that should be minimized by sending information to more powerful servers, which will be responsible for processing and sending the results to the device mobile [15]. The computational requirements involved in security must also be considered, by adhering to the most modern application development paradigms.

This solution was chosen based on government IT parameters. However, the prototype is expected to be validated through research by survey, to assess its navigability, ease of use and adherence to the parameters identified, in addition to the opinion of Inmetro IT experts.

Acknowledgement

The authors acknowledge the support of Inmetro, which enabled us to carry out this research.

References

[1] Brasil. Lei nº 12.527, 18 de novembro de 2011 Diário Oficial da República Federativa do Brasil
[2] Colesca, S. et al. 2009 Understanding trust in e-Government *Inzinerine Ekonomika* **63** 7-15
[3] Ciconello, A.; Moroni, J.. Participação social no governo Lula. Avançamos? *ABONG A Abong nas Conferências 2005: Criança e Adolescente – Assistência Social, Brasília*
[4] Teixeira, A.C.; Dagnino, E.; Silva, C. A. 2002 La constitución de la sociedad civil en Brasil. Sociedad civil, espacios públicos y democratización: Brasil. *México: Fondo de Cultura Económica*, p. 21-76
[5] Naves, R. 2003 Novas possibilidades para o exercício da cidadania. *História da cidadania*. (São Paulo: Contexto) p. 563
[6] Fishkin, J. S. 2002. Possibilidades democráticas virtuais: Perspectivas da democracia via internet.

"*Internet e política: teoria e prática da democracia eletrônica*". (Belo Horizonte: UFMG)

[7] Ferguson, M. 2002 Estratégias de governo eletrônico: o cenário internacional em desenvolvimento.. *Internet e política: teoria e prática da democracia eletrônica.* (Belo Horizonte: UFMG) p 103-140.

[8] Kushchu, I.; Kuscu, H. 2003 From E-government to M-government: Facing the Inevitable. *In: the 3rd European Conference on e-Government. MCIL Trinity College Dublin Ireland.* 253-260

[9] Eugster, P. et al. 2014 Debugging the Internet of Things: The Case of Wireless Sensor Networks *IEEE Software* **32** 38-49

[10] Satyanarayanan, M. et al. 2001 Pervasive Computing: Vision and Challenges *IEEE Personal Communications* **8** 10-17

[11] Bernstein, P. 1996 Middleware: An Architecture for Distributed System Services *Communications of the ACM* **39** 86-98

[12] Armbrust, M. et al. 2010 A View of Cloud Computing *Communications of the ACM* **53** 50-58

[13] Welke, R. et al. 2011 Service-Oriented Architecture Maturity *IEEE Computer* **44** 61-67

[14] Anthony, D. et al. 2007 Privacy in Location-Aware Computing Environments *Pervasive Computing* **6** 64-72

[15] Kumar, K. et al. 2013 A Survey of computation offloading for mobile systems *Mobile Networks and Applications* **19** 129-140

[16] Kambourakis, G. et al. 2014 Editorial: Developments in Security and Privacy-Preserving Mechanisms for Future Mobile *Mobile Networks and Applications* **19** 61-63 2014.

[17] SurveyMonkey®. Disponível em: http://pt.surveymonkey.com/.

[18] Rahimi, M. et al. 2014 Mobile cloud computing: c curvey, state of art and future directions *Mobile Networks and Applications* **19** 133-143

Metrology applied to ultrasound characterization of trabecular bones using the AIB parameter

D S Braz[1], C E Silva[1], A V Alvarenga[1], D S Junior[1] and R P B Costa-Félix[1]

[1] Laboratory of Ultrasound (Labus), Directory of Scientific and Industrial Metrology (Dimci), National Institute of Metrology, Quality and Technology (Inmetro);

E-mail: douglassbraz@gmail.com

Abstract. Apparent Integrated Backscattering (AIB) presents correlation between Apparent Backscatter Transfer Function and the transducer bandwidth. Replicas of trabecular bones (cubes of 20 mm side length) created by 3D printing technique were characterized using AIB with a 2.25 MHz center frequency transducer. A mechanical scanning system was used to acquire multiple backscatter signals. An uncertainty model in measurement was proposed based on the Guide to the Expression of Uncertainty in Measurement. Initial AIB results are not metrologically reliable, presenting high measurement uncertainties (sample: 0.5_0.2032/AIB: -15.1 dB ± 13.9 dB). It is noteworthy that the uncertainty model proposed contributes as unprecedented way for metrological assessment of trabecular bone characterization using AIB.

1. Introduction

The Quantitative Ultrasound (QUS) was introduced around the 1980s to clinical evaluation of bone health conditions [1]. The acoustic properties of trabecular bone have been widely investigated in vitro. Using pulse-echo methods, there are many parameters that can be determined. In this paper, the parameter that will be considered is AIB – Apparent Integrated Backscatter. Previous studies indicate a significant potential of this parameter to reflect structure, density, composition and mechanical properties of trabecular bones [2]. However, it cannot be found in literature metrological reliable results.

Therefore, standard samples were developed to imitate healthy human trabecular bones and also with osteoporosis with propose to obtain a reference material to be used to ensure measurements with metrological quality.

The measurement uncertainty has a great importance in the characterization of the samples because it is possible to detect how much the value scatters around the mean.

2. Materials and Methods

2.1. Samples

Isotropic bone replicas were produced in acrylonitrile butadiene styrene polymer (ABS) using 3D printing technique from CTI Renato Archer, Campinas. They are cubes of 20 mm × 20 mm × 20 mm (Figure 1).

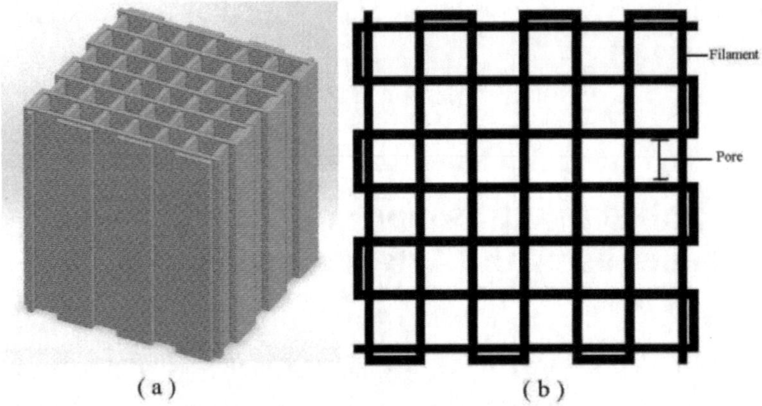

Figure 1. Bone replica form and structure.

The samples were distinguished by pore size and filament size (in mm); see Table 1.

Table 1. Isotropic bone replica structure.

Sample	Pore size [mm]	Filament size [mm]
0.5_0.2032	0.5	0.2032
0.5_0.5782	0.5	0.5782
2.5_0.2782	2.5	0.2782
2.5_0.5282	2.5	0.5282

2.2. Measurement System.

To the measurement system was used a 2.25 MHz central nominal frequency transducer, an arbitrary waveform generator (Agilent 33250A), an oscilloscope (Agilent DSO 6032A), a water tank (240 mm × 240 mm × 240 mm) filled with deionized water and a polished steel plate reflector (Ø 63 mm × 10 mm) (Figure 2).

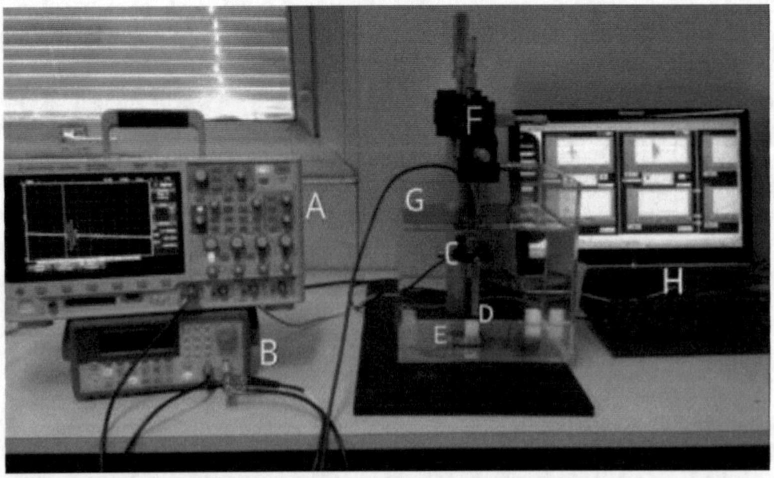

Figure 2. Measurement System: A- Oscilloscope, B- Arbitrary waveform Generator, C- Transducer, D- Bone replica sample, E- Reflector, F- Linear and angular stages, G- Water tank, H- Computer.

Measurements were made using a system with three linear stages; 25.4 mm travel each, responsible for transducer movement in three different axes (x, y, z) and two goniometric stages, ± 5° travel that are responsible for transducer alignment with reflector.

2.3. Methodology

The sample was degassed under vacuum and placed on the bottom of water tank at 22°C [4].

The transducer was positioned one near field from the front surface of reflector to acquire a reference signal. A time window gated was selected by the operator in LabView's software acquiring the initial of backscattering signal situated after sample surface signal until initial reflector signal. FFT (Fast Fourier Transform) was performed and it was determined the transducer bandwidth, above the -6 dB of the reference spectrum. The sample was positioned with the surface parallel to transducer surface, within the near field produced by the transducer. After that, the power spectrum of the signal backscattered from the sample was obtained through a similar procedure, except that the gate was delayed to verify that no energy from the surface reflection was incorporated in the sample-backscattering window. This signal is called sample signal. So, using these two signals, the parameters can be calculated.

The measurements were performed in the four sides of cube, which have the same physical architecture. It was carried out 9 measurements per side.

2.4. Characterization parameters

The AIB was determined by integrating the Apparent Backscatter Transfer Function (ABTF), over the -6 dB bandwidth of the transducer, this can be viewed in the equations (1) and (2) [2]:

$$AIB = \frac{1}{\Delta f}\int_{\Delta f} ABTF\, df \tag{1}$$

Where:
Δf – The transducer bandwidth.

And,

$$ABTF = 10 \times log\left(\frac{A_b}{A_r}\right) \tag{2}$$

Where:
A_b – Power spectrum of the sample in the acoustic path;

A_r – Power spectrum of the reference signal.

The AIB informs the mean value of apparent backscattering in the bandwidth of transducer used.

2.5. Measurement Uncertainty

It was used the Guide to the expression of uncertainty in measurement (GUM) to express the uncertainty of measurement result [5].

The measurement uncertainty was obtained considering Type A standard uncertainty and Type B standard uncertainty. Type A standard uncertainty was obtained from standard deviation divided by √9. Type B evaluation of standard uncertainty was obtained from the amplitude expanded uncertainty, in volts, from oscilloscope certificate divided by k=2.

Type A and Type B standard uncertainties were combined as can be viewed in equation (3).

$$u_c = \sqrt{(u_{type\,A} \times c.s_A)^2 + (u_{type\,B} \times c.s_B)^2} \tag{3}$$

Where:
u_c – combined uncertainty;

$c.s_A$ - Sensitivity coefficient to Type A standard uncertainty;

$c.s_B$ - Sensitivity coefficient to Type B standard uncertainty.

The coverage factor k was determined from effective degrees of freedom and the level of confidence equal to 95%. Expanded uncertainty U can be given by equation (4).

$$U = u_c \times k \tag{4}$$

Thus, this expanded uncertainty relates one side of one sample result, so there are four expanded uncertainty for each sample. The final combination of uncertainties was assessed with side combined uncertainties by their effective degrees of freedom according to equation (5).

$$u_{C_G} = \frac{v_1 u_1 + v_2 u_2 \ldots v_n u_n}{v_1 + v_2 + \cdots v_n} \tag{5}$$

Where:
u_{C_G} – global combined uncertainty;

$v_1, v_2 \ldots v_n$ - effective degrees of freedom;

$u_1, u_2 \ldots u_n$ – combined uncertainty.

So the coverage factor k was determined and the global expanded uncertainty U_G was calculated.

To compare the results obtained by each side, the measurements were compared with the global result of sample using the standardized error E_n as defined in equation (6). Values are considered statistically equal if the E_n is less or equal to 1.

$$E_n = \frac{X_G - X_S}{\sqrt{U_G^2 + U_S^2}} \tag{6}$$

Where:
E_n – normalized error;

X_G – global mean (sample mean);

X_S – side mean;

U_S – side measurement uncertainty;

3. Results

Table 2 provides Apparent Integrated Backscatter results accompanied by coverage factor and measurement uncertainty values of trabecular bones samples measurements and normalized error between each side.

Table 2. Results of trabecular bones samples measurements.

Sample	Side	X (dB)	k	U (dB)	E_n $Satisfactory?$ 1	2	3	4
0.5_0.5782	1	-17.2	2.14	0.22	X	Not	Yes	Yes
	2	-16.1	2.20	0.29	Not	X	Not	Not
	3	-17.1	2.26	0.54	Yes	Not	X	Yes
	4	17.6	2.31	0.63	Yes	Not	Yes	X
2.5_0.5282	1	-13.6	2.31	0.59	X	Yes	Yes	Not
	2	-13.5	2.31	0.47	Yes	X	Yes	Not
	3	-13.7	2.26	0.45	Yes	Yes	X	Not
	4	-12.4	2.09	0.12	Not	Not	Not	X
0.5_0.2032	1	-16.7	2.31	0.68	X	Not	Yes	Not
	2	-13.9	2.00	0.20	Not	X	Not	Yes
	3	-15.9	2.26	0.76	Yes	Not	X	Not
	4	-14.1	2.31	0.68	Not	Yes	Not	X
2.5_0.2782	1	-13.5	2.31	0.32	X	Not	Not	Not
	2	-12.6	2.26	0.24	Not	X	Not	Not
	3	-13.0	2.20	0.17	Not	Not	X	Not
	4	-17.3	2.20	0.34	Not	Not	Not	X

Analysing Table 2 it is possible to realize that there are many results not equal statically. To solve this, it was created a multiplier factor *MF* used to increase the global expanded uncertainty and make equal results, as can be seen in Table 3.

Tabela 3. Results with multiplier factor.

Sample	Side	MF	X_G (dB)	k_G	U_G (dB)	E_n	$Satisfactory?$
0.5_0.5782	1	2.22	-17.0	2.02	4.2	0.32	Yes
	2					1.00	Yes
	3					0.06	Yes
	4					0.30	Yes
2.5_0.5282	1	5.45	-13.3	2.01	12.3	0.07	Yes
	2					0.05	Yes
	3					0.12	Yes
	4					1.00	Yes
0.5_0.2032	1	4.45	-15.1	1.99	13.9	0.35	Yes
	2					1.00	Yes
	3					0.16	Yes
	4					0.24	Yes
2.5_0.2782	1	6.70	-14.1	2.02	8.0	0.21	Yes
	2					0.68	Yes
	3					0.68	Yes
	4					1.00	Yes

Table 4 is responsible to show E_n results combining each sample to demonstrate if the characterization method can detect differences between samples used.

Table 4. E_n of samples combination.

Samples Combination	E_n	$Satisfactory?$
0.5_0.5782/2.5_0.5282	0.3	Yes
0.5_0.5782/0.5_0.2032	0.1	Yes
0.5_0.5782/2.5_0.2782	0.3	Yes
2.5_0.5282/0.5_0.2032	0.1	Yes
2.5_0.5282/2.5_0.2782	0.1	Yes
0.5_0.2032/2.5_0.2782	0.1	Yes

4. Discussion and Conclusion

The isotropic bones replica should provide results statistically similar for all the sides, but the results in Table 2 show that the samples 0.5_0.5782 and 2.5_0.5282 present different results for sides 2 and 4, respectively.

The samples 0.5_0.2032 and 2.5_0.2782 have 0.2 mm-thickness filaments, which are, as expected, thinner than the ability of the transducer frequency perceive. This can explain why they are not statistically equal.

Assuming that the samples were structurally isotropic, the multiplier factor was applied in order to make all samples side results statistically equal, and the bone replica global uncertainty possible to be estimated (Table 3). High values of expanded uncertainty were obtained for all samples. Therefore, these results are not metrologically reliable. Moreover, Table 4 displays that all results are statistically equal showing that this method cannot be able to identify and classify different bone replica architecture, as it is expected for. Besides, the results associated with its respective uncertainties demonstrated that it is not possible to relate the AIB with BMD (Bone mineral density), as it was established in the literature [3].

References

[1] Langton C M, Palmer S B and Porter R W 1984 The measurement of broadband ultrasonic attenuation in cancellous bone *Eng Med* pp 89-91

[2] Matusin D P 2009 Estudo preliminar de caracterização ultrassônica da pseudoartrose Rio de Janeiro, RJ Dissertação (Mestrado em Engenharia Biomédica) – Universidade Federal do Rio de Janeiro, UFRJ.

[3] Laugier P and Haïat G 2011 *Bone Quantitative Ultrasound* Springer Dordrecht Heidelberg London New York

[4] Hoffmeister B K et al 2008 Ultrasonic Characterization of Human Cancellous Bone In Vitro using Three Different Apparent Backscatter Parameters in the Frequency Range 0.6–15.0 MHz, IEEE Trans *Ultrasound Ferroelectr Freq Control* v. **55**, n. 7, pp. 1442-1452.

[5] JCGM 2008 Guide to the expression of uncertainty in measurement JCGM 100:2008.

Analytical validation applied to simultaneous determination of solvents dichloromethane (DCM), methyl isobutyl ketone (MIBK), tetrahydrofuran (THF) and toluene (TOL) in urine by headspace extraction and injection on chromatographic system with a flame ionization detector

E D M Muna[1] and R P Pereira[1]
[1] Instituto Militar de Engenharia, IME – Praça Gen. Tibúrcio, 80 - Urca - Rio de Janeiro - RJ - Brazil

E-mail: al_dobloni@hotmail.com, dobloni@ime.eb.br

Abstract. The determination of the volatile organic solvents dichloromethane (DCM), methyl isobutyl ketone (MIBK), tetrahydrofuran (THF) and toluene (TOL) is applied on toxicological monitoring of employees in various industrial activities. The gas chromatography technique with flame ionization detector and headspace injection system has been applied. The analytical procedure developed allows the simultaneous determination of the above-mentioned solvents and the accuracy of the method was tested following the INMETRO guidelines through the DOQ-CGRE 008 Rev.04- July/2011.

1. Introduction

The importance of determination of the volatile organic compounds such as dichloromethane (DCM), methyl isobutyl ketone (MIBK), tetrahydrofuran (THF) and toluene (TOL) is associated with the risk they offer to the health of workers exposed to those substances during their journey labor. The biological exposure index (BEI) values, originates from studies conducted by the American Conference of Governmental Industrial Hygienists (ACGIH), where the analyte concentration levels found in urine were correlated with their symptomatology. The BEI values are related to exposures of eight hours per day, five days per week, with concentrations in urine of 300 μg/L for DCM, 1000 μg/L for MIBK, 2000 μg/L for THF and 30 μg/L for TOL [1].

DCM is associated to cancer predisposition, heart diseases, central nervous system and liver failure, skin and eyes irritation. Exposure occurs primarily by inhalation and skin absorption [2]. The MIBK is irritating to the eyes, nose, throat and skin, causing headaches and nausea, and also effects on liver and kidneys. Exposure to high concentrations can be life-threatening due to the narcotic effects of this substance [3]. The THF has similar toxicological effects to MIBK, however, with less intensity [4]. TOL is the most toxic of the mentioned species and is the one presenting greater use, being the major substituent of benzene in various industrial activities where the use of this compound was prohibited. It is a strong central nervous system depressor, leading to mental confusion, loss of coordination and unconsciousness, which may lead to death in cases of acute exposure [5].

In the present work, the technique utilized was gas chromatography with flame ionization detector, and injection system by headspace extraction. This approach was adopted due to the chemical characteristics of the analytes and the complexity of the matrix. Despite the theoretical and methodological discussion, the method was developed experimentally. The main goal of the present work involves the analytical validation aiming to assign metrological reliability to the measurements made, which requires a set of experiments planned to compose statistical representation of the chemical assay [6, 7].

2. Materials and Methods

2.1. Materials

Reverse-osmosis type quality water (purified with a Milli-RO plus Milli-Q station from Millipore Milford, MA) and pure solvents for the composition of the patterns.

2.2. Sample and standards preparation

Sample preparation consists of collecting an aliquot of 8 mL, adding 1 mL of internal standard (fluorobenzene in aqueous solution) and sealing the headspace vial, subsequently submitting to extraction programming of the headspace equipment.

The small number of steps in sample manipulation can be considered a great advantage, which considerably reduces the contribution of uncertainty sources associated with the sample preparation step.

The standards were prepared from pure substances diluted appropriately in water containing 5% methanol. In the case of reproducing the sample behavior, a spiked urine sample was prepared, ensuring its homogeneity, which was used in the execution of matrix experiments.

2.3. Chromatographic conditions

A headspace injection system (Perkin Elmer Turbo Matrix 40 Trap) coupled to gas chromatograph with flame ionization detector (Perkin Elmer Clarus 500) were used to quantify the analytes in urine. The analyses were performed under the following chromatographic conditions: Column, DB-624 with 60 m length, 250 µm in diameter and 1.4 µm of inner film.

The temperature of the FID was 220ºC, and the injector temperature was 200°C. The oven temperature was programmed to 45°C for 5.5 minutes, ascending 15°C/min to 140°C, standing for 4 minutes, subsequently rising 40°C/min to 200°C and following for the cleaning stage, ascending 100°C/min to 240°C waiting for 2 minutes. Helium was the carrier gas with a flow of 1.4 mL/min for 17 minutes followed by an increase of 100 mL/min to 1.8 mL/min.

The parameters of headspace injection were: incubation time of 60 minutes, temperature of 80 ºC, needle temperature 200 ºC and time of pressurization of 1 minute. The analytes exhibited retention times of 7,441 min (DCM); 9,699 min (THF); 12,528 min (MIBK); 12,823 min (TOL) and 10,813 min for the internal standard fluorobenzene.

3. Results and Discussion

3.1. Analytical validation

3.1.1. Analytical Range

The measurement range was selected based on values indicated by the BEI/ACGIH (American Conference of Governmental Industrial Hygienists).

The IBE values applies to exposures of eight hours per day, five days a week, with values in urine, 300 µg/L of DCM, 1000 µg/L of MIBK, 2000 µg/L of THF, and 30 µg/L for TOL.

Table 1 shows the analytical range selected in accordance with toxicological limits.

Table 1. Analytical curve in µg/L (ppb).

Levels	Dichloromethane	Methyl isobutyl ketone	Tetrahydrofuran	Toluene
1	150	450	900	15
2	300	900	1800	30
3	450	1350	2700	45
4	600	1800	3600	60
5	750	2250	4500	75

3.1.2. Selectivity

The selectivity evaluates the ability of an analytical method of providing specific responses for each analyte, respecting the variation of concentrations.

This evaluation was performed through the t test applied to the angular coefficients of matrix without curve (standards in water) and curve with matrix (standards in urine).

Table 2 lists t_{calc} and t_{tab} values considered for evaluation.

Table 2. Values of t_{calc} and t_{tab}.

t	Dichloromethane	Methyl isobutyl ketone	Tetrahydrofuran	Toluene
t_{calc}	-0,107	0,317	-0,017	0,577
t_{tab}		2,160		

The t_{tab} values refer to a total number of 15 replicates (3 per level) and confidence interval of 95%. All values were observed to be lower than the reference, pointing to the inexistence of matrix effects.

3.1.3. Linearity

The linearity is the capacity of a analytical method to provide specific responses for each concentration within a stipulated range. To make this assessment, it is used the calculus of the coefficient of determination (R Squared).

For the linearity evaluation, it was used the analytical curve without matrix addition, with 3 replicates per level, being evaluated the values for R^2 and dispersion of the residues through the Cochran's test. Table 3 shows the values obtained for R^2 and Cochran.

Table 3. Values of R^2 and Cochran's test.

R^2 / C	Dichloromethane	Methyl isobutyl ketone	Tetrahydrofuran	Toluene
R^2	0.9954	0.9966	0.9959	0.9930
C_{calc}	0.5455	0.4753	0.4999	0.3395
C_{tab}		0.6838		

These results indicate linearity within the studied range and the homocedasticity of the residues (free residues trend over the analytical range) which is observed for C_{calc} lower than C_{tab}.

Figures 1 to 4, below, show the residues dispersion obtained from the calibration curves for each analyte.

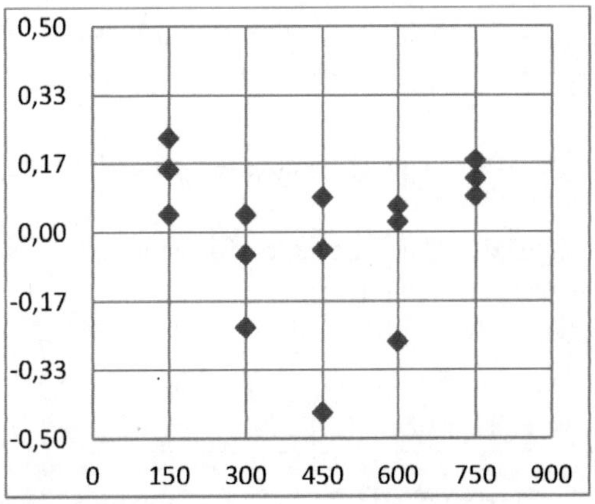

Figure 1: Dichloromethane residues dispersion.

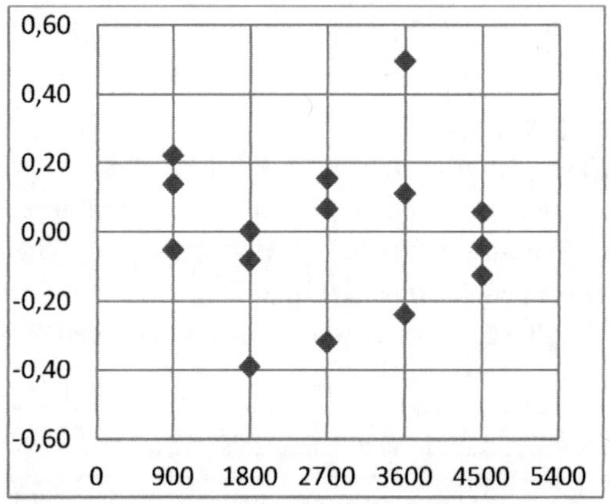

Figure 2: Tetrahydrofuran residues dispersion.

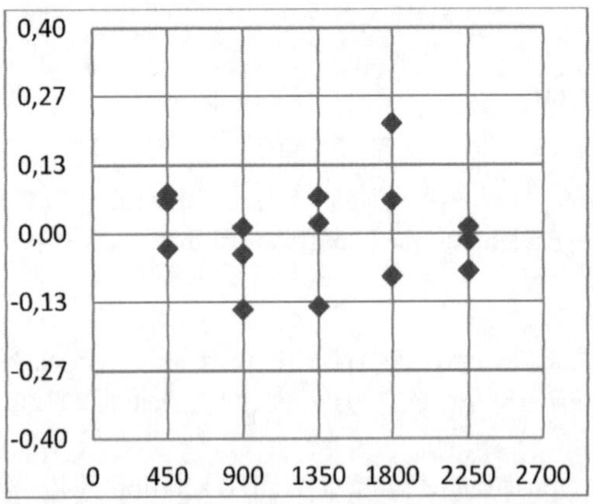

Figure 3: Methylisobutylketone residues dispersion.

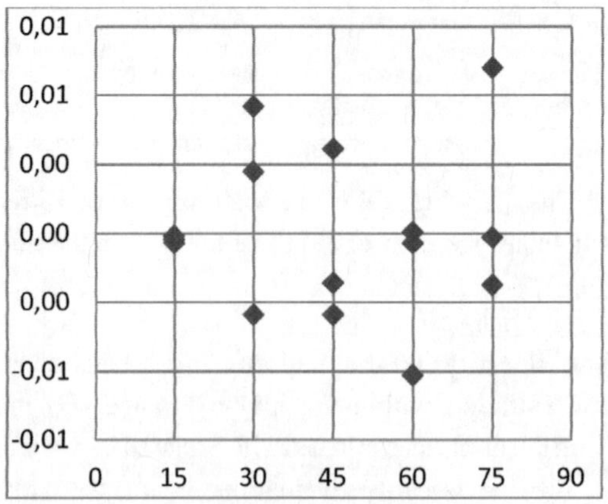

Figure 4: Toluene residues dispersion.

3.1.4. Limits of detection (LOD) and quantification (LOQ)

The detection limit relates the minimum signal from which can be considered a positive signal related to the presence of the analyte.

The importance of determining the limits of detection and quantification, allows the recognition of the metrological intervals of the analytical method. Allowing to evaluate the capacity of the method to attend a particular application.

For the determination of LOD and LOQ, 7 replicates were obtained of the lowest level of the curve applying the standard deviation method. Table 4 shows the LOD and LOQ obtained statistically.

Table 4. LOD and LOQ in µg/L (ppb).

Limits	Dichloromethane	Methyl isobutyl ketone	Tetrahydrofuran	Toluene
LOD	25.75	68.86	155.12	1.63
LOQ	85.83	229.54	517.07	5.44

The theoretical values obtained from the quantification limit were found below the lowest point of the curve for the analytes studied, ensuring metrological reliability in measurements close to the first point of the analytical curve.

3.1.5. Recovery

The recovery represents the study of the analytical system performance compared to samples containing known concentrations of the target analyte. The limitation of this procedure is that the added analyte is not necessarily in the same way as in the sample. The presence of analytes added into a more easily detectable form may result in overly optimistic evaluation of retrieval.

The success of the evaluation of recovery is the simulation capability of the actual conditions of the samples. In this case, the characteristics of the analytes and low concentrations studied, ensured the prospect of recovery with the expected success.

For the recovery evaluation, 3 replicates of fortified samples in the low, medium and high ranges were obtained. Table 5 presents the average recovery values obtained for the series of substances studied.

Table 5. Recovery (%).

Rec	Dichloromethane	Methyl isobutyl ketone	Tetrahydrofuran	Toluene
%	97.60	99.48	99.54	99.21

Journal of Physics: Conference Series **733** (2016) 012013 doi:10.1088/1742-6596/733/1/012013

3.1.6. Repeatibility

Repeatability is the relative standard deviation (RSD) evaluate of the replicates used in the composition of the curve. It has relation to the evaluation of the residue dispersion. The RSD values have great importance in the evaluation of homocedasticity, showing tendencies along the calibration curve. The repeatability must be uniform across the analytical range, indicating that responsiveness is equivalent at all concentrations.

Repeatability study was conducted from the matrix without curve and in triplicate per level. The relative standard deviation (RSD) for each concentration level was calculated for all analytes under study, there was no transposition of the acceptability criterion limit, which in this case is 10%.

4. Conclusions

The values obtained in the statistical analysis ensure metrological confidence in the measurements performed and the applicability of this methodology for occupational toxicological monitoring of the mentioned substances.

The framework of the values in the BEI/ACGIH within the analytical curve allows the anticipated deflagration of toxic conditions which individuals may be subjected. The observation of quantized values, even lower than the limit, presents an important toxicological monitoring tool.

The substances studied have toxicological effects observed in the short-term when there is transposition of the reference values, however, there are still uncertainties about the effects associated with long-term exposure to lower doses of the referenced values.

The technique presents considerable complexity, however, the operating cost is relatively low compared to other analytical techniques. CRMs (certified reference materials) can be utilized aiming to expand the metrological traceability to attend national and international quality standards.

Acknowledgments

We thank Serviço Social da Indústria – Federação das Indústrias do Estado do Rio de Janeiro (FIRJAN) for the support during the development of the present work.

References

[1] American Conference of Governmental Industrial Hygienists (ACGIH) 2010 *Threshold limit values and biological exposure indices for 2010* (Cincinnati)

[2] Osha analytical methods manual 1995. *Methylene Chloride: Method 80* (Salt Lake)

[3] Osha analytical methods manual 2000. *2-Butanone (MEK) Hexone (MIBK): Method 1004* (Salt Lake)

[4] CDC – Centers for Desease Control and Prevention in <www.cdc.gov/niosh/ipcsneng/neng0578.html> accessed in April 2015

[5] Osha analytical methods manual 1998. *Toluene: Method 111* (Salt Lake)

[6] Kolb B and Ettre L S 2006 *Static headspace-gas chromatography* 2nd ed (New Jersey: Wiley-Interscience)

[7] DOQ-CGCRE-008 2011 *Orientações sobre validação de métodos analíticos* (Rio de Janeiro: Inmetro)

[8] Diego C N *et al* 2008 *J. Braz. Chem. Soc.* São Paulo, v. **19,** n. 6, 1201 – 1206

[9] E D M Muna *et al* 2015 *J. Phys.: Conf. Ser.* **575** 012031

Effects of heat treatments on the thermal diffusivity of Uranium-Molybdenum alloy

D M Camarano[1], F A Mansur[1], A M M Santos[1], W B Ferraz[1] and T A Pedrosa[1]

[1]Nuclear Technology Development Center, CDTN, 31270-901, Belo Horizonte, MG, Brazil

E-mail: dmc@cdtn.br

Abstract. U-Mo alloys are the most investigated nuclear fuel material to be used in research reactors. The addition of molybdenum stabilizes the gamma phase of uranium and increases its melting point. A research program under development at Nuclear Technology Development Center (CDTN) aims the obtaining of uranium-molybdenum alloys to enable the high enriched uranium (HEU) to low enriched uranium (LEU) conversions. U-Mo ingots with 10 % by weight were induction melted and heat treated at 300 °C for 72 h, 120 h and 240 h. Thermal diffusivity was determined by the laser flash method and thermal quadrupole method, from room temperature to 300 °C and 400 °C. It was observed that the thermal diffusivity tends to increase with increasing temperature.

1. Introduction

A research program is in progress at CDTN - Nuclear Technology Development Center to develop a fuel consisting of a metallic alloy of uranium and molybdenum, that may eventually in future be used in the RMB - Multipurpose Brazilian Reactor, and which will enable the use of low enriched uranium (LEU). The use of metallic uranium is not feasible due to the occurrence of intense swelling related to the presence of the α-phase with orthorhombic crystal structure [1], the uranium stable phase at room temperature.

The uranium-molybdenum system was initially investigated in 1945 by Ahmann, Snow and Wilson [2] and has been one of the most investigated due to the retention capacity of the γ-phase, which rises as the Mo content is increased [3], and their higher densities. The aim of the present work is to investigate the thermal diffusivity of U-Mo with Mo content of 10% by weight, at room temperature, 300 °C and 400 °C.

2. Materials and methods

2.1. Production of the alloys

U-10 wt.% Mo alloys were melted in an induction furnace under a protective atmosphere of argon. Pieces of uranium and molybdenum were placed in a graphite crucible and, after purging under vacuum, the protective atmosphere of argon was established and the furnace was heated up to a temperature of about 1500 °C. This temperature was maintained for 12 min to provide the complete fusion and homogenization of the two elements. This melted alloy was poured into a cylindrical mold of copper cooled in air, resulting on U-Mo ingots with diameter of 25 mm and length of 150 mm [4-6].

Discoid samples with 8 mm diameter and 2.5 mm thickness were prepared from these ingots and its thermal diffusivity were determined by a laser flash method at room temperature and thermal quadrupole method at 300 °C and 400 °C.

2.2. Density determination of the alloys

The density ρ (gcm^{-3}) of each sample was determined geometrically from measurements of its diameter D (cm), its thickness L (cm) and its mass m (g) using a micrometer and a calibrated analytical balance, by means of the following equation (1):

$$\rho = \frac{m}{\frac{\pi \cdot D^2 \cdot L}{4}} \tag{1}$$

2.3. Thermal diffusivity determination of the alloys

2.3.1. Laser flash method. For thermal diffusivity measurements at room temperature, the laser flash method [7] was adopted using a bench developed by CDTN researchers, and applied according to ASTM-E-1461 [8]. Figure 1 presents schematically the experimental apparatus of CDTN for thermophysical properties measurements based on the laser flash method.

By this method, the front face of a small disk-shaped sample is subjected to a very short burst of radiant energy. The source of the radiant energy is a CO_2 laser and the irradiation times are of the order of milliseconds. The resulting temperature rise of the rear surface of the sample is registered and from the obtained thermogram, the sample thermal diffusivity is calculated by the following equation (2):

$$\alpha = \frac{1 \cdot 37 \cdot L^2}{\pi^2 \cdot t_{1/2}} \tag{2}$$

Where α is the sample thermal diffusivity (m^2s^{-1}), L is the sample thickness (m) and $t_{1/2}$ is the half time (s).

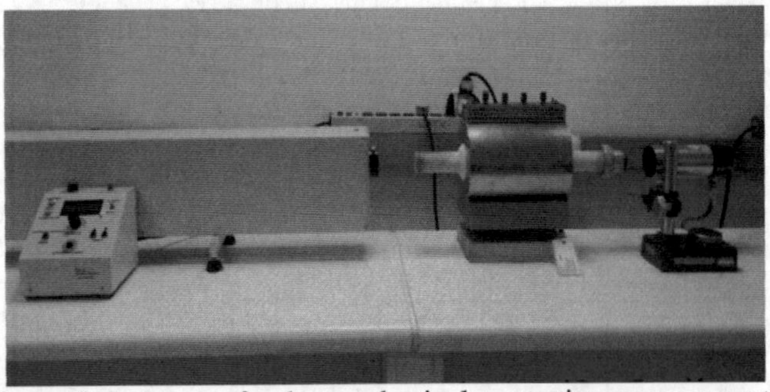

Figure 1. Apparatus for thermophysical properties measurements.

The used sealed CO_2 laser (10,6 µm wave length), model BL 80i provided by Bioluz Company has a maximal power output of 100 W and it is usually set to limit the sample rear face temperature rise in a maximum of 3 °C. The temperature of the sample rear face is measured by an infrared thermometer, model MMLTSSF1L provided by Raytek. For the processing of the thermometer signal, the data acquisition and calculations it is used the LabVIEW plataform and a 16 bits A/D card. The in house made sample oven has a heat element of Platinum/ 30 % Rodium that can be heated up to 1700 °C, but the diffusivities measurements were made at room temperature.

Journal of Physics: Conference Series **733** (2016) 012014 doi:10.1088/1742-6596/733/1/012014

2.3.2. Thermal quadrupole method. The thermal quadrupole method was developed by Degiovanni [9] and can be seen as an extension of the concept of thermal resistance at steady state to transient conditions. It is an analytical method whose resolution is approximated by the least squares method, leading to identification of the thermal diffusivity. The sample gets a pulse of light on one side, raising its temperature. From the temperature field over the distance, it is possible to compare the theoretical curve and the experimental curve by the method of least squares. The modeling for the method is illustrated in figure 2 [9,10].

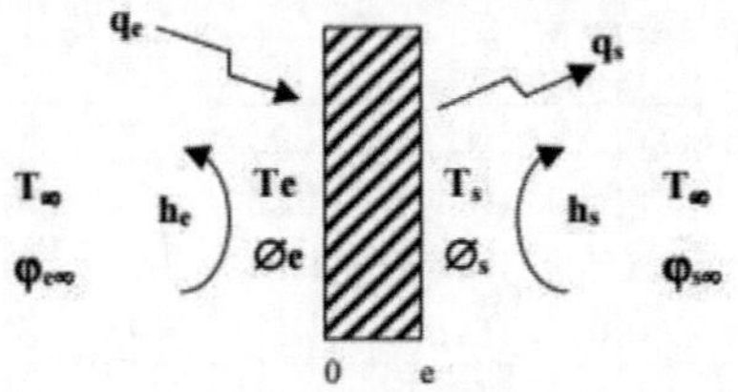

Figure 2. Heat transfer through a unidirectional plane wall.

The corresponding equations are:

$$\varphi_e = -\lambda \cdot \frac{\partial T}{\partial x}|_0 = q_e - h_e \cdot S \cdot (T_e - T_\infty) \quad (3)$$

$$\varphi_s = -\lambda \cdot \frac{\partial T}{\partial x}|_0 = q_s + h_s \cdot S \cdot (T_s - T_\infty) \quad (4)$$

Where: φ corresponds to the density of heat flow per unit of time, λ is the thermal conductivity ($Wm^{-1}K^{-1}$), *T* is the temperature (K), *S* is the surface (m), *h* is the heat exchange coefficient ($Wm^{-2}K^{-1}$) and *q* is the heat flow (Wm^{-2}).

To determine the thermal diffusivity, it was used the diffusivimeter Protolab model QuadruFlash 1200 (figure 3) of Thermophysical Properties Measurement Laboratory (LMPT) of CDTN. The thermal diffusivity was measured 10 times for each sample, and the data averaged.

This diffusivimeter, manufactured in Brazil, is constituted by a xenon lamp (1200 J), responsible for the energy pulse, three K-type thermocouples of special class, an InSb infrared detector, an oven for heating the sample and a signal processing unit. A short thermal excitation is generated by the xenon lamp. The flash is directed inside a resistive furnace, allowing heating of the tested specimen from room temperature to 1200 °C and the temperature rise of the specimen is measured by three K-type thermocouples.

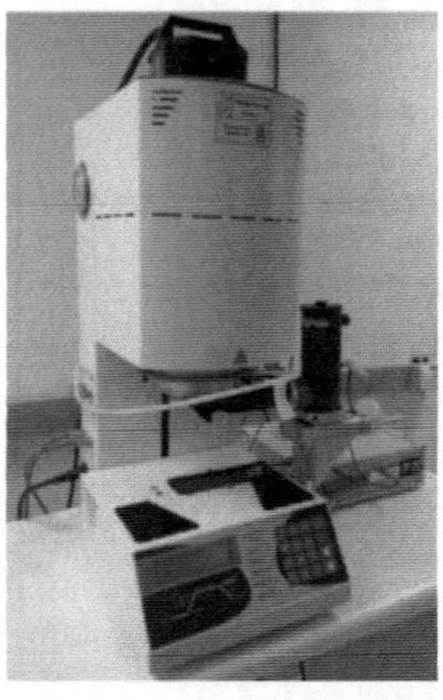

Figure 3. Diffusivimeter QuadruFlash 1200.

3. Results and discussion

Table 1 shows the results of the density and thermal diffusivity of the U-10 wt.% Mo alloys.

Table 1. Thermal diffusivity and density of the samples.

U-10%Mo	Density (gcm^{-3})	Thermal Diffusivity ($x10^{-6} m^2s^{-1}$) Temperature /°C		
		25	300	400
As cast	16.38	3.90	7.72	8.22
300 °C 72h	16.45	3.83	7.74	8.17
300 °C 120h	16.26	4.18	7.27	8.29
300 °C 240h	16.19	4.17	7.43	8.39

There was no significant variation in the densities of the specimens, remaining around 16 gcm^{-3}. From room temperature to 300 °C and 400 °C, the thermal diffusivity increased significantly, but from 300 °C to 400 °C it increases modestly for all the samples. The results obtained in this study are similar to the work of Burkes et al. [11] at both temperatures evaluated.

It can be seen that the heat treatment at 300 °C for 72 h, 120 h and 240 h do not have major influence at the results. Calculated average thermal diffusivity as a function of temperature for the U-10 wt.% Mo alloy is provided in figure 4. The maximum standard deviation obtained was 9 %.

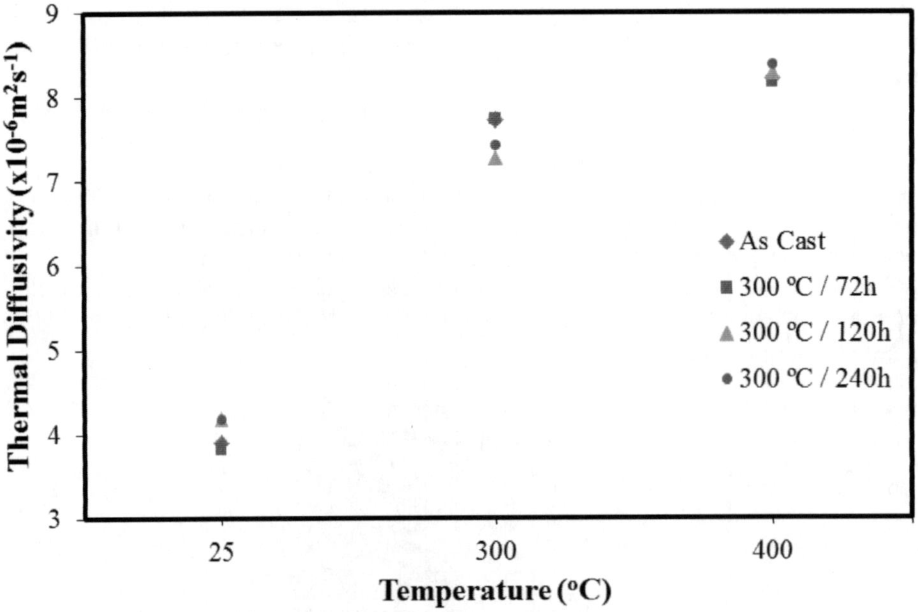

Figure 4. Average thermal diffusivity as a function of temperature of U-10 wt.% Mo alloys.

It is clearly seen that the increase of temperature increases the thermal diffusivity of the U-10 wt.% Mo alloy.

4. Conclusions

Thermophysical properties are an important role of any fuel development campaign. The present work investigated the thermal diffusivity of U-Mo with Mo content of 10% by weight, at room temperature, 300 °C and 400 °C.

It can be seen that the thermal diffusivity tends to increase with increasing temperature. Further studies may be done to evaluate different temperatures as well as heat treatments, to improve the thermophysical properties of U-10 wt.% Mo alloys.

Acknowledgments

The authors thanks the financial support of *Rede* GTD SIBRATEC – *Rede de Serviços Tecnológicos em Geração, Transmissão e Distribuição de Energia Elétrica* and INCT – *Instituto Nacional de Ciência e Tecnologia de Reatores Nucleares Inovadores*.

References

[1] Pugh S F 1961 Swelling in alpha uranium due to irradiation *J. of Nuclear Materials* **4** 177-199

[2] Ahman D, Snow A I and Wilson A S 1945 The Uranium-Molybdenum Binary System *Report CT-2946*

[3] Sinha V P, Hegde P V and Prasad G J 2010 Phase transformation of metastable cubic γ-phase in U-Mo alloys *J. of Alloys and Compounds* **506** 253-262

[4] Ferreira R A N et al. 2011. Thermophysical properties determination of uranium-molybdenum fuel by the flash laser method. *Proc. of the International Nuclear Atlantic Conference* (Belo Horizonte)

[5] Pedrosa T A 2014 **Transformações de fases em ligas U-Mo no estado bruto de fusão envelhecidas artificialmente** Thesis (CDTN) Minas Gerais Brazil

[6] Pedrosa T A *et al.* 2013. Investigation of the microstructure influence in the thermo-physical properties of U-Mo alloys through the laser flash method. *Proc. of the International Nuclear Atlantic Conference* (Recife)

[7] Parker W J, Jenkins R J, Butler C P and Abbott G L 1961 Flash Method of Determining Thermal Diffusivity, Heat Capacity and Thermal Conductivity *J. Appl. Phys.* **32** 1679-84

[8] ASTM - American Society for Testing and Materials. ASTM E-1461: Standard test method for thermal diffusivity of solids by the flash method. West Conshohocken, 2011

[9] Degiovanni A, Maillet D, André S, Batsale J C and Moyne C 2000. Thermal Quadrupoles: Solving the Heat Equation Through Integral Transforms. New York, *John Wiley & Sons,* 388p

[10] Guisard-Restivo T A 2003 *Proc. of the 47th Annual Meeting of the Brazilian Ceramic Society* (João Pessoa)

[11] Burkes D E, Papesch C A, Maddison A P, Hartmann T and Rice F J 2010 Thermo-physical properties of DU–10 wt.% Mo alloys *J. of Nuclear Materials* **403** 160-166

Thermophysical properties of ethylene glycol mixture based CNT nanofluids

D M Camarano[1], F A Mansur[1], T L C F Araújo[1], G C Salles[1] and A P Santos[1]

[1]Nuclear Technology Development Center, CDTN, 31270-901, Belo Horizonte, MG, Brazil

E-mail: dmc@cdtn.br

Abstract. Nanofluids are produced by dispersing nanometer-scale solid particles into base liquids such as water, ethylene glycol, etc. The thermal quadrupole method is utilized to determine the thermophysical properties of materials. By this technique, the thermal diffusivity and conductivity of different nanofluids containing surfactants (humic acid, sodium salt of humic acid and sodium carboxymethyl cellulose and multi-wall carbon nanotubes were evaluated at room temperature and at 75 °C. Values of thermal diffusivity varying in the range from 9.60×10^{-8} m^2s^{-1} to 1.46×10^{-7} m^2s^{-1} and thermal conductivity from 0.26 $Wm^{-1}K^{-1}$ to 0.41 $Wm^{-1}K^{-1}$ were obtained. As main conclusions, it was noted that nanofluids exhibit superior heat transfer characteristics than the conventional heat transfer fluid and the thermal conductivity is enhanced by 50 % for the nanofluid containing 0.0275 mg/mL of sodium salt of humic acid + ethylene glycol, at the temperature of 25 °C.

1. Introduction

Dispersions of colloidal particles of nanometric dimensions (below 100 nm), added to conventional heat transfer fluids, including oil, water and ethylene glycol, have received great attention in constituting a new class of engineering fluids, the nanofluids. Nanoparticles can be as nanopowders, such as Al, Cu, SiC and CuO, or carbon nanotubes (also known as CNT). The excellent thermal, chemical and electrical properties of nanomaterials point these materials as the main additives for the production of high performance and multifunctional composite systems [1,2].

Preparation of nanofuids is the first key step in applying nanophase particles to changing the heat transfer performance of conventional fluids. The nanofluid does not simply refer to a liquid/solid mixture and some requirements are necessary, such as stable suspension, low agglomeration of particles and no chemical change of the fluid [3].

The aim of the present work is to investigate the thermal diffusivity and thermal conductivity of ethylene glycol with additions of multi-wall carbon nanotubes (MWCNT) and surfactants. For this purpose, the thermal quadrupole method was applied, and the tests were realized at room temperature and at 75 °C.

2. Materials and methods

2.1. Preparation of nanofluid

For the preparation of nanofluids was used ethylene glycol (EG) as a base fluid with addition of surfactants and nanomaterials. As surfactants were used humic acid (HA), sodium salt of humic acid

(HAS) and sodium salt of carboxymethyl cellulose (CMC), all from Sigma-Aldrich, and as nanomaterials multi-wall carbon nanotubes (MWCNT), from Nanocyl (3100 series). Six different samples with surfactants and MWCNT were studied. Table 1 shows information of each sample in detail. The preparation method of nanofluids has been implemented and conducted by the Laboratory of Carbon Nanostructures Chemistry (LQN) of CDTN [4].

The preparation of nanofluids was as follows: first, the surfactant and the carbon nanotubes were weighed and placed in a beaker. After that, ethylene glycol was added, and then the samples were placed in ultrasonic bath for 3 hours, to ensure the homogenization and dispersion stability. After 1 month of preparation, no visually observable sedimentation was found, indicating good stability of the mixture.

Table 1. Nanofluids composition.

ID in the graphs	Nanofluids and base fluid
1	EG (base fluid)
2	0.0275 mg/mL HA + EG
3	0.0275 mg/mL HAS + EG
4	0.0275 mg/mL HA + 0.05 mg/mL MWCNT + EG
5	0.0275 mg/mL HAS + 0.05 mg/mL MWCNT + EG
6	0.03 mg/mL HAS + 0.03 mg/mL MWCNT + EG
7	0.0275 mg/L CMC + 0.03 mg/mL MWCNT + EG

2.2. Density determination of the alloys

The densities ρ of the nanofluids were determined by weighing a known volume of the fluid in a glass graduated cylinder with an analytical balance Ohaus AV264CP at room temperature. The procedure was repeated three times, and the collected data are averaged. The density was determined using the equation (1):

$$\rho = \frac{m}{v} \tag{1}$$

Where m and v are the mass and the volume of the fluid, respectively. The densities of the nanofluids at 75 °C were estimated according to the variation of the density of ethylene glycol.

2.3. Determination of Thermal diffusivity – Thermal Quadrupole Method

The thermal quadrupole method was developed by Degiovanni [5] and can be seen as an extension of the concept of thermal resistance at steady state to transient conditions. It is an analytical method whose resolution is approximated by the least squares method, leading to identification of the thermal diffusivity. The sample gets a pulse of light on one side, raising its temperature. From the temperature field over the distance, it is possible to compare the theoretical curve and the experimental curve by the method of least squares. The modeling for the method is illustrated in figure 1 [5,6].

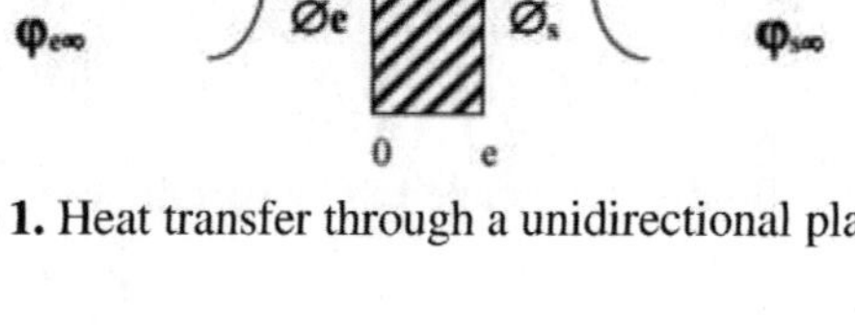

Figure 1. Heat transfer through a unidirectional plane wall.

The corresponding equations are:

$$\varphi_e = -\lambda \cdot \frac{\partial T}{\partial x}|_0 = q_e - h_e \cdot S \cdot (T_e - T_\infty) \quad (2)$$

$$\varphi_s = -\lambda \cdot \frac{\partial T}{\partial x}|_0 = q_s + h_s \cdot S \cdot (T_s - T_\infty) \quad (3)$$

Where: φ corresponds to the density of heat flow per unit of time, λ is the thermal conductivity ($Wm^{-1}K^{-1}$), T is the temperature (K), S is the surface (m), h is the heat exchange coefficient ($Wm^{-2}K^{-1}$) and q is the heat flow (Wm^{-2}).

To determine the thermal diffusivity, it was used the diffusivimeter Protolab model QuadruFlash 1200 (figure 2) of Thermophysical Properties Measurement Laboratory (LMPT) of CDTN. The thermal diffusivity was measured 10 times for each sample, and the data averaged.

This diffusivimeter, manufactured in Brazil, is constituted by a xenon lamp (1200 J), responsible for the energy pulse, three K-type thermocouples of special class, an InSb infrared detector, an oven for heating the sample and a signal processing unit. A short thermal excitation is generated by the xenon lamp. The flash is directed inside a resistive furnace, allowing heating of the tested specimen from room temperature to 1200 °C and the temperature rise of the specimen is measured by three K-type thermocouples.

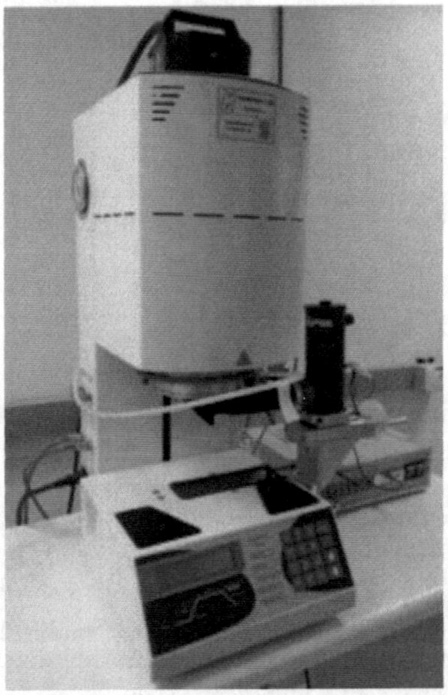

Figure 2. Diffusivimeter QuadruFlash 1200.

2.4. Determination of Thermal conductivity

The thermal conductivity k ($Wm^{-1}K^{-1}$) of each sample was calculated by the equation (4):

$$k = \alpha \cdot \rho \cdot C_p \quad (4)$$

Where: α is the sample thermal diffusivity (m^2s^{-1}), ρ is the sample density (kgm^{-3}) and C_p is the specific heat of the material ($Jkg^{-1}K^{-1}$).

The specific heat C_p of all the fluids was assumed to be specific heat of the ethylene glycol at both temperatures.

3. Results and discussion

3.1. Density determination

Table 2 presents the densities of the nanofluids. The maximum standard deviation obtained was 0.32%. It can be seen that the density decrease with increasing temperature. As expected, the densities of nanofluids are in good accordance with the density of the base fluid.

Table 2. Nanofluids densities.

ID	Density (kgm^{-3})	
	25 °C	75 °C
1	1112.43	1069.86
2	1102.13	1059.96
3	1108.63	1066.21
4	1099.77	1057.68
5	1121.80	1078.87
6	1121.33	1078.42
7	1141.37	1097.69

3.2. Thermal diffusivity and thermal conductivity

Tables 3 and 4 present the thermal diffusivity and thermal conductivity of the nanofluids.

Table 3. Thermal diffusivity of nanofluids.

ID	Thermal Diffusivity ($x10^{-7}\ m^2s^{-1}$)	
	25 °C	75 °C
1	0.97	0.96
2	1.41	1.41
3	1.46	1.44
4	1.33	1.34
5	1.33	1.34
6	1.26	1.27
7	1.10	1.12

Table 4. Thermal conductivity of nanofluids.

ID	Thermal Conductivity ($Wm^{-1}K^{-1}$)	
	25 °C	75 °C
1	0.26	0.27
2	0.37	0.39
3	0.39	0.41
4	0.35	0.37
5	0.36	0.38
6	0.34	0.36
7	0.30	0.32

As expected, the pure ethylene glycol showed lower values than the fluids with additives. It can be observed that with the temperature increase the thermal diffusivity did not change significantly and the thermal conductivity increased slightly for all the studied fluids (figure 3). The maximum standard deviation obtained was 9 %.

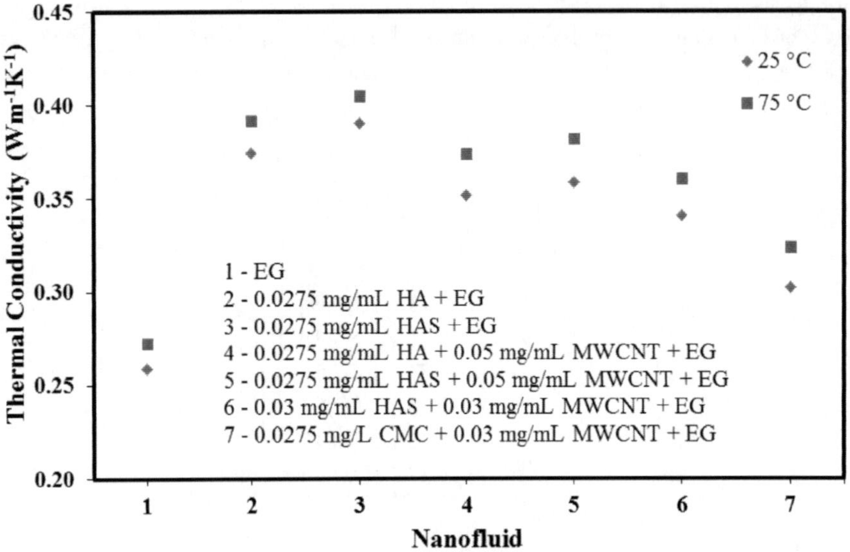

Figure 3. Thermal conductivity of the nanofluids.

The fluids with only the surfactants HA and HAS (0.0275 mg/mL each) exhibited the highest increases of thermal conductivity. However, there was no substantial increase in the thermophysical properties of the nanofluids with addition of the surfactants plus MWCNT. The probable cause of this difference is due to the lower volumetric concentration of MWCNT investigated in this study in comparison with other studies [7]. The nanofluid with CMC exhibited the lowest thermal conductivity increase in comparison to the other fluids.

4. Conclusions

The present work concerns the determination of the thermophysical properties of MWCNT/EG based nanofluids. The following conclusions are made based on the experimental results.

- Nanofluids exhibit superior heat transfer characteristics than the conventional heat transfer fluid;
- The thermal conductivity is enhanced by 50 % for the nanofluid containing 0.0275 mg/mL HAS + EG, at the temperature of 25 °C and 52 % at 75 °C.

More studies will be done to investigate the thermophysical properties of nanofluids applying higher concentrations of MWCNT.

Acknowledgments

The authors thanks the financial support of CNPq - *Conselho Nacional de Desenvolvimento Científico e Tecnológico*, *Rede* SisNANO and *Rede* GTD SIBRATEC – *Rede de Serviços Tecnológicos em Geração, Transmissão e Distribuição de Energia Elétrica.*

References

[1] Kakaç S, Pramuanjaroenkij A 2009 Review of convective heat transfer enhancement with nanofluids *Int. J. Heat and Mass Transfer* **52** 3187-96

[2] Kumaresan V, Velraj R 2012 Experimental investigation of the thermo-physical properties of water–ethylene glycol mixture based CNT nanofluids *Thermochimica Acta* **545** 180-6

[3] Xuan Y and Li Q 2000 Heat Transfer Enhancement of Nanofluids *Int. J. Heat Fluid Flow* **21** 58–64

[4] Araújo T, Salles G, Furtado C and Santos A 2014 *Proc. of the 37th Annual Meeting of the Brazilian Chemical Society* (Natal)

[5] Degiovanni A, Maillet D, André S, Batsale J C and Moyne C 2000. Thermal Quadrupoles: Solving the Heat Equation Through Integral Transforms. New York, *John Wiley & Sons*, 388p

[6] Guisard-Restivo T A 2003 *Proc. of the 47th Annual Meeting of the Brazilian Ceramic Society* (João Pessoa)

[7] Liu M-S, Lin M C-C, Huang I-T and Wang C-C 2005 Enhancement of thermal conductivity with carbon nanotube for nanofluids *Int. Commun. Heat Mass Transfer* **32** 1202-10

Fundamental studies to develop certified reference material to calibrate spectrophotometer in the ultraviolet region

F C da Conceição[1], P P Borges[1] and J F S Gomes[2]

[1]National Institute of Metrology, Quality and Technology-Inmetro, Chemistry Metrology Division, Electrochemistry Laboratory. Av. Nossa Senhora das Graças, 50, Duque de Caxias, RJ, CEP 25250-020, Brazil.
[2]National Institute of Metrology, Quality and Tecnology - Inmetro, Optical Metrology Division, Radiometry and Photometry Laboratory.

E-mail: jfgomes@inmetro.gov.br

Abstract. Spectrophotometry is the technique used in a great number of laboratories around the world. Quantitative determination of a high number of inorganic, organic and biological species can be made by spectrophotometry using calibrated spectrophotometers. International standards require the use of optical filters to perform the calibration of spectrophotometers. One of the recommended materials is the crystalline potassium dichromate ($K_2Cr_2O_7$), which is used to prepare solutions in specific concentrations for calibration or verification of spectrophotometers in the ultraviolet (UV) spectral regions. This paper presents the results concerning the fundamental studies for developing a certified reference material (CRM) of crystalline potassium dichromate to be used as standard of spectrophotometers in order to contribute to reliable quantitative analyses.

1. Introduction

The absorption molecular spectrometry (also called spectrophotometry) in the ultraviolet (UV) spectral regions is very used for quantitative determining of a high number of inorganic, organic and biological species in chemical laboratories, petrochemical, environmental, forensic and clinic around the world. It is based on the transmittance or absorbance measurements carried out in spectrophotometer, which is an instrument that relates the concentration of determined substance with the quantity of the absorbed or transmitted radiation in specific wavelength, that traversed transparent container with length of path defined [1,2] according to the Lambert-Beer´s law as equation (1).

$$A_a = a_a bc \quad (1)$$

Where, A_a is the apparent absorbance; a_a, apparent absorptivity; b, length of path and c, the concentration. The term named apparent was introduced together with the parameters absorbance and absorptivity because the effect related to the buoyancy correction for mass and also internal reflections of light in the quartz cuvette were not considered in the measurements.

For guaranteeing of reliability and metrological traceability from measurement results, the ISO/IEC 17025 standard states that all equipment, which has an important effect in the accuracy and reliability of the measurement results of testing, calibration or sampling, shall be calibrated before being used [3], on the other hand, one of the statements concerning the spectrophotometers in the regional standard ASTM E925[4] says that it is of own analyst´s responsibility verify and validate either if the instrument is running correctly or if it is capable to provide acceptable analytical results. The verification of the performance of the instrument shall be done with the use of certified reference materials (CRMs). These CRMs must have optical properties very well characterized to be used for that purpose.

Currently, the National Institute of Standards & Technology (NIST) provides the SRM 935a [5]. This standard reference material (SRM) is the crystalline potassium dichromate, from which solutions with specific concentrations to the wavelengths in the spectral regions of UV are prepared. These standard solutions are used to evaluate the performance of the spectrophotometers in the UV regions.

The National Institute of Metrology, Quality and Technology (Inmetro) was granted the calibration and measurement capability (CMC) for high purity potassium dichromate. Due to the importance of verifying the performance of the spectrophotometers to regular time interval, the Electrochemistry Laboratory from the Chemistry Metrology Division (Dquim) together with the Radiometry and Photometry Laboratory from the Optical Division (Diopt) are developing the CRM of crystalline potassium dichromate.

The main goal of this work is to present the fundamental studies needed to develop the CRM of crystalline potassium dichromate in Brazil, since this CRM will contribute to guarantee the accuracy and linearity of the absorbance scale to the spectrophotometers, through which analytical measurement results will be provided with metrological traceability and reliability.

2. Feasibility studies, stability and batchpreparation of the candidate of CRM

2.1. Feasibility studies

Initially, solutions of potassium dichromate ($K_2Cr_2O_7$) in 0.001 mol L^{-1}perchloric acid ($HClO_4$) medium were prepared. The concentrations of the $K_2Cr_2O_7$ were 20, 40, 60, 80 and 100 mg kg^{-1}. These concentrations will provide absorption values in the specific wavelength range needed to verify the performance of the spectrophotometers in the UV regions. With the measurement results of absorbance for each solution, there was calculated the apparent absorptivity as described in equation (1) and the results were compared to the values shown in the SRM 935a certificate.

2.2. Stability of the solutions

It was prepared, in addition, 1 L of two new solutions of $K_2Cr_2O_7$ in 0.001mol $L^{-1}HClO_4$ either in the lowest level of concentrations, 20 mg kg^{-1}, or in the highest level, 100 mg kg^{-1}, for the stability study. This study is needed to verify the bias in the parameter of apparent absorptivity during the time. This study is very important for the consumer to know if the dichromate solutions will remain their optical properties constant along the time.

2.3. Batch preparation

A batch contained 64 bottles of the candidate of CRM was prepared. Firstly, 1 kg of the salt of $K_2Cr_2O_7$ was mixed and manually homogenized; after, 15 g of the material was weighed in an analytical balance and was filled in each glass bottle, which was numbered in the order of filling.

3. Experimental

3.1. Reagents

High purity $K_2Cr_2O_7$ (99.5%) from Sigma-Aldrich and $HClO_4$analytical purity (70-72%) from Merck were used. All solutions were prepared using deionized water from Elga® model Option-Q with electrolytic conductivity lower or equal to 1.0 μS cm^{-1}.

3.2. Equipment

The apparent absorbances were measured by UV/VIS Spectrophotometer Lambda 25, using quartz cell both from Perkin Elmer® with length of path of 1.0 cm. The $K_2Cr_2O_7$ solutions were prepared weighing the material using analytical balance both from Mettler Toledo® with resolution 0.001 g or a microbalance with resolution of 0.1μg, when necessary.

3.3. Methodology
The utilized methodology was as follows: Initially, $K_2Cr_2O_7$ solutions were prepared in 0.001 mol L^{-1} $HClO_4$ using the CRM of potassium dichromate from Inmetro (MRC 8477) in the following concentrations: 20 mg kg^{-1}, 40 mg kg^{-1}, 60 mg kg^{-1}, 80 mg kg^{-1}, and 100 mg kg^{-1}. After then, measurements of apparent absorbance in triplicate were done in the specific wavelengths (λ) in the UV regions: 235 nm (1 nm = 10^{-9} m), 257 nm, 313 nm, 345 nm and 350 nm. Finally, from the apparent absorbance values, the apparent absorptivities were calculated by using equation (1) for each wavelength measured.

4. Results and discussion
Figure 1 shows the absorption spectrum for specific concentrations of the solutions (range 20-100 mol kg^{-1}). It can be seen that the apparent absorbance increases in each wavelength measured as soon as the concentration is raised. The lowest values measured were obtained for λ = 313 nm, as showed in the literature [4,5].

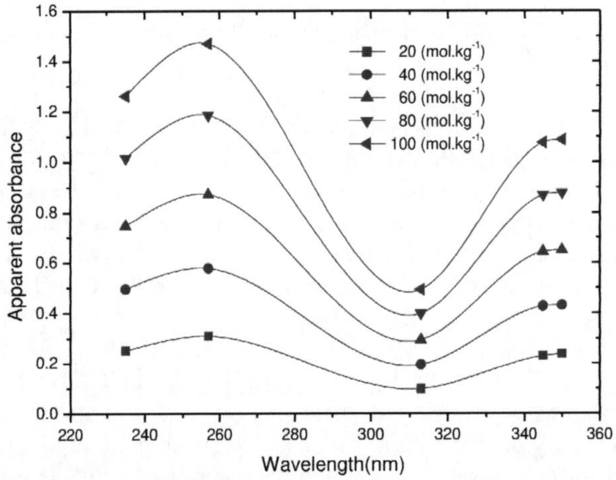

Figure 1. Absorption spectra of $K_2Cr_2O_7$ solutions in $HClO_4$ medium.

Using these values of apparent absorbance measured, it was calculated the correspondent apparent absorptivity values by equation (1). The values of apparent absorptivity reached in this work were compared to the values of absorptivity presented in the certificate of analysis of the SRM 935a as shown in table 1.

Table1. Apparent absorptivity and the relative error considering this work and the SRM 935a.

Concentration (mg kg^{-1})	λ (nm)	a_a (kg cm^{-1}g^{-1})	Error[1] (%)
20	235	11.879	3.1
	257	14.659	-2.8
	313	4.644	3.3
	345	10.868	-2.5
	350	11.231	-5.3
40	235	12.329	-0.20
	257	14.387	-0.48
	313	4.848	-0.76
	345	10.643	-0.38
	350	10.726	-0.41
60	235	12.306	0.33
	257	14.366	0.05
	313	4.823	-0.16
	345	10.592	0.09
	350	10.696	-0.04
80	235	12.418	-0.22
	257	14.461	-0.22
	313	4.857	-0.75
	345	10.610	-0.08
	350	10.720	-0.18
100	235	12.462	-0.23
	257	14.527	-0.29
	313	4.861	-0.62
	345	10.628	-0.26
	350	10.743	-0.30

[1] Apparent absorptivity of this work divided by the apparent absorptivity from NIST.

Although the spectral bandwidth (SBW) used in this work was constant and equal to 1 nm, the results were comparable do the results showed in the NIST's certificate. In the certificate SRM 935a the SBWs varied in the range of 0.8 to 1.2 nm. On the other hand, the ASTM E-925 standard recommends that the SBW should not be higher than 2 nm, therefore the comparability of the measurement results can be acceptable.

From table 1, it can be seen that the higher relative error was around 5%, due to the lower concentration used, since this concentration has lower absorptions and, thus, can present deviations from Lambert-Beer law. Therefore, these results show that it is viable to develop this new CRM of crystalline potassium dichromate in Brazil.

For 5 months, the stability of the $K_2Cr_2O_7$ solutions in 0.001 mol L^{-1} $HClO_4$ medium in the concentrations of 20 mg kg^{-1} and 100 mg kg^{-1} is showed in figure 2.

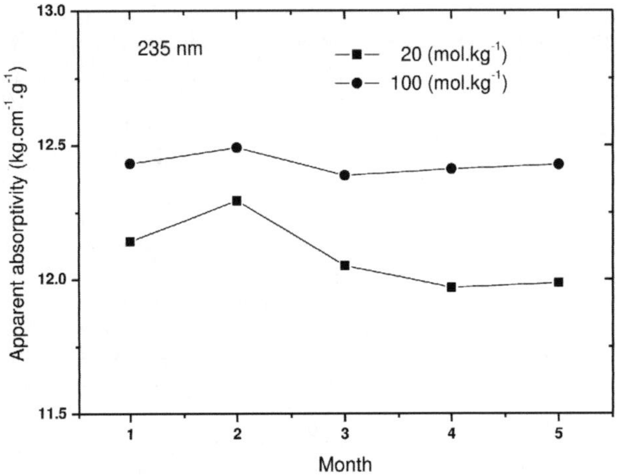

Figure 2. Stability of the $K_2Cr_2O_7$ solutions related to the apparent absorptivity parameter.

The wavelength of 235 nm was considered as an example. All the data of apparent absorptivity was considered normally distributed by Shapiro-Wilk test. By regression analysis, the values of apparent absorptivity measured in the solutions were considered stable (p-value > 0.05) for a confidence level of 95% [6]. Because of that, the $K_2Cr_2O_7$ solutions can be considered stable for 5 months.

Considering the results of feasibility studies, a batch of a candidate CRM of crystalline potassium dichromate was prepared. Four bottles were at random selected from the batch for the studies of homogeneity, characterization and stability of the solutions. Both homogeneity and characterization studies are being performed. The uncertainty related to the both studies will be added to the uncertainty of the CRM. The absorbance of the solutions in five different concentrations of $K_2Cr_2O_7$ was chosen as the property value. The inhomogeneity of the batch will be inferred by using the statistical test of One-way ANOVA. The characterization of the candidate CRM will be carried out by a comparison test performed by two laboratories.

The process of certification of the batch of crystalline $K_2Cr_2O_7$ will be subject to a next work.

5. Conclusions

Spectrophotometry is an analytical technique very well used in diverse laboratories in the country. With the development of the CRM of crystalline potassium dichromate, the laboratories can guarantee better accuracy in the spectrophotometric analyses in diverse areas, for example, the petrochemical laboratories for analyzing some quality parameters of biofuels needed to be in accordance with the specifications of National Agency of Oil, Natural Gas and Biofuels (ANP).

Aknowledgements

One of the authors (F C da Conceição) wants to thank to Petrobras for granting a fellowship (PRH-PB nº 103).

References

[1] Holler FJ, Skoog DA andCrouch SR 2009 *Princípios de Análise Instrumental*Trad Célio Pasquini 6ªed (São Paulo:Bookman)

[2] Ewing G W 1969*Intrumental methods of Chemical Analysis* 3rd Edition (McGraw-Hill:New York)

[3] ISO/IEC 17025 2005 *General requirements for the competence of testing and calibration laboratories*

[4] ASTM E925-09 2014 *Standard practice for monitoring the calibration of ultraviolet-visible spectrophotometers whose spectral bandwidth does not exceed 2 nm*
[5] SRM 935a NIST *certificate of analysis crystalline potassium dichromate for use as an ultraviolet absorbance standard*
[6] ISO GUIDE 35 2006 *Reference materials – General and statistical principles for certification*

Development of a hand- transmitted vibration measurement instrument to perform tests in medical equipment according to the international standard IEC 60601-1-2005 Ed.3

F. F. de Lima , H. T. Moriya, J. C. T. B. Moraes

Divisão de Ensaios e Calibração, Laboratório de Engenharia Biomédica, Escola Politécnica, Universidade de São Paulo, São Paulo, SP, Brazil

E-mail: felipe@leb.usp.br

Abstract. According to the sub clause 9.6.3 of the international standard IEC 60601-1-2005, medical electrical equipment must provide means of protection against hazardous hand-transmitted vibrations. Compliance of this sub clause is checked by using a vibration measurement instrument in accordance with ISO 5349-1-2001 standard. The present article describes the development of a vibration measurement instrument to perform vibration measurements tests in medical equipments.

1. Introduction

Exposure to hazardous hand vibrations for long periods of time can result in a health issue known as vibration syndrome. This syndrome has adverse circulatory and neuronal effects and can cause pain, loss of light touch, loss of grip strength and bone cysts in fingers and wrists [1].

Some medical equipment can produce and transmit harmful vibrations to the operator or patient hands during normal operation. According to the sub clause 9.6.3 of the international standard of medical electrical equipment IEC 60601-1-2005 Ed.3[2] (equivalent to the Brazilian standard NBR IEC 60601-1-2010 Ed.3 [3]), these equipment must provide means of protection against hand-transmitted hazardous vibrations. To verify if an electrical medical equipment (ME) complies with this sub clause a vibration measurement instrument is used as described in the international standard ISO 5349-1-2001 [4].

2. Objectives

Aiming the construction and evaluation of the vibration measurement instrument the following objectives were outlined:

- The instrument hardware must be developed following the measurement requirements of IEC 60601-1-2005 and ISO 5349-1-2001. The hardware must be composed of a vibration sensor, an anti-aliasing filter and an analog to digital converter (A/D).
- The instrument software must be developed to implement a digital filter (ISO 5349-1-2001), to convert the signal to the desired unit (m/s^2 in this case) and to show the results to the operator.
- The developed instrument must be verified by comparison with a certified instrument.
- The used test procedure must be verified by five experienced professionals in a real test scenario

3. Materials and methods

The instrument hardware was based on the diagram shown in Figure 1.

Figure 1. Hardware diagram

The vibration sensor employed was a small 4mm x 4mm packaged analog 3-axes accelerometer with maximum sensing range of 2g and bandwidth of 0.5 Hz to 1600 Hz (ADXL327, Analog Devices, USA). .

The chosen accelerometer measures accelerations in the range of 1g (9.8m/s²) and frequencies in the range of 1 Hz to 1000Hz. These ranges are justified since 1 g is approximately four times greater than the maximum allowed vibration (IEC60601-1-2005) and the frequency range of 1 Hz to 1000Hz is consistent with the attenuation of 40dB caused by the required ISO 5349-1-2001 filter.

The anti-aliasing filter used to filter the signal from the accelerometer was a low-pass unity gain 4th order Sallen Key Chebyshev filter (Fcutoff= 1400Hz).

The anti-aliasing filter and the accelerometer were assembled in different circuit boards, resulting in a small vibration sensor easy to attach to a ME.

The analog to digital converter connected to the output of the anti-aliasing filter was a multipurpose interface (PCI – 6052E - National Instruments, USA).

A virtual instrument (VI) software was based on the diagram shown in Figure 2.

Figure 2. Software diagram

The VI has four main functions:

• Digitally filter the acquired data with the specified ISO 5349-1-2001 filter.
• Convert the data to the standard m/s² unit.
• Compute the root square of the sum of the square of the data.
• Show the results of the measurement to the operator.

Based on the ISO 5349-1-2001 filter (continuous 6th order frequency weighting and band-limiting filter) a digital filter was designed and used to filter the digital data.

After the digital filter, gain is applied to the data to convert it to the desired unit (m/s²). Finally, the root square of the sum of this data is computed, producing a final result in m/s², shown in graphical user interface (GUI).

To verify the measurements obtained by the new assembled instrument, a mechanical setup (Figure 3) was mounted.

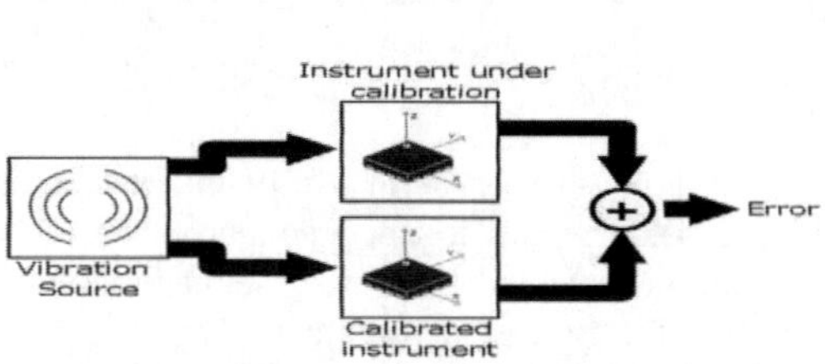

Figure 3. Instrument measurement verification

A 1-axis shaker (LDS V780, LDS Test and Measurement, UK) was used as vibration source. It was configured to produce sinusoidal oscillations in different frequencies/magnitudes. On the shaker vibration platform, a calibrated 1-axis accelerometer (LDS 353B31, LDS Test and Measurement, UK) and the developed instrument sensor were fixed as close as possible to ensure that the same vibration disturbance was acting in both sensors. The verification procedure was repeated three times, once for each accelerometer axis. In each repetition, the shaker was programmed to produce sinusoidal oscillations in the range of 3Hz to 1200Hz (10 s in each frequency, 10 frequencies per decade), and the magnitude was set to produce the maximum permitted shaker output (limited to 15 m/s²). For each frequency data from the calibrated accelerometer and from the developed instrument was processed with the same algorithm and compared through an error plot.

To verify the procedure adopted to execute the vibration compliance test, five qualified professionals with proven experience in execution of tests of the IEC 60601-1-2005, performed the vibration test in a simulated ME (a conventional table fan CM Premium - Delta - CN operating in its maximum intensity). As patient hand contacts were adopted two physical points: the fan transport handle and the middle of the rod responsible to mechanically support the electrical motor of the fan. Each of the five professionals was responsible to attach the vibration sensor to the simulated ME and to collect the vibrations measurements.

4. Results

The relative error between the calibrated accelerometer and the developed instrument was obtained for each axis, as shown in figure 4.

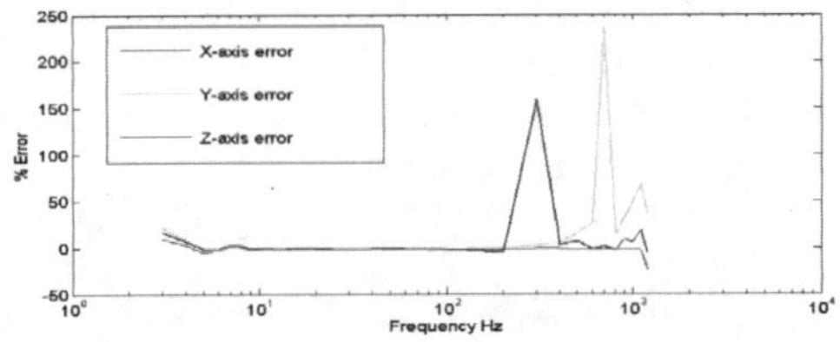

Figure 4. Instrument error comparison

In the frequency range of 3 Hz to 100 Hz, in all three axes, relative errors were below 5%. In frequencies above 200 Hz some resonant frequencies were observed.

Table 1 shows the results obtained by the five experienced professionals when executing the vibration measurement in the simulated ME.1.

Table 1. Test procedure results

	Fan handle (m/s²)	Fan rod (m/s²)
1	6.28 ± 0.86	1.49 ± 0.18
2	5.34 ± 0.66	1.92 ± 0.23
3	9.12 ± 1.08	2.13 ± 0.26
4	5.77 ± 0.79	1.87 ± 0.26
5	6.28 ± 0.74	2.09 ± 0.25

5. Discussion

During the instrument verification was observed that resonant frequencies were achieved in the range of 200Hz to 1200Hz (Figure 4). These resonant frequencies are the result from the mechanical assembly and could be reduced or canceled with more sophisticated mechanical elements. As these resonant frequencies are produced in frequencies attenuated by at least 20dB through the ISO 5349-1-2001 filter the assembled instrument can be accepted to be used in ME vibrations measurements tests, since it is guaranteed that vibrations frequencies in this specific range are not significant in the measurement. This guarantee can be easily verified providing to the operator a frequency spectrum estimator of the raw vibration signal in the software GUI.

ISO 5349-1-2001 provides a continuous time transfer function of the filter to be used. To reduce hardware complexity, this filter was digitally implemented by transforming the continuous transfer function to a discrete time form using bilinear transformation. The sampling frequency of the analog to digital converter was chosen so as to the differences between the analog and digital filter in the bandwidth of interest (3 Hz - 100Hz) were less than 0.01%.

Although relative error between calibrated accelerometer and assembled instrument remained as low as 5%, this measurement error could be lower by using a digital channel equalization filter in the assembled instrument to compensate electrical and mechanical system differences between both devices.

Although the professional number 3 has measured a fan handle vibration of 9.12 ± 1.08 m/s^2 (Table 1), differing expressively from the others professionals, all measurements were homogenous. This difference, probably, can be explained by the fact that this test can be sensitive to the attachment position and orientation of the sensor in the ME.

6. Conclusion

A vibration measurement instrument, following requirements of IEC 60601-1-2005 and ISO 53492-2001, was assembled and both instrument measurement verification and test procedure verification were done, indicating that the usage of the assembled instrument in real ME tests is possible, as long as some limits are respected. Moreover, some improvements to the instrument can be done, as the channel equalization filter and the spectrum estimator.

References

[1] Millar Donald J D, the *National Institute for Occupational Safety and Health (NIOSH)* Publications (1983) 38

[2] *International Electrotechnical Commission (IEC),* 2005, IEC 60601-1, Ed. 3: Medical electrical equipment - Part 1: General requirements for basic safety and essential performance (2005);

[3] *Associação Brasileira de Normas Técnicas (ABNT),* 2010, ABNT NBR IEC 60601-1_2010 - Equipamento eletromédico Parte 1: Requisitos gerais para segurança básica e desempenho essencial , (2010).

[4] *International Organization for Standardization (ISO),* 2001, Mechanical vibration – measurement and evaluation of human exposure to hand-transmitted vibration – Part 1 – General Requirements (2001)

Determination of Al in cake mix: Method validation and estimation of measurement uncertainty

G Andrade[1], O Rocha[1] and R Junqueira[2]

[1] Centro de Inovação e Tecnologia SENAI FIEMG - Campus CETEC, Brazil
[2] Faculdade de Farmácia, Universidade Federal de Minas Gerais, Brazil

E-mail: giselle.andrade@fiemg.com.br

Abstract: An analytical method for the determination of Al in cake mix was developed. Acceptable values were obtained for the following parameters: linearity, detection limit – LOD (5.00 mg·kg^{-1}) quantification limit – LOQ (12.5 mg·kg^{-1}), the recovery assays values (between 91 and 102%), the relative standard deviation under repeatability and within-reproducibility conditions (<20.0%) and measurement uncertainty tests (<10.0%) The results of the validation process showed that the proposed method is fitness for purpose.

1. Introduction

Aluminum (Al) is the third most abundant element in the earth's crust and is also present in water and food. For over a century it has been recognized as a neurotoxic element [1].

One of the main routes of exposure to aluminum in the general population is food consumption, especially of those with food additives containing Al [2]. The Joint FAO/WHO Expert Committee on Food Additives evaluated the toxicity of Al and established the Provisional Tolerable Weekly Intake (PTWI) 2 mg Al / kg body weight per week, from all sources, including additives. As reported in sixty-seventh meeting, some population groups could exceed Al, especially children, due to the consumption of food additives containing aluminum added in its composition [3]. The aluminum content in food has been studied in several countries. However, in Brazil, there are few data addressing the Al concentration.

Considering the relevance of the Al presence in the diet of the population, the aim of this work is to present a detailed procedure for in-house method validation and evaluate the uncertainty sources for determining the levels of Al in cake mix by ICP OES. The analytical performance of the entire procedure, such as linearity, detection, limit, quantification limit, precision and accuracy were assessed statistically to evaluate the developed procedures.

2. Material and methods

2.1. Instrumentation

The acid digestion of cake mix samples was performed using a closed-vessel microwave system (Model D Start, Milestone, Italy). The device can work up to the power of 1000 W during the time set by the user, providing that a vessel does not exceed 100 bars pressure or 300°C.

The Table 1 summarizes the operating conditions used in ICP OES with radially viewed configuration (Optima 3000, Perkin Elmer, USA).

Table 1. Instrumental conditions of the ICP OES

Parameter	Value
RF incident power	1500
Nebulizer	Cross-flow
Nebulizer argon flow rate	1.0 L·min^{-1}
Sample flow rate	1.5 mL·min^{-1}
Auxiliary argon flow rate	0.5 L·min^{-1}
Plasma argon flow rate	15 L·min^{-1}
Flow pumping	1.5 mL·min^{-1}
Diameter of the injector tube	2.0 mm
Height of view radial	10 mm
Wavelength of the Al (atomic line)	396.153 nm
Ionization Energy	9.13 eV[1]

2.2. Reagents and solutions
High purity deionized water (resistivity 18 MΩ cm) was obtained from a Milli-Q Plus water purification system (Millipore Corp, EUA). Analytical reagent grade HNO_3 65% (w·w^{-1}) (Merck, Germany) was previously purified by sub-boiling (Berghof, Germany). For the preparation of standards was used certified multielement solution (SPEX, Certipur, Lot 27200JB), containing 500 mg·L^{-1} of Al.

2.3. Contamination care
In order to prevent contaminations all operations were performed in a clean room ISO Class 7 equipped with islands of laminar flow ISO class 5. The materials used in the test consists of Teflon or polyethylene and were decontaminated with 10 % HNO_3 for 24 hours, then washed with ultrapure water, dried and kept in laminar flow.

2.4. Digestion procedure
The digestion procedure for all samples was performed as in [4], [5] and [6]. Aliquots (0.2-0.3 g) of homogenized sample were digested with 3 mL of 65 % nitric acid (w·w^{-1}) and 3 ml of high purity deionized water in a closed microwave. The heating program for digestion of the samples was previously optimized [6]. The samples were transferred to 50 mL volumetric flasks.

2.5. Calibration and sample analysis
Two sets of standards were used for calibration: one for low concentrations of Al (20 µg·L^{-1} to 200 µg·L^{-1}). Analytical blanks were also prepared in order to evaluate the noise and zero adjustment instrumental, but were not included in the calculations. The calibration curves were assessed through ordinary least squares estimation, outlier detection and lack of fit tests [7].

2.6. Validation
The performance characteristics of the method were established by assays employing standards solutions, sample blanks and spiked samples. Linearity, matrix effects, lack of trueness (recovery), precision under repeatability and within-reproducibility conditions, detection and quantification limits were the parameters evaluated [8].

All the tests for linearity assumption were also done for these curves [7]. The slopes and intercepts were compared by t-test for a 0.05 after evaluation the assumption of homogeneity of variances. The standard addition method was used to check the matrix effects. Due to the absence certified reference material for matrix studied, the parameters lack of trueness and precision were estimated through recovery.

The recovery and precision were established by spiked samples in four different days, with three independent replicates at the same levels of linearity test. Blank samples were also prepared, in triplicate, for each curve. The values obtained were analyzed. The repeatability and within-reproducibility were estimated by ANOVA [7].

The theoretical detection limit (LD) and quantification limit (LQ) were estimated using the background equivalent concentration, and after were confirmed by testing the recovery and precision.

2.7. *Measurement uncertainty*

Uncertainties were estimated according to the Guide to the Expression of Uncertainty in Measurement [9]. Relevant sources for the uncertainty assessment have been divided into four groups: uncertainties associated with a) sample preparation, b) preparation of standard solutions, c) the analytical curve, d) method precision. To calculate the expanded uncertainty (U), a coverage factor (k) of 2 was used which corresponds to a 95 % interval for a normal probability distribution.

3. Results and discussion

3.1. *Validation of method*

Outliers were removed before statistical tests. The results obtained for the linearity assessment are shown in Table 2. The assumptions that the errors are normally distributed, homoscedastic and independent were confirmed as proposed by [7].

All the regression assumptions were confirmed for the solvent and matrix-matched calibration curves. No significant matrix effects were detected when the slopes of the solvent and matrix-matched curves were compared.

Table 2. Evaluation of Linearity for curves in nitric acid and the matrix in the concentration range studied.

Statistic	Curve in nitric acid	Curve in matrix
n^a	18	18
Outilers	0	2
R^b	0.9937	0.9894
$R_{critic}{}^c$	0.9567	0.9529
p	>0.10	>0.10
d^d	1.516	2.116
p	>0.10	>0.10
tL^e	1.101	0.732
p	0.2870	0.4764
$F_1{}^f$	$2.545 \cdot 10^3$	$1.710 \cdot 10^3$
p	$4.57 \cdot 10^{-19}$	$4.89 \cdot 10^{-16}$
$F_2{}^g$	1.224	2.083
p^h	0.352	0.158

[a] Number of observations;
[b] Ryan–Joiner correlation coefficient.
[c] Ryan–Joiner critics correlation coefficients;
[d] Durbin–Watson statistic.
[e] Levene t-statistic.;
[f] ratio of variances of the regression;
[g] Variance ratio;
[h] significance.

The Table 3 presents the estimated recovery, relative standard deviations under repeatability (RSDr) and within-reproducibility (RSDR).

Table 3. Mean recovery, relative standard deviation under repeatability and within-reproducibility conditions.

Concentration ($\mu g \cdot L^{-1}$)	MR[a] (%)	RSDr[b] (%)	RSDR[c] (%)
20	93.16	16.88	19.39
50	101.98	4.29	5.59
80	101.98	3.10	4.13
120	97.71	3.24	6.46
160	93.72	7.33	7.33
200	91.70	4.81	6.20

[a] mean recovery
[b] repeatability relative standard deviation
[c] within reproducibility relative standard deviations.

These recovery results were considered acceptable and demonstrated no lack of trueness of the validated method. For the first level were obtained RSDr and RSDR above than the respective critical values (14.7%) [8], these results indicate accuracy and precision of the method in the range above of 12.5 mg $\cdot kg^{-1}$.

The theoretical detection and quantification limits were estimated in 5.0 mg kg^{-1} and 12.5 mg kg^{-1}, respectively. They were confirmed by recovery experiments.

The measurement uncertainty of the method was estimated using the bottom-up approach. The Ishikawa diagram with the uncertainty sources of the measurement uncertainty of the analysis shown in Figure 1.

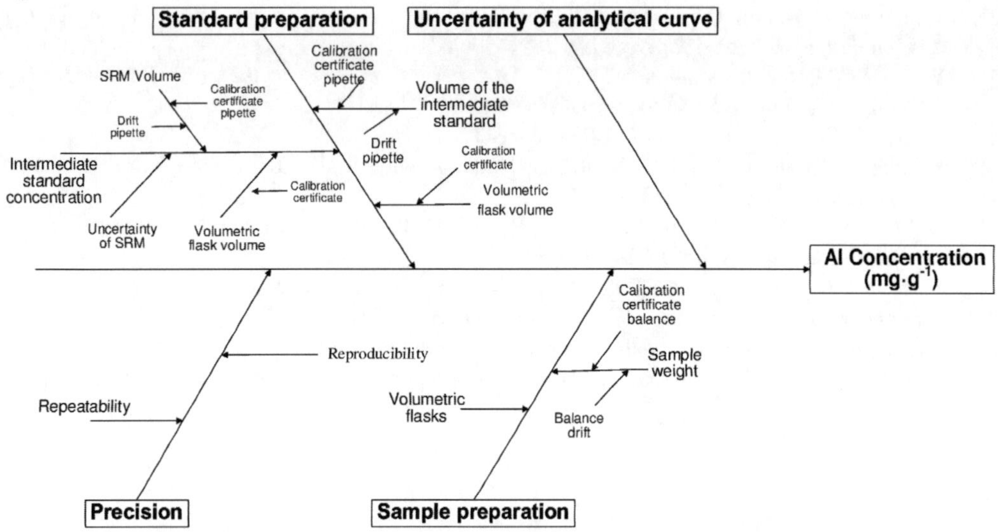

Figure 1. Cause and effect diagram for the determination of Al.

Percent values of the standard uncertainty contribution from each group of uncertainties to the standard uncertainty are shown in Figure 2. The uncertainties associated with the precision are the main contribution in the method.

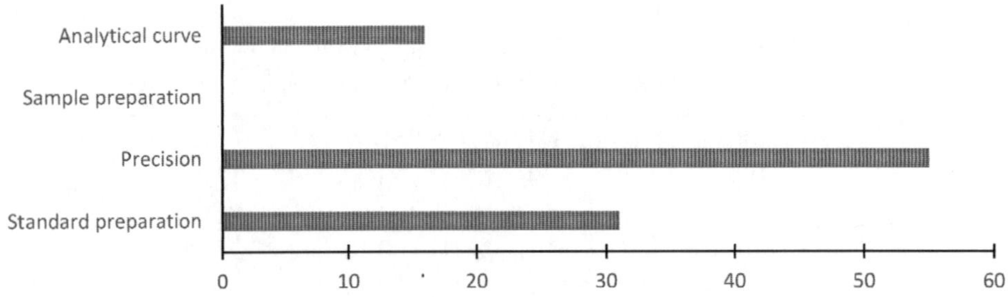

Figure 2. Relative contribution from each group of uncertainties.

For the analyzed sample, whose Al concentration determined, was 34.6 $mg \cdot kg^{-1}$, the value of expanded uncertainty (U) was 2.0 $mg \cdot kg^{-1}$, which corresponds less than 10 % of the concentration value.

4. Conclusions

The method presented good results of selectivity, linearity, precision (repeatability and reproducibility), recovery and the expanded uncertainties. These results showed that the proposed method is fitness for purpose.

References

[1] Zatta P, Ibn-Lkhayat-Idrissi M, Zambenedetti P, Kilyen M and Kiss T 2000 *Brain Res. Bull.* **59** 41

[2] Who (World Health Organization) 1989 *Thirty-third Report of the Joint FAO/WHO Expert Committee on Food Additives*. **776** 28

[3] Who (World Health Organization) 2012 *Sixty-seventh meeting of the Joint FAO/WHO Expert Committee on Food Additives (JECFA)*. **65**

[4] Guérin T, Noël L 2009 *Journal of Reference*. **1**

[5] Millour S, Noël L, Chekri R, Vastel C, Kadar A, Guérin T 2010 *Accreditation and Quality Assurance*. **15** 503

[6] Millour S, Noël L, Kadar A, Chekri R, Vastel C and Guérin T 2011 *Journal of Food Composition and Analysis*. **24** 111

[7] Souza S V C, Junqueira R G 2005 *Analytica Chimica Acta*. **552** 25

[8] Thompson M, Ellison S L R, Wood R 2002 *Pure Appl. Chem.* **74** 835

[9] INMETRO (Instituto Nacional de Metrologia, Qualidade e Tecnologia) 2012 *Avaliação de dados de medição - Guia para a expressão de incerteza de medição - GUM 2008*. **2012**

A novel design for a primary measurement standard for the quantity torque

G M Taccola[1] and R J Leão[2]

[1] Center for Mechanical, Electrical and Fluid Flow Metrology, Institute for Technological Research, São Paulo, SP, 05508-901, Brazil

[2] Brazilian Synchrotron Light Laboratory, Brazilian Center for Research in Energy and Materials, Campinas, SP, 13083-970, Brazil

Abstract. Primary standards are established using a reference measurement procedure. Through the realization of the definition of a given quantity they provide traceability in the metrology chain. Most torque standards use calibrated weights linked to a lever arm to provide known torques in the rotational axis of the device. These standards do not succeed to provide continuous torque values because of limitation in weights combination and the intrinsic incremental steps of the technique. In automatic machines this can be overcome but it requires a complex mechanical system to operate. In designs that use movable weights along the lever arm the determination of the length of the lever arm is a large source of errors. Zeroing the torque standard also represents a major problem. An innovative and functional design that aims to solve the aforementioned disadvantages is presented.

1. Introduction

In the torque metrological chain, torque transducers are used to calibrate torque wrenches, and torque transducers are normally traceable to base quantities of mass and length, through torque standard machines. An example of SI traceability of torque is given in [1].

By definition, primary standards are established using reference measurement procedures [2]. The most widely solution for torque realization is the use of calibrated weights and a lever arm. Most National Metrology Institutes – NMI have such machines [3-6]. Complete metrological characterizations are performed [7-9] and also comparisons are made to assure metrological confidence [10]. Despite some variations, the main concept involves a known length levelled with a known mass in its end, applying a torque value in the main axis of the device [11]. The transducer to be calibrated is coupled to the machine. In order to change the applied torque the weights combinations are varied. Thus, the loading cycle achievable with deadweight machines usually follows a discrete pattern, due to limitations of the system [1, 12].

The Jockey-weight concept [13] works by sliding a movable weight along the lever arm. The weight can be gradually moved, and the torque can be linearly raised, but accuracy and repeatability of weight placement in a certain location requires complex mechanisms. As for any type of machine, a great deal of effort need to be done in order to know precisely the length of the lever arm, as this will affect directly the applied torque [14]. At the National Metrology Institute of Japan – NMIJ a detailed analysis of a 10 Nm dead weight torque standard machine was performed, with an emphasis on thermal expansion characteristics [15].

Designs should account for ease in automation, due to possible operator influences in hand operated machines. The need for automation of torque standard machines is investigated in the literature [16].

The bending of the arm and the thermal influences on its size are critical, as they cause undesirable changes in the applied torque. The reference line, where the weight is connected to the arm also plays an important role in torque metrology [14].

Especially in the case of the lower part of the working range, friction in the bearings affects negatively the torque standard. Some authors focus on the study of this specific matter, studying these effects and finding ways of determining the static friction torque [4, 17].

The industry is constantly pushing the development of more reliable torque standards [6-7, 10]. The construction of both small and large torque ranges face problems related to the realization of torque and reducing the uncertainty of the system.

The purpose of this paper is to introduce a new design for the realization of torque with potential to serve as a new primary torque standard machine [18].

2. Concept Description

The new design for torque standard is based on the angular variation of the position of a reaction arm. Current designs rely on the application of a known weight in the end of a lever arm of known length, which should be well aligned to the horizontal axis.

Since the torque will change according to the change of the length of the arm, rotating the arm will cause torque to change with the projection of the arm length.

The zero position of this new design is the vertical position, where no torque is applied to its main axis. By gradually rotating machine's angle, torque is applied with the sine of the angle with the vertical axis, measured by means of a precision level or encoder. Figure 1 explains the basic idea of this concept.

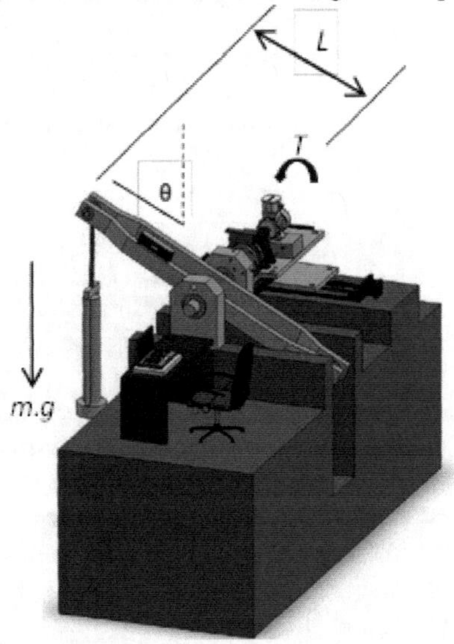

Figure 1. Scheme of the new standard (Source: authors).

A single calibrated weight can be applied in the construction of the torque standard. We propose a symmetric arm to act as a counter-torque. The design of the arm accounts for the machine to operate in two different modes, calibrating torque transducers in the clockwise and anticlockwise direction. Both calibrations can be performed in only one setup.

Two separate level indicators can be applied to redundancy, one in each side of the double arm. As the new design do not use weight stack as main operational principle, automation is easier to be done.

One of the main advantages of the proposed design is the possibility of application of continuous torque, without discrete ramp-up. The mechanism of applying gradually increasing or decreasing torque with discrete weights is much more complex, involving a drive mechanism and torque transducers or strain gauges to continuously measure the torque and compensate during weight stacking. Stabilization and control issues of the new system are perhaps not simple to achieve, and future prototypes will allow the validation of the standard for continuous torque application.

3. Operation

The torque can be modelled as T (m, g, L, θ), where m is the applied mass, g is the local gravity acceleration, L is the length of the arm and θ is the angle between the arm and the vertical axis, as in Equation 1.

$$T = m.g.L.sin\theta \tag{1}$$

The maximum torque will be achieved when the angle is 90 degrees. A control loop for the operation of the standard can be seen in figure 2.

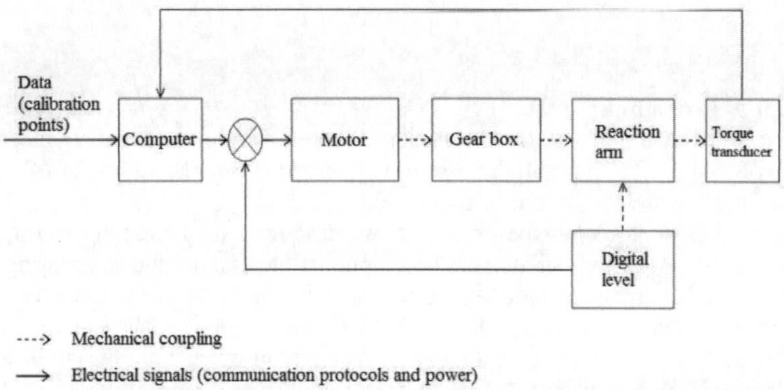

Figure 2. Control diagram (Source: authors).

Accounting for the exemption of weight changes, a simulation of the time involved in a torque transducer calibration (considering weights standardized by OIML, more commonly used by laboratories) shows that the new standard will be able to save time and therefore costs in calibration services (figure 3).

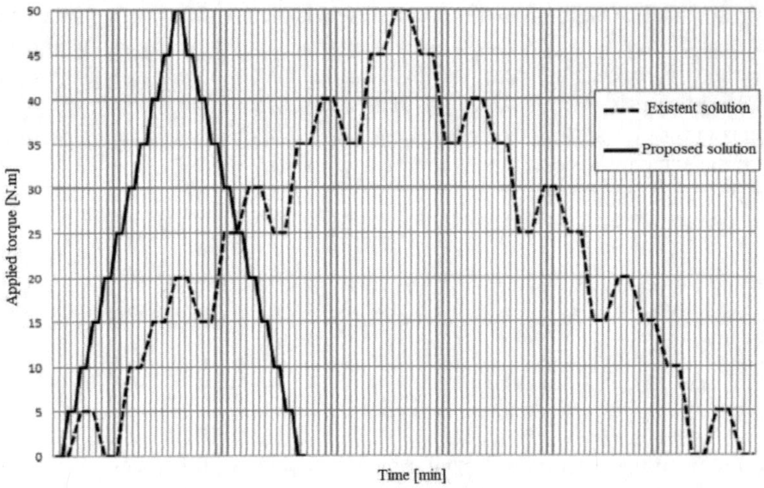

Figure 3. Comparison between times required to perform a calibration of a torque transducer (Source: authors)

4. Uncertainty Estimation

Besides parasitic influences, such as bearing friction, bending moments and transversal forces, uncertainty of the new design, as in the case of any torque standard, is a matter of how well one knows the length of the arm and the weight being applied.

In the particular case of the new design being described, the length of the arm is measured by the inclination and length of the bar itself. The employed electronic level measuring system should be reliable and accurate, and have enough resolution. A detailed review about uncertainty evaluation for torque primary standards is done by Merlo [17]. The lower limit for the relative expanded uncertainty is referred to as 1 x 10-5. Other interesting articles about this subject are the studies conducted by Park et al [4], Röske [7], Koji et al [8].

Equation 1 can be extended to include the influences of buoyancy acting on the dead-weight, unbalancing and inhomogeneities in the double reaction arm and friction on the rotating axis as follows in Equation 2:

$$T_{tot} = m.g.L.sin\theta.(1-\rho_a/\rho_w) - T_{bal} - T_{fri} \tag{2}$$

Uncertainty propagation law [19] to combine the effects of each component and related sensitivity coefficients can be easily deduced. Table 1 details the calculation of the expected total uncertainty of the proposed torque standard (simulating a 100 Nm applied torque).

The dimensions of the sensibility coefficients come from the partial derivatives of Equation 1. The probability distributions were assumed to be normal for all the sources of uncertainty except for Buoyancy, Unbalance and Friction (considered uniform), and the effective degrees-of-freedom were considered large enough to assume a coverage factor of two.

Components of Angle, Length and Mass are current Calibration and Measurement Capability – CMC of the Institute for Technological Research – IPT in São Paulo, Brazil and other Brazilian accredited laboratories. Local gravity uncertainty was obtained by measurements made by National Observatory – ON at IPT site. Buoyancy, Friction and thermal expansion of the reaction arm components are estimated by using reference values [4]. Arm flexure effects were adopted also according to literature [8]. Unbalance of the reaction arm may be corrected and the residual difference in mass between the arm and counter-arm parts (in this case 0.1 g was assumed) might generate a Torque and the uncertainty related to that effect would be equal to $(\Delta m.g.L.sin\theta)/\sqrt{3}$.

Table 1. Uncertainty estimation (Source: authors).

	Estimate	*c*	u	*c.u (Nm)*
Angle (rad)	0.52359	173.206	2.18E-6	0.000378
Length (m)	1.02	98.040	2.00E-6	0.000196
Arm Flexure (m)	0	98.040	2.04E-6	0.000200
Thermal Growth (m)	0	98.040	1.77E-6	0.000174
Mass (kg)	20.039	4.990	1.65E-6	0.000082
Gravity (m/s²)	9.78643	10.218	6.00E-6	6.13E-7
Air density (kg/m³)	1.225	-0.013	3.00E-2	-0.000380
Weight density	7906	1.96E-6	4.56E+1	0.000089
Balancing (Nm)	0	1	2.88E-4	0.000288
Friction (Nm)	0	1	1.73E-3	0.001730
Torque Applied (Nm)				100.0007
Combined Uncertainty – u (Nm)				0.0019
k				2.00
Expanded Uncertainty – U (Nm)				0.0038

5. Conclusion

This work describes a new torque standard machine concept, intended to improve the current technique used to calibrate torque transducers. The state-of-the-art in terms of torque quantity is the use of weight stacks, despite the discrete application of torque and the difficulty in automation.

The concept of angular variation of the arm open up the possibility to expand the calibration procedure to a more realistic situation, where the transducer to be calibrated is used in continuous torque slope. Performance estimation has been done and a relative expanded uncertainty in the order of 4E-5 is expected. This uncertainty fulfils recommended CMC for torque primary standards and would allow the calibration of torque transducers to 0,05 class [20].

Future work may evaluate a full prototype of the new torque standard and calculate the CMC for the new primary torque standard, as well as intercomparison studies. A CMC in the order of the results found in this paper would be compatible with those found in the literature. For instance, NMI torque standards CMC range from 2E10-5 to 10E-5 [4, 7].

6. References

[1] Atsuhiro N, Koji O and Kazunaga U 2015 Measurement 61 1–8
[2] Joint Committee for Guides in Metrology 2012 "International Vocabulary of Metrology – Basic and General Concepts and Associated Terms" BIPM 200
[3] Koji O, Takashi O, Kazunaga U and Furuta E 2002 "Design and development of 20 kNm deadweight torque standard machine" Proceedings of the IMEKO World Congress
[4] Park Y, Kim M and Kang D 2007 Meas. Sci. Technol 18 3273–3278
[5] Zhimin Z, Yue Z, Bin G, Feng M, Tao L, Honglei J and Ming D 2009 Proceedings of the IMEKO World Congress 339-342
[6] Peschel D, Mauesrberger D, Schwind D and Kolwinski U 2005 Proceedings of the IMEKO World Congress 001-007
[7] Röske, D 2014 Metrologia 51 87–96
[8] Koji O, Takashi O and Kazunaga U 2007 Measurement 40 797–802
[9] Oliveira R, Cabral L, Kolwinski U and Schwind D 2005 "Performance of the new primary torque standard machine of INMETRO, Brazil" Proceedings of the IMEKO World Congress
[10] Atsuhiro N, Koji O and Kazunaga U 2014 Measurement 49 77–90

[11] Gassmann H, Allgeier T and Kolwinski U 2000 Proceedings of the IMEKO World Congress 063-074
[12] Atsuhiro N, Koji O and Kazunaga U 2012 Measurement 45 2393–2399
[13] Allgeier T, Kolwinski U and Schwind D 2002 "Jockey-weight lever machines for force and torque" Proceedings of the IMEKO World Congress
[14] Röske D 1997 Measurement 20 (1) 23-32
[15] Atsuhiro N, Koji O and Kazunaga U 2010 Measurement 43 1318–1326
[16] Oliveira R, Guilherme R, Cabral L 2012 Measurement 45 2413–2418
[17] Merlo S 2001 Measurement 29 279–285
[18] Taccola G and Leão R 2013 Instituto Nacional da Propriedade Industrial BR1020130327786
[19] Joint Committee for Guides in Metrology 2008 "Evaluation of measurement data – Guide to the expression of uncertainty in measurement" BIPM 100
[20] Comitê Brasileiro de Máquinas e Equipamentos Mecânicos 2013 "Metallic materials - Calibration of static torque measuring devices" ABNT 12240

Acknowledgments
The authors would like to thank the Institute for Technological Research for the support during the development of this work and the colleagues from the Mechanical Metrology Laboratory for the useful discussions regarding the subject of this paper.

Journal of Physics: Conference Series **733** (2016) 012020 doi:10.1088/1742-6596/733/1/012020

Historical overview of Ramsey spectroscopy and its relevance on Time and Frequency Metrology

M M Amaral [1,4], L V G Tarelho [1], M A de Souza [1], A C Baratto[1], G A Garcia [1], S T Muller [2], J De Martin Jr [2], A S Rodriguez [2], A Bebeachibuli[2], D V Magalhães [2,3]

[1] National Institute of Metrology, Quality and Technology - INMETRO; Av. Nossa Senhora das Graças, 50, 25250-020, Duque de Caxias- RJ – Brazil.
[2] Institute of Physics of São Carlos– University of São Paulo - IFSC-USP, Av. Trabalhador São-Carlense, 400, 13566-590, São Carlos - SP – Brazil.
[3] School of Engineering of São Carlos – University of São Paulo - EESC-USP, Av. Trabalhador São-Carlense, 400, 13566-590, São Carlos - SP – Brazil.
[4]Institute of Nuclear and Energy Research – IPEN, Avenida Lineu Prestes, 2242, 05508-000, São Paulo - SP -Brazil

E-mail: lvtarelho@inmetro.gov.br

Abstract. A brief overview of the historical evolution of the method of successive oscillatory fields developed by Norman Ramsey, and some different implementations of the decurrent methodology are presented. We use time and frequency standards, from Cs atomic beams to optical standards, as examples. The scientific progress and the technological implementation achieved through a partnership between USP-SC and INMETRO are shown on the characterization of each time and frequency standard.

1. Introduction

The 1989 Physics' Nobel Prize was divided between Normal F. Ramsey, Hans G. Dehmelt and Wolfgang Paul. Ramsey was awarded "for the invention of the separated oscillatory fields method and its use in the hydrogen maser and other atomic clocks" [1].

Ramsey started to work on Rabi's group at Columbia University in 1937, right after the invention of molecular-beam magnetic resonance. The first experiments in the group were made with lithium fluoride molecules, hydrogen and deuterium. In fact, very good spectra were obtained from lithium fluoride molecules, but the results from hydrogen and deuterium were not as good. This problem was turned over to Ramsey, whereas other researchers in the group pursued, then, more promising and productive goals [2].

The Rabi's method setup employed a uniform magnetic field provided by a coil in which atoms and molecules pass through. When a "correct" frequency of a RF field is applied to the coil, the nuclear spin of an atom flips. The spectral line width in Rabi's resonance is a function of the time of flight throughout the coil, so the longer the time the sharper the line.

After joining Harvard University in 1947, Ramsey started to design his new molecular-beam magnetic resonance apparatus. He faced the challenge of maximizing the spectral resolution. One idea was to making the RF coils as long as possible to obtain extremely uniform magnetic field, but it became extremely hard to achieve.

Inspired by the Michelson [3] stellar interferometer, that has almost twice the resolution of a perfect telescope with mirror diameter of the same length of the interferometer, Ramsey discovered that using two coils to apply an oscillatory field to the atoms, the resolution of the spectral lines would be twice the resolution obtained in Rabi's method [2,4,5].

In this article we will present a brief overview of the method of successive oscillatory fields developed by Norman Ramsey, its application on the main atomic clock models (Cesium Atomic Beam, Compact Atomic Clock, Fountain Clock, Hydrogen Maser and Optical Atomic Clock) for primary standards of time and frequency, and its importance on other physics experiments. As illustration, we present the progress of the scientific development and technology implementation on IFSC-USP and INMETRO.

2. The Ramsey Method [5]

Ramsey developed a method, designed originally for measurements of nuclear magnetic moments, which can be applied for both atomic and molecular beams. In this methodology, instead of applying a coherent field over a period of time *T*, as in Rabi's method, the atoms or molecules are exposed to the field twice by a short period of time τ, separated by a period *T* with no field applied (Figure 1). The atoms are prepared on an initial state and enter the region of interaction on the time $t = 0$ and interact with a microwave radiation with frequency ν during the time interval τ. In $t = \tau$ the atoms pass throughout a microwave free region and their quantum states evolve freely during the time interval *T*. Finally, on $t = \tau + T$, the atoms interact again with the microwave during a new period of time τ. In each region of microwave field, the atoms feel a $\pi/2$ pulse. At the end of this process, with the right condition of frequency and of field power, that maximize the transition probability, the atoms must be at the state$|F = 4$. Parallel to the atomic beam direction, in the region of interaction, there is a static magnetic field (*C-field*) that permits to lift the degeneration of the hyperfine steady state. After that, the atoms pass through a detection system that measures the transition probability by an induction signal proportional to the number of atoms in the upper state.

The Full Width at Half Maximum (FWHM) of the interference pattern is $dn = 1 \quad 2T$. So, the longer the time spent by the atoms between the regions of interaction, the narrower the FWHM, improving the sensitivity.

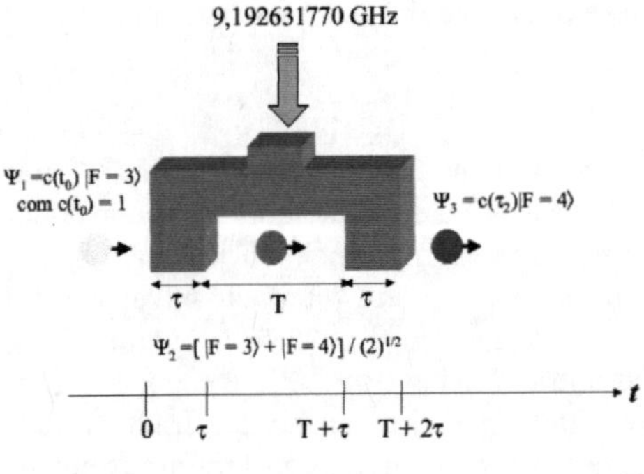

Figure 1. Schematic representation of Ramsey spectroscopy for Cesium Atomic Beam.

The temporal separation between the oscillatory fields can be achieved by a spatial separation of the two regions of field interaction (e.g. Cesium Atomic Beam and Hydrogen MASER), or by the interaction with the field in the same spatial location, turning on and off the field (e.g. Compact

Atomic Clock and Optical Atomic Clock), or making the atoms to pass twice the same region (e. g. Fountain Clock).

In this method of successive oscillatory fields the resonance line is about 1.7 times narrower than the Rabi's methodology for the same length of the apparatus, the sharpness of the resonance is not affected by non-uniformities of the constant field, leading to an increase in precision by a factor of 20, it is more efficient at high frequencies, and the first-order Doppler shifts are eliminated. Notwithstanding those advantages, some precautions must be taken, e.g., the variation on the amplitude of the oscillating field may change the shape of the resonance line.

3. Primary Standards of Time and Frequency

The BIPM establishes the hyperfine transition |6 $^2S_{1/2}$, F=3, m_f=0>→|6 $^2S_{1/2}$, F=4, m_f=0> of the non-perturbed Cesium (Cs) fundamental state, at the microwave frequency region (9.192 631 770 GHz), as the definition of the second (Figure 2).

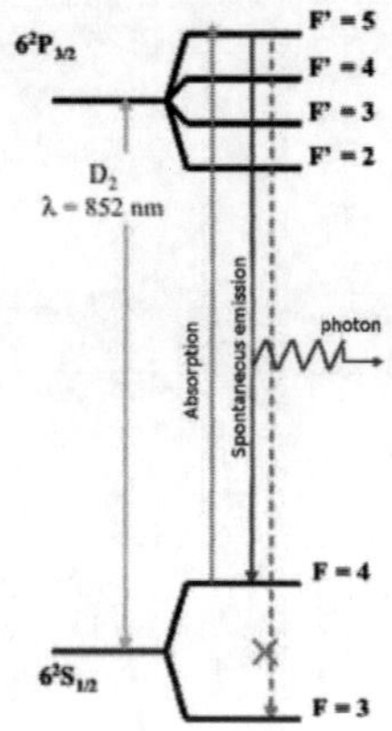

Figure 2. Cesium Energy Levels for D_2 transition at 852 nm.

A primary standard of time and frequency possess the best possible accuracy and stability and it finds applications on local representation of time scale, telecommunications, base stations for satellites, geo positioning, among others, emphasizing the economic impact of the methodology developed by Ramsey, which gave rise to the atomic clock technology.

3.1. Cesium Atomic Beam

Most of the standard atomic clocks in operation around the world are commercial Cs atomic beam clocks employing magnetic selection, notwithstanding the facts that their metrological characterization are not so easy, and that there are some difficulties to correct all uncertainty factors. The assembling of an optically pumped Cs beam clock is a good alternative that allows a complete metrological characterization of the effects that can affect its operation and of the correction factors. Thus, an accurate and stable operation can be achieved. The main differences between the magnetic selected and optically pumped beam clocks are related to the way the atoms of Cs are prepared on the quantum state F = |3 and how the transition probability to |F = 4 is measured.

3.1.1 Optically Pumped Atomic Beam. The atoms of Cs that emerge from a heated oven with velocities of about 200 m/s on the states $|F = 3$ and $|F = 4$, are collimated by a nozzle. The atoms are optically prepared to obtain a population inversion, putting the most part on the $|F = 4$ state. They pass then through spatially separated oscillatory fields (figure 1) and the Ramsey spectroscopy is performed, obtaining the interference pattern shown in figure 3. Locking the microwave frequency at the maximum of the interference pattern allows operating it as a primary frequency standard (or a local "atomic clock"). The Ramsey fringe shown in figure 3 was obtained with a primary frequency standard developed at the Physics Institute of São Carlos [6] since 1998. This standard was allocated to Inmetro at the end of 2013 [7] to be commissioned as a national primary standard.

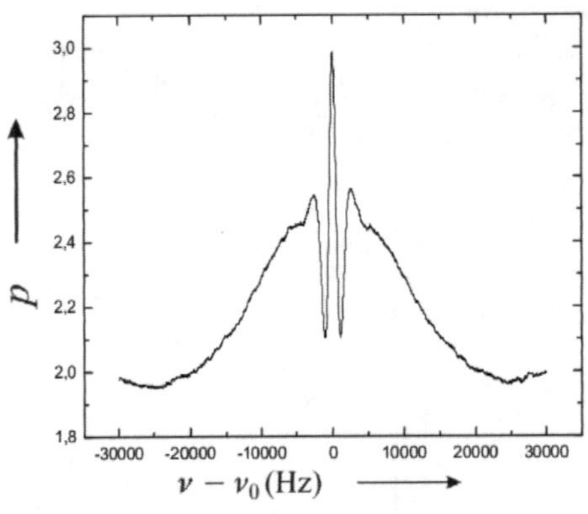

Figure 3. Ramsey fringe for cesium atomic beam.

3.2. *Compact Atomic Clock*

The compact atomic clock is an evolution of the thermal beam clock. It uses cold atoms trapped by laser to interrogate the clock transition, improving its performance.

Figure 4 shows a setup of the compact atomic clock. Three-axis laser beams and a magnetic field trap the atoms (magneto-optical trap) and cool them down at the center of the setup. The atoms are prepared on the state |F = 3 in the same way described at Cs beam section. In sequence, the magnetic field is turned off and the atoms are exposed to the microwave oscillatory field for a time period τ, the microwave oscillatory field is turned off for a time period T and the atoms are exposed again to the microwave oscillatory field to a new period of time τ as described on section 2.

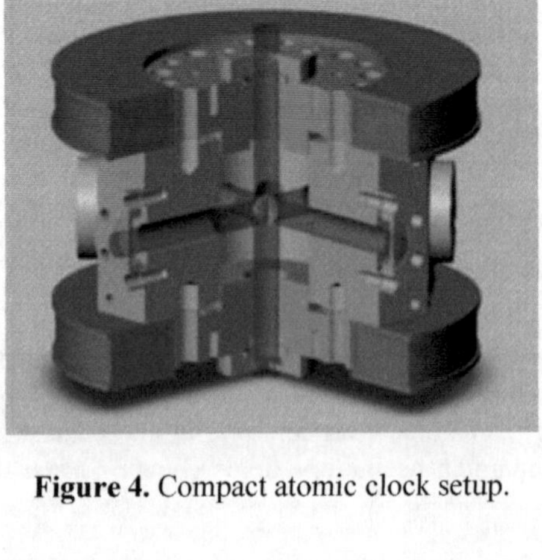

Figure 4. Compact atomic clock setup.

The main advantage of this setup is its compact size compared to the size of the thermal beam clock, allowing its use at base stations, satellites and for high performance signal dissemination.

Figure 5 presents the Ramsey interferometry pattern, provided by a compact clock, measured at São Carlos laboratory [8]. It is possible to observe an improvement on the spectral resolution when it is compared to that of the Cs beam.

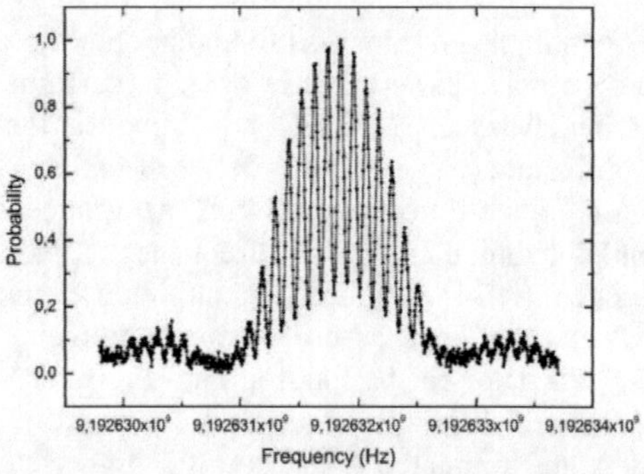

Figure 5. Ramsey fringe obtained with a compact atomic clock

3.3. Fountain Atomic Clock

The fountain atomic clock also uses cold atoms to measure the hyperfine (clock) transition [9]. The process of trapping the atoms is similar to that of the method described for the compact clock, the difference being that it follows a ballistic trajectory concept introduced by Zacharias [5].

After preparing the atoms on the state |F = 3 , the cold atoms are launched up vertically with a few meters per second velocity and pass through a region of oscillatory microwave field, for the first time, in a time period τ. The gravitational force slows the atoms down until them fall back and, after a time period T, the atoms pass again at the same region of oscillatory microwave field by another time period τ (see figure 6). The fringe shown in the figure was obtained in São Carlos [9].

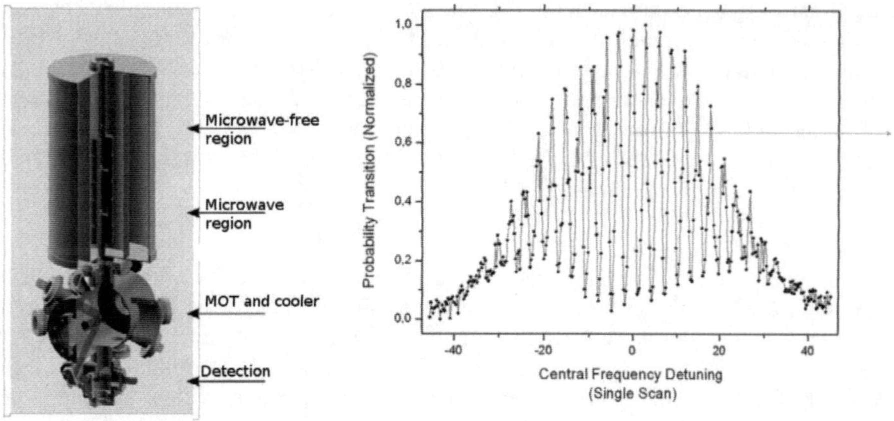

Figure 6. Fountain atomic clock setup and Ramsey fringe.

It is interesting to note that, due to the different constructive characteristics, the time interval *T* between the oscillatory microwave fields is bigger at Fountain clock, followed by Compact Clock and then by the Atomic Cs Beam and thus the spectral resolution and accuracy are better at the same order.

4. Other Frequency Standards

4.1. Hydrogen MASER

The microwave amplification by stimulated emission of radiation, whose acronym is MASER, has been built using various atomic or molecular species in order to perform high-resolution microwave spectroscopic investigations. The Hydrogen MASER has achieved the most widespread use, its stability being superior to that of Cesium (Cs) clocks for periods from about 10 s to a day

The atomic hydrogen MASER uses a transition between two ground state levels $|F = 1, m_f = 0$ and $|F = 0, m_f = 0$ of atomic hydrogen, with a frequency separation of 1 420 405 752 Hz (1.42 GHz). The atomic hydrogen (H) is produced in an intense electrical discharge from the molecular hydrogen gas (H_2). The atoms emerge from the source into an evacuated region and enter a state-selecting magnet. The $|F = 1$ state will be focused onto the small aperture of the storage cell, whereas the $|F = 0$ state is defocused. When these atoms are exposed to microwave radiation at the hyperfine frequency, more atoms are stimulated to go from the higher-energy state to the lower one. Thus, this device is an amplifier [4,10].

Hydrogen MASER has also been operated with the microwave power confined in two small cavities that function as separated oscillatory field devices (Ramsey's method). The atoms that are stimulated to emit radiation move randomly into and out of these cavities with oscillatory fields and spend the intermediate time in the large container with no such fields. Due to the large size of the storage box, there are longer flight times and less frequent wall collisions, so the resonances are narrower and the wall shifts are smaller than for a normal hydrogen MASER (with just microwave cavity).

4.2. Optical Atomic Clock

The natural evolution of these atomic clocks points to the most promised measurements in the optical range frequencies (THz), instead of using microwave excited transitions (GHz) range. Even at this newfangled range, Ramsey method can be of further assistance. Single laser pulse techniques are similar to Rabi spectroscopy and multiple pulses techniques can be related to Ramsey spectroscopy. Multi pulse Ramsey spectroscopy offers several ways to significantly manipulate some frequency shifts present in the excitation, interrogation or detection systems. The most promising method is called Hyper Ramsey spectroscopy scheme [11]. It is based on time-separated pulses that can have different durations, frequencies and phases. The pulses manipulate the induced frequency shifts of the spectroscopic signals due to interference effects. The signature Ramsey fringes are present in Hyper Ramsey fringes under some conditions, and the capability of reducing frequency shifts is very important for uncertainty characterization of high precision optical clocks. Inmetro and Physics Institute of São Carlos have established a cooperation agreement to develop this class of frequency standards based on Sr (Strontium) atoms [12].

5. Conclusion

The field of time and frequency standards received, and continues to receive, important contributions from the Ramsey method. This method turned to be fundamental to the evolution of the atomic clock apparatuses. Brazil is now in the route to construct an infrastructure to build, operate and maintain primary frequency standards that should be, in a near future, capable of contributing to the TAI of the BIPM. It is interesting to note that, due to the different constructive characteristics, the time T between the oscillatory microwave fields is, at Fountain clocks, bigger than the Compact Clock and the Atomic

Cs Beam Clock. Mainly due to these characteristics, its spectral resolution and its accuracy are better than these last clocks.

References

[1] "The Nobel Prize in Physics 1989". Nobelprize.org. Nobel Media AB 2014. Web. 24 Jul 2015. <http://www.nobelprize.org/nobel_prizes/ physics/laureates/1989/>

[2] Kleppner, D. "Norman Ramsey and his method". Phys. Today, 66, 1, 25 (2013)

[3] Michelson, A. A., and Pease, F. G. ,"Measurement of the diameter of Alpha-Orionis by the interferometer", Astrophys. J. 53, 249–259(1921).

[4] Ramsey, N. "The method of successive oscillatory fields". Phys. Today 66, 1, 36 (2013)

[5] Ramsey,Norman F., "Molecular Beams",Oxford University, 1990.

[6] Bebeachibuli, A. "Relógio Atômico a Feixe Efusivo de Césio: Estudo Da Estabilidade e da Acurácia Como Função do Deslocamento Da Frequência Atômica Devido ao Efeito Zeeman de Segunda Ordem, ao Cavity Pulling e ao Rabi Pulling". M.Sc. dissertation, São Paulo: USP, 2003.

[7] Tarelho, L.V.G.; Garcia, G.A. ; Baratto, A.C.; de Souza, R.S.L. ; de Martin Junior, J.; Muller, S.T. ; Magalhães, D.V.; Bebeachibuli, A.; Bagnato, V.S.; Rovera, G.D.; "Joint Effort to Commissioning a Thermal Cesium Beam with Optical Pumping as Primary Frequency Standard to Brazilian NMI", Annals of 29th Conference on Precise Electromagnetic Measurements (CPEM), Rio de Janeiro , Brazil, 2014

[8] Muller, S T, "Padrão de frequência compacto", DSc. Thesis, São Carlos, 2010

[9] Alves, R F, "CHAFARIZ ATÔMICO DE Cs 133", M.Sc. Dissertation, São Carlos, 2012

[10] Riehle, F. "Frequency standards basics and aplications", Wiley-VCH, Weinheim, 2004

[11] Yudin, V.I.; Taichenachev, A.V.; Oates, C. W. ; Barber, Z. W.;Lemke, N. D.; Ludlow, A. D.; Sterr, U.; Lisdat, Ch.; and Riehle, F. ,"Hyper-Ramsey spectroscopy of optical clock transitions", Physical Review A 82, 011804(R) (2010).

[12] Magalhães, D V ; Müller, S T ; Pechoneri, R D; de Martin Junior, J ; Bueno, C; Bagnato, V S; Rodriguez, S.A., Courteille, P.W.; Paiva R.R. ; Garcia G.A.; Tarelho L.V.G.; Amaral M.M.; Baratto, A.C.; Souza, R.S.L; Atomic Frequency Standards in Brazil -Research and ongoing applications, Annals of XXXIII Encontro de Físicos do Norte e do Nordeste, Natal,Brazil, 2015.

Necessary and convenient number of indications in a measurement

A C Baratto[1], L V G Tarelho[1], G A Garcia[1]

[1] Divisão de Metrologia em Tecnologias da Informação e Telecomunicações, Inmetro, Av. Nossa Senhora das Graças, 50, 25250-020, Duque de Caxias- RJ – Brazil

E-mail: acbaratto@inmetro.gov.br

Abstract. The work addresses the problem of how to properly stipulate a suitable number of indications or repetitions to be used in a measurement. This appropriate and adequate number of indications is understood and is proposed as the minimal number that induces, when the resulting Type A uncertainty is included in the calculation, a relative change in the preexisting combined uncertainty lower than some parameter Q. In its turn, the parameter Q is anticipated by the metrologist based on an analysis of the experimentational context in which the measurement is being made.

1. Introduction

There is a widespread and recurrent idea that getting a large number of indications is a positive and safe procedure in measurement processes, and that it is always better than taking a small number of them. But the cost of a specific process, which derives from superfluous overwork, from extra time spent in measuring and in storage, transmission and processing of data, and from the waste of energy expended to maintain the measurement system in a steady state, grows with the number of indications. Time and energy are money, but the matter will not be a significant problem for you if you have an abundance of time, energy and money. However, the subject is of great importance, and particularly delicate, in cases in which the measurement procedure is cumbersome and tiring, leading to abnormal results when extensively repeated, and when the integrity of the measurand is in some way sensitive to certain parts of the process.

The decision about the best number of indications to be used in a measurement involves a careful consideration of the "*experimentational perspective that dictates the uses and applications one intends to give to* (the) *results*" [1]. Before the beginning of a measurement it is necessary to define a measurand, or to make a careful exegesis of the elected preexisting definition. After, we need to select the method and the instruments that will be used in the measurement, all in accordance to the use and application that is intended for the results. At this time we already know (or we are in conditions to know) the values of all Type B uncertainty components. The combination of these Type B uncertainties in the "law of propagation of uncertainties" results in a minimum value for the combined uncertainty (u_{cB}). Now, the next step is deciding how many indications will be acquired. For this, the metrologist needs to know the minimum acceptable number of indications and, additionally, the most convenient number of indications to be used so that the measurement is carried out based on the wisest decision. The use of a number smaller than the necessary implies an undesirable large uncertainty. The use of a number greater than the most convenient (in a "*lato sensu*") implies an unwanted large cost of the measurement process.

The number of indications delivered by the physical measurement process will result in a Type A uncertainty that will increase the value of the minimum combined uncertainty u_{cB} reported above by some relative factor q. The metrologist wants to establish a parameter Q as an upper limit for the value of q. The problem treated in this work is to calculate the minimum number n of indications that satisfies the condition $q < Q$. The value of Q, thought as a tolerable or desirable value, is chosen by the metrologist based on experimentational requirements. Of course, the number (n) that results from the treatment is in general not a natural one, and should be rounded up to the next integer. In practice, the use of *this next integer* will result in a relative variation factor q slightly smaller than Q. This is valid for the expanded uncertainty as well.

2. Calculation of the minimum number of indications

For a type A evaluation, the **experimental standard deviation of the mean** $s(\overline{q})$ of a series of n observations ([2, B.2.17 NOTE 2]) in a measurement process is given by equation (1):

$$s(\overline{q}) = \frac{s(q_i)}{\sqrt{n}} = \frac{1}{\sqrt{n}} \frac{\sqrt{\sum_{k=1}^{n}(q_k - \overline{q})^2}}{\sqrt{(n-1)}} = \frac{\sqrt{\sum_{k=1}^{n}(q_k - \overline{q})^2}}{\sqrt{n(n-1)}} \qquad (1)$$

where $s(q_i)$ is the experimental standard deviation of the observations and $\overline{q}$ is the mean of n effective q_i indications.

By (1) we see that $s(\overline{q})$ decreases when the number of indications increases. For $n >> 1$ the dependence of $s(\overline{q})$ is just inversely proportional to n. In the present treatment, and by supposition, the other uncertainty components $u(x_i)$ are evaluated by Type B uncertainty evaluation methods.

We will consider the evaluation of a measurand Y, with estimate y and combined uncertainty $u_c(y)$, which is composed by N-1 identified Type B uncertainty components ($u_i(y) = c_i u(x_i), i = 1,\ldots,N-1$), and just only one Type A ($u_N(y) = c_N u(x_N) = s(\overline{q})$) component. The associated combined variance $u_c^2(y) = \sum_{i=1}^{N} c_i^2 u^2(x_i) = c_N^2 u^2(x_N)) + \sum_{i=1}^{N-1} c_i^2 u^2(x_i)$ may be written as in equation (2):

$$u_c^2(y) = c_N^2 u^2(x_N)) + \sum_{i=1}^{N-1} c_i^2 u^2(x_i) = s^2(\overline{q}) + \sum_{i=1}^{N-1} c_i^2 u^2(x_i) = \frac{1}{n(n-1)} \sum_{k=1}^{n} (q_k - \overline{q})^2 + \sum_{i=1}^{N-1} c_i^2 u^2(x_i) \qquad (2)$$

For simplicity we will, from now on, write u_c in place of $u_c(y)$. We consider in principle that, before starting the procedures for the physical measurement, we had made a study of all the Type B uncertainties, we know all their values and, therefore, we also know the value of the Type B combined standard uncertainty u_{cB}. The value of the combined variance of the measurement will be the sum of the corresponding combined B variance (u_{cB}^2) and the Type A variance component ($c_N^2 u^2(x_N)$) when the last one is included in the calculation. Considering all N-1 Type B uncertainty components $u_B(x_i)$ associated with the measurement, the combined B variance u_{cB}^2 is given by equation (3) below:

$$u_{cB}^2 = \sum_{j=1}^{N-1} c_j^2 u^2(x_j) \qquad (3)$$

The introduction of any new component $u_N(y) = c_N u(x_N)$ will change the combined uncertainty, given in equation (3), from u_{cB} to a new value u_*, given by equation (4):

$$u_*^2 = u_N^2(y) + u_{cB}^2 = c_N^2 u^2(x_N) + u_{cB}^2 = c_N^2 u^2(x_N) + \sum_{j=1}^{N-1} c_j^2 u^2(x_j) \tag{4}$$

The relative change δu_c, produced in the preexisting u_{cB}, will be given, according to Baratto and Bezerra [3, formula (7)], by equation (5) below.

$$\delta u_c(N) = \frac{u_* - u_{cB}}{u_{cB}} = \frac{u_*}{u_{cB}} - 1 = \frac{\sqrt{c_N^2 u^2(x_N) + \sum_{j=1}^{N-1} c_j^2 u^2(x_j)}}{\sqrt{u_{cB}^2}} - 1 = \sqrt{\frac{c_N^2 u^2(x_N)}{u_{cB}^2} + 1} - 1 \tag{5}$$

We look for the number of indications that is necessary and sufficient to be acquired in the measurement. This number of indications will determine the value of the Type A uncertainty, and then the final uncertainty. In the case discussed here, the new N^{th} component $u_N(y)$ included in the calculation, as in equation (4), is the uncertainty associated to the dispersion of the repeated observations q_i, more specifically the **experimental standard deviation of the mean**, $s(\overline{q})$, given in (1) above. The change of the term $c_N^2 u^2(x_N)$ by the experimental variance of the mean $s^2(\overline{q})$ results in equation (6):

$$\delta u_c(N) = \sqrt{\frac{s^2(\overline{q})}{u_{cB}^2} + 1} - 1 \tag{6}$$

We want to find the number n of indications that, when used in a measurement, induces a relative increase δu_c in the combined standard uncertainty equal to some appropriate chosen value Q. The Q value is previously decided by the metrologist based on the final quality planned for the results (the value of Q could be, for example, 0.3, or 0.1, or 0.05). This is done, supposedly, when all Type B components have already been optimized for the intended purposes. Equations (7) and (8) develop the equality $Q = \delta u_c$, and equation (9) gives the expression for the desired value of n.

$$Q = \delta u_c(N) = \sqrt{\frac{s^2(\overline{q})}{u_{cB}^2} + 1} - 1 \Rightarrow (Q+1)^2 - 1 = \frac{s^2(\overline{q})}{u_{cB}^2} = Q(Q+2) \tag{7}$$

$$Q(Q+2) = \frac{1}{u_{cB}^2} \frac{\sum_{k=1}^{n}(q_k - \overline{q})^2}{n(n-1)} = \frac{1}{n.u_{cB}^2} \frac{\sum_{k=1}^{n}(q_k - \overline{q})^2}{(n-1)} = \frac{1}{n.u_{cB}^2} s^2(q_i) \tag{8}$$

Finally, the value of n is given by:

$$n = \frac{s^2(q_i)}{u_{cB}^2 . Q(Q+2)} \tag{9}$$

The value of n delivered by equation (9) gives a value to the Type A uncertainty that, when included in the uncertainty calculation (equation (4)), will provide the elect relative increase Q in the preexisting Type B combined uncertainty u_{cB} (given by equation (3)). This means that a measurement made with n indications will result in a final combined uncertainty that is Q times greater than the preexisting Type B combined uncertainty ($u_* = Q \times u_{cB}$).

3. Discussion and conclusion

For some combinations of the values of u_{cB} and $s(q_i)$, and the values stipulated for Q, the result given by equation (9) may not be an integer. In these cases the result will be round up to the next integer n. In practice, it is acceptable (in most cases it is mandatory) to use a number n_u greater than the number n derived from equation (9). The number n_u to be used can be obtained from the knowledge of the experimental standard deviation $s(q_i)$ associated with the measurement. The uncertainty associated to this knowledge depends on the number of indications used in the determination of $s(q_i)$.

If we have a well-characterized measurement and we know a pooled estimate of the experimental standard deviation $s(q_i)$ with good confidence from m_p indications obtained in previous measurements (for $m_p = 50$ observations the relative uncertainty in $s(\overline{q})$ is 10 percent, for example [2, GUM, E.4.3, Table E.1]), and if we had collected m_q indications ($m_q \geq n$) in the current measurement, then m_q will be used to calculate $u_c^2(y)$ in equation (2), and m_p will be used to calculate the effective number of degrees of freedom using the Welch-Satterthwaite formula ([2], GUM, G.4). If the measurement is being taken for the first time (in which case we do not have a pooled variance), and we had made m observations, than the value of n calculated in (9) (or other value $m > n$) will be used for the two calculations.

The value n_u to be actually used in the measurement should certainly be somewhat greater than the value n given by equation (9). It may be applied to n an increase factor consistent with the degree of confidence by which the experimental standard deviation $s(q_i)$ is known. This degree of confidence can be given by the **relative standard deviation** of the **experimental standard deviation of the mean,** "*which is given by the ratio* $\sigma[s(\overline{q})]/\sigma(\overline{q})$ *and which can be taken as a measure of the relative uncertainty of* $s(\overline{q})$" (of K independent observations q_k) [2, GUM E.4.3].

From equation (1), we can see that the relative uncertainty deviation of $s(q_i)$ is equal to the relative uncertainty of $s(\overline{q})$. As the dependence of n to $s(q_i)$ is quadratic (see equation (9)), the relative uncertainty of n will be given by twice the relative uncertainty of $s(q_i)$ that, as we have supposed, had previously been obtained. So, if $s(q_i)$ is obtained from $K = 10$ indications, this "uncertainty of the uncertainty" is 24 % (= 0.24) [2, GUM, Table E.1]. The value of n given by equation (9) would then be multiplied by a factor equal to 1.48 (that is, the value to be effectively used will be $n_u = 1.48\ n$). This way, the consequent relative increase (q) in the value of the combined uncertainty u_{cB} will likely satisfy the condition $q < Q$, as anticipated.

The present derivation is valid for a measurement of a single measurand that has only one uncertainty component of Type A. When the measurement involves a measurand that depends on other measurands (**input quantities in a measurement model** [4, VIM3: 2.50]) we need to use equation (9) above to calculate a specific value of n associated to each input quantity. This will be done, of course, only for those input quantities that have a concomitant Type A uncertainty component. And only for those that contribute the most to u_c. The "contribution" of each component to u_c may be evaluated using the results (7) or (12) given in [3].

We need to consider that this treatment is correct only in the (very restricted) GUM context, in which the definitional uncertainty is considered to be zero (or negligible), the measurand being characterized by a unique value. In the more general context addressed in [1] the inherent uncertainty is not necessarily zero. When this is the case, in general it will make little sense to spend considerable effort in the intent of limiting the relative influence of the Type A contribution to a predetermined value Q. In a sense, this is equivalent to the procedure of increase the number *n* of indications in an attempt to make irrelevant the Type A uncertainty. This is not possible, because the inherent uncertainty establishes a deterministic minimum limit to the value of the uncertainty associated to the dispersion of the indications: this value cannot be smaller than the inherent or definitional uncertainty. So, it is not correct to reduce it artificially to zero. In these cases, in which the inherent uncertainty is dominant, we need to consider the use of the standard deviation, in place of the standard deviation of the mean, in the evaluation of the dispersion associated to the definitional uncertainty. This is the value to be used in the calculation of the uncertainty of the measurement. Of course, there will be some difficulties in separating the dispersion due to the inherent uncertainty from that dispersion linked to the fluctuations of the influence quantities, which are the exclusive responsible for the genuine Type A uncertainty.

Electronic Supplementary Material can be solicited through the e-mail listed in the first page. It contains a self-explanatory Excel Worksheet with a comprehensive calculation of the value of *n*, according to equation (9), for hypothetical measurement examples. The worksheet treats general measurement contexts with six Type B uncertainty components, whose values are chosen by the operator. It also includes the calculation of the degree of freedom and the relative increase in the expanded uncertainty which, in most cases, is the most relevant information the operator wants.

4. References

[1] Baratto A C 2008 Measurand: a cornerstone concept in metrology Metrologia **45** 299

[2] JCGM BIPM, IEC, IFCC, ILAC, ISO, IUPAC, IUPAP and OIML 2008 Evaluation of measurement data — Guide to the expression of uncertainty in measurement JCGM 100:2008 - GUM 1995 with minor corrections

[3] Baratto A C and Bezerra I L 2013 Relative sensitivity of combined standard uncertainty to changes induced in uncertainty components Accred Qual Assur **18** p 129

[4] JCGM BIPM, IEC, IFCC, ILAC, ISO, IUPAC, IUPAP and OIML 2012 International vocabulary of metrology – Basic and general concepts and associated terms (VIM) 3rd edition 2008 Version with minor corrections

Acknowledgments: The authors wish to thank Tatiana C S Baratto for the valuable comments and stylistic formulations concerning the English language usage.

Comparison of direct mercury analyzer and FIA-CV-AAS in determination of methylmercury in fish

J C Ulrich, M A Hortellani, J E S Sarkis and M A Nakatsubo

Instituto de Pesquisas Energéticas e Nucleares – Ipen/CNEN-SP, São Paulo, Brazil

E-mail: jculrich@ipen.br

Abstract. Methylmercury (MeHg) has been determined in fish reference materials by direct mercury analyzer (DMA 80) and FIA-CV-AAS. In order to evaluate accuracy, certified reference materials (Fish protein, NRCC - Dorm 4 and fish material, Ipen - Dourada 1) were analyzed after extraction and separation of mercury species. Good agreement of the results have been obtained (relative error of the determination between the methods varied from 1.5 % to 39 %). The repeatability of the results varied from 4 % to 26 %.

1. Introduction

Monomethylmercury (MeHg) is the most commonly occurring organo-mercury compound and one of the most toxic, and it is recognized as a major environmental pollution issue and health hazard for humans. Contaminated seafood is the major route of exposure for humans to MeHg. It represents, on average, 85 % of the total mercury present in fish [1].

Several methods for determining the concentration of inorganic mercury and organomercury species have been developed [2-6].

The wet digestion procedure generally is used, but it involves a number of reagents both for acidic digestion and mercury reduction and is therefore time-consuming and presents risks of mercury loss due to volatilization and significant manipulation of the samples. In contrast, the systems that combine sample combustion (thermal decomposition in the presence of O_2), Hg amalgamation, and atomic absorption spectrometry has already been proven to be an effective method to obtain reliable results [7-9].

Currently, we are investigating the MeHg concentration in fish materials. An important feature of these studies was that measurements were conducted by two different methods, direct mercury analyzer (DMA) and atomic absorption spectrophotometry with cold vapor generation and flow injection (FIA-CV-AAS), and its results were compared.

2. Experimental

2.1. Samples preparation

Two certified reference materials (fish protein, NRCC-Dorm 4, and fish material, Ipen-Dourada 1 [10]) were prepared and analyzed. The method is based on the acid leaching with hydrochloric acid solution, HCl 6 mol L^{-1} (by volume), and mercury separation of the organic and inorganic ion in exchange resin (Dowex 1×8 100–200 mesh). The methodology was based in Horvat and May [11,12].

Once separated, MeHg was decomposed to inorganic Hg^{2+} by ultraviolet (UV) irradiation and the final solution was diluted to approximately 30g with demineralised water. The solutions are ready to analyzed by both methods, FIA-CV-AAS and direct mercury analyzer.

2.2. FIA-CV-AAS
The samples solutions were inserted into the sample introduction system of an atomic absorption spectrophotometer (FS-SpectrAA220 Varian Australia Pty Ltd.) and methylmercury (such as mercury) is determined by the technique of atomic absorption spectrophotometry with cold vapour generation and flow injection (FIA-CV-AAS). A tin II chloride solution ($SnCl_2$ 25 % in HCl 25 % (by volume)) was used as reducer of the Hg. Argon was used as carrier gas at constant flow at 200 mL min^{-1}. Before analysis, the equipment was calibrated with Hg standard solutions in the range 2 to 12 μg kg^{-1} Hg.

2.3. Direct mercury analyzer
The same solutions analyzed by FIA-CV-AAS were introduced in the direct mercury analyzer (DMA-80, Milestone, Sorisole, Italy). An aliquot of the 300 μL each sample was added in the quartz boats. This equipment contains an automatic sampler, a quartz furnace, a cobalt-manganese oxide catalyst, a gold-coated sand amalgamator and an atomic absorption detection cell. The different steps of the analysis are controlled by software. Similarly, the equipment was calibrated with Hg standard solutions in the range 0.5 to 100 ηg Hg.

3. Results and discussion
Basic parameters obtained during validation of two analytical methods are presented in table 1.

Table 1. Quality assurance of both analytical procedures.

Parameter	DMA		FIA-CV-AAS	
	Dorm 4	Dou[a] 1	Dorm 4	Dou[a] 1
Repeatability, %	9	26	4	12
Recovery, %	61	88	70	98
Expanded uncertainty[b], %	23	46	28	48

[a]Dou 1 is Dourada 1
[b]Uncertainties expressed as 95% of level of confidence and k=2

The results indicate that both methods represents similar results. FIA-CV-AAS has a better repeatability of the results compared to DMA method. The recovery rates in both methods were at a similar level but to Dorm 4 was lower than Dourada 1. Probably it occurred due composition of the Dorm 4 (fish protein). To evaluated the measurement uncertainty, all possible sources of uncertainty, to both methods, were carefully identified. Afterwards, the uncertainty components were quantified and the combined standard uncertainty was calculated. Finally, using the equation $U = k.u_c$ where u_c is the combined standard uncertainty and k is a coverage factor equal to 2.

3.1. Comparison of methods with CRM values

The analytical performance of the methods was evaluated by the analysis of two certified reference materials (Dorm 4 and Dourada 1). The concentrations obtained are presented in table 2 and showed in the figures 1 and 2.

Table 2. Results of MeHg (expressed in $\mu g\ g^{-1}$ as Hg), with their uncertainties, in certified reference materials by different methods.

CRM	Certified value	DMA	FIA-CV-AAS
Dorm 4	0.354 ± 0.031	0.215 ± 0.049 (n = 5)	0.250 ± 0.070 (n = 5)
Dourada 1	0.245 ± 0.053	0.215 ± 0.099 (n = 15)	0.241 ± 0.115 (n = 15)

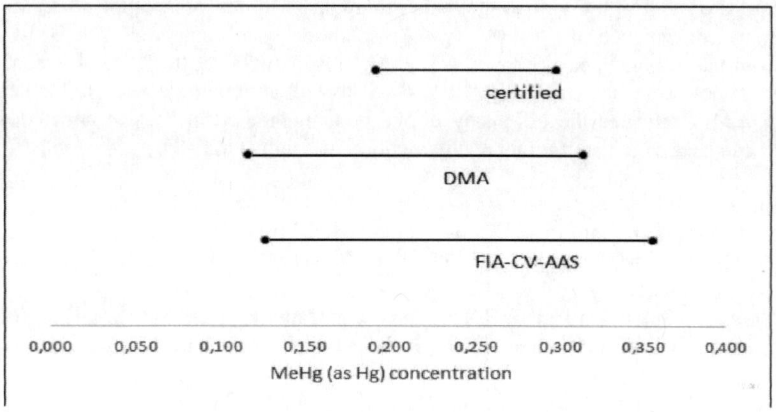

Figure 1. Results obtained by DMA and FIA-CV-AAS for MeHg (as Hg) in the Dourada 1.

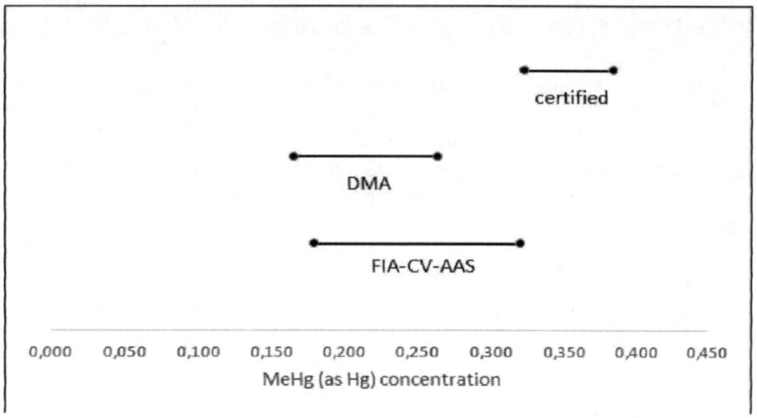

Figure 2. Results obtained by DMA and FIA-CV-AAS for MeHg (as Hg) in the Dorm 4.

The concentrations obtained show agreement in both methods, mainly by DMA method (showed in table 2 and figure 1). But the results obtained in the Dorm 4, when compared with the certified values were, on average, 34 % lower. Probably the composition of the Dorm 4 (fish protein) and the acid leaching method used, influenced the results obtained. Others studies has been necessary to improve this results for Dorm 4.

In the DMA method, relative error (R) of determination varied from 12 % to 39 % and in the FIA-CV-AAS method varied from 1.5 % to 29 %.

A paired Student's t-test, applied to compare the analytical results of the samples analyzed by both methods, showed that MeHg (as Hg) concentrations were not significantly different (t calculated < t tabulated, $\alpha = 1$ %) when used DMA or FIA-CV-AAS.

It should be noted that the DMA method is under development and the FIA-CV-AAS method is routine in our laboratory and accredited by CGCRE/INMETRO.

4. Conclusions

The analytical methods were characterized by good agreement of the results. Both methods showed sufficient sensitivity and considered to be robust. The good performance obtained with DMA method when compared with FIA-CV-AAS method encouraging with regards to application on a routine basis. The main inconvenience of the FIA-CV-AAS method, when compared with DMA method, is the quantity of the reagents and generated waste. The results obtained to Dorm 4, when compared with certified values, were not satisfactory. It will be studied, although the results obtained by DMA and FIA-CV-AAS confirmed the efficiency of the both methods. Finally, we concluded that the both methods can be used to determining methylmercury in fish materials.

References

[1] Bisinoti M C and Jardim W F 2004 Quím. Nova **27** 593
[2] Zabaljauregui M, Delgado A, Usobiaga A, Zuloaga O, Diego A de and Madariaga J M 2007 *J. of Chromatography A* **1148** 78
[3] Clough R, Belt S T, Evans E H, Fairman B and Catterick T 2003 *Anal. Chim. Acta* **500** 155
[4] Sannac S, Fisicaro P, Labarraque G, Pannier F and Potin-Gautier M 2009 *Accred. Qual. Assur.* **14** 263
[5] Lee S H and Suh J K 2005 *Microchemical Journal* **80** 233
[6] Hortellani M A, Sarkis J E S, Bonetti J and Bonetti C 2005 *J. Braz. Chem. Soc.* **16** 6A 1140
[7] Barst B D, Hammerschmidt C R, Chumchal M M, Muir D C G, Smith J D, Roberts, A P, Rainwater T R and Drevnick P E 2013 *Environ. Toxic. and Chem.* **32** 6 1237
[8] YanHua P, Yao S, Lu G, Nan W, XinTing P, HaiJia S and JiJuan C 2015 *J. of Food Safety and Quality* **6** 1 54
[9] Ferlin S, Fostier A H and Melendez-Perez J J 2014 *Anal. Methods* **6** 4537
[10] Ulrich J C and Sarkis J E S 2013 *Accred. Qual. Assur.* **18** 511
[11] Horvat M, May K, Stoeppler M and Byrne A R 1988 *Appl. Organomet. Chem.* **2** 515
[12] May K, Stoeppler M and Reisinger K 1988 In: Merian E, Frei R W, Hardi W and Schlatter C H (ed) Carcinogenic and mutagenic metal compounds 2. Gordon and Breach Science Publishers, NY

Acknowledgments

I wish to acknowledge assistance and financial support from Instituto de Pesquisas Energéticas e Nucleares – Ipen.

Uncertainty Estimation of Metals and Semimetals Determination in Wastewater by Inductively Coupled Plasma Optical Emission Spectrometry (ICP-OES)

J R Marques, S M Villa-Soares, T B Stellato, T B S C Silva, M G Faustino, L R Monteiro, M A F Pires, M E B Cotrim

Centro de Química e Meio Ambiente, Instituto de Pesquisas Energéticas e Nucleares (IPEN-CNEN/SP), Brazil

joyce.marques@usp.br

Abstract. The measurement uncertainty is a parameter that represents the dispersion of the results obtained by a method of analysis. The estimation of measurement uncertainty in the determination of metals and semimetals is important to compare the results with limits defined by environmental legislation and conclude if the analytes are meeting the requirements. Therefore, the aim of this paper is present all the steps followed to estimate the uncertainty of the determination of amount of metals and semimetals in wastewater by Inductively Coupled Plasma Optical Emission Spectrometry (ICP-OES). Measurement uncertainty obtained was between 4.6 and 12.2% in the concentration range of $mg.L^{-1}$.

1. Introduction

According to the International Vocabulary of Metrology (VIM) measurement uncertainty is a non-negative parameter characterizing the dispersion of the quantity values that is attributed to a measurand [1]. To carry out this estimation, it is necessary to follow some steps such as: to determine the measurement model; evaluate the sources of uncertainty; estimate the uncertainty in each step of the method and determine the combined and expanded uncertainty, regarding the confidence interval wanted [2].

The estimation of measurement uncertainty is important to express how precise is the method used and whether the results obtained are really meeting the needs of environmental legislations or other limits.

2. Objective

This work aims to demonstrate the steps used to estimate the measurement uncertainty in the determination of metals and semimetals (B, Sn, Cd, Ba, Cr, Cu, Fe, Mn, Ni, Pb and Ni) in wastewater, after acid digestion and analysis by ICP-OES.

3. Measurement Uncertainty

It is necessary to perform the entire method study, in order to estimate the measurement uncertainty. In this case, the method validation was performed, and all the calculations needed are described below.

Journal of Physics: Conference Series **733** (2016) 012023 doi:10.1088/1742-6596/733/1/012023

3.1. Measurement Model

The first step was to determine the measurement model used to obtain the measurand. From the equation 1, it was found feasible to raise the sources of uncertainty that influences in the quantification of metals and semimetals in wastewater, using the method studied [3, 4].

$$C = \frac{(CA\ x\ V_{final})}{V_{inicial}} \cdot \frac{1}{R} \quad (1)$$

Where: C = Concentration in the sample (mg.L-1); CA = Concentration of the element according to the calibration curve (mg.L^{-1}); $Vfinal$ = Final volume of the sample (50 mL); $Vinitial$ = Initial volume of the sample (45 mL); R = Recovery of the method.

3.2. Sources of Uncertainty

The uncertainty sources are presented with a quality tool called Ishikawa Diagram in Figure 1. All stages of Sample preparation and analysis were considered [5].

The contribution of each source of uncertainty is described in the next items.

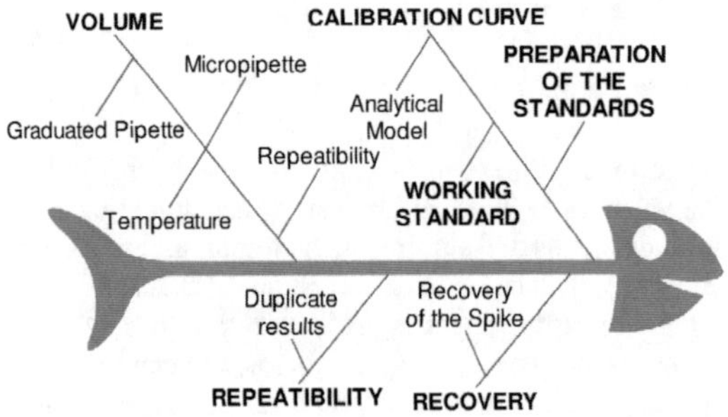

Figure 1: Ishikawa Diagram to the studied method

3.3. Volume Uncertainty

In this step, all sources of uncertainty that contribute to metal concentration dispersion were considered. Part of the uncertainty come from the volume measurement in the acid digestion process. The initial volume of sample is measured using a graduated pipette (two aliquots - 25 mL and 20 mL), after that, two acids are add to the sample, measured with a micropipette. The first source of uncertainty considered was the maximum variation of 5°C of temperature in the laboratory. This uncertainty was calculated according to equation 2 [3, 4].

$$\mu(V_T) = \frac{\Delta T.Q.V}{\sqrt{3}} \quad (2)$$

Where: ΔT = Variation of temperature in the laboratory (°C); Q = Expansion coefficient of water (0.000124°C^{-1}); V = Volume and $\sqrt{3}$ = considering rectangular distribution.

The second source of uncertainty considered was the uncertainty of the glassware, calculated according to equation 3.

$$\mu(V_P) = \frac{\mu(cert)}{\sqrt{6}} \quad (3)$$

Where: $\mu(cert)$ = Uncertainty informed by the producer and $\sqrt{6}$ = considering triangular distribution.

The last source of uncertainty of volume considered was the repeatability of the measure. In this case, ten measures were carried out, using all glassware, and the uncertainty was calculated according to equation 4 [3, 4].

$$\mu(V_{repe}) = \left(\frac{s}{\sqrt{n}}\right) \tag{4}$$

Where: s = Standard deviation of the measures and n = Number of measures.
The volume uncertainty were combined as presented in equation 5.

$$\mu(V_X) = \sqrt{\mu(V_T)^2 + \mu(V_P)^2 + \mu(V_{repe})^2} \tag{5}$$

3.4. Calibration Curve: Preparation of Standards
All the standards used to build the calibration curves were prepared from a multielemental working standard (WS), prepared from monoelemental certified standards. To estimate the uncertainty of the preparation of WS, firstly was estimated the uncertainty of volume, as presented before, and the other sources are presented in equation 6 [3, 4].

$$\mu(C_{WS}) = \left[\sqrt{\left(\frac{\mu V_P}{V_P}\right)^2 + \left(\frac{\mu V_{VF}}{V_{VF}}\right)^2 + \left(\frac{\mu C_{std}}{C_{std}}\right)^2}\right] . C_{WS} \tag{6}$$

Where μVP = Uncertainty of the volume used of the certified standard; VP = Volume used of the certified standard; μVVF = Uncertainty of the volumetric flask; VVF = Volume of the volumetric flask; $\mu Cstd$ = Uncertainty informed in the certificate of the standard; $Cstd$ = Concentration of the standard (informed in the certificate); CWS= Final concentration of the working standard.

In equation 7, it is presented the method used to estimate the uncertainty of each standard used to build the calibration curve [3, 4].

$$\mu(C_{PX}) = \left[\sqrt{\left(\frac{\mu V_P}{V_P}\right)^2 + \left(\frac{\mu V_{VF}}{V_{VF}}\right)^2 + \left(\frac{\mu C_{WS}}{C_{WS}}\right)^2}\right] . C_{SX} \tag{7}$$

Where: CSX = Final concentration of each standard.
Results were presented as a percentage, and the greater one was found to be the preparation of the calibration curve uncertainty.

3.5. Calibration Curve: Analytical Model
The uncertainty of the analytical model was estimated using the calculation presented in the EURACHEM Guide (equation 8) [3, 5].

$$\mu(C_X) = \frac{S}{B_1}\left[\sqrt{\frac{1}{p} + \frac{1}{n} + \frac{(C_X - \bar{C})}{S_{xx}}}\right] \tag{8}$$

Where: S = Residual standard deviation; $B1$ = slope; p = Number of measures to determine CX; n = Total number of replicates; CX = Sample concentration; $\bar{C}$ = Calibrations average and $S_{xx} = \sum_{i=1}^{n}(C_i - \bar{C})^2$ where Ci = Concentration obtained from the calibration curve.

3.6. Repeatability Uncertainty
To estimate the repeatability uncertainty was used the equation 9 [3, 4].

$$\mu(Repe) = \left(\frac{Máx\ RSD}{\sqrt{n}}\right) \tag{9}$$

Where: $Máx\ DPR$ = Greater value of relative standard deviation of the interval;
In this work, data from the analysis performed in 2013 and 2014 applying this method were used.

3.7. Recovery Uncertainty
To estimate the recovery uncertainty the equation 10 was used [3, 4].

$$\mu(Rec) = \left(\frac{\overline{Rec}}{\sqrt{s}}\right) \tag{10}$$

Where: $\overline{Rec}$ = Recovery average and s = Standard deviation of the recovery.
In this work data from the spikes analyzed in 2013 and 2014 applying this method were used to estimate the recovery of the method.

3.8. Combined and Expanded Uncertainties
The standard uncertainties were combined using the equation 11 [3, 4].

$$\mu(Conc_{Am}) = \left[\sqrt{\left(\frac{\mu V}{V}\right)^2 + \left(\frac{\mu PX}{100}\right)^2 + \left(\frac{\mu C_X}{Cx}\right)^2 + \left(\frac{\mu Repe}{100}\right)^2 + \left(\frac{\mu Rec}{100}\right)^2}\right].C_S \tag{11}$$

Where: CS = Concentration of the element in the sample.
To obtain the result of the expanded uncertainty, the results of combined uncertainty have to be multiplied by the expansion factor (k), to results with 95% of confidence interval k = 2.

4. Results and Discussion

Using the calculations described above the measurement uncertainty was estimated. Results of each source of uncertainty, combined uncertainty and expanded uncertainty are presented in table 1. The uncertainty source that contributes the most to all expanded uncertainties is the recovery. However, as was considered a study of 2 years to this source, the results can be considered more conservative and comprehensive.

In table 1 all the standard uncertainties (μ(x)/x) are without measurement unit, due to it is adimensional it is possible combine it after all the individual steps.

Table 1: Results of standard, combined and expanded uncertainty in the studied method.

	μ (x)/x							Expanded Uncertainty	
	V	Prep.	C_o	Rep	Rec	C_o (mg.L^{-1})	μ comb. (mg.L^{-1})	U (mg.L^{-1})	U (%)
B	0.00122	0.00792	0.01200	0.00463	0.03014	0.2499	0.0084	0.0169	6.75
Sn	0.00122	0.00847	0.01116	0.00701	0.03181	1.5230	0.0540	0.1081	7.10
Cd	0.00122	0.00803	0.00792	0.00726	0.02756	0.1260	0.0039	0.0077	6.13
Ba	0.00122	0.00893	0.00875	0.00378	0.01888	2.9710	0.0683	0.1366	4.60
Cr	0.00122	0.00851	0.01603	0.01844	0.02713	0.8110	0.0304	0.0608	7.50
Cu	0.00122	0.00807	0.00869	0.00705	0.02975	0.9210	0.0302	0.0604	6.56
Fe	0.00122	0.00841	0.00621	0.00492	0.06001	3.0610	0.1871	0.3742	12.23
Mn	0.00122	0.00794	0.01101	0.00391	0.02226	0.2725	0.0072	0.0144	5.28
Ni	0.00122	0.00892	0.00391	0.00385	0.02605	0.2555	0.0072	0.0144	5.62
Pb	0.00122	0.00807	0.01425	0.01077	0.04630	0.7720	0.0388	0.0776	10.06
Zn	0.00122	0.00839	0.00617	0.00170	0.03210	0.3243	0.0110	0.0219	6.76

5. Conclusion

Evaluating the results is possible to conclude that using data from validation process, EURACHEM guide [3, 5] and the analyst knowledge was possible to estimate, firstly, the sources of uncertainty involved in the determination of metals in wastewater, and the expanded uncertainty of each element.

The uncertainty measured is in between 4.6% and 12.2% of the concentration determined, what is coherent with the method used.

References

[1] JCGM, International Vocabulary of Metrology - Basic and General Concepts and Associated Terms (VIM), 2012.

[2] S. YENISOY-KARAKAS, "Estimation of uncertanties of the method to determine the concentration of Cd, Cu, Fe, Pb, Sn and Zn in tomato paste samples analysed by high resolution ICP-MS," Food Chemistry, pp. 1555 - 1561, 2011.

[3] S L R Ellison and A Williams (Eds). Eurachem/CITAC guide: Quantifying Uncertainty in Analytical Measurement, Third edition, (2012) ISBN 978-0-948926-30-3. Available from www.eurachem.org.

[4] H. A. FURUSAWA, "Estimativa de incerteza em análise química," 2007 São Paulo.

[5] INMETRO, Avaliação de dados de medição: Guia para expressão de Incerteza de medição - GUM, Duque de Caxias: INMETRO/CICMA/SEPIN, 2008, p. 141.

Study for standardization of the lighting system in fruit sorting

J F S Gomes[1], F O Baldner[2], P B Costa[3], M B Guedes[1], I A A Oliveira[1], F R Leta[2,4]

[1] National Institute of Metrology, Quality and Tecnology - Inmetro, Optical Metrology Division, Radiometry and Photometry Laboratory. Av. Nossa Senhora das Graças, 50, Duque de Caxias, RJ, CEP 25250-020, Brazil.
[2] Universidade Federal Fluminense, Programa de Pós-Graduação em Eng. Mecânica.
[3] National Institute of Metrology, Quality and Tecnology - Inmetro, Mechanical Metrology Division, Dimensional Laboratory.
[4] Universidade Federal Fluminense, Departamento de Engenharia Mecânica, Laboratório de Metrologia Dimensional e Computacional.

E-mail: jfgomes@inmetro.gov.br

Abstract. Sorting is a very important step in the fruit processing. The attributes definition and characterization are important for both marketing and end user, making it necessary to establish regulations for classification and standardization in order to unify the language of the market and enabling a more efficient market and also increase consumer awareness. For this end, it is necessary to standardize the technical criteria that can change the perception of the product. Studies have been developed in order to standardize a methodology to determine the subclass of fruit ripening, evaluating the influence of different light sources in the subclass evaluation.

1. Introduction

Brazil, following the changes in the global economy, improves its processes to ensure quality of various products and to meet international requirements, and to position itself consistently in the global market. Brazil's agricultural sector is the one that has shown greater international success, occupying the 3rd place in the world in fruit production.

However, some production problems still hinder the sector's growth according to its capacity. It can be inferred from the displayed volume production a loss reaching 40 % domestically. Misuse of soil management techniques and planting, lack of storage infrastructure and logistics, inadequate packaging and misinformation of the producer are all factors that contribute to increasing losses in fruit production [1].

Parameters such as appearance, taste and typical fruit color, as well as uniformity of size and shape, are just some of the demands from international consumers. To meet all requirements and to deal with the high levels of perishability and fragility that are intrinsic, the fruit production requires a complex system of planting, harvesting, post-harvest treatment, storage, transport and display in retail outlets.

To meet international requirements, especially from Europe, Brazil created regulations and standards that would guarantee quality and standardization in fruit production. The main purpose of this classification is to separate fruit into uniform lots, characterizing fruits by group (genome classification), class (size), subclass (ripening stage), presentation and category (quality).

Among the parameters to be evaluated, the one showing most complexity is the determination of the subclass, which evaluates fruit ripening. This evaluation is performed visually, according to the fruit coloration. However, there is no technical definition regarding different colors because the classification is tied to only a fixed number of categories. The banana, for example, has ripening subclasses defined from C1 to C7, corresponding to the following stages: completely green; green with yellow traces; greener than yellow; more yellow than green; yellow with green tip; yellow; and yellow with brown spots [2].

This comparative method is very subjective, since the visual analysis of color depends on several factors, such as room light, observer, viewing angle, etc., thus undermining the classification process' reliability.

This paper presents a study to define a scientific methodology to standardize the classification of fruit based on colorimetry and photometry techniques. The ripening scale was associated with the peel's spectral distribution, with the study of the influence of illumination on the characterization of the ripening stages.

This research was conducted in the Division of Optical Metrology (Diopt) of the National Institute of Metrology, Quality and Technology (Inmetro) in collaboration with the Dimensional and Computational Metrology Laboratory of Universidade Federal Fluminense (UFF), Rio de Janeiro, Brazil.

2. Methodology

A fruit's ripening and quality can be indicated well enough by the color of its peel. Each fruit has different types and concentrations of pigments which absorb and reflect electromagnetic radiation in the visible wavelength and, for each fruit, color difference may be determined on the basis of the reflected radiation on the surface. As a study to define the methodology for characterization of color in fruits, the spectrocolorimetric system from Inmetro's Optical Metrology Division (Diopt) was used to analyse the spectral feature of some fruits bought in common market. The fruits analysed were: banana, guava, orange, lemon, apple, papaya, mango and passion fruit.

2.1. Materials and methods

Diopt's spectrocolorimetric system is basically composed by a spectrocolorimeter, an electrical current source, a lamp holder and a sample holder. Measurements of the peel's spectral radiance were performed by frontally illuminating it, and acquiring its reflection at 45°, by the spectrocolorimeter - is the so called 0°: 45° geometry. Nine commercial lamps were used to illuminate the fruit and check for differences in spectral distribution and final color perception [3].

Using the light spectra reflected by the fruit's peel both the radiance factor and the tristimulus values XYZ were calculated according to CIE recommendation [4]. From this result, the chromaticity coordinates were calculated in CIE xy and CIE L*a*b* systems as well as the color difference ΔE^* for the different lighting systems. The lamps used were characterized by their correlated color temperature (CCT) and color rendering index (CRI) [4].

3. Results and Discussion

An example of the lighting influence in the color perception of two ripening stages of the Papaya can be seen in Table 1.

In figures 1 to 3 it's shown the spectral distribution in two ripening stages of the analyzed fruits. For each fruit, distinct features and change in visual perception are represented in the spectrum change. To analyse the influence of lighting in the characterization of peel color and of each ripening subclass, the banana was chosen because of its importance in Brazilian production [1]. From the radiance factor the CIE L*a*b* was calculated for each ripening stage. The value of a* increases with ripening to C7 class, with no overlap between classes of values, which is in accordance with the

Journal of Physics: Conference Series **733** (2016) 012024 doi:10.1088/1742-6596/733/1/012024

visually perceived. This behaviour does not occur with other parameters, and the value of C7 class coincides with those of other classes. Thus, the value of a* proves to be a good indicator for ripening [2].

Table 1. Papaya perception using four different lighting systems [1].

Source	Green papaya	Ripe papaya
Dichroic	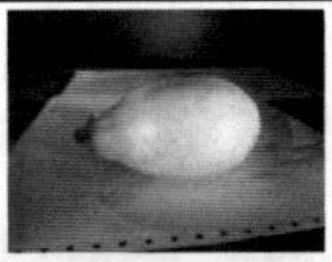	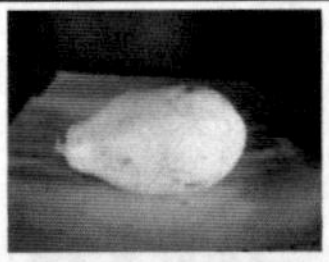
LED	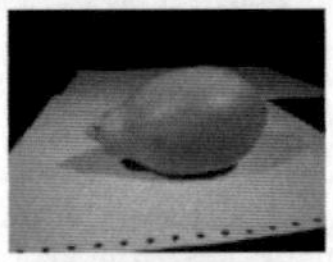	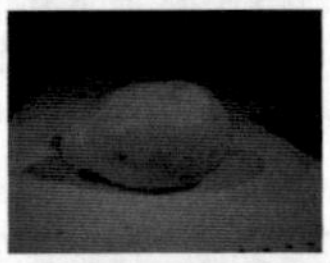
Metal halide	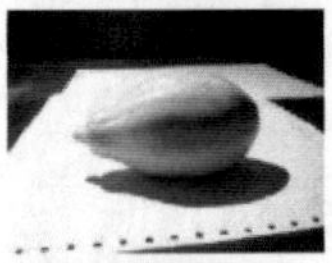	
High Pressure sodium	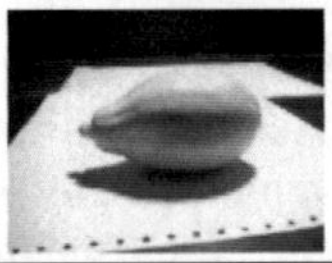	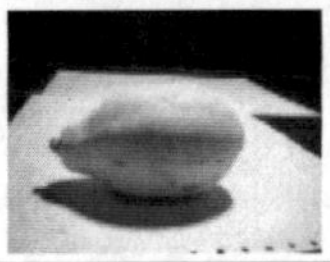

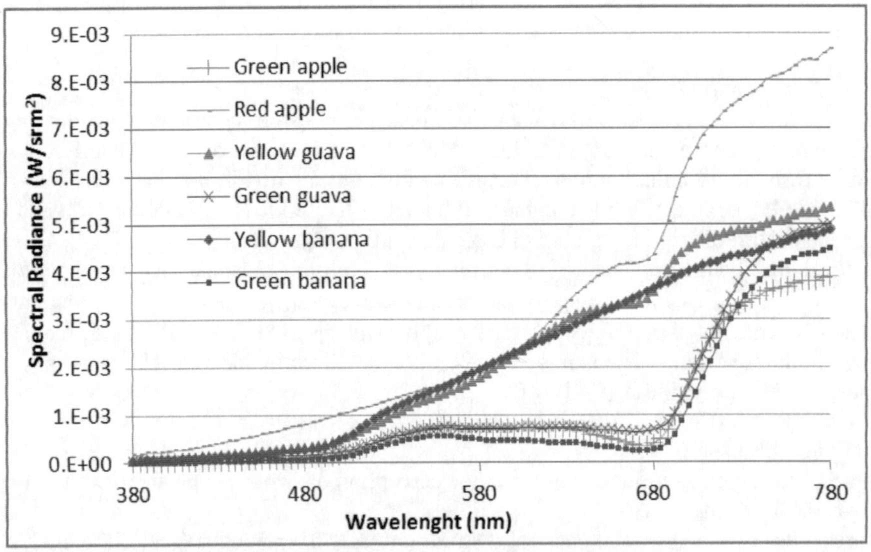

Figure 1. Measurements of two ripening subclasses: apple, guava and banana.

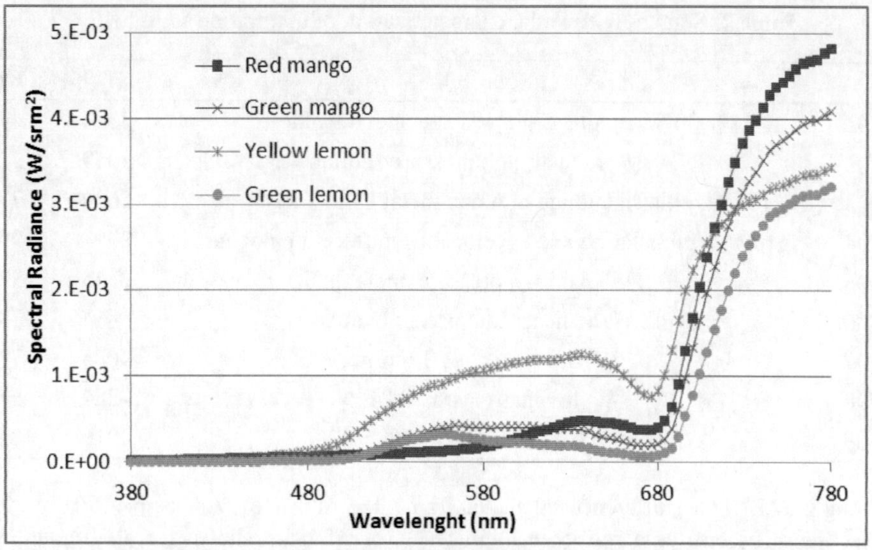

Figure 2. Measurements of two ripening subclasses: mango and lemon.

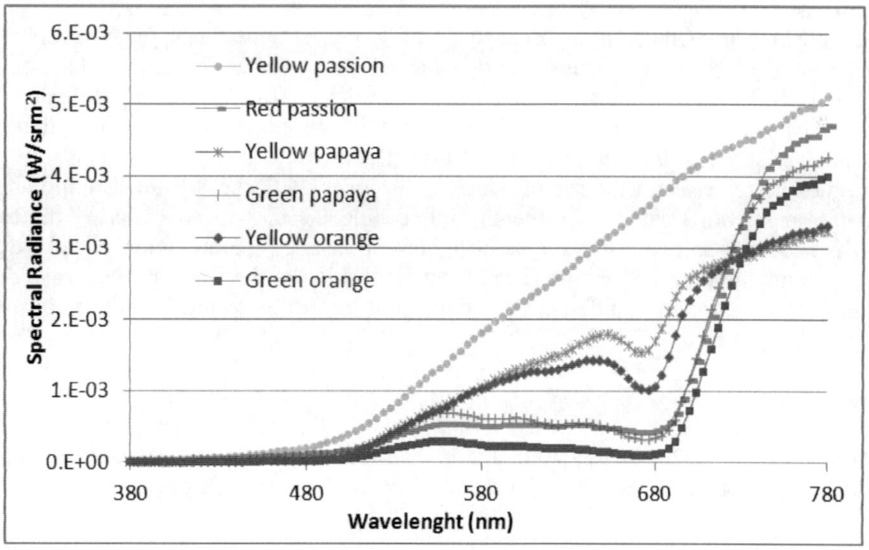

Figure 3. Measurements of two ripening subclasses: passion, papaya and orange.

Nine commercial lamps (F01 to F09) were used to analyze the influence of illumination on the peel color, with their CCT and CRI values shown in table 2. The banana peel's colors were measured in each ripening under each subclass for the nine types of lighting (F01 to F09) [3].

Table 2. Sources used in the characterization of maturation steps [3].

	Source	CCT (K)	CRI
F01	400 W, tubular high pressure metal halide	3834	49.5
F02	250 W, tubular high pressure sodium	2059	22.5
F03	400 W, ellipsoidal high pressure metal halide, daylight	5680	77.0
F04	1000 W, tubular double envelope incandescent halogen	2991	99.6
F05	150 W, tubular high pressure metal halide	3545	97.1
F06	42 W, compact fluorescent Dulux	3864	78.8
F07	54 W, fluorescent T5 Lumilux	4085	78.2
F08	73 W, luminaire with 49 LED	4284	67.3
F09	32 W, fluorescent T8	4010	83.3

Analysing the ΔE* color difference calculated from CIEL*a*b* for each source (F01 to F09) using source F04 (incandescent) as a reference, high levels of ΔE* are shown for all sources, with the exception of F05. The ΔE* values of sources F01 to F08 increase as the banana ripens, i.e. the color difference is smaller in the evaluation of the green banana, than for the yellow. The color difference is large (ΔE* > 8.0) for lamps F06, F07 and F09, regardless of the assessed subclass.

4. Conclusions

This paper presents a methodology for analysis of a fruit's appearance, with the objective of standardize the characterization stage of ripening classes. The influence of lighting in this classification is evaluated by comparing nine different commercial lamps. During the analysis it was found that the chromaticity coordinate a*, of the CIEL*a*b* system, proved to be a good indicator of ripening and was strongly influenced by the different sources in the lighting system. A greenish appearance was perceived with the use of lamps with higher TCC and a yellowish appearance was noticed with lamps with smaller CCT thereby influencing the classification during the packaging process. It is necessary to establish the type of lighting to be used mainly in automated fruit sorting systems. This work highlights the importance of standardization of the fruit ripening scale, due to the diversity undertaken among the different exporting countries in this sector, in order to guarantee the quality of the final product.

Acknowledgments

The authors would like to acknowledge Inmetro and Faperj, with projects under grants n° 52600.031066/2014, n° E-26/171.362/2001 and n° E-26/103.618/2012, respectively. We would also like to acknowledge Rafaela Rezende Vieira for collaboration with the measurement

References

[1] Gomes J F S 2013Padronização de metodologia para caracterização de cor por imagem aplicada à seleção de frutas *Doctoral Thesis from Universidade Federal Fluminense*.

[2] Gomes J F S, Vieira R R and Leta F R 2013 *Scientia Horticulturae* 150, 201–205.

[3] Gomes J F S, Vieira R R, de Oliveira I A A and Leta F R 2014 *Journal of Food Engineering*, 120, 215–222.

[4] CIE 15.3 2004 Colorimetry; *CIE Publication.*

Spectral measurements of light sources with a goniophotometer with and without mirror arrangement

Juliana Freitas Santos Gomes[1], Ivo Antônio Ázara de Oliveira[1], André da Silva Sardinha[1], Domingos David Viana[1], Rafaela Rezende Vieira[2], Amanda Hang Barbosa[2], Ana Paula Dornelles de Alvarenga[3]

[1]National Institute of Metrology, Quality and Tecnology - Inmetro, Optical Metrology Division, Radiometry and Photometry Laboratory. Av. Nossa Senhora das Graças, 50, Duque de Caxias, RJ, CEP 25250-020, Brazil.
[2]Universidade Católica de Petrópolis – UCP.
[3]National Institute of Metrology, Quality and Tecnology - Inmetro, Optical Metrology Division, Colorimetry and Spectrophotometry Laboratory.

E-mail: jfgomes@inmetro.gov.br

Abstract. This paper presents a comparative study of the relative spectral radiance measurements of incandescent and LED light sources performed in two different setups with a rotating–mirror-goniophotometer. In one configuration the light is reflected by the goniophotometer mirror, while in the other the mirror is not used. This study permits to determine a spectral correction factor at each measured wavelength in order to obtain a more precise value for the Correlated Color Temperature (CCT) for any light sources with different spectral power distributions.

1. Introduction

The Inmetro's spectrocolorimetric system performs calibration and tests services of color tiles (radiance factor and chromaticity coordinates) and light sources (correlated color temperature (CCT) and color reproduction index (CRI)).

This system is assembled onto an optical table, and to measure CCT the lamp lights a standard white tile and the reflected light is collected and analyzed at the spectroradiometer. High wattage lamps or the measuring of CCT in several directions made it necessary to assemble the spectrocolorimetric system together with the mirror goniophotometer of the Radiometry and Photometry Laboratory (Laraf).

There are two possible different setups for the measurements after the lamp is mounted in the goniophotometer arm: in the first the light reflected by goniophotometer's mirror lights the standard white tile and is collected by the spectroradiometer; in the second the mirror is not used and the lamp lights the tile directly. Comparing the results of both methods it was noticed a discrepancy in the obtained CCT values.

In order to characterize for LEDs colorimetric measurements this assembled system a series of varied incandescent and LED lamps were measured to obtain a correction factor to be applied in the measurements with the mirror.

The main goal of this study is to quantify the discrepancies in the CCT measurements of the different light sources considering both setups and to show the comparison of the lamps spectral radiance measurements obtained with and without the goniophotometer mirror reflection.

2. Methodology

The characterization of both setups was based on the comparison of the spectral distribution of a calibrated standard lamp together with a reflectance standard white tile in order to assess the radiance scale systematic error.

2.1. Setup without mirror reflection

Figure 1 shows this setup where the spectroradiometer is fixed on a tripod at a 0°:45° geometry where the lamp illuminates the standard white tile at 0° incidence and the light is reflected at 45° to the spectroradiometer. A standard FEL lamp (F-603, calibration certificate NIST 844/274563-07) was positioned at 50 cm from the tile, and the spectroradiometer focusing the tile at 71 cm and 45° to the tile surface normal.

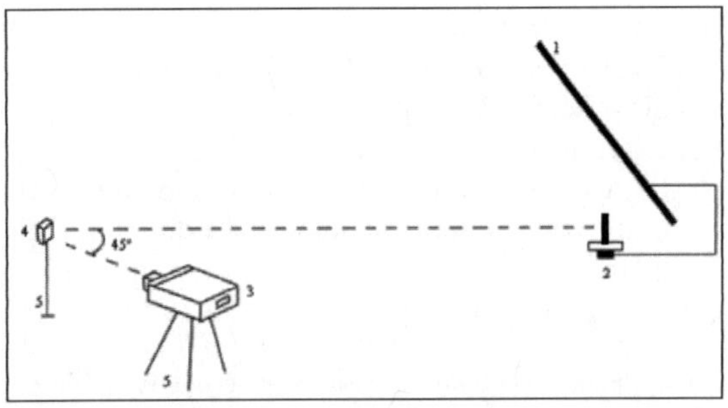

Figure 1. Setup without mirror reflection: 1: mirror; 2: lamp mounted on the goniophotometer arm; 3: spectroradiometer; 4: standard white tile; 5: tile mount and tripod.

2.2. Setup with the lamp radiation reflected from the mirror.

In this setup depicted in figure 2 the same standard lamp is mounted on the goniophotometer arm distant 1.95 m from the mirror. The lamp reflected radiation insides perpendicularly to the tile positioned at 3.6 m from the mirror. The spectroradiometer is positioned at the same 0°:45° geometry as before. In this setup there is a black variable closure between the mirror and the rest in order to avoid spurious scattered light. In both setups the whole laboratory is maintained under dark conditions.

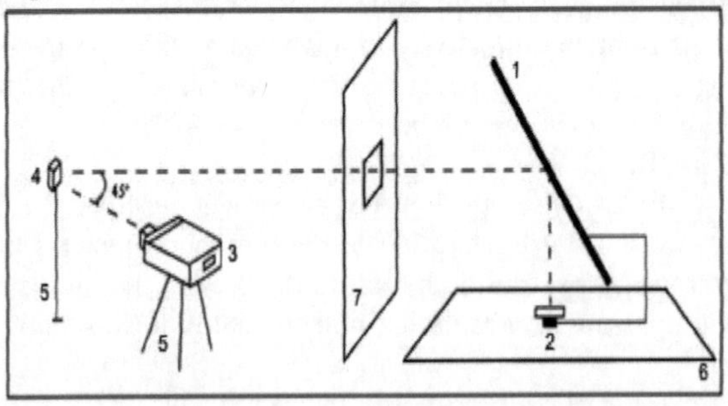

Figure 2. Setup using the mirror – 1: elliptical mirror (axis 2.83 m x 2 m); 2: lamp socket in the goniophotometer arm; 3: spectroradiometer; 4: standard white tile; 5: tile mount and tripod.

3. Results and discussion

Six light sources (source 01 incandescent lamp and the others LED sources) were measured using both setups described before in order to compare their CCT measured values shown in table 1.

Table 1. Calculated CCT values for the six measured lamps in both setups.

Lamp	Without mirror		With mirror		Difference
	CCT (K)	U(K) (k = 2)	CCT (K)	U(K) (k = 2)	CCT (K)
Source 01	2857	26	2963	26	-106
Source 02	2960	28	3022	28	-62
Source 03	4478	99	4494	93	-16
Source 04	3065	30	3105	30	-40
Source 05	3634	48	3732	50	-98
Source 06	3327	38	3335	35	-8

One can see a large difference on the CCT values calculated from the measurements results from both experimental setups (except source 06). This discrepancy is due to the light absorption by the mirror used in the setup described in 2.2 and is clearly seen in figures 3, 4 and 5 that show the measured spectral distribution from sources 01, 02 and 06 respectively.

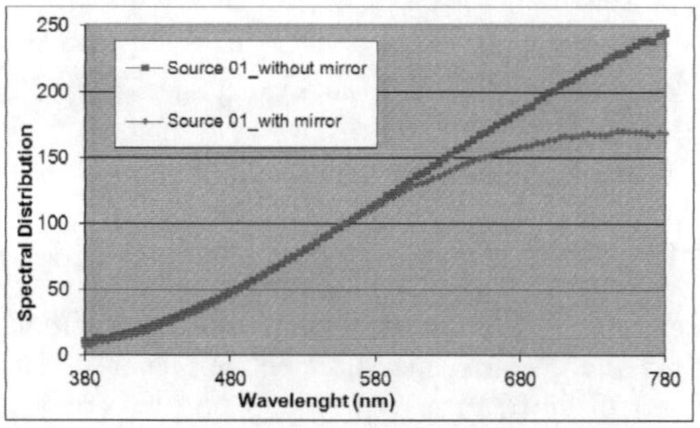

Figure 3. Spectral distribution from source 01 with and without mirror.

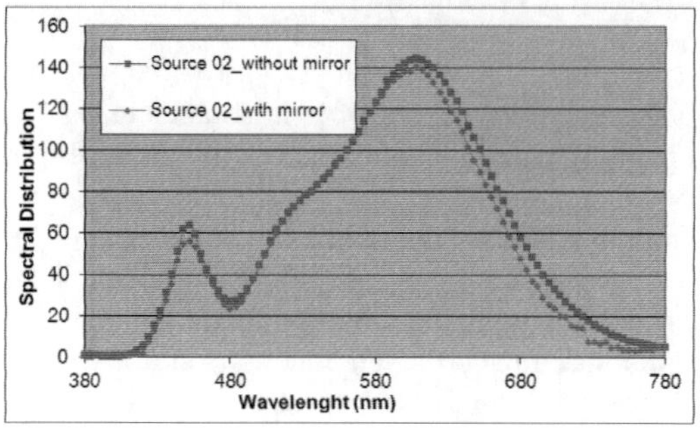

Figure 4. Spectral distribution from source 02 with and without mirror.

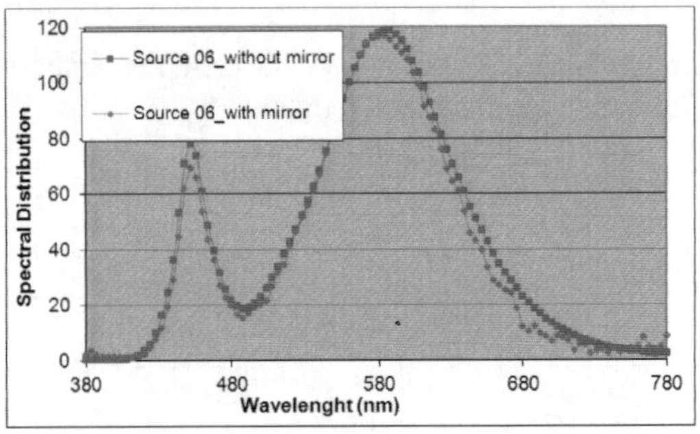

Figure 5. Spectral distribution from source 06 with and without mirror.

Using the measurements from the standard lamp described in 2 a correction factor for the mirror absorption was established for application in all measurements using the setup in 2.2. Table 2 shows the calculated CCT values after applying the correction factor to the measured spectral distributions.

Table 2. CCT values for the six sources calculated after applying the correction factor to the measurements with mirror.

Lamp	Without mirror		With mirror		Difference
	CCT (K)	U(K) (k = 2)	CCT (K)	U(K) (k = 2)	CCT (K)
Source 01	2857	26	2856	26	1
Source 02	2960	28	2937	28	23
Source 03	4478	99	4475	98	3
Source 04	3065	30	3032	29	33
Source 05	3634	48	3678	51	-44
Source 06	3327	38	3271	35	56

After applying the correction factor the differences in CCT from measurements with and without mirror became significantly smaller being lesser than the declared measurement uncertainty. The exception is source 06 where the correction enlarged the difference. This lamp had a very low intensity generating noisy spectra that may have caused in the spectra integration.

4. Conclusions

This article shows a comparative study of spectral radiance measurements of light sources using two experimental setups, with and without reflection from the goniophotometer mirror. Incandescent as well as LED lamps were measured in both setups.

The conclusion is that the mirror absorbs a large amount of the incident radiation at the ends of the spectra and thus modifying the spectral radiance measured spectra and consequently the calculated CCT value. An experimental correction factor was defined and applied generating good results for the incandescent and the majority of the LED lamps. This spectral correction function diminishes the systematic error introduced by the wavelength dependence reflectance of the mirror.

Further studies are necessary for obtaining mirror spectral correction factors using a source that presents a more uniform spectral distribution from the ultraviolet to the near infrared.

Aknowledgments
A.D. Alvarenga, R.R. Vieira, D. D. Viana and A.H. Barbosa acknowledge CNPq. I. A. Á. de Oliveira would like to acknowledge Inmetro with project under grants n◦ 52600.031066/2014.

References

[1] CIE 15.3 2004 Colorimetry *CIE Publication.*
[2] CIE 13.3 1995 Method of Measuring and Specifying colour rendering properties of light sources *CIE publication.*

Stability study of standards used for calibration of the spectrophotometer wavelength scale

L F Debossan[1], E M S Carvalho[1], M A Souza[1], J F S Gomes[1]

[1]Laboratório de Radiometria e Fotometria, Divisão de Metrologia Óptica, Inmetro, Av. Nossa Senhora das Graças, 50, Xerém, Duque de Caxias, RJ, Brasil

laraf@inmetro.gov.br

Abstract. The calibration of spectrophotometers is a procedure recommended by international standards to provide quality assurance of results and traceability. Due to its intrinsic properties, holmium oxide filters are indicated as reference standards for calibrating the wavelength scale of such equipment. This paper presents a study aiming to assess the repeatability and drift of holmium oxide standard filters calibrated in the Radiometry and Photometry Laboratory (Laraf) of Inmetro in order to verify their stability.

1. Introduction

Spectroscopy is an analysis methodology studies the interaction between substances and electromagnetic radiation, through sophisticated optical instruments. This technique has been used in increasing proportions to quantify substances and to determine the concentrations because it is simple, fast and provides reliable results. It is widely used in the areas of water analysis, biology, physics, chemistry, and pharmaceuticals, in the polymer and petrochemical industry, among others.

All equipment used for tests and /or calibrations, which have significant effect on the accuracy or validity of the result must be calibrated and checked periodically in order to maintain traceability [1]. For these periodic checks and characterizations, there are international standards that determine the types of tests that must be made and that the reference should be used. The holmium oxide is one of the most recommended spectrophotometry filters for calibration of wavelength scale, because of its characteristic absorption peaks.

In studies developed by NIST (National Institute of Standards and Technology), holmium oxide standards were compared among national metrology institutes of several countries, all of them have control of environmental conditions and storage techniques and handling of filters. These studies show results obtained from analysis of the absorption bands of holmium oxide standards [2, 3, 4]. These standards have intrinsic characteristics and remained stable for many years. It is important to note that this stability is related only to the wavelength scale (abscissa) as the photometric scale (ordinate) may be subject to changes due to environmental conditions and contamination on the glass surface.

The Laboratory of Radiometry and Photometry (Laraf) at the Division of Optical Metrology (Diopt) of Inmetro has calibration services for optical filters used in the calibration of spectrophotometers.

Once the lab has measured a lot of wavelength standards, comparative studies of data will be made to verify the behavior of these standards between themselves, since they are from different manufacturers, used and handled in different ways. Such analysis will allow conclusions about the holmium oxide behavior, since those standards analyzed were kept in unknown conditions which could influence the final result.

This paper describes the comparative analysis of the absorption bands values of different holmium oxide filters measured in Laraf.

2. Materials and methods

The spectrophotometer is an instrument used for measuring the amount of light absorbed or transmitted through a substance, whether solid, liquid or gas. With this technique, unknown components present in a sample can be identified, because their absorption bands are sensitive to ultraviolet, visible and infrared radiation.

The holmium oxide filter used as a reference standard to verify the wavelength scale can be found in two forms: liquid solution in cuvette cells and glass within a cuvette mount [5].

Although both filters are based on the same substance (holmium oxide), glass filters have near infrared region bands that are not present in the solution filters due to absorption by water. Consequently, the characteristic peaks values used are different in each case [3].

In the case of solution filters, when measured for a spectral bandwidth of 1 nm, characteristic peaks used for measurements at the spectrophotometric system were: (241; 250; 278; 287; 333; 345; 361; 386; 416; 451; 468; 485; 537; 641) nm, according to the articles published by NIST [2, 4] used as reference in the study, using the minimum presented uncertainty (0,2 nm).

For glass filters, the certified wavelengths of minimum transmittance for a spectral bandwidth of 1 nm are: (241; 279; 287; 333; 360; 385; 418; 453; 460; 536; 637) nm [3] with an uncertainty of 0,2 nm (k = 2). Some differences in the found peaks can be seen mainly in the band of 241 nm because some filters have an absorption variation in this region, depending on the base glass used.

In the measurements analyzed, it was used the scanning method, which there is a scan from the wavelength with greater value (initial) towards the lower value (final), presenting the peak values of the holmium oxide. This spectrum is exemplified in figure 1.

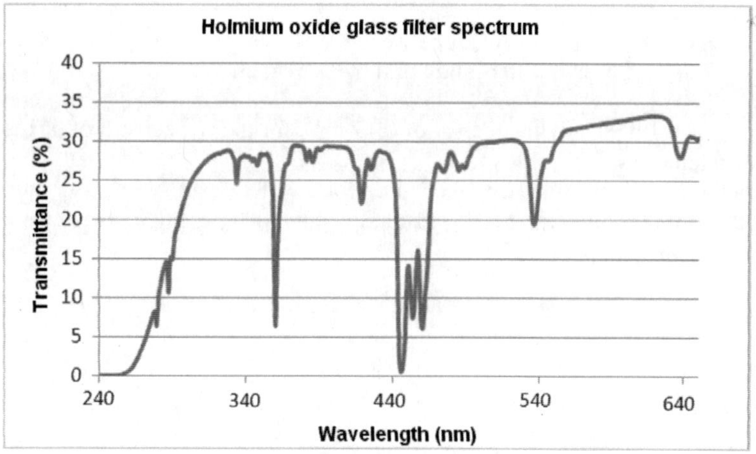

Figure 1. Holmium oxide glass filter spectrum.

The main parameter in the spectrum is the position of the found peaks, since the transmittance value can be considered irrelevant.

The analyzes were made using as base holmium oxide filters measured in the laboratory with the following specifications: analysis scale of 230 nm to 680 nm and scanning speed of 30 nm/min to 60 nm/min, being a total of 5 cycles or more with spectral bandwidths of 1 nm. The analysis was presented as percentage transmittance (%).

3. Results and discussions

In total, 90 measurements of holmium oxide filter data, among the years of 2000 and 2015 (between calibrations and recalibrations) were analyzed. Among the 90 holmium oxide filter measurements made by the laboratory in the midst of the years cited, 87 % of these measurements were made with glass filters doped with holmium oxide and 13 % with solution filters.

For holmium oxide solution filters, the results from 12 filters measurements were compared with each other (2001-2014 among calibration and recalibration). The peaks were compared with the values provided by NIST [2]. Errors are shown in the figure 2. It was used the lowest measurement uncertainty, provided by the NIST certificate, 0,16 nm, as the maximum and minimum limits of acceptance of the variation among the peak values provided by NIST and the measurement results.

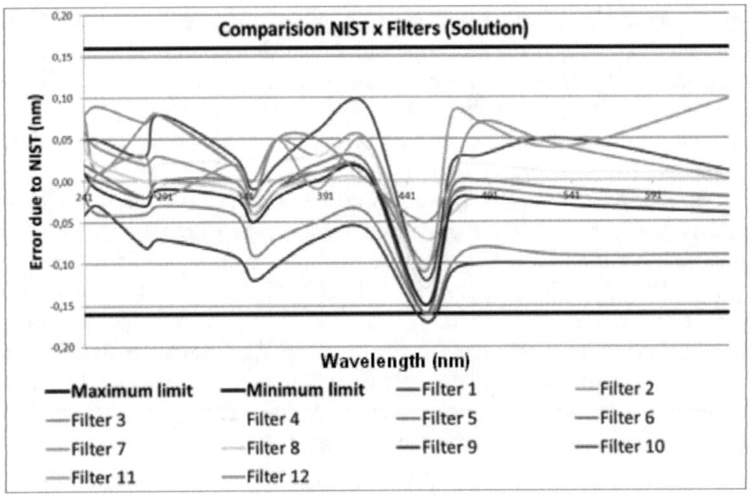

Figure 2. Comparison chart between the values provided by NIST for filters and Laraf measurements.

It is noted that only filters 1 (with an error of 0,17 nm), for the wavelength of 451 nm, ranged 6 % (0,01 nm) of the reference value.

Regarding holmium oxide glass filters, the values for the 78 glass filters measurements made in the laboratory (a total of 12 different manufacturers) from 2000 to 2015, between calibration and recalibration, were compared (figure 3).

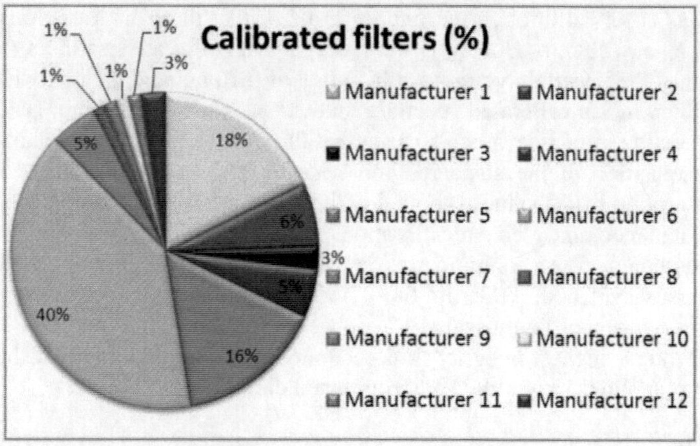

Figure 3. Ratio among manufacturers and filters calibrated in the laboratory, since 100 % is the total of 78 glass filters measurements.

As there was no data on the types of glass used in each measurement, it was adopted that filters provided by the same manufacturer had the same composition. For this reason, filters of the same manufacturer data were compared, getting an average of peaks by manufacturers.

The curves shown in the figure 4 represent the comparisons among such manufacturers and peak values supplied by NIST. It was used the measurement uncertainty of 0,20 nm as a parameter for the maximum and minimum limits of the analysis.

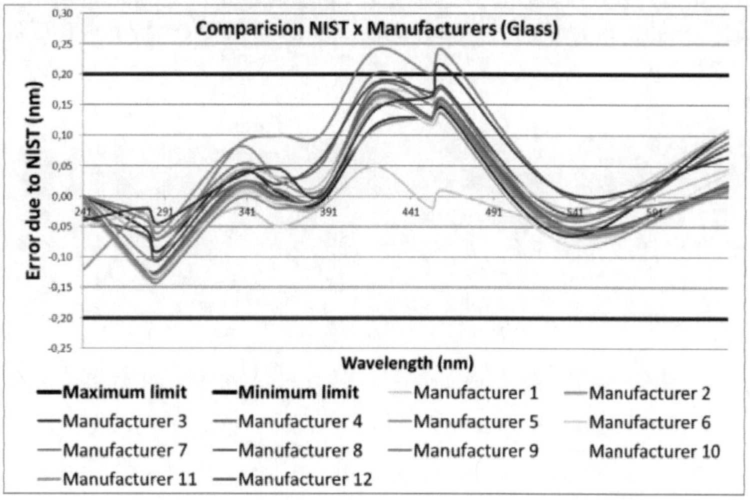

Figure 4. Comparison chart between manufacturers peak values (their mean) and the values provided by NIST onto glass filters.

The manufacturer 11 has a deviation of 20 % of the uncertainty value for the wavelength of 419 nm. This manufacturer also has a 20 % deviation for peak 460 nm, along with manufacturer 12, which deflects 10 % of the reference value. It is noteworthy that the manufacturer 11 is 1 % of the analyzed filters, equivalent to only a filter.

4. Conclusion

In this study it was observed through the figures 2 and 4, that there are variations in peak values for some filters and manufacturers. However, the vast majority of analysis filters (± 96 %) maintained their drift lower than the uncertainty provided by NIST of 0,16 nm and 0,2 nm, which is acceptable for holmium oxide filters when calibrated spectral bandwidth smaller than 2 nm. Based on the extensive study of holmium oxide standards, it can be seen that there is the need to make an initial calibration as soon after the acquisition of the standard, and periodic checks by the client comparing previous certificates or absorption bands values supplied with those measured in equipment. If the values found are not expected, it is recommended a recalibration.

It should be noted that holmium oxide standards should be used for verification of spectrophotometers wavelength scale in the UV/VIS region, it is not recommended its use in verification/calibration of a different scale.

To ensure the filters' highest integrity, it is recommended the proper use of the filter with gloves, storage free of dust and under controlled environmental conditions.

Acknowledgments

Thanks to Petrobras for the the financial support through the Programa Petrobras de Formação de Recursos Humanos (PFRH) and to the Curso Técnico em Metrologia from Inmetro.

References

[1] ABNT NBR ISO/IEC 17025:2005 2005 *ABNT.*

[2] Travis JC *et al.* 2005 *J. Phys. Chem.* Ref. Data Vol. **34** 41-56.

[3] Allen DW 2007 *J. Res. Natl. Inst. Stand. Technol.* **112** 303-306.

[4] Travis JC and Duewer DL 2003 *Fourth Oxford Conference on Spectroscopy*, Art Springsteen, Michael Pointer Editors, Proc. SPIE **4826** 113-122.

[5] ASTM E925 - 09 2014 Standard Practice for Monitoring the Calibration of Ultraviolet-Visible Spectrophotometers whose Spectral Bandwidth does not Exceed 2 nm *ASTM.*

Primary certification of reference material for electrolytic conductivity of bioethanol

L F da Silva, M R F Gomes, G C Cassini, A C V Faria, I C S Fraga

Inmetro – National Metrology Institute, Quality and Technology, Chemical Metrology
Av. Nossa Senhora das Graças, 50, Prédio 4, Xerém, Rio de Janeiro, RJ, Brazil

E-mails: leonardo.enq@gmail.com and sertafraga@gmail.com

Abstract. Nowadays the preservation of the planet is spreading into the international scene with the use of renewable energy sources such as bioethanol. The challenge is to guarantee the quality of produced bioethanol, and the electrolytic conductivity (EC) is one of the specified parameters for this purpose. However, is necessary to demonstrate the metrological traceability of the measurement results for EC in this matrix. This study presents the certification of a reference material for EC in bioethanol by using only primary measurements. The value of primary certified reference material (CRM) is (0.77 ± 0.06) μS.cm^{-1}, and its use will provide the metrological traceability needed for measurement results in laboratories.

Keywords: bioethanol, electrolytic conductivity, metrology

1. INTRODUCTION

The production of biofuels has attracted the attention of society due to social, economic, and environmental factors. This increased attention has also highlighted the possibility of using these energy sources to replace fossil fuels, as they can facilitate the reduction of global warming and the greenhouse effect [1,2]. In addition, biofuels constitute a viable alternative in relation to fuel derived from petrol, and they are considered renewable energy sources because obtaining them involves the processing of raw materials (e.g., sugarcane, vegetable oils, etc.) that can be continuously produced through agriculture or from animal fats. The participation of biofuel in the energy matrix depends on a complex array of factors involving many aspects of production (i.e., raw material and technology) and the sustainability (e.g., social, economic, and environmental) of this energy source.

Since the start of the PROALCOOL program, the quality of bioethanol has been connected with the performance of engines working with hydrated alcohol. Playing an important role in bioethanol's economic sustainability, Brazil is one of the leaders in the production and development of technologies related to its use [3-5]. Therefore, efforts to ensure the quality of this matrix are required, such as achieving the operation of automotive engines without corrosion damage. Aiming for this control are international and national specifications [6-8] for quality parameters in bioethanol (e.g., pH, electrolytic conductivity (EC), etc.) and standards for measurement methodologies [9-11] to harmonise the results of these measurement parameters in Brazil and other countries.

The validation of biofuel is markedly dependent on three basic factors: normalised specifications; official methods of analysis, and certified reference materials (CRMs). The normalised specifications are a set of characteristics required to guarantee the quality of the biofuels. The official methods of analysis are the tools necessary to construct a correct evaluation of the normalised specifications of the fuel. The CRMs are tools utilised in the validation and official method control of fuel analysis [5].

The National Institute of Metrology, Quality and Technology (Inmetro) in Brazil has been active since 2006 in the production of CRMs [12-15]. In 2007, Inmetro implemented the primary system of electrolytic conductivity (PSEC); its objectives are to provide metrological traceability and to characterise (i.e., define the property value of) a CRM at a primary level. Inmetro participates in several international comparisons within the ambit of the Consultative Committee for Amount of Substance (CCQM): Metrology in Chemistry and Biology and the Inter-American Metrology System (SIM) in order to compare the results to those of its peers [16-20]. In 2014, Inmetro coordinated an international comparison (via pilot study) within the ambit of the CCQM to determine the EC of bioethanol. In preparation for the study, a batch of bioethanol was submitted to studies of certification (specifically, homogeneity, stability, and property value) [21,22] by PSEC. The main results of these studies are presented and discussed in this work.

2. MATERIALS AND METHODS

2.1 Bioethanol batch production

The bioethanol, from a national supplier, was collected at the Sugarcane Technology Center (CTC) in the city of Piracicaba in the state of São Paulo. The bioethanol was homogenised by magnetic stirring for 24 hour in a container made of polyvinyl chloride (PVC) at room temperature (20 ± 1) °C. The bioethanol was bottled in 500 mL amber bottles in a batch of 120 units individually sealed with Parafilm®, identified, and stored in the laboratory at the same temperature.

2.2 Equipment

The PSEC used in the measurements has a piston-type cell with a cylindrical ceramic body made with aluminum oxide (Al_2O_3) and two platinum electrodes (Pt). In addition, the system is composed of a series of equipment for measuring the quantities that are used to determine EC [13-20].

2.3 EC measurement

The EC is defined (1) as being inversely proportional to the electrical resistance of the solution and directly proportional to the constant of the cell used. This constant is defined by its geometric dimensions, and for a cylinder with parallel electrodes at each side, it is equal to its length (or the distance between the electrodes, ΔL) divided by the cross-sectional area (A) inherent.

$$k = \frac{1}{R}.K_{cel} = \frac{1}{R}.\frac{\Delta L}{A} \qquad (1)$$

Where k is the EC (S.m^{-1}), R is the electrical resistance from the solution (Ω), and K_{cel} is the cell constant (m^{-1}).

In the PSEC measurement, the solution resistance is given by the resistance variation between two electrode positions ($\Delta R = R_i - R_f$) with the length of the cell given by the variation ($\Delta L = L_i - L_f$), while the cell diameter is fixed (0.049998 m) and determined through calibration. The most complete representation for determination of EC (2) takes into account the correction of the values to the reference temperature of 25 °C (at which all the results of this study are expressed). Usually, 30 repetitions are performed to obtain the EC value of a single sample, and the final value is the average of these results.

$$\kappa = \left[\frac{4.(L_i - L_f)}{(R_i - R_f).\pi.D^2}\right].\left[\frac{1}{1 + \alpha\,(T_m - T_o)}\right] \qquad (2)$$

Where α is the temperature coefficient of the solution (°C^{-1}), T_m is the temperature of the measurement (°C), and T_o is the reference temperature (25 °C).

3. RESULTS

3.1 Study of the impedance spectrum

Prior to the certification studies (i.e., the homogeneity, characterisation, and long-term stability study), it is relevant to obtain a profile of the electrical behaviour of the material by studying the electrochemical impedance spectroscopy (EIS) of the solution. This profile is related to the fact that for electric current applied (or potential difference) with different electrical frequency bands, the electrical resistance may assume different values.

This occurs because, in the cell model used, the total resistance to the passage of electrical current is subdivided into components, such as the electrical resistance (the real component of impedance), the reactance (the imaginary component of the impedance), and the capacitance originating from the cell geometry, and also into parasitic effects such as the polarisation of the electrodes and effects of adsorption and desorption of ions on the surface of the electrodes used. Therefore, it is important to choose a model in which these effects can be minimised, such that the impedance values observed experimentally are dominated by its real component (electrical resistance) that is used in the determination of the electrolytic conductivity of the solution.

Experimentally, this occurs when, for a wide range of frequencies studied (wide frequency study), some regions of the impedance spectrum indicate stable values for the real component of the impedance within specified ranges of electrical frequencies. In these regions, it is observed that the values of the imaginary component are minimal and, therefore, the spectrum of the solution is dominated by the electrical resistance of the solution. Another important aspect is that around this particular 'ideal' frequency range, the parasitic effects can be minimised by extrapolating the measured values of the electrical resistance in this region to the limit of high frequency (or its inverse to zero). In this situation, the system can be modeled simply as an equivalent circuit in which there is electrical resistance of the solution in parallel to the capacitance of the electrodes [13,23,24].

The impedance spectrum of bioethanol samples is shown in Figure 1, in the Nyquist plot (a) that presents the variation of the reactance values and electrical resistance for different associated electrical frequencies. The bioethanol samples were subjected to a wide range of frequencies (20-100000 Hz).

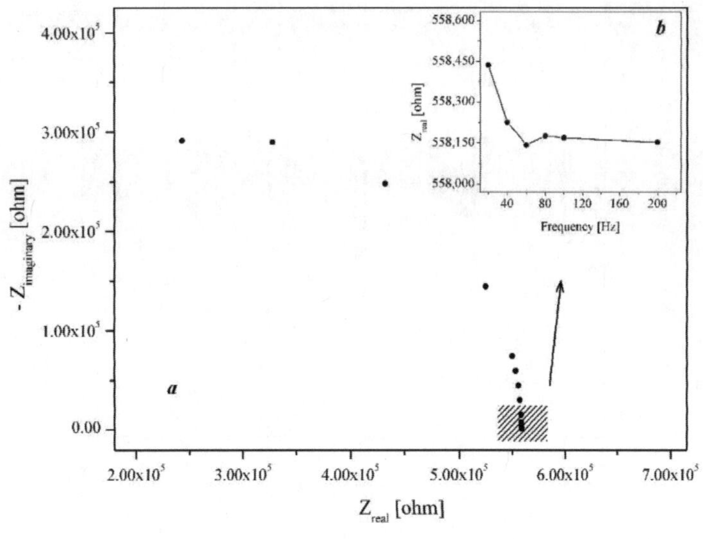

Figure 1. Impedance spectrum for the bioethanol samples.

For the samples of bioethanol, it was observed that the minimum reactance values (a) were found for an associated frequency band that comprises a range between 20 and 200 Hz. For this same range, the observed electrical resistance values are of the magnitude of 558.000 Ω (b). These values must be extrapolated linearly to infinite frequency values in order to minimize parasitic effects, as mentioned above. For this work, all subsequent measurements were performed in the electrical frequency range between 60 and 100 Hz. This is a more specific working range in which there is an intersection with the optimum working range observed in the study of the impedance spectrum of the samples.

3.2 Homogeneity study

The homogeneity of bioethanol was verified through the determination of the EC in six samples randomly chosen from the produced batch. This study aimed to obtain the standard uncertainty associated with the degree of heterogeneity inherent in the material and contributing to the overall uncertainty of the material [25,26].

The estimation (3) of this uncertainty (u_{hom}) follows the concepts defined in ISO Guide 35 [22], being equal to the variance observed between the bottles (S_{bb}).

$$u_{hom} = s_{bb} = \sqrt{\frac{(MS_{among} - MS_{within})}{no}} \quad (3)$$

Where MS_{among} is the mean square between the bottles studied, S_{bb} is the between-bottle (in) homogeneity standard deviation, MS_{within} is the mean square inside the bottles studied, and, no is the number of repetitions (30) done.

The EC measurement results of this study are presented in the graph in Figure 2. They were statistically analysed by analysis of variance (ANOVA) for the bottles numbered 3, 30, 54, 56, 113, and 116.

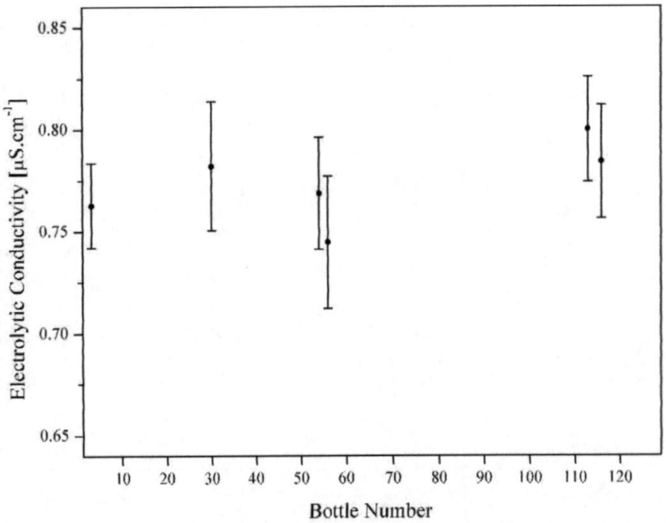

Figure 2. Bioethanol homogeneity study.

The results demonstrate the convergence of values because, for each individual measurement, there is an intersection with the subsequent measurement. The estimated value of the standard uncertainty was 0.02 $\mu S.cm^{-1}$.

3.3 Stability study

The stability of the batch was determined with EC measurements on samples randomly chosen more than one year after the preparation of the batch. The results of these measurements are presented in the graph in Figure 3.

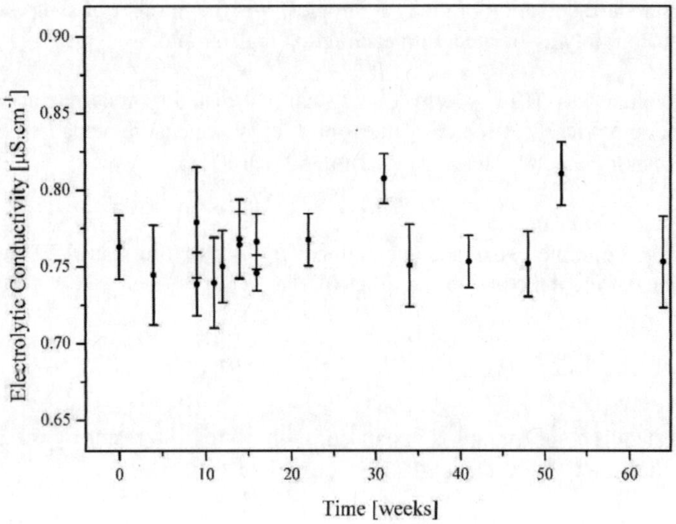

Figure 3. Bioethanol stability study.

The results presented a weak correlation (fitting a simple linear regression model) for EC variation with time (correlation coefficient, R = 0.44) that indicates good stability of the material. Furthermore, applying a significance test, the $f_{statistics}$, with 95% confidence level (CL), is 1.19 with an associated p_{value} of 0.29 (> 0.05), which does not allow the rejection of the null hypothesis that the slope is zero and, therefore, the material could be considered stable.

Despite this, a slight variation in the EC value was observed, which could be explained by the evaporation of the matrix over time, leading to an increase in the ionic strength of the solution. This effect must be incorporated into the evaluation of the standard uncertainty due to long-term stability (u_{lts}), multiplying the slope from the linear fit obtained (0.0003 μS.cm^{-1}.week^{-1}) by the time studied (4).

$$u_{lts} = a \times t \quad (4)$$

Where a is the slope from the linear regression model (μS.cm^{-1}.week^{-1}), and t is the time studied (week).

The uncertainty associated with the instability of the material can be estimated as 0.02 μS.cm^{-1}. As the material was studied over 64 weeks (1 year and 3 months), this period also could be estimated as the shelf life of the material.

3.4 Property value

The characterisation, or the determination of the property value of the material, was performed in duplicate with randomly chosen samples. The associated uncertainty determined by characterisation (u_{char}) takes into account the repeatability and the individual uncertainties of each measurement (5).

$$u_{char} = \sqrt{\frac{s}{\sqrt{n}}^2 + \sum_{i=1}^{i=2} u_i^2} \tag{5}$$

Where s is the standard deviation of the two samples, n is the number of samples (two bottles), and u_i is the standard uncertainty associated with each individual sample.

The property value was 0.77 μS.cm^{-1}, with an estimated standard uncertainty [25,26] of 0.02 μS.cm^{-1}. The electrical resistance values in the EC measurements were in the order of 1.4×10^5 Ω, with an associated frequency range from 60 to 100 Hz.

3.5 Certified reference material value

The evaluation of the combined standard uncertainty (u_{CRM}) for the material takes into account the standard uncertainties of the three previous studies (6).

$$u_{CRM} = \sqrt{u_{hom}^2 + u_{lts}^2 + u_{char}^2} \tag{6}$$

This combined standard uncertainty is expanded, multiplying the result by the convergence factor (k) = 2 (95% CL), then the CRM value is (0.77 ± 0.06) μS.cm^{-1}.

4. CONCLUSION

The EC primary measurements indicate the possibility of establishment and the production of CRM in alcoholic matrices. However, the final uncertainty from the material (0.06 μS.cm^{-1}) should be improved, because this value is approximately 8% of the property value of the material. It is important to emphasise that this value is very low (0.77 μS.cm^{-1}) and is therefore not easy to measure.

This improvement will involve the investigation of the matrix effects (alcoholic media) and the methodology of the measurements where smaller associated uncertainties should be studied. However, the obtained values do not invalidate the use of this type of CRM, because it allows the establishment of a traceable and metrological chain. As this is the first work of primary certification of a CRM for measurements of the bioethanol matrix, the purpose was achieved with the highest level of accuracy possible. Laboratory analysis will confirm reliability and metrological traceability of the issued results of the measurements of EC.

REFERENCES

[1] Goldemberg J, Coelho S T, and Guardabassi P 2008 The sustainability of ethanol production from sugarcane *Ene Poli* **36** 2086-97

[2] A comprehensive analysis of biodiesel impacts on exhaust emissions. EPA Office of Transportation and Air Quality, Environmental Protection Agency, Washington DC 2007 118 Draft Technical Report, EPA420-P-02–001

[3] Crago C, Khana M, Barton J and Giuliani E 2010 Competitiveness of Brazilian sugarcane ethanol compared to US corn ethanol *Ene Poli* **38** 7404-15

[4] Rodrigues J M, Guimarães E F, Sousa M V B, Silva V F, Fraga I C S, Souza V, Cunha V S 2012 Certified reference material to water content determination in bioethanol fuel *Quím Nova* **35** n. 5

[5] Stradiotto N R, Zanoni M V B, Fraga I C S, Borges P P, Cortez L A B 2010 Sugarcane bioethanol R&D for productivity and sustainability fourth part 21 ed Blücher (São Paulo) 813-828 http://www.blucher.com.br/livro.asp?Codlivro=05302 Accessed 28 Mar 2016

[6] Agência Nacional do Petróleo, Gás Natural e Biocombustíveis (ANP), Resolução N°19, 20.4.2015

http://nxt.anp.gov.br/NXT/gateway.dll/leg/resolucoes_anp/2015/abril/ranp%2019%20-%202015.xml Accessed 28 Mar 2016

[7] Automotive fuels – Ethanol as a blending component for petrol EN 15376:2011 (European Community), http://www.analiit.ee/index.php?option=com_content&view=article&id=58&lang=et Accessed 28 Mar 2016

[8] ASTM International D4806 (U.S.A.) ASTM D4806 - 13a Standard Specification for Denatured Fuel Ethanol for Blending with Gasolines for use as Automotive Spark-Ignition Engine Fuel, http://www.astm.org/Standards/D4806.htm. Accessed 28 Mar 2016

[9] ABNT NBR 10547:2016 Etanol combustível - Determinação da condutividade elétrica http://abnt.org.br/noticias/4618-etanol-combustivel-determinacao-da-condutividade-eletrica Accessed 28 Mar 2016

[10] ISO 17308:2015 Petroleum products and other liquids - Ethanol - Determination of electrical conductivity http://www.iso.org/iso/iso_catalogue/catalogue_tc/catalogue_detail.htm?csnumber=59558 Accessed 28 Mar 2016

[11] Spitzer P, Fisicaro P, Seitz S, Champion R 2009 *Accred Qual Assur* **14** 671-676.

[12] http://www.inmetro.gov.br/metcientifica/formularios/form_mrc.asp Accessed 28 Mar 2016

[13] Fraga I C S, Lopes J C, Cordeiro L R, Silva L F, Borges P P 2015 Evaluation of the Stability of Solutions of Low Electrolytic Conductivity by Primary Measurements *J Solution Chem* **44** 1920-1936

[14] Fraga I C S, Ribeiro C M, Sobral S P, Dias J C, Gonçalves M A, Borges P P, Gonzaga F B 2012 Certified reference material of bioethanol for metrological traceability in electrochemical parameters analyses *Talanta* **99** 99-103

[15] Fraga I C S, Lopes J C, Sobral S P, Ribeiro C M 2013 Certification of a low value electrolytic conductivity solution using traceable measurements *Accred Qual Assur* **18** 99-104

[16] Vyskocil L *et al* 2013 Final report on key comparison CCQM-K92: Electrolytic conductivity at 0.05 S m^{-1} and 20 S m^{-1} *Metrologia* **50** Tech. Suppl. 08013

[17] Durbiano F *et al* 2014 Final report of key comparison CCQM - K105 Electrolytic conductivity at 5.3 S.m^{-1} *Metrologia* **51** Tech. Suppl. 08016

[18] Gonzaga F B, Fraga I C S, Lopes J C, Costa A O 2015 Final report of key comparison SIM.QM-K92 Electrolytic conductivity at 0.05 S m^{-1} *Metrologia* **52** Tech. Suppl. 08008

[19] Jensen HD 2009 Final Report on CCQM–K36.1 *Metrologia* **46** Tech. Suppl. 08004

[20] Jensen HD 2010 Final Report on CCQM–K36 *Metrologia* **47** Tech. Suppl. 08025

[21] ISO Guide 34, General requirements for the competence of reference material producers, Geneva 2009

[22] ISO Guide 35, Reference materials - general and statistical principles for certification, Geneva 2006

[23] Barsoukov E, MacDonald J R 2005 *Impedance Spectroscopy, Theory, Experiment, and Applications* ed John Wiley (New York)

[24] Seitz S, Manzin A, Jensen H D, Jakobsen P T, Spitzer P 2010 Traceability of electrolytic conductivity measurements to the international system of units in the sub mS m^{-1} region and review of models of electrolytic conductivity cells *Electrochim Acta* **55** pp 6323-6331

[25] BIPM, IEC, IFCC, ILAC, ISO, IUPAC, IUPAC and OIML 2008 Guide to the Expression of Uncertainty in Measurement–GUM 1995 with minor corrections JCGM 100:2008 http://www.bipm.org/utils/common/documents/jcgm/JCGM_100_2008_E.pdf. Accessed 22 Mar 2016

[26] EURACHEM/CITAC 2000 Guide quantifying uncertainty in analytical measurement 2nd edn http://www.eurachem.org/guides/pdf/QUAM2000-1.pdf. Accessed 22 Mar 2016

Transmittance Variations Analysis in Sunglasses Lenses Post Sun Exposure

A D Loureiro, L M Gomes, L Ventura

LIO – Laboratory of Ophthalmic Instrumentation - Dept. of Electrical Engineering, Sao Carlos School of Engineering, University of Sao Paulo

lilianeventura@usp.br

Abstract. The hypothesis that sunglass ultraviolet (UV) protection can degrade with Sun exposure has never been proven experimentally. No sunglasses standards take into account UV transmittance changes after long Sun exposure. We selected 12 sunglass lenses and measured transmittance values from 280 nm to 780 nm. After 50 hours of exposure, new transmittance measurements were taken and transmittance variations inferior to 0.2% were observed. The exposition continues longer and more lenses will be tested to obtain conclusive results. We hope to obtain experimental data to confirm UV protection loss hypothesis and obtain a relation between Sun and solar simulator exposition.

1. Introduction

Sunglasses are widely used fundamentally for visual comfort, fashion and protective issues. There is a strong association between UV light exposure and at least 6 different diseases involving 3 different ocular structures (eyelid, cornea and crystalline lens) [1]. Eye's natural response to intense visible light is to reduce its exposure by closing the eyelid and constricting the pupil. When wearing dark lenses, visible light is attenuated and the natural response is not activated, with less visible light reaching the eye, the pupil does not constrict increasing eye's exposure, thus sunglasses with improper UV protection are dangerous [2]. There is a hypothesis that after long exposition to Sun sunglasses lenses may lose their UV protection.

The brazilian standard NBR ISO 12312-1:2015 specifies physical characteristics (mechanical, optical etc) for sunglasses and sunglare filters for general use [3].

The luminous transmittance or visible transmittance, τ_V, is defined as the mean of the spectral transmittance, $\tau_F(\lambda)$, between 380 nm and 780 nm weighted with the spectral distribution of radiation of day light (CIE Standard illuminant D65), $S_{D65\lambda}(\lambda)$, and relative spectral visibility function of daylight vision, $V(\lambda)$, and it is calculated using Equation (1) [3].

$$\tau_V = \frac{\int_{380}^{780} \tau_F(\lambda) V(\lambda) S_{D65\lambda}(\lambda) d\lambda}{\int_{380}^{780} V(\lambda) S_{D65\lambda}(\lambda) d\lambda} \quad (1)$$

The solar UV transmittance, τ_{SUV}, is the mean of the spectral transmittance, $\tau_F(\lambda)$, between 280 nm and 380 nm weighted with the solar radiation, $E_{s\lambda}(\lambda)$, at sea level for air mass 2 and the relative spectral effectiveness function for UV radiation, $S(\lambda)$, and it is calculated using Equation (2) [3].

$$\tau_{SUV} = \frac{\int_{280}^{380} \tau_F(\lambda) E_{s\lambda}(\lambda) S(\lambda) d\lambda}{\int_{280}^{380} E_{s\lambda}(\lambda) S(\lambda) d\lambda} \quad (2)$$

The solar UVA transmittance, τ_{SUVA}, and the UVB transmittance, τ_{SUVB}, are calculated using Equation 2 but using integration interval between 315 nm and 380 nm, and between 280 nm and 315 respectively.

Depending on visible spectral transmittance, lenses are classified into different categories and for each category there is a minimum required UV protection (table 1) [3]. For being improper for general use, lenses with visible transmittance less than 3% are not classified into any category.

Table 1. Transmittance requirements for sunglass lenses for general use [3].

	Visible spectral range		UV spectral range	
Lens category	Range of luminous transmittance (τ_V)		Maximum value of solar UVB transmittance (τ_{SUVB})	Maximum value of solar UVA transmittance (τ_{SUVA})
	from over (%)	to (%)	280 nm to 315 nm	315 nm to 380 nm
0	80	100	$0.05\ \tau_V$	τ_V
1	43	80		
2	18	43	1.0% absolute or $0.05\ \tau_V$ (the greater)	$0.5\ \tau_V$
3	8	18	1.0% absolute	1.0% absolute or $0.25\ \tau_V$ (the greater)
4	3	8		

Brazilian standard proposes a resistance-to-irradiation test in which lenses' category and UV requirements could not change after 50 uninterrupted hours of irradiation by 450 W solar simulator. There are not any sunglasses standards that take into account UV transmittance changes after long Sun exposure or any previous experimental studies about transmittance variation by solar exposition.

Our team has developed an online survey to evaluate the sunglasses use conditions by the Brazilian population. Over 2000 people have been surveyed. Some questions were asked, for example where the person lives, person's degree of education and the time one takes to change or buy new sunglasses. Another important asked question was about how long was the daily wearing time of sunglasses. The results have shown that people used to wear the sunglasses for 22 days and 8 hours in summer, and 52.6% buy new sunglasses every 2-4 years. Based on the answers and estimations, it is reasonable to assume that the UV protection of sunglasses should last at least 2 years [4].

Our team has also developed a device capable of automatically expose sunglasses lenses to the Sun [5]. The project consists of a panel with cover, housing 100 lenses fixed on a rotating axis, which will be irradiated by the Sun from sunrise until sunset. The lid opens and turns the panel towards the Sun automatically, so that the lens will always be facing the Sun. Humidity, dust, time and UV index sensors, as well as a video camera are part of the system. The exposure time and UV index will be recorded and automatic opening or closing the lid may also be controlled by a PC using online software.

In this study, we analyze UV and visible spectral range transmittance variation in sunglass lenses after long Sun exposition. We aim to obtain experimental data to confirm the hypothesis that sunglass lenses lose their UV protection after solar exposition. As secondary goal, we aim to compare Sun exposition and solar simulator exposition from transmittance variations.

2. Methodology

Twelve dark sunglasses (categories 3, 4 and without category) were selected for this study. Once our goal is to analyze variation in UV (280 nm - 380 nm) and visible transmittance (380 nm - 780 nm), initial transmittance values from 280 nm to 780 nm of the left lenses were measured using VARIAN Cary 5000 spectrophotometer. The right lenses were saved for future exposition on solar simulator. An acrylic frame and metal base panel as shown in figure 1 were mounted to hold left lenses for exposition.

Figure 1. Lenses during Sun exposure.

Every 50 hours of solar exposition new transmittance measures are performed. The transmittance measures were performed in 5 different points of each lens. Lenses transmittance variation analyses are calculated using the free software GNU Octave v.3.8.2 [6].

3. Results

After 50 hours of exposition small transmittance variations were observed. These variations were inferior to 0.2% of initial value for all wavelengths from 280 nm to 780 nm. Figure 2 illustrates percentage transmittance variation for measured wavelengths. This figure shows one lens' percentage transmittance variation for every wavelength from 280 nm to 780 nm. A negative value means transmittance increase in some wavelength and a positive one means transmittance decrease.

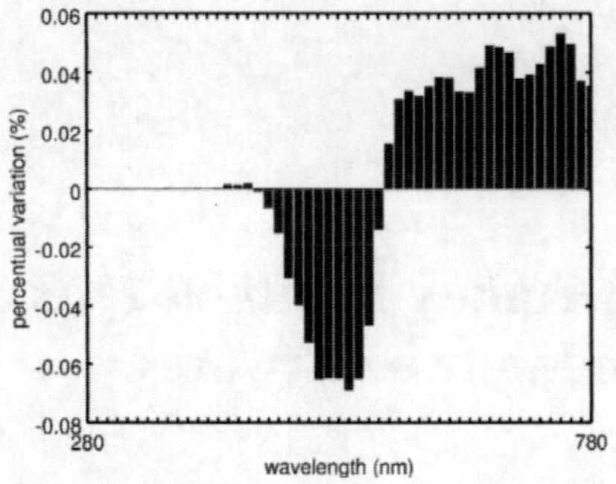

Figure 2. Percentage transmittance variation from 280 nm to 780 nm.

In 7 lenses, transmittance values increased for all wavelengths. In 4 lenses, transmittance values increased for some wavelengths and decreased for others. In only 1 lens, transmittance values decreased for all wavelengths. This short solar exposition (50 hours) produced small transmittance variations.

4. Conclusion

In almost all lenses, transmittance values increased slightly for every wavelength. However, we could not trace a variation behavior trend because not only the variations measured were little and number of lenses small but also we encountered different responses.

With more exposition time and more lenses tested, we hope acquire enough data to confirm the hypothesis that lenses lose their UV protection after long natural solar exposition. Furthermore, with these data and spectroscopy measures after solar simulator expositions of right lenses, we will be able to compare experimentally solar exposition and solar simulator exposition. Therefore we will obtain experimental data to analyze whether UV protection is lost after 2 years (1460 hours of use).

We are also developing an automated panel [5] to facilitate the exposure of the lens and the control over exposure.

Acknowledgements

The authors acknowledge FAPESP grant no. 2014/17252-5 and 2014/16838-0, Pro-Reitoria de Cultura e Extensao – USP and ABIOPTICA (Associacao Brasileira da Industria Optica).

References

[1] Yam J C S and Kwok A K H. Ultraviolet light and ocular diseases. International Ophthalmology, 34(2):383–400, 2014.

[2] Sliney D H. Intraocular and crystalline lens protection from ultraviolet damage. Eye & contact lens, 37(4):250–258, 2011.

[3] ABNT NBR ISO 12312-1. Proteção dos olhos e do rosto – Óculos para proteção solar e óculos relacionados Parte 1: Óculos para proteção solar para uso geral. Associação Brasileira de Normas Técnicas, 2015.

[4] Masili M, Schiabel H and Ventura L. Contribution to the radiation protection for sunglasses standards. Radiat Prot Dosimetry, 164 (3): 435-443, 2015.

[5] Gomes L M and Ventura L. Development of automated prototype for studying the effect of solar aging on sunglasses. Proc. SPIE 9531, Biophotonics South America, 95314D, June 19, 2015, Rio de Janeiro, RJ, Brazil.

[6] Eaton J W, Bateman D, and Hauberg S. GNU Octave version 3.0.1 manual: a high-level interactive language for numerical computations. CreateSpace Independent Publishing Platform, 2009. ISBN 1441413006.

Comparison of uncertainty in fatigue tests obtained by the Monte Carlo method in two softwares

Lisiane Trevisan[1], Daniel Antonio Kapper Fabricio[2], Afonso Reguly[3]

[1] Instituto Federal de Educação, Ciência e Tecnologia do Rio Grande do Sul – Câmpus Farroupilha. Av. São Vicente, 785, Farroupilha, RS, Brasil
[2] Instituto Federal de Educação, Ciência e Tecnologia de Santa Catarina – Câmpus Chapecó. Rua Nereu Ramos, 3450 D, Chapecó, SC, Brasil
[3] Laboratório de Metalurgia Física, Universidade Federal do Rio Grande do Sul. Avenida Bento Gonçalves, 9500, prédio 43820, Porto Alegre, RS, Brasil

E-mail: lisiane.trevisan@farroupilha.ifrs.edu.br

Abstract. The Supplement 1 to the "Guide to the expression of uncertainty in measurement" indicates the Monte Carlo method for calculating the expanded measurement uncertainty. The objective of this work is to compare the measurement uncertainty values obtained via Monte Carlo method through two commercial softwares (Matlab® and Crystal Ball®) for the parameter 'adjusted strain', obtained from fatigue tests. Simulations were carried out using different number of iterations and different levels of confidence. The results showed that there are short differences between the measurement uncertainty values generated by different software.

1. Introduction

The fatigue phenomenon represents more than 90% of failures in service components built with metallic materials [1]. Fatigue can be defined as the process of progressive localized permanent structural change occurring in material subjected to conditions which produce fluctuating stresses and strains at some point or points and may culminate in cracks or complete fracture after a sufficient number of fluctuations [2].

The value of the adjusted strain is determined from the proof load considering the surface roughness of the metallic material, and its determination is critical for indicating the tension effectively suffered by the material during cyclic loading.

The Monte Carlo method consists in applying an artificial sampling technique that operates numerically complex systems with independent input variables [3]. Different software may be used to generate pseudo-random numbers and used for the application of Monte Carlo simulation [4].

The Monte Carlo method is especially recommended for the input variables of uncertainty calculations with arbitrarily large values, measurement models with high nonlinearity and complexity, asymmetric distributions of the input quantities and probability distribution function associated with non-Gaussian a dominant component [5]. In such cases, the uncertainty estimation by the Monte Carlo method tends to be more representative than the ISO/GUM.

The number of iterations is an important factor in obtaining reliable results of the Monte Carlo method, such that the greater the number of simulations, the better the analysis of the results.

However, the larger the number of iterations, the greater the computational work, and sometimes the use may be very slow or even infeasible. In general, results to one or two significant figures [5].

Thus, the main objective of this study is to determine the expanded uncertainty to the 'adjusted strain' fatigue tests determined through two commercial softwares: Matlab® and Crystal Ball®, for two different confidence levels in different numbers of iterations.

2. Methodological Procedures

In this study, 36 test specimens of AISI 316L material were prepared for testing at three stress levels (480 MPa, 500 MPa and 520 MPa) and three different levels of roughness (Ru1, Ru2 and Ru3, according to table 1).

Table 1. Preparation of the surface of the specimens.

Roughness	Surface Preparation
Ru1	Sanding with sandpaper water # 800 and polishing with diamond paste 6 microns
Ru2	Sanding with sandpaper water # 500, free polishing
Ru3	Sanding with sandpaper water # 320, free polishing

The fatigue specimens were tested on a servo hydraulic testing machine with 100 kN load capacity. For the tests, we used a frequency of 20 Hz, fatigue ratio 0.1 and as failure criterion to complete rupture of the specimen.

The measures of surface roughness were made on a roughness tester, being held five measures in the area of interest, with the help of a fixing system for the specimens. The Brinell hardness test (HB) were conducted five measurements for each specimen.

2.1. Mathematical model to determine the 'adjusted strain'

In the fatigue test, the value of the proof load is changed due to the influence of surface roughness. Thus, the 'adjusted strain' (σ_{corr}) is determined by multiplying the proof load (σ_{test}) for a correction factor ($C_{\sigma,R}$), as described in equation (1).

$$\sigma_{corr} = C_{\sigma,R}\sigma_{test} \tag{1}$$

One of the simplest methods for determining the correction factor ($C_{\sigma,R}$) is the method described by FKM-Guideline guide [6] as shown in equation (2).

$$C_{\sigma,R} = 1 - a_r \log(R_z) \log\left(\frac{2S_{t,u}}{S_{t,u,min}}\right) \tag{2}$$

The values of a_r and $S_{t,u,min}$ depend on the material and heat treatment to which the material is subjected and the value of $S_{t,u}$ is the actual resistance of the mechanical component. Thus, considering the values for stainless steel, the equation (2) can be rewritten as equation (3), a function of only surface roughness (R_z) and hardness (HB).

$$C_{\sigma,R} = 1 - 0.22 \log(R_Z) \log\big(0.014973(\mathrm{HB})\big) \tag{3}$$

The 'adjusted strain' is then determined based on a normal probability distribution and the estimation of measurement uncertainty is accomplished through the Monte Carlo method, based on the values of the lower limit (L_i) and the upper limit (L_s) of each variable [7], whereas the standard deviation for the construction of the probability distribution curve.

Finally, Monte Carlo simulations were performed for the estimation of measurement uncertainty of the test, comparing the values obtained by two commercial softwares (Matlab® and Crystal Ball®), two levels of confidence (95.45% and 99.73%) and two iterations numbers (10^5 and 10^6).

3. Results and discussion

Table 2 shows the results obtained for 10^5 iterations and table 3 show results for 10^6 iterations, were μ represents the mean values of adjusted stress and U represents the expanded measurement uncertainty.

Table 2. Measurement uncertainty values for 'adjusted strain', with 10^5 iterations.

		Confidence level of 95.45%				Confidence level of 99.73%			
Software		Matlab®		Crystal Ball®		Matlab®		Crystal Ball®	
Parametric		*μ*	*U*	*μ*	*U*	*μ*	*U*	*μ*	*U*
480 MPa	**Ru1**	491.10	7.38	491.09	7.37	491.11	11.05	491.09	11.06
	Ru2	480.48	2.79	480.48	2.79	480.49	4.18	480.48	4.18
	Ru3	469.65	3.34	469.66	3.34	469.65	5.04	469.66	5.01
500 MPa	**Ru1**	501.79	7.70	501.78	7.75	501.80	11.58	501.78	11.62
	Ru2	494.05	8.56	494.03	8.59	494.00	12.78	494.03	12.88
	Ru3	489.39	6.91	489.41	6.94	489.40	10.36	489.41	10.41
520 MPa	**Ru1**	526.11	5.69	526.13	5.67	526.11	8.50	526.13	8.51
	Ru2	514.16	13.07	514.19	12.99	514.16	19.61	514.19	19.48
	Ru3	506.81	6.65	506.81	6.65	506.80	9.96	506.81	9.98

Table 3. Measurement uncertainty values for 'adjusted strain', with 10^6 iterations.

		Confidence level of 95.45%				Confidence level of 99.73%			
Software		Matlab®		Crystal Ball®		Matlab®		Crystal Ball®	
Parametric		*μ*	*U*	*μ*	*U*	*μ*	*U*	*μ*	*U*
480 MPa	**Ru1**	491.12	7.40	491.12	7.40	491.12	11.09	491.12	11.10
	Ru2	480.49	2.79	480.49	2.79	480.49	4.18	480.49	4.18
	Ru3	469.65	3.35	469.65	3.35	469.65	5.03	469.65	5.03
500 MPa	**Ru1**	501.78	7.70	501.78	7.69	501.79	11.55	501.78	11.54
	Ru2	494.04	8.55	494.03	8.53	494.04	12.82	494.03	12.80
	Ru3	489.40	6.91	489.41	6.93	489.41	10.37	489.41	10.39
520 MPa	**Ru1**	526.12	5.68	526.11	5.67	526.11	8.52	526.11	8.51
	Ru2	514.16	13.20	514.12	12.80	514.16	19.75	514.12	19.20
	Ru3	506.79	6.66	506.80	6.65	506.80	9.98	506.80	9.97

The influence of the level of confidence in the uncertainty calculation was expected, because the higher the confidence level, the larger the covered area under the probability curve. For greater confidence level, the greater the uncertainty of the measurement value. It was also observed that for higher confidence levels, the greater the difference between the results obtained by comparing the different software.

The group of specimens with strain of 520 MPa and Ru2 presented great dispersion of the experimental results of surface roughness. It emphasizes the importance of the surface preparation, which must be suitable in order to avoid errors due to the machining marks. Surface roughness is a key factor in resistance to dynamic forces, this happens because of the fact that the indents act as surface tension concentrators.

It was used a numerical routine for each software to calculate the expanded uncertainty. Furthermore, there is a difference in the methodology for the generation of random numbers used by each software by changing the shape of the probability distribution curve of the measurand. Still, the difference between software was much lower than the difference caused by other significant factors such as the condition tested and the confidence level.

It can be experimentally checked that the computational time to perform the simulations is different between the two softwares. The implementation way of each numerical routine, allows Crystal Ball® running faster than Matlab®. Using a desired level confidence of 95% confidence, the execution of 200,000 iterations can be considered enough [4].

4. Conclusions

The results showed that the higher the confidence level, the greater will be the expanded uncertainty of measurement values. Proportionally, these values suffer major influences of the software when used by larger magnitudes confidence level.

Differences between the measurement uncertainty of results were observed between the software when compared under the same simulation conditions. Most likely, these differences are due to the generation of random numbers and also due to the parameters initially determined based on the probability distribution curve.

However, even with the differences caused by each software, in most strain conditions, the values have been within the acceptable error as described in the calibration certificate of the cell load class 1 (acceptable error is lower than 1%) used in the fatigue test. Nevertheless, these differences can be considered or not, depending on the project requirements to run or standard used in the development of fatigue test.

References

[1] Callister W 2007 Materials Science and Engineering. 7th ed, New York

[2] ASTM International 2013 ASTM E1823: Standard Terminology Relating to Fatigue and Fracture Testing. Pennsylvania

[3] Bruni A L 2008 Avaliação de Investimentos, 1st ed, Atlas, São Paulo

[4] JCGM 2008 Evaluation of measurement data: guide to the expression of uncertainty in measurement (GUM). Joint Committee for Guides in Metrology, 1st ed, Geneva

[5] JCGM 2008 Evaluation of measurement data: Supplement 1 to the Guide to the expression of uncertainty in measurement – Propagation of distributions using a Monte Carlo method. Joint Committee for Guides in Metrology, 1st ed., Geneva

[6] FKM Guideline 2003 Analytical strength assessment of components in mechanical engineering. 5th ed

[7] Couto P R G 2008 A estimativa da incerteza de medição pelos métodos do ISO GUM 95 e de simulação de Monte Carlo. Instituto Nacional de Metrologia, Qualidade e Tecnologia.

Validation of the method of quantitative phase analysis by X-ray diffraction in API: case of Tibolone

R P Silva[1,2], M F S Ambrósio[1,5], E K Epprecht[2], R R Avillez[3], C A Achete[4], A Kuznetsov[4], L C Visentin[5]

[1] Universidade Federal do Rio de Janeiro, Campus de Xerém, 25245-390, Duque de Caxias, RJ – Brasil
[2] Pontifícia Universidade Católica do Rio de Janeiro, Centro Técnico-Científico, Departamento de Engenharia Industrial, 22451-900, Rio de Janeiro, RJ - Brasil
[3] Pontifícia Universidade Católica do Rio de Janeiro, Centro Técnico-Científico, Departamento de Engenharia Química e de Materiais, 22451-900 - Rio de Janeiro, RJ – Brasil
[4] INMETRO, Instituto Nacional de Metrologia, Qualidade e Tecnologia, 25250-020, Duque de Caxias, RJ - Brasil
[5] NanoBusiness Informação e Inovação Ltda, 25250-020, Duque de Caxias, RJ – Brasil

E-mail: lorenzo.visentin@nanobusiness.com.br

Abstract. In this study, different structural and microstructural models applied to X-ray analysis of powder diffraction data of polymorphic mixtures of known concentrations of Tibolone were investigated. The X-ray data obtained in different diffraction instruments were analysed via Rietveld method using the same analytical models. The results of quantitative phase analysis show that regardless of the instrument used, the values of the calculated concentrations follow the same systematics with respect to the final errors. The strategy to select a specific analytical model that leads to lower measurement errors is here presented.

1. Introduction

The quantitative phase analysis by X-ray diffraction (XRD) of organic solids which compose active pharmaceutical ingredients (API) is susceptible to many factors (instrumental, knowledge of materials properties, and method of data treatment) that influence the final result.

The Rietveld method is accepted as the most accurate approach to extract from X-ray data the values of concentrations of phases in phase mixtures [1]. However, the choice of structural and microstructural model in this method of analysis becomes a critical factor from the point of view of its contribution to the errors of calculated concentrations. This situation represents a serious problem for the pharmaceutical industry and regulatory organizations in production and distribution of safe and efficient solid drugs.

This study investigates, in a systematic manner, different assumptions about structural and microstructural properties of phases of Tibolone during the data modelling. Tibolone is a synthetic steroid used to relieve hypo-estrogenic symptoms and protect against bone loss in post-menopausal women [2]. The general and specific constraints for the reliable quantitative phase analyses were established on the base of multiple experimental data and variable concentrations of phases of Tibolone.

2. Materials and methods

2.1. Sample Preparation

Three standard mixtures of known concentrations, referred hereafter as M1, M2 e M3, were prepared using Tibolone pure Form I (commercial) and Form II (prepared according to Araujo, et.al.) [2]. The concentrations of the mixtures are shown in the Table 1. A calibrated analytical balance with the error of indication of -0,01 mg and the expanded measurement uncertainty (coverage factor k = 2) of 0.01 mg was used to establish the real phase proportions in the studied samples.

The overall mass of each prepared sample was approximately 1 g. The mixtures were placed in eppendorfs and stirred in vibratory mixer mill for 30 minutes. Table 1 presents the acceptable lower limits (M1, preferably M2) concentrations for a crystalline pure Form I [3], and the expected guess for a detection limit of Form II in Tibolone.

Table 1. Sample identification and phase composition.

Tibolone (API)	Nominal [F1-F2] (%)	Real [F1-F2] (%)
M1	90-10	90.09-9.91
M2	95-5	94.99-5.01
M3	99-1	98.73-1.27

F1= Tibolone Form I and F2 = Tibolone Form II

2.2. X-ray data collection

The instruments used for X-ray data collections, their basic configurations, and sample preparation aspects are listed in the Table 2.

Table 2. Instruments and sample preparation types used in X-ray experiments.

Instruments	Geometry	Tube	Detector	Powder Sample
X'Pert Pro Panalytical[1]	Reflection, Bragg-Brentano	CuKα	PiXcel	Mixed and pressed *Mixed, grinded and pressed
D8 Focus Bruker[2]	Reflection, Bragg-Brentano	CoKα	Scintillator	Mixed and pressed
Agilent Technologies[3]	Transmission, Debye Scherrer	CuKα	Charge-coupled device (CCD)	Mixed powder in capillary

[1] NanoBusiness Informação e Inovação Ltda.
[2] Inmetro - Laboratório de Difração e Espectroscopia, Divisão de Metrologia de Materiais (DIMAT).
[3] Inmetro - Diretoria de Metrologia Aplicada às Ciências da Vida (DIMAV).

2.3. Structural and Microstructural models

Eight models of Form I and Form II of Tibolone were employed to calculate the concentrations of these phases in the samples. Different constraints for structural and/or microstructural properties of phases distinguish these models (see Table 3).

Model 1 assumes the fixed structural parameters for Form I and Form II of Tibolone taken from data base of Cambridge Crystallography Data Centre (No.: Form I 1126814 and Form II 1126815), and isotropic crystallite size peaks broadening (Voigt function) for diffraction peak shape fitting. Subsequent models include other refinement options, as shown in Table 3.

Table 3. Description of models.

Structural and Microstructural Models	Model Constraints							
	1	2	3	4	5	6	7	8
1. Isotropic crystallite size broadening	●	●	●	●	●	●	●*	●
2. Lattice Parameters	○	●	●	●	●	●	●	●
3. Texture (spherical harmonics)	○	○	●	●	●	●	●	●
4. Hydrogen atoms inserted in Form I	○	○	○	●	●	●	●	●
5. Isotropic atomic displacement parameter, *Beq*	○	○	○	○	●	●	●	○
6. Atomic coordinates (constrained to maximum displacement of 0.0016)	○	○	○	○	○	●#	●	●
7. Anisotropic crystallite size broadening	○	○	○	○	○	○	●#	○
8. Anisotropic atomic displacement parameters, *Uij*	○	○	○	○	○	○	○	●

● Applied or variable

○ Not applied or not variable

* Applied only to Form II

Applied only to Form I

3. Results and Discussion

In Figure 1 it is shown the selected diffraction patterns of samples M1, M2 and M3 obtained on instrument 1. Figure 2 shows the experimental X-ray diffraction pattern of M1 sample and the one calculated from the Rietveld refinement of model 8.

The use of the latter model results in the excellent agreement between calculated and real concentrations of phases in the sample. Figure 3 summarises the relative errors of the calculated concentrations of Form II for all samples (Table 1) and experimental conditions (Table 2) used in this study.

As one can expect, the values of relative errors increase with decreasing concentration of Form II. This is a general trend, which can vary for a particular structural and microstructural model. One can also note that the results obtained on instruments 1 and 2 have markedly better agreement with the real values of concentrations than the ones of instrument 1* and 3.

The main distinction between these sets of conditions is the 2Θ resolution of diffraction peaks. This resolution is worse in case of instruments 3 and 1* due to a low angular resolution of CCD detector and additional crystallite size peak broadening induced by grinding, respectively.

The inset of Figure 3 compares the values of real relative errors (the relative differences between the real and calculated concentrations) and standard deviations for concentrations of Form II obtained from Rietveld refinement of a particular model. The calculated and real errors are comparable for models that result in a weighted profile residual (R_{wp}-factor) bellow 12 and a goodness of fit (GoF) bellow 2 (see Figure 4).

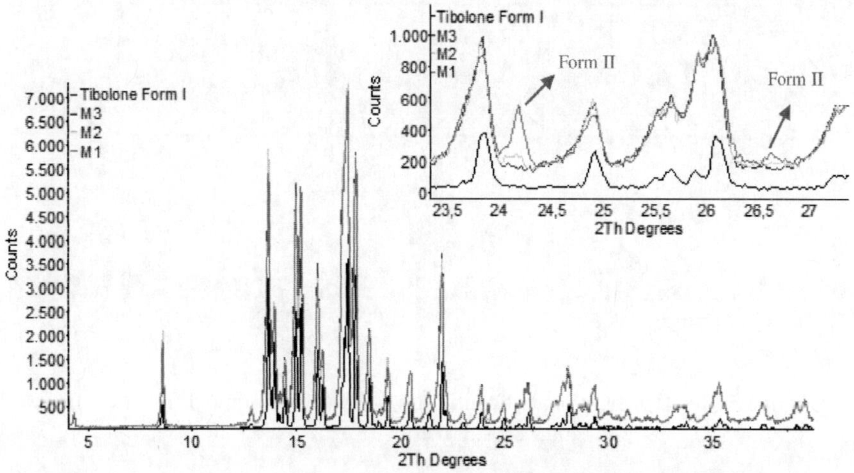

Figure 1. Diffraction patterns of M1, M2 and M3 samples. Inset emphasises the 2Θ regions of two not overlapped peaks of Form I and Form II.

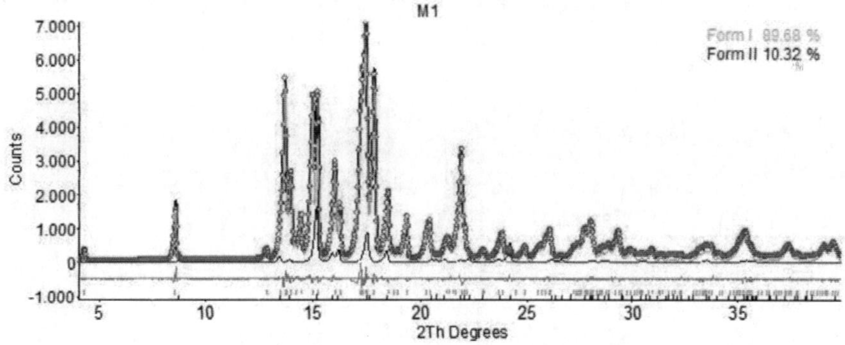

Figure 2. Experimental (black curve), calculated (red curve) and difference (grey) from Rietveld refinement (model 8) XRD patterns of the sample M1. Form I (green curve) and Form II (blue curve)

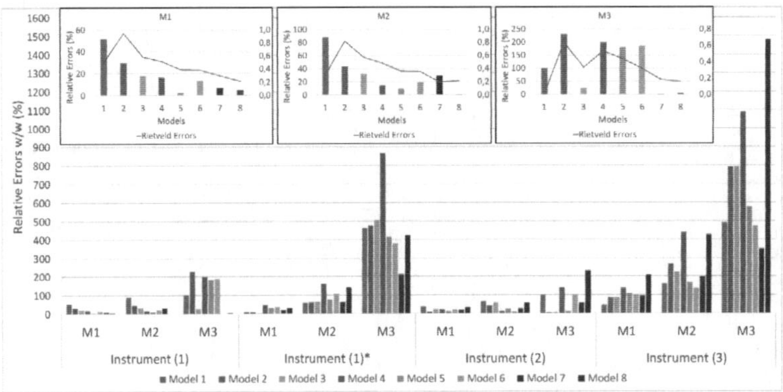

Figure 3. Relative errors in concentration. The three extended figures is from measures in the instrument 1.

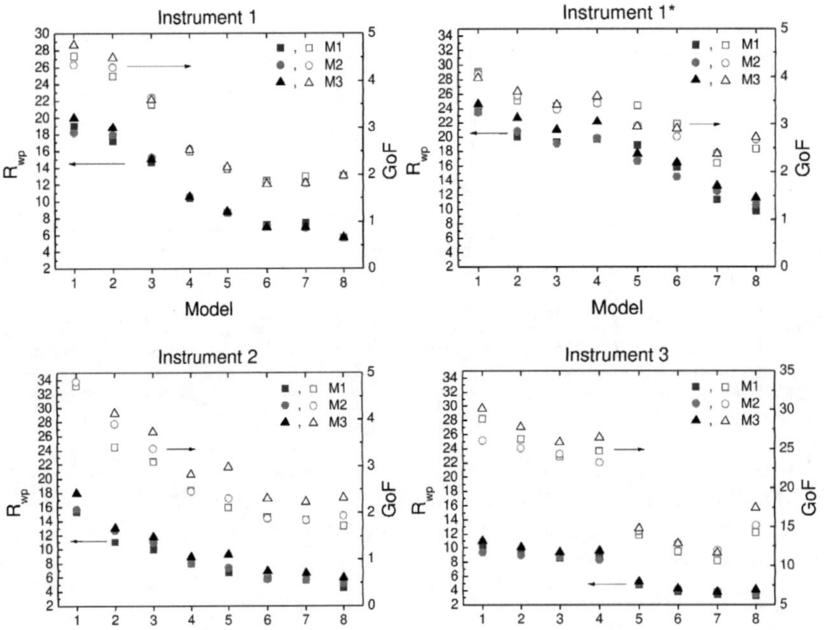

Figure 4. R_{wp} and GoF indexes ($R_{wp} = \sqrt{\sum w_m(Y_{o,m}-Y_{c,m})^2/\sum w_m Y_{o,m}^2}$; $GOF = chi2 = Rwp/Rexp$).

4. Conclusion

The obtained results showed that a good resolution of X-ray diffraction data is a necessary condition for reliable quantitative analysis of polymorphism in Tibolone. The Rietveld method of analysis of diffraction patterns gives reliable estimations of concentrations of Form II with respective errors for concentrations above 5 % w/w. The modelling of microstructural (texture) and structural (H atoms, ADPs) features was required to achieve this result. Additional models sophistications, such as consideration of anisotropy of crystallite size broadening and anisotropy of ADPs may be necessary to obtain reliable results in case of lower concentrations of Form II.

Acknowledgments

The e-Diffraction Lab from NanoBusiness Company is grateful to the agencies CNPq, FAPERJ and to INMETRO for the collaboration in this research. An especially acknowledgment to Daniel Giacometti by management support in this project.

5. References

[1] Young R A 1993 *Introduction to the Rietveld method. - The Rietveld Method* (Oxford: Oxford University Press - IUCr Book Series) pp 1-39.

[2] Araujo G L B, Faria D L A, Zaim M H, Carvalho F M S, Andrade F R D and Matos J R 2010 *J. Therm. Anal. Calorim.* **102** 233

[3] Akzo Nobel N.V. G A J M T Sas *et. al.* Pharmaceutical composition which contains a pharmaceutically suitable carrier and the compound having the structure (7α,17α)-17- hydroxy-7-methyl-19- nor-17- pregn-5(10)- en-20- yn-3- one. EP0389035A1, 12 mar. 1990, 26 sep. 1990.

Proficiency testing pilot for determination of total mercury in fresh fish

[1]L V de Santana, [1]J E S Sarkis, [1]J C Ulrich and [1]M A Hortellani.

[1] Institute of energy and nuclear research-National (IPEN/CNEN-SP), Av. Professor Lineu Prestes 2242 05508-000 São Paulo, SP, Brazil

E-mail: santana-luciana@ig.com.br

Abstract: A proficiency-testing scheme concerning total mercury determination in fish tissue involved 10 laboratories as participants, who used their regular in-house analytical methods, and the assigned value and the standard deviation used in proficiency testing program was derived from calibration against the certified reference values of the CRMs. The majority of participants obtained satisfactory Z-scores, and laboratories that need to revise their procedures were singled out. The objective these exercises were makes a useful contribution towards the production of proficiency test in Brazil. The uncertainty expanded calculated for the reference material was 22%.

1. Introduction

Proficiency tests are inter-laboratory exercises and the results of the particular laboratory are compared with reference value or with the results obtained by other laboratories. A proficiency item test is required for the effective execution of these tests and it can be a sample product or device, reference material, equipment, a standard set of data or other information [1].

The statistical treatment data produced during an inter-laboratory exercise is the most useful for population analysis data and adds information on its statistical properties technical scrutiny enables discussion of the possible sources of disagreement between laboratories [1].

Nowadays there are not proficiency providers essay in Brazil for toxic metals such as mercury in fishes tissues. Furthermore, the production of the RM was explored by gamma irradiation process to maximize the shelf life of it. In additional it was examined the homogeneity and stability by atomic absorption spectrometry with hydride generation.

A proficiency pilot test quality assurance program was also successfully performed using this material. The developed RM candidate in this study has demonstrated its suitability for quality assurance in the analysis of the total arsenic in fish tissue [1, 2].

2. Methodology

The preparation of the reference material used in the proficiency test for determination of mercury in fish tissues followed the recommendations of ISO Guide 35: 2012 [3].

2.1. Study of homogeneity and stability

For the study of homogeneity were randomly assigned to ten sachets temperature of 23 ± 0.2 ° C. The statistical tool ANOVA was used to evaluate the possible heterogeneity of the sachets and the sachets inside, as described in ISO Guide 35 [3].

The stability study was performed with 16 randomly selected sachets for a period of 0, 7, 15, 30 and 45 days in 4 different temperatures ranges 5 ° C, 23 ° C, 40 ° C and 60 ° C using model isochronous. The evaluation results of the stability study of the material were performed by linear regression analysis [3].

3. Results and discussions

3.1 Homogeneity study and stability study and uncertainty of the assigned value

The characterization of a reference material is defined as the complete set of measurements, which establishes, as such, the property values, not necessarily their uncertainties. In this work the characterization results were obtained from a method of measuring single (FIA-CV-AAS) where each parameter was calculated by Excel spreadsheet, developed in the laboratory for the determination of mercury in fish (test accredited in CGCRE / INMETRO second ABNT: ISO / IEC 17025: 2005) [4, 5].

The homogeneity study was conducted within the sachet or between sachets to assess possible differences of concentration which can be displayed during the production stage [1, 2]. The Shapiro-Wilk test (W) was used to verify the results showed normal behavior at a significance level of 5%. (Wcalc.= 0.78 Wcrit.(0.05;5) = 0.762).

The ANOVA evaluates the value of the result $F_{critical}$ and $F_{calculated}$ for a confidence level of 95%,

The stability study was performed in different temperature ranges (5-60 °C) for a period of 0-45 days using an isochronous model. The results were submitted to Grubbs test for the presence of outliers and were then submitted to linear regression analysis and significance testing.

The significance of the linear regression slope was evaluated using the "regression analysis" implemented in MS Excel ® at a level of 95% [2, 3, 5].

The expanded uncertainty U_{pt} is obtained by multiplying the combined standard uncertainty, u_{char} = uncertainty for characterization, u_{bbr} = uncertainty for homogeneity, and u_{sts} = uncertainty for stability by a coverage factor k (k= 2), the result was expressed by equation 1. [2, 3, 5].

$$U_{pt} = k\sqrt{u_{char}^2 + u_{bb}^2 + u_{sts}^2} \quad (1)$$

The reference results obtained for total mercury was 0.69 ± 0.15 µg g^{-1} (table 1).

The value proficiency assigned samples presented were calculated by calibration against the certified reference values of the CRMs (DORM-2 - "Dogfish" – National ReserchConciul of Canada); the process is recommended by ISO/IEC 13528:2015[2].

Table 1. Estimation of uncertainties in µg g^{-1} for the element Hg-total determined by FIA-CV-AAS (wet basis).

Element Hg	(µg g^{-1})	%
u_{char}	0.074	10.6
u_{bb}	0.0002	0.03
u_{sts}	0.003	0.43
[a]Average ± U_{pt}	0.69 ± 0.15	

[a]Average-obtained in the characterization, represents the average of 45 replicates of each material

3.3 Proficiency test pilot

The proficiency testing was relied on the ten participation governmental and private laboratories. The reference material for proficiency testing was prepared according to the recommendations of the ISO/IEC 17043:2010 [1].

The participating laboratories received one samples of 15 g (recommended size to be about 1 g each). They were analyzed in three different days in order to check the robustness of their results. The laboratories were instructed to proceed with the test item following the same procedure used in their routine analysis. They were allowed to use any suitable method(s) of measurement. The analytical techniques were used absorption spectrometry atomic- AAS (CV-AAS, HG-AAS, GF-AAS, TDA-AAS), Hg analysis Thermal decomposition mass spectrometry (ICP-MS) and spectroscopy optical (ICPOES).

All results were statistically evaluated using procedures based on standard ISO/IEC 17043:2010 (Table 2) [1].

Table 2. Results reported by participants

Laboratories	Average ± U (µg g^{-1})
1	0.50 ± 0.09
2	X
3	0.65 ± 0.10
4	0.82 ± 0.02
5	0.84 ± 0.04
6	0.69 ± 0.02
7	**0.23 ± 0.02**
8	**4.03 ± 0.06**
9	0.56 ± 0.04
10	**0.18 ± 0.02**

X = result not reported.

The assigned value and the standard deviation used in proficiency testing program was derived from calibration against the certified reference values of the CRMs. A good agreement was observed among the values calculated and results obteined by (Table 3).

Table 3. Estimate for the assigned values and standard deviation, in µg g^{-1} for the results of pilot proficiency testing (wet basis)

Element	Assigned values (µg g^{-1}) %
Hg	0.693 ± 0.15

The analytical performances were evaluated by calculating the Z-score values characterizing individual laboratories [1], which is the most conventional way of performance evaluating a laboratory participating in a PT scheme [1, 2].
These values were calculated by using as assigned value (X_{ref}) the corresponding standard deviation unit (s) according to the following formula:

$$Z\text{-}score = \frac{(X_{lab} - X_{ref})}{s} \tag{2}$$

Where: X_{lab} = average result of the laboratory, X_{ref} = reference value, and s = standard deviation for proficiency testing.
The z-score, helps identify outlying results as z-scores are standardized values that give it a score note for each result on the others in the group. Thus, a Z value close to zero means that the findings are consistent with those of other participants. A disperse value in proficiency testing is any results that present an absolute value of z-score greater than three.
The Z-score classifies laboratory performance as follows:

$|Z| \leq 2$ is "satisfactory"

$2 < |Z| \leq 3$ is "acceptable"

$|Z| > 3$ is "unsatisfactory".

The evaluated laboratory performance is summarized in the Figure 1. Considering the results obtained to date, the majority laboratories obtained a satisfactory z-score.
The laboratory 7, 8 and 10 showed result for z-score considered unsatisfactory, because were above 3 and -3. The z-score for laboratory 8 was 9.04, in this case, the graph was presented with open bar and axis to -5 and 5 for better visualization.

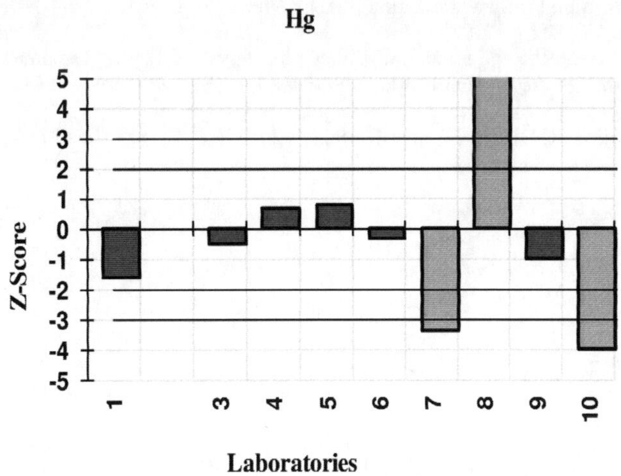

Figure 1: shows the results of z-scores for each participating laboratory proficiency test pilot program for mercury in fish tissue

4. Conclusions

The studies of homogeneity and stability at 23 °C indicated no significant variations that could affect the intended use of this material. Thus, the produced sample is suitable to be used in a program of proficiency testing of laboratories. Therefore, there is no need for freezing during transport.
The results obtained by the collaborating laboratories were in agreement with the results determined by the producer laboratory using the homogeneity test. Each participant sent along with the concentrations found information on the analytical methods used by the analyzes.
Laboratories with unsatisfactory results had informed by proficiency completion testing meeting that they had sent the wrong results of the sample analysis moreover just one laboratory does not report the result. Laboratories with unsatisfactory results had informed by proficiency completion testing meeting that they had sent the wrong results of the sample analysis moreover just one laboratory does not report the result. In addition laboratories that have sent the correct data, there are no inferences or deviation from the reference value, z-score ≤ 2.
In the first analysis the number of laboratories was small and for this reason we have considered this comparison pilot study safety for proficiency testing development in the future program with larger group laboratories.

5. Aknowledgements

We are indebted to the IPEN-CNEN/SP and CNPq for financial support in the preparation of this work and CTR-IPEN by irradiation of samples.

6. References

[1] ISO/IEC Guide 17043(2010) Conformity assessment- General requirements for proficiency testing. Internacional Organization for Standardization (ISO), Geneva, Switzerland.

[2] ISO/IEC Guide 13528(2005) Statistical Methods for use in proficiency testing by interlaboratory comparisons. Internacional Organization for Standardization (ISO), Geneva, Switzerland

[3] ISO/IEC Guide 35. 3rd (2006) Certification of reference materials—general and statistical principles. Internacional Organization for Standardization (ISO), Geneva, Switzerland.

[4] ISO/IEC 17025(2005) General Requirements for the Competence of Testing Laboratories and Calibration. Internacional Organization for Standardization (ISO), Geneva, Switzerland.

[5] L V Santana, J E S Sarkis, J C Ulrich and M A Hortellani 2015 *J. Phys.* ***575*** *012041.*

Uncertainty estimation in the determination of metals in superficial water by ICP-OES

Mainara G. Faustino[1], Joyce R. Marques[1], Lucilena R. Monteiro[1], Thamiris B. Stellato[1], Sabrina M. V. Soares[1], Tatiane B. S. C. Silva[1], Douglas B. da Silva[1], Maria Aparecida F. Pires[1] and Marycel E. B. Cotrim[1]

[1]Instituto de Pesquisas Energéticas e Nucleares (IPEN-CNEN/SP), Avenida Lineu Prestes, 2242, Cidade Universitária, São Paulo - SP 05508-000, Brazil.

mainarag@usp.br

Abstract. From validation studies, it was possible to estimate a measurement uncertainty of several elements such as Al, Ba, Ca, Cu, Cr, Cd, Fe, Mg, Mn, Ni and K in water samples from Guarapiranga Dam. These elements were analyzed by optical emission spectrometry with inductively coupled plasma (ICP-OES). The value of relative estimated uncertainties were between 3% and 15%. The greatest uncertainty contributions were analytical curve, and the recovery method, which were related with elements concentrations and the equipment response. Water samples analyzed were compared with CONAMA Resolution #357/2005.

1. Introduction

The uncertainty of measurement according to VIM (2012), are all "*non-negative parameter that characterizes the dispersion of values of a measurand, based on the information used*" [1]. According to GUM (Guide to the Expression of Uncertainty in Measurement, 2008), the uncertainty of the result of a measurement reflects the lack of exact knowledge of the value of the measurand. Whereof the correction of recognized systematic effects, the result of measurement is only an estimation of the measurand value, originating from the random effects and imperfect correction of the results for systematic effects [2].

The result of a measurement after correction may be too close to the value of the measurand, and yet, have a negligible error, although it may have a large uncertainty. Therefore, should not confuse the uncertainty of the measurement result with the remaining unknown error occurred [2].

The EURACHEM/CITAC (2012) guide summarizes in four steps how the uncertainty estimation should be executed:

Step 1 – Specify measurand: declare clearly, what is being measured, with the inclusion of measuring and its input quantities, of which it depends;

Step 2 – Identify uncertainty sources: the inclusion of sources that will contribute to the uncertainty of the parameters in relation to Step 1, but may include other sources and also include sources resulting from chemical assumption;

Step 3 – Quantify uncertainty components: measure or estimate the size of the uncertainty component associated with each potential source of uncertainty identified;

Step 4 – Calculate combined uncertainty: with the information obtained in step 3 will contribute to the quantification of total uncertainty, whether associated with individual sources or with the combined effects of several sources. Contributions must be expressed as standard deviations, and combined according to the appropriate rules for the combined standard uncertainty. The appropriate coverage factor should be applied to achieve an expanded uncertainty [3].

When issuing a result, it is necessary the use of a measurement uncertainty, an example, the CONAMA's Resolution 357/2005, demands that the results should be statistically analyzed taking into consideration the necessary measurement uncertainties [4]. In Table 1 shows the permissible maximum values of the analyzed elements in this work in the CONAMA's Resolution.

Table 1. Permissible Maximum Values of the analyzed elements in the CONAMA's Resolution.

	CLASSE 1 Fresh Water
Chemical Elements	Permissible Maximum Values ($mg.L^{-1}$)
Dissolved Aluminum	0.1
Barium	0.7
Cadmium	0.001
Dissolved Copper	0.009
Chromium	0.05
Dissolved Iron	0.3
Manganese	0.1
Nickel	0.025
Potassium*	-
Calcium*	-
Magnesium*	-

*There is not permissible maximum concentration in CONAMA's Resolution [4].

Uncertainties were applied in the concentration results from Guarapiranga Dam's water samples. The selection of the water collection points in Guarapiranga Dam was aiming the influence of its tributaries and sub-basins in water quality. The selection was carried out with the participation of representatives of the Botanical Institute - SP, Energy and Nuclear Research Institute (IPEN) and the Center for Integration and Coordination of Information Management Environmental Planning Department of the Environment (SMA), they selected 14 points [5]. Fourteen collection points are shown in Figure 1.

Water samples were collected on 07 and 09 October 2014. The collection and sampling procedures were performed according to the procedures described in "National Guide Collection and Sample Preservation" of the National Water Agency (ANA) / Environmental Company the State of São Paulo [6] and Standard Methods for the Examination of Water and Wastewater [7].

This work intends to explain the steps of the uncertainty estimation applied to water samples collected in Guarapiranga Dam and compare to the CONAMA's Resolution 357/2015.

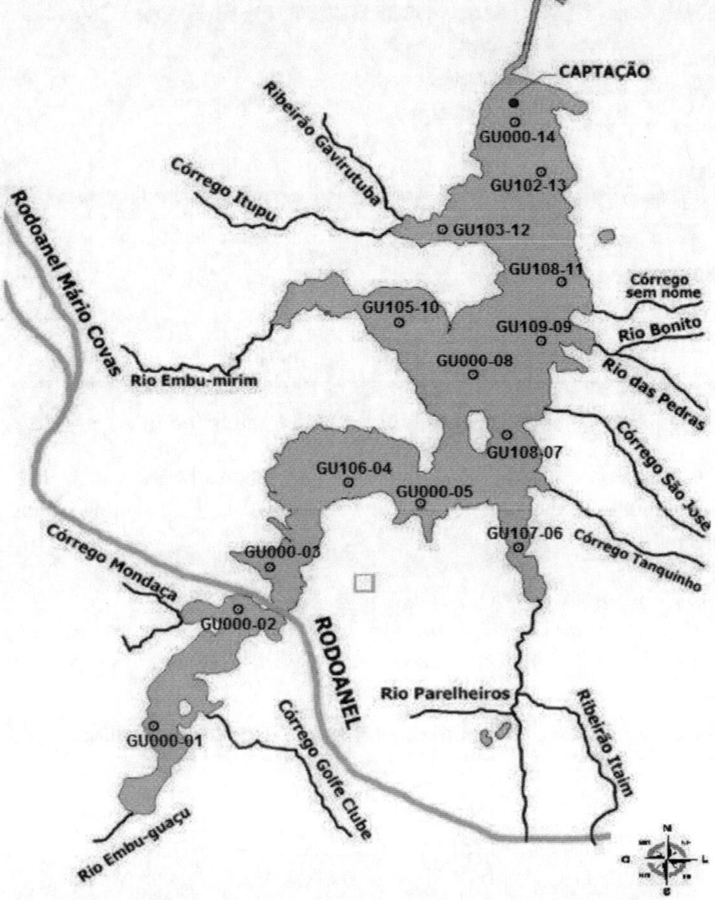

Figure 1. Distribution of collection points in Guarapiranga Dam and its tributaries [5].

2. Material and methods

Data from method development and validation of metal analysis in water was used, as follows the Ishikawa diagram (Figure 2). Uncertainties from sample and standards volume measurement, analytical curve preparation, measurement and linear regression, precision (repeatability) and recovery were estimated.

The mathematical model used to quantify the measurand is presented in (1), shows the variable that can influence the measurement of the concentration of metals in water.

$$C = \frac{C_O \times V_f}{V_i} \times \frac{1}{R} \quad (1)$$

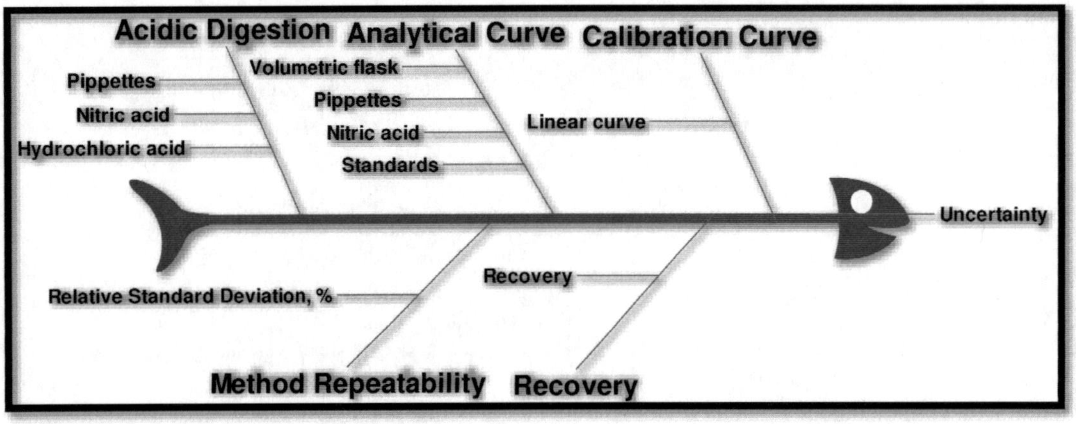

Figure 2. Ishikawa Diagram considered for metal determination in natural waters by ICP-OES.

Where C = element concentration, mg.L^{-1}; C_O = element concentration according to analytical curve, mg.L^{-1}; V_f = final volume of the sample after digestion, 50 mL; V_i = initial volume of sample, 45 mL; R = recovery from the method [3] [5].

2.1 Uncertainty associated with acidic digestion.

The sources of uncertainty due digestion in microwave were: temperature, sample, nitric and hydrochloric acid volumes (25 mL, 4mL and 1 mL, respectively), besides pipettes repeatability used in volume measurement.

The pipettes uncertainties estimation according the sample temperature and acid collection are present in (2) [3].

$$\mu(V_t) = \frac{V.Q.\Delta T}{\sqrt{3}} \tag{2}$$

Where $\mu(V_t)$ = uncertainty volume pipette according to temperature, mL; Q = expansion coefficient, °C; Δ= temperature variation, °C; V = collected volume of sample/acid, mL; $\sqrt{3}$ = rectangular distribution.

To estimate sample volume uncertainty and the acid volume uncertainty, several sources of uncertainties were combined as presented in (3).

$$\mu(V_s/V_a) = \sqrt{\mu(V_p)^2 + \mu(V_t)^2 + \mu(V_{repe})^2} \tag{3}$$

Where $\mu(V_{s/a})$ = Sample volume Uncertainty (s) and/or volume of the acid (a), mL; $\mu(V_p)$ = uncertainty of volume measurement in function of variations of Glassmaking and / or micropipette, mL; $\mu(V_{repe})$ = uuncertainty volume according repeatability, ml;

To estimate the volume combined uncertainty due acid digestion in microwave, it was used the equation (4).

$$\mu(V_f) = \sqrt{\mu(V_s)^2 + \mu(V_a)^2} \tag{4}$$

Where $\mu(V_f)$ = combined volume uncertainty due acid digestion, mL.

2.2 Uncertainty estimation on analytical curve

Uncertainties sources of analytical curve were considered as: Volume from volumetric flask and pipettes used to prepare secondary standards, nitric acid volume measurement and pipette repeatability. The uncertainty estimation was performed as presented at (5).

$$\mu(C_{PC}) = \left(\sqrt{\left(\frac{\mu V_p}{V_p}\right)^2 + \left(\frac{\mu V_{bv}}{V_{bv}}\right)^2 + \left(\frac{\mu C_{SE}}{C_{SE}}\right)^2}\right) x\, C_{PC} \quad (5)$$

Where $\mu(C_{PC})$ = Concentration uncertainty of each secondary standard of analytical curve (PC); μV_p = standard uncertainty of the pipetted volume of the SE (Stock Solution); V_p = Pipette volume of the SE, mL; μV_{bv} = uncertainty from the volumetric flask; V_{bv} = volumetric flask of volume, mL; μC_{SE} = standard uncertainty of SE; C_{SE} = elements concentration in the SE; C_{PC} = final concentration of the element on the spot.

2.3 Uncertainty associated with the calibration curve

The uncertainty estimation due to analytical curve, $\mu(C_o)$, was performed using the intensity data of each element measurement. [3] [5]. An example of the analytical curve used from aluminum (Al) is present in Figure 3. For calculation was used (6).

$$\mu C_0 = \frac{S}{B_1}\left(\sqrt{\frac{1}{p} + \frac{1}{n} + \frac{(C_0 - \bar{C})^2}{S_{xx}}}\right) \quad (6)$$

Where $\mu(C_O)$ = Standard uncertainty of the analytical curve at the point 0, mg.L^{-1}; S = residual standard deviation; B_1 = slope; p = number of replicate to determine Co; n = total number of sample; C_O = sample concentration; $\bar{C}$ = average concentration of the analytical calibration curve; S_{XX} = $\sum_{i=1}^{n}(C_i - \bar{C})^2$, where Ci = concentration value obtained from the calibration curve.

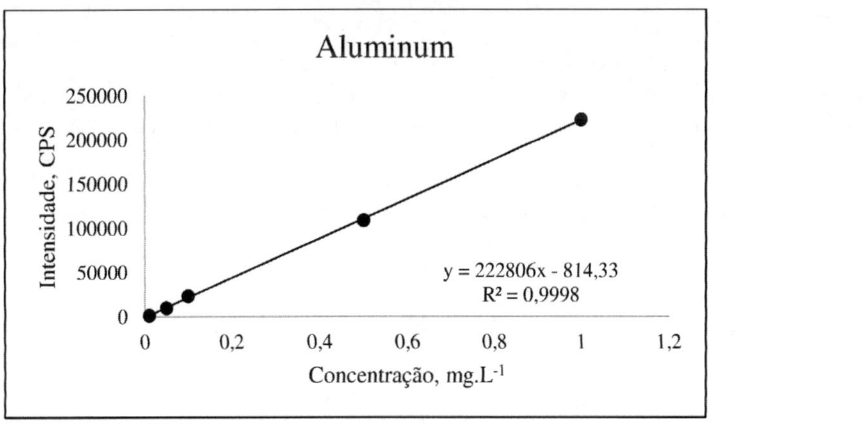

Figure 3. Al Analytical curve.

2.4 Method Repeatability Uncertainty

To estimate the uncertainty due method repeatability, it was used the relative standard deviation (RSD%) of the samples with standard addition measurement of the analytical curve intermediate point each studied element. For the calculation, was used (7).

$$\mu(Repe)=\left(\frac{Máx\ DPR}{\sqrt{n}}\right) \tag{7}$$

Where $Máx\ DPR$ = Higher deviation of replicate values; n = number of replicates.

2.5 Recovery Uncertainty

The recovery uncertainty was obtained from recovery tests performed on method validation, in which standard additions were made in seven independent samples in ten replicates for each element. Thereafter was the average of the replicates used for calculating the measurement uncertainty associated recovery according to (8).

$$\mu(Rec) = \left(\frac{\overline{Rec}}{\sqrt{s}}\right) \tag{8}$$

Where $\overline{Rec}$ = average method of recovery for the element; s = standard deviation of recovery.

2.6 Combined and expanded uncertainty

As step 4 of the EURACHEM/CITAC (2012) guide was performed combining all sources of uncertainty using (9).

$$\mu(C_s) = \left(\sqrt{\left(\frac{\mu V}{V}\right)^2 + \left(\frac{\mu C_{PC}}{C_{PC}}\right)^2 + \left(\frac{\mu C_O}{C_o}\right)^2 + \left(\frac{\mu Repe}{Rep}\right)^2 + \left(\frac{\mu Rec}{Rec}\right)^2}\right) x\ C_s \tag{9}$$

Where μV_f = standard uncertainty of volume, mL; V_f = total volume, mL; μC_{PC} = higher standard uncertainty of the points in the calibration curve, %; μC_O = standard uncertainty adjusting the calibration curve, mg.L^{-1}; Co = concentration of a point of the calibration curve, mg.L^{-1}; $\mu Repe$ = standard uncertainty of repeatability, %; μRec = standard uncertainty of recovery, %; C_S = sample concentration, mg.L^{-1}.

The relative extended uncertainty estimation was performed using an expansion factor of uncertainty k = 2, using a 95% confidence interval of the results obtained, presented in (10).

$$\mathrm{U} = \mu(C_S)\ x\ k \tag{10}$$

Where U = relative expanded uncertainty; $\mu(C_S)$ = final uncertainty of the concentration of the sample; k = expansion factor = 2

3. Results and Discussion

As an example, Aluminum expanded relative uncertainty estimation (%) was 12% and are presented in Table 2. The other elements such as Ba, Ca, Cd, Cu, Cr, Fe, K, Mg, Mn, e Ni, presented values of 3%, 3%, 5%, 13%, 4%, 5%, 15%, 8%, 4%, and 4%, respectively.

Journal of Physics: Conference Series **733** (2016) 012032 doi:10.1088/1742-6596/733/1/012032

Aluminum (Al), Ba, Ca, Cd, Cu, Fe, K, Mg, Mn and Ni showed high values of contributions to the uncertainty associated with analytical curve μ (Co), as they relate to the analyte concentration and instrument response. Some elements, such as Cu, K and Al showed high values in uncertainty associated with recovery μ (Rec) as equal to analytical curve element is related to the concentration in the sample and may have variations in concentration. Samples from Guarapiranga Dam were analyzed by ICP-OES. Data are presented in Table 3.

Table 2. Uncertainties components used to estimate Aluminum expanded uncertainty.

Source	Value (x)	μ(x)	μ(x)/x
$\mu(V_f)$, mL	50	0.061	0.00122
$\mu(C_{PC})$, %	100	1.421	0.01421
$\mu(C_o)$, mg.L^{-1}	0.098	0.003	0.02670
μ(Rep), %	100	0.017	0.00017
μ(Rec), %	98	5.194	0.05300
$\mu(C_s)$		0.006	
U (k=2)		0.012	

Table 3. Elements concentration with the uncertainty (mg.L^{-1}).

Collection points	Mg	K	Ca	Al	Cr	Mn	Fe	Ni	Cu	Cd	Ba
G000-01 S	0.61 ±0.05	1.49 ±0.22	1.70 ±0.04	0.23 ±0.014	<LQ	0.044 ±0.002	1.23 ±0.06	<LQ	<LQ	<LQ	<LQ
G000-02 S	0.74 ±0.06	1.82 ±0.27	2.47 ±0.06	0.28 ±0.017	<LQ	0.037 ±0.002	1.12 ±0.05	<LQ	<LQ	<LQ	<LD
G000-03 S	0.92 ±0.07	2.37 ±0.35	3.78 ±0.10	0.16 ±0.010	<LQ	0.029 ±0.001	0.45 ±0.02	<LQ	<LQ	<LQ	LD
G000-04 S	1.16 ±0.09	2.96 ±0.44	5.49 ±0.14	0.10 ±0.006	<LQ	0.039 ±0.002	0.21 ±0.01	<LQ	<LQ	<LQ	<LQ
G000-05 S	1.56 ±0.12	3.83 ±0.57	8.20 ±0.21	0.04 ±0.003	<LQ	0.075 ±0.003	0.17 ±0.01	<LQ	0.024 ±0.003	<LQ	<LQ
G107-06 S	1.68 ±0.13	4.13 ±0.62	9.28 ±0.24	0.09 ±0.006	<LQ	0.092 ±0.004	0.29 ±0.01	<LQ	0.023 ±0.003	<LQ	<LQ
G108-07 S	1.61 ±0.12	4.02 ±0.60	8.64 ±0.22	0.058 ±0.004	<LQ	0.094 ±0.004	0.19 ±0.01	<LQ	0.032 ±0.004	<LQ	<LQ
G000-08 S	1.63 ±0.13	4.01 ±0.60	8.75 ±0.23	0.044 ±0.003	<LQ	0.072 ±0.003	0.12 ±0.01	<LQ	0.026 ±0.003	<LQ	0.012 ±0.032
G109-09 S	1.61 ±0.12	4.02 ±0.60	8.70 ±0.22	0.043 ±0.003	<LQ	0.072 ±0.003	0.13 ±0.01	<LQ	0.021 ±0.003	<LQ	<LQ
G105-10 S	1.73 ±0.13	4.32 ±0.64	9.14 ±0.23	0.08 ±0.003	<LQ	0.081 ±0.003	0.24 ±0.01	<LQ	0.016 ±0.002	<LQ	0.01 ±0.03
G104-11 S	1.68 ±0.13	4.20 ±0.61	9.01 ±0.23	0.10 ±0.002	<LQ	0.047 ±0.002	<LQ	<LQ	0.038 ±0.005	<LQ	<LQ
G103-12 S	1.64 ±0.13	4.23 ±0.63	8.78 ±0.23	0.016 ±0.002	<LQ	0.055 ±0.002	0.17 ±0.01	<LQ	0.035 ±0.005	<LQ	0.01 ±0.03
G102-13 S	1.60 ±0.12	4.06 ±0.61	8.50 ±0.22	0.062 ±0.002	<LQ	0.046 ±0.002	<LQ	<LQ	0.039 ±0.005	<LQ	<LQ
G000-14 S	1.59 ±0.12	4.03 ±0.60	8.52 ±0.22	0.25 ±0.002	<LQ	0.050 ±0.002	0.13 ±0.01	<LQ	0.041 ±0.005	<LQ	<LQ

Elements such as Fe, Al and Cu showed values above CONAMA Resolution 357/2005. Fe is related to geological characteristic of the area; Al in some points is associated with aluminum sulfate also used as flocculant in the dam, in water treatment, once Guarapiranga is used as one main source of water public supply; and Cu is used as copper sulfate, to prevent algae blooms that are frequent in the region.

Other evaluated elements such as Cd, Cr Ba, Mn, Mg, K and Ni are present below permitted values by Brazilian regulation and were considered satisfactory.

4. Conclusion

By using method development and validation procedures it was possible to identify the most significant uncertainties sources and with EURACHEM/CITAC guide (2012) it was possible to estimate expanded uncertainties of all elements measures in water samples collected at Guarapiranga Dam and compare its values with appropriated regulation.

References

[1] VIM INMETRO 2012 *Instituto Nacional de Metrologia, Qualidade e Tecnologia (*Brasil: Duque de Caxias)

[2] INMETRO/CICMA/SEPIN 2012 *Avaliação de dados de medição: Guia para a expressão de incerteza de medição – GUM 2008* (Brasil)

[3] EURACHEM/CITAC Guide 2012 *Quantifying Uncertainty in Analytical Measurentent* (United Kingdom)

[4] BRASIL *CONAMA 357/2005 17 Março 2005* [Online] Available: http://www.mma.gov.br/port/conama/res/res05/res35705.pdf. [Acesso em 15 Janeiro 2013].

[5] H. M. SHIHOMATSU 2015 *Desenvolvimento e validação de metodologia SPE-LC-MS/MS para a determinação de fármacos e droga de abuso nas águas da represa Guarapiranga-São Paulo/SP* (São Paulo / Instituto de Pesquisas Energéticas e Nucleares)

[6] AGÊNCIA NACIONAL DAS ÁGUAS and CIA SANEAMENTO DO ESTADO DE SÃO PAULO 2011 *Guia nacional de coleta e preservação de amostras: água, sedimento, comunidades aquáticas e efluentes líquidos* (Brasil)

[7] American Public Health Association (APHA), American Water Works Association (AWWA), and Water Environment Federation (WEF) 2005 *Standard Methods for the Examination of Water and Wastewater 21th Edition* (Baltimore Maryland: United Book Press Inc)

[8] H. A. FURUSAWA 2007 *Estimativa de incerteza em análise química* (Brasil: Instituto de Pesquisas Energéticas e Nucleares)

The importance of metrological metadata in the environmental monitoring

Márcio A A Santana[1], Patrícia L O Guimarães[1], Eugênio S Almeida[2] and Tero Eklin[3]

[1] Environmental Metrology, Meteorological Instrumentation Laboratory, Center for Weather Forecasting and Climate Research, National Institute for Space Research, Rodovia Presidente Dutra km 39, Cachoeira Paulista, SP, Brazil
[2] Supercomputing Group, Center for Weather Forecasting and Climate Research, National Institute for Space Research, Rodovia Presidente Dutra km 39, Cachoeira Paulista, SP, Brazil
[3] Laboratory Centre, Finnish Environment Institute, Hakuninmaantie 6, 00430 Helsinki, Finland

E-mail: marcio.santana@cptec.inpe.br, profmarciosantana@hotmail.com

Abstract. The metrological metadata propagation contributes significantly to improve the data analysis of the meteorological observation systems. An overview of the scenarios data and metadata treatment in environmental monitoring is presented in this article. We also discussed the ways of use of the calibration results on the meteorological measurement systems as well as the convergence of the methods used in the corrections treatment and estimation of the measuring uncertainty in metrological and meteorological areas.

1. Introduction

The measurement results of the instruments used in observation systems reveal important features of the environment. The environmental data are often uncertain and require additional information that provides the context of who, what, where, how and why these measurements were performed [1, 2].

Metadata is additional information about the measurement that improves the results quality [2]. Beyond the geographic coordinates and meteorological measurements, users of meteorological observations require information about the type of instrumentation used. They need details about the installation, maintenance and historical calibration, to outline scientific conclusions and decision-making [3]. Efforts to collect, store and process the metadata should be implemented. They improve the reliability of measurement results by detecting problems, analyzing the data quality and verifying the compatibility between data sets.

In this article we denominate metrological metadata the additional information related to the science of measurements applied to meteorological instrumentation.

2. Monitoring and measurement

The climate change has become the greatest threat to humanity and the environment. Meteorology, and especially the climate community are support instruments in this scenario to minimize the impacts [4]. On the other hand, they require observation systems and reliable information.

The meteorological instrumentation used in observation systems follows the technological innovations. However, many manufacturers still do not show the traceability patterns used for calibration. Instead, they execute only one performance test, since they consider their sensors and meters a remarkable scientific equipment. Another important information that should compose the metadata database is the experimental site classification for each metrological variable. The classification takes into account topography, land cover, obstacles distance, artificial heat sources, reflective surfaces, distance from water sources, shading in relation to the solar elevation, distance relationship and obstacle height, protection devices (shield), latitude, etc. Depending on the classification, it is recommended the addition of significant uncertainties, such as ± 5° C for the air temperature, > ± 25% for rain and > ± 50% for wind [3].

For example, if rainfall measurements are carried out for an extended period it is likely that local conditions have changed, such as vegetation cover (growth and appearance of trees), construction works and relocation of the meter. It can also occur meter replacement due to maintenance, recalibration or updates. These changes affect measurements. In the case of climate studies, the changes must be analyzed and approved carefully, and the metadata must be updated [5].

Having in place observational networks that can take and record regular climate observations is a necessary, but not sufficient, condition for ensuring the climate record can adequately support climate monitoring and service provision. The data must also be properly quality-controlled, archived and easily accessible [4].

In many countries, great part of the environmental data is inadequately treated alone. As a consequence, the degree of reliability is unknown [6]. A common protocol has not yet been established between Metrology, Instrumentation and Meteorology areas.

3. Metrology and environmental monitoring

The greatest challenge is making important decisions quickly from systems that contain huge variety of data decisions, especially in strategic sectors such as health, civil defense, energy and the environment. The consequences of inadequate decision-making can lead to economic problems, legal implications and negative environmental impacts, and have as the direct cause the data quality.

Figure 1 illustrates the environmental monitoring scenarios: with and without the use of metadata. In the first scenario (in color), the information is based on reliable data guided by measurement results with traceability to the International System of Units (SI) and to quality control, leading to truly wise decision-making.

In the second scenario (without color), there is no metadata or they are insufficient or poor. It leads to decision-making that cause decline in quality of life and economic losses. On the other hand, the lack of measurements traceability leads to nowhere.

Metrology is a horizontal scientific-technical field, which underpins almost all subject fields in natural sciences and engineering [7]. The Environmental Metrology is considered an emerging multidisciplinary field and this interaction with other areas of human knowledge causes major impact on improving the quality of life. The propagation of information inherent to the metrological activities together with the meteorological instrumentation underpins the quality of the results of environmental monitoring. An adequate metrological infrastructure in the environmental area is essential to the development of the entire global economy [7]. In practice, calibration and periodic verification activities are the basis of quality. All data becomes questionable without metrological traceability.

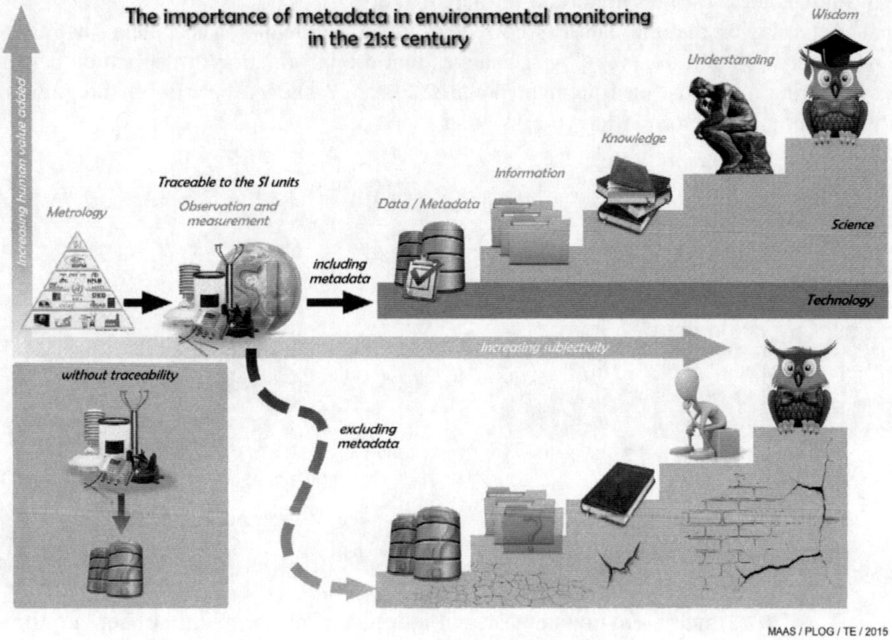

Figure 1. Ladder of knowledge – scenarios

4. Metrological traceability

The modern society requires reliable measurements that give the same answer wherever they are made. This is achieved by ensuring traceability to the SI [7]. The meteorological traceability in the **Environmental** Metrology can be represented by figure 2.

The smallest pyramid in the center represents the Regional Instrument Centres (RIC) from the World Meteorological Organization (WMO), a BIPM member since April 2010. They were established aiming to perform the meteorological instruments calibration to address the growing need for meteorological and hydrological data with high quality, the meteorological instruments standardization and international comparisons [3, 8, 9].

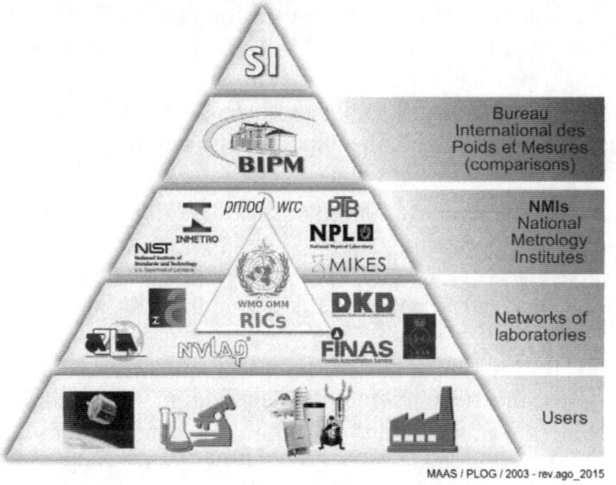

Figure 2. Traceability diagram applied to the Environmental Metrology area.

5. Use of metrological data as metadata in meteorology

The final user believes that the data has corrected, the instrumentation technician believes that raw data is post processed and the metrologist believes that the information of calibration certificates are propagated during all of the measurement process. Figure 3 shows the lack of integration between Metrology, Instrumentation and Meteorology areas.

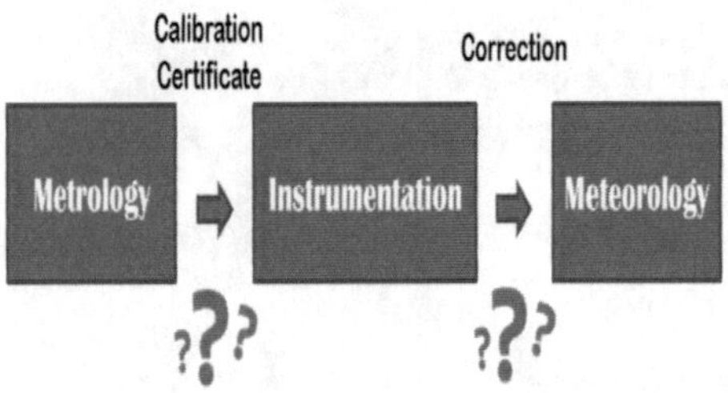

Figure 3. The three environmental monitoring knowledge areas.

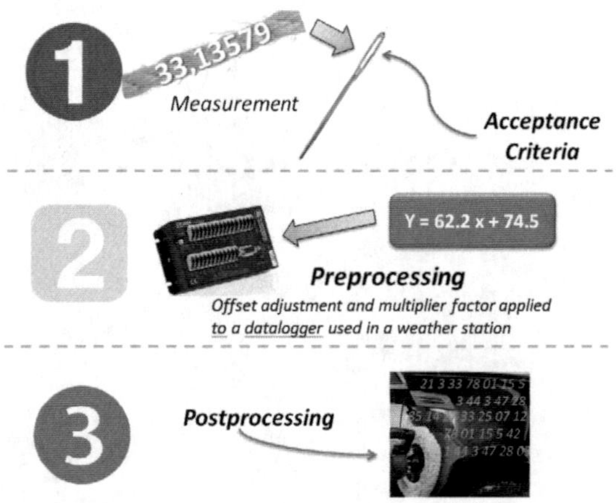

Figure 4. Preprocessing, offset adjustment and processing.

Many times the environmental data are initially qualified but not systematically corrected. Because of a misinterpretation, many people understand that the calibrated instrument do not have or do not require a correction factor.

Normally, the qualification data consists of validation tests to verify if the georeferenced environmental data are within maximum and minimum range, and if there are temporal and spatial consistency. As a result, the absence of a metadata of the calibration results leads to the lack of knowledge of the reliability degree of measurements.

The use of metadata related to the instrumentation calibration can occur in three different situations: before, during or after the collection of the environmental data. The first case uses the Acceptance Criteria, which are the values that serve as limits to prove that the instrument is suitable for use. The Acceptance Criteria is defined before the calibration and results analyzed to verify if they fit within a desirable range. Normally, it uses a maximum error (E_{MAX}) where the sum of the error module or combined correction of a measurement system (C) with the module of the uncertainty associated (U) must be less than or equal to the maximum permissible value, as in equation (1).

$$E_{MAX} \geq |C| + |U| \qquad (1)$$

In the second case (pre-processing), a correction equation or calibration factor is introduced in the data acquisition system of the weather station (Figure 4). Then, the corrected data are inserted on the database server.

Two situations occurs in this case: (1) data acquisition systems without features like an offset adjustment and/or a multiplier factor to correct the data, and (2) data acquisition systems with these features but requires technicians on the field to configure the systems individually for each calibration. The uncertainty values U should be treated later. In the third case (post-processing), the Calibration Certificate information (Table 1) is treated as metadata, and the data can be corrected and qualified.

Table 1. Calibration certificate (partial data).

Nominal Value [°C]	Conventional value, T_{ITS-90} [°C]	Measured value T_{IND} [°C]	Correction T_{ITS-90} - T_{IND}, C [°C]	Expanded measurement uncertainty, U [°C]	Coverage factor @ 95.45%, k
25.0	25.1	25.2	0.1	± 0.2	2

Suggestions of meteorological data correction procedures can be obtained from the Guide of Meteorological Instrumentation. An example is the case of precipitation measurements that are significantly affected by exposure conditions, wind and topography. In this case, the metadata are particularly important for users of the data, because the metadata describes the circumstances of the measures [3].

With the use of the metadata (C and U) which may correspond to systematic and random errors the measurement of the variable (M) obtained by an instrument or measurement system becomes a measurement range (RM), according with equation (2). In the case of meteorological observations, more than one source of uncertainty has significant contribution in the measurement process.

$$RM = M + C + U \tag{2}$$

The combined correction (C) is the correction value to be applied to the measurement process and is the result of simultaneous action of systematic contributions from all sources of uncertainty [10]. The expanded measurement uncertainty (U) is estimated by combined standard uncertainty (u_c), which is the set of uncertainties type A with type B [11].

The methodologies used to estimate the measurement uncertainty on metrology area started in recent years in the meteorology area, in particular the *Guide to the expression of uncertainty in measurement* [11].

6. Discussion and conclusion

We conclude that the results from the measurements by themselves rarely are sufficient to help answer questions about meteorological observations. Therefore the metadata becomes as important as the meteorological data.

Systems or environmental monitoring networks require investment, as well as current information processing systems. So, to achieve success in this investment, the quality and reliability must be guaranteed through the use of the metrological metadata in the area of meteorology.

Workshops, seminars and conferences that brings together meteorology, instrumentation and metrology professionals should be encouraged in order to accelerate the integration process between areas, to improve the computational models and to get improvement on the information reliability level.

References

[1] Günther O 1997 Environmental information systems *ACM SIGMOD Record* **26**(1) 3-4

[2] OTT Hydromet. The importance of metadata in water quality monitoring programs [Internet]. Kempten (Germany): OTT Hydromet2015 Feb 18 [cited 2015 Nov 29]. Available from: http://www.ott.com/blog/2015/02/water-quality-monitoring-programs/

[3] World Meteorological Organization 2008 (updated in 2010) *Guide to eteorological Instruments and Methods of Observation* Seventh edition WMO-No. 8 Geneva http://www.wmo.int/pages/prog/www/IMOP/CIMO-Guide.html

[4] Wright W 2008 Observing the climate-Challenges for the 21st century *Bulletin of the WMO*, **57**(1), 29-34

[5] Strangeways I 2010 *Precipitation: theory, measurement and distribution* Cambridge Univ. Press

[6] Trenberth K E 2008 Observational needs for climate prediction and adaptation *Bulletin of the WMO* **57**(1) 17-21

[7] EURAMET/European Union 2008 *European metrology research programme (EMRP)* implementing Metrology in the European Research Area (iMERA)

[8] BIPM, IEC, IFCC, ILAC, ISO, IUPAC, IUPAP and OIML 2012 *International Vocabulary of Metrology - Basic and General Concepts and Associated Terms, 3rd edition* JCGM 200:2012 (JCGM 200:2008 with minor corrections)

[9] Associação Brasileira de Normas Técnicas 2005 *ABNT NBR ISO/IEC 17025: 2005-requisitos gerais para a competência de laboratórios de ensaio e calibração. General requirements for the competence of testing and calibration laboratories* ABNT

[10] Junior A A Gand Sousa A R 2008 *Fundamentos de Metrologia científica e industrial*Manole

[11] BIPM, IEC, IFCC, ILAC, ISO, IUPAC, IUPAP and OIML 2008 *Evaluation of measurement data - Guide to the expression of uncertainty in measurement (GUM)* JCGM 100:2008 (GUM 1995 with minor corrections)

Surface evaluation of orthopedic hip implants marketed in Brazil

M M Souza [1], R M Trommer [2], M M Maru [2], C R M Roesler [3], W S Barros [1] and M S Dutra [3]

[1] Dimensional Metrology Laboratory, Mechanical Division Inmetro. Brazil
[2] Materials Metrology Division, Inmetro. Brazil
[3] BioMechanical Engineering Laboratory, Mechanical Engineering Department. UFSC, Brazil
[4] Mechanical Engineering Program. UFRJ, Brazil

mmsouza@inmetro.gov.br

Abstract. One of the factors that contribute to the quality of total hip prostheses is the degree of accuracy in the manufacturing of the joint surfaces. The dimensional control of joint components is important because of its direct influence on the durability and, consequently, in the patients' life quality. This work presents studies on the form and roughness of orthopedic hip prostheses marketed in Brazil. The results provide data for quality control of the surfaces of the femoral heads and acetabular components of hip prostheses and indicate the need of improvement in the procedures used to this control.

1. Introduction

The technology of orthopedic implants allowed significant improvement in the life of patients suffering from muscle-skeletal disorders. However, the failure of implants in clinical use is not rare, and the poor quality of products, this concerning several requirements, is regarded as one of the major causes of adverse events. In order to ensure the safety and efficacy of implants, manufacturers of medical devices have to meet biomechanical requirements related to the biological, physical, chemical and tribological properties. In parallel, suitable geometric design, selection of materials and manufacturing are important factors to obtain a final product which fulfills such requirements [1].

Regarding the geometrical features of surface metrology, the past half century has been extremely rich in events related to equipment design, data treatment and development of standards. Consequently, the manufacturing and metrology of orthopedic hip implants have been also improved.

There are two essential areas of metrology that guide the evaluation of orthopedic implant design: geometry measurement and surface finish assessment. Both techniques are used as quality monitoring methods of the manufacturing process, and are useful tools in the assessment of the performance of orthopedic implants. Geometrical metrology encompasses techniques such as the assessment of form by sphericity measurement in submicron resolution. In contrast, surface texture assessment is focused on nanometer measurement [2].

Particularly in Brazil, the Brazilian Health Surveillance Agency (Anvisa) is the official agency responsible for regulating the orthopedic implants trading. Additionally, the Brazilian government has been investing in several projects aiming to control the quality of orthopedic hip prostheses produced by national and international manufacturers. Another important point to be considered is the fact that the Brazilian population is currently 195 million, of whom 10% are over 60 years old (i.e., around 20 million). It is estimated that in 2050 the Brazilian population will be approximately 215 million, with 30% over 60 years old; that is, over 64 million people [3]. The increase of elderly population will cause an increase in chronic and degenerative diseases, surgical requirements, and thus, increase in orthopedic hip prosthesis surgeries is therefore predicted.

One of the main clinical problems limiting the functional life of a total hip implant is the aseptic loosening, which is related to the osteolysis resulting from the reaction of the host tissue with the particles (debris) produced by wear during the sliding contact between the femoral head and the polyethylene liner [4, 5]. Decreasing the generation of debris is a challenging objective in the design and selection of materials of a hip implant. To reach this goal, sphericity and roughness of both femoral head and acetabular liner are key factors, since these parameters can potentially affect the wear behavior. In the case of polymer bearing surfaces running against smooth counterfaces, the resulting polymer wear will involve adhesion wear mechanism [6, 7]. On the other hand, with high roughness counterfaces, the asperities will provoke deeper deformations in the polymer, and the wear will result from polymer abrasion and plastic deformation, as well as severe damages by fatigue mechanism. Thus, the counterface roughness characteristics play an important role regarding the transition from smooth to severe wear mechanisms and the associated wear rate.

Currently in Brazil, the measurement of the dimensional parameters of a hip implant is usually performed by the manufacturer, based on the procedure recommended in ABNT NBR ISO 7206-2 standard [8].

This work reports the findings of measurements carried out on national and imported acetabular liners and femoral heads with respect to their surface roughness and sphericity.

2. Material and Methods

2.1. Hip implant specimens

The specimens studied in this work were: 03 national femoral heads (NFH) and 02 imported metallic femoral heads (IFH), both made of stainless steel in accordance to ASTM F138 standard [9], 03 national acetabular liners (NAL) and 03 imported acetabular liners (IAL), all made of ultra-high molecular weight polyethylene (UHMWPE) following the ASTM F648 standard [10]. In the field of orthopedics, the UHMWPE material is still considered as the gold standard for articulating surface for total joints [11].

2.2. CMM and Roughness equipment

Sphericity measurements were carried out on a calibrated Coordinate Measurement Machine (CMM), based on the ABNT NBR ISO 7206-2 [8] standard recommendations. A total of 25 data points were taken for all hip components, being 8 points equally spaced along a circumference in 3 planes (one in the equatorial plane and the others, dislocated in 30° and 60° from the polar axis) and a point on the pole.

The sphericity was calculated using the 25 points measured, according to equation (1):

$$Sd = \text{distance } OP - 0{,}5D \tag{1}$$

Where "Sd" is the sphericity deviation, "D" the mean diameter, "O" the femoral head center and "P" the measured points.

Roughness measurements were performed through a calibrated surface profilometer with 0.8 nm height resolution and 8mm height measurement range. Three measurements were taken for each femoral head, in two orthogonal directions. The Ra roughness parameter was obtained with cut-off value of 0.08 mm, Gaussian filter and least square arc fitting over 15 mm length. For the UHMWPE liners, the same measurement procedure was applied, but at 4 quadrants, 0°, 90°, 180° and 270°.

The expanded uncertainty values (U) of the measurements were determined according to the equation (2) [12].

$$u^2(z_g) = u^2(Pt_n) + u^2(b) + u^2(z_0) \quad (2)$$

Where $u^2(z_g)$ is the overall profile uncertainty, $u^2(Pt_n)$ the uncertainty of reference standard, $u^2(b)$ repeatability of tracing profile measurements and $u^2(z_0)$ the uncertainty due to background noise of the measurement device

3. Results

The results of sphericity deviation for the national femoral heads and UHMWPE liners are shown in Table 1.

Table 1. Sphericity deviation values for the NFH and NAL components.

	NFH	NAL
Sd (μm)		
1	30.65	91.61
2	27.47	81.02
3	17.73	86.13
U	1.40	1.50

According to the requirements described in the ABNT NBR ISO 7206-2 standard [8], the maximum value for the sphericity deviation of the femoral head is 10 μm. It can be noted from the data in table 1 that all of the national femoral heads (NFH) do not meet the requirements of the standard. This was associated to the existence of a deep valley in the pole of the femoral heads, detected in the measurements done.

Concerning the UHMWPE liners (NAL), the ABNT NBR ISO 7206-2 standard [8] describes that the sphericity deviation shall not be greater than 100 μm. By observing the data presented in table 1, it is concluded that all measured specimens fulfill this requirement.

The roughness results of the NFH and IFH are shown in table 2.

Table 2. Roughness values of the NFH and IFH specimens.

	NFH	IFH
cut-off (mm)	0.08	0.08
Ra (μm)		
1	0.006	0.005
2	0.007	0.005
3	0.006	-
U	0.005	0.005

Based on the requirements established in the ABNT NBR ISO 7206-2 [8] standard, which specifies that the Ra roughness value of femoral heads shall not exceed 0.05 μm, it is seen that all the NFH and IFH specimens were in agreement with the criteria according to the ABNT NBR ISO 7206-2 [8].

A discussion should be made concerning the procedure taken for obtaining the roughness parameters for the acetabular liner specimens (NAL and IAL). The ABNT NBR ISO 7206-2 standard [8] recommends 0.08 mm cut-off value, since most of the measurement equipment applies this cut-off.

In opposite, ISO 4288 standard [12] describes that 0.8 mm cut-off shall be used if the Ra values are between 0.1 and 2 µm.

The results obtained for the roughness of the acetabular liners, using 0.08 mm and 0.8 mm cut-off, are shown in table 3.

Table 3. Roughness values of the IAL and NAL specimens.

	IAL		NAL	
cut-off (mm)	0.08	0.8	0.08	0.8
Ra (µm)				
1	0.362	0.968	1.267	2.506
2	0.320	0.652	1.373	2.861
3	-	0.646	1.112	3.012
U	0.018	0.030	0.018	0.030

According to the ABNT NBR ISO 7206-2 standard [8], the Ra value for UHMWPE acetabular liners shall be equal or lower than 2 µm. Applying 0.08 mm cut-off, it can be seen that all of the IAL and NAL specimens were approved in terms of roughness measurements.

However, carrying out the analysis based on the ISO 4288 [13] with 0.8 mm cut-off, the NAL do not comply the ABNT NBR ISO 7206-2 requirements [8], because the Ra values exceed the maximum accepted criterion of 2 µm. In opposite, the Ra values for the IAL specimens did not exceed 1.00 µm with the same 0.8 mm cut-off, being them even so approved.

The above discussion suggested that the cut-off value recommended in the ABNT NBR ISO 7206-2 standard [8] is not suitable for adequate control of the surface roughness quality of the acetabular liners. Therefore, it is important that manufacturers in Brazil carefully adopt the method with the larger cut-off value of 0.8 µm for the Ra roughness evaluation; as it is more conservative. Also, it must be observed that it do reveal the existing high roughness in the liner surface that can lead to early failures, especially with respect to wear, because the roughness values can be 25 to 50% higher than those considered suitable [14].

4. Conclusions

The results of the measurements of the national femoral head indicate that all of them did not fulfill the criteria for sphericity according to the ABNT NBR ISO 7206-2 standard.

Based on the roughness measurements using 0.08 mm cut-off, all the femoral heads and acetabular liners were approved.

Nevertheless, using 0.8 mm cut-off for the roughness analyses, all the national acetabular liners would be rejected due to the values be higher than the limit established in ABNT NBR ISO 7206-2 standard. On the other hand, all of the imported acetabular liners were still in conformity despite the use of a larger cut-off value.

References

[1] Mathia T G, Pawlus P, Wieczorowski M Recent trends in surface metrology, Wear 271 494–508, 2011.

[2] Brown L T, The use of 3d Surface Analysis Techniques to Investigate the Wear of Matt Surface Finish Femoral Stems In Total Hip Replacement, Ph.D. Thesis, University of Huddersfield, Huddersfield, UK, 2006.

[3] IBGE, 2011 "Instituto Brasileiro de Geografia e Estatística". Available from www.ibge.gov.br. Accessed in: 19 March. 2011.

[4] Wright T, Goodman S Implant wear in total joint replacement, AAOS, 2001.

[5] Maal, N A et al. The Incidence of Acetabular Osteolysis in Young Patients with Conventional versus Highly Crosslinked Polyethylene, Clin Orthop Relat Res 469:372–381, 2010.

[6] Hutchings I M: *Tribology, Friction and Wear of Engineering Materials*, Bodmin, Cornwall, 1992.

[7] Wang A et al. *Orientation Softening as a Mechanism of Ultra-High Molecular Weight Polyethylene (UHMWPE) Wear in Artificial Hip and Knee Joints*, 1998.

[8] ABNT NBR ISO 7206-2 Implantes para cirurgia – Próteses parcial e total de articulação de quadril. Parte 2: Superfícies de articulação feitas de materiais metálico e plástico. Set 1999.

[9] ASTM F138 Standard Specification for Wrought 18 Chromium-14 Nickel-2.5 Molybdenum Stainless Steel Bar and Wire for Surgical Implants (UNS S31673), 2008.

[10] ASTM F648 - 10a Standard Specification for Ultra-High-Molecular-Weight Polyethylene Powder and Fabricated Form for Surgical Implants, 2010.

[11] Kurtz, S UHMWPE Biomaterials Handbook. 2nd ed. Burlington, MA: Academic Press; 2009.

[12] BIPM, IEC, IFCC, ISO, IUPAC, IUPAP and OIML (1995) Evaluation of measurement data—guide to the expression of uncertainty in measurement JCGM 100:2008. BIPM, Sèvres.

[13] ISO 4288:1996 Geometrical Product Specifications (GPS) – Surface Texture: Profile method – Rules and procedures for the assessment of surface texture, 1996.

[14] Souza M M, Estudo da Forma e da Rugosidade em Próteses Ortopédicas de Quadril, D.Sc. Thesis, UFRJ, RJ, Brazil, 2011.

Development of benzene, toluene, ethylbenzene and xylenes certified gaseous reference materials

M C Brum [1], D C G Sobrinho [1], F A Fagundes [1], R J Oudwater[1], C R Augusto[1]
1 Divisão de Metrologia Química (Dquim), Instituto Nacional de Metrologia, Qualidade e Tecnologia, (Inmetro), 25250-020, Xerém-RJ, Brasil

mariana.brum@pq.cnpq.br

Abstract. The work describes the production of certified gaseous reference materials of benzene, toluene, ethylbenzene and xylenes (BTEX) in nitrogen from the gravimetric production up to the long term stability tests followed by the certifying step. The uncertainty in the amount fractions of the compounds in these mixtures was approximately 4 % (relative) for the range studied from 2 to 16 µmol/mol. Also the adsorption of the BTEX on the cylinder surface and the tubing were investigated as potential uncertainty source.

1. Introduction

Benzene, toluene, ethylbenzene, m-xylene, p-xylene and orto-xylene (BTEX) compounds can be released to the atmosphere through emissions from motor vehicles, cigarette smoke and oil and gas emissions. BTEX compounds are generated during the processing of petroleum products and also when paints, thinners, adhesives are produced. These compounds can cause damage to the health, since they may lead to lethal diseases such as lung cancer and therefore are classified as carcinogenic. They are monitored in the atmosphere though the World Meteorological Organization, Global Atmospheric Watch Program (WMO-GAW) [1]. These volatile organic compounds (VOCs) components need to be quantified in a high confidence level and the traceability to the International System of Units (SI) has to be assured. Therefore, the use of certified reference material is necessary for this purpose. These certified gaseous reference materials (CRM) are commercialized within various concentrations from some National Metrology Institutes (NMI) and companies with this type of expertise since they are indispensable in the analysis of gaseous media ranges [2,3,4].

BTEX in nitrogen reference materials are usually produced for environmental purposes at ppb range but there is also a demand for higher concentration standards, at levels that can be used in the case of occupational exposure.

The Brazilian NMI, National Institute of Metrology, Quality and Technology (Inmetro), has the knowledge and facilities on producing certified gas mixtures, both binary and multicomponent CRM [5]. However, for BTEX there is no gaseous reference material produced in Inmetro till now and since there is a demand for this CRM this study is very promising and will also help to reduce the importation costs.

These reference materials are usually commercialized in cylinders that have a special surface treatment to minimize wall adsorption. However, recent studies show that depending on the treatment used for the internal cylinder wall this can influence the final concentration of the mixture and has to be taken into account [6].

In the present work the development of a BTEX/N_2 certified reference material is presented and also additional testing to determine the magnitude of the losses due to adsorption.

2. Experimental

Before the gravimetric preparation of the mixture, the cylinder is cleaned by evacuation in a Pfeiffer turbo molecular pumping station followed by filling with high purity nitrogen (N_2) 99.9999 % mol/mol. New 5L aluminium gas cylinders treated with a proprietary process Aculife IV, by Scott Specialty Gases, were used. Before the actual preparation the clean cylinder was evacuated overnight. Afterwards, the empty cylinder is weighed in a mass comparator, Mettler balance, where it is compared against an identical (reference) cylinder. Then the pre-mixture with the BTEX components is transferred via a filling station to the empty cylinder. This process of weighing and filling is repeated for the matrix, N_2. Once the mixture is produced the final composition is calculated by using the added masses of each component that is expressed as mole fraction µmol/mol (ppm) according to ISO 6142 standard [7].

The BTEX in nitrogen mixtures (BTEX/ N_2) were prepared in the range of 2 to 16 µmol/mol from the dilution of a 50 µmol/mol BTEX/ N_2 certified reference material purchased from NPL (National Physical Laboratory). Finally, after the gravimetric concentration of the mixtures produced were determined, the value was validated by comparing it with other primary standards (5,10 and 15 µmol/mol), also purchased from NPL in accordance with the ISO 6143 standard [8]. The analytical method employed was Gas Chromatography by using a Variant GC-SP3800 equipment with a CP Sil-5 CB column, FID, helium carrier and with the following oven program:temperature: 60 °C, hold 1 min, 10 °C/min to 140 °C .

Afterwards, the long term stability studies were performed with five measurements during a period of 15 months after the production of the mixtures. After all the criteria are satisfied the mixtures can be certified. Finally, the certificate value is the gravimetric value and the uncertainty is the contribution of gravimetry, stability and analytical uncertainty.

For the additional tests that evaluate the influence of the cylinder surface treatment on the adsorption of BTEX compounds, a 6 µmol/mol BTEX/ N_2 was transferred to an evacuated cylinder. Gas transferring was performed twice and the mixtures were analyzed the next day, after being homogenized.

3. Results and Discussion

The gravimetric preparation data with the respective uncertainty is obtained by the Gas Metrology software and afterwards the analytical verification takes place to confirm these results.

In order to verify the mixtures produced, a lot of effort was done to optimize the analytical method since ethylbenzene and xylenes are challenging components.

The stability tests were performed with a 5 µmol/mol NPL standard along with the reference material candidates 2,4,6,10 and 16 µmol/mol showed that the difference between the NPL certificate value and the analytical one obtained for the standard is lower for benzene and toluene, while for the other components it was higher than 3% (table 1).

Table 1. The difference between the certificate value and the analytical value for a 5 µmol/mol standard used in the stability tests.

Component	Δ(%)
Benzene	-0.49
Toluene	0.11
Ethylbenzene	3.32
Meta + para - xylene	3.03
Orto-xylene	3.45

The reference material candidates are produced from the dilution of a 50 ± 1.5 μmol/mol standard performed in a filling station, so there could be some adsorption taking place in the line while transferring the gas mixture. In order to minimize this effect the use of extra heating of the line could help, since the pressure involved is high and some condensation could be happening due to the presence of needle valves that are used to reduce the flow, so the exact amount of gas can be introduced precisely into the cylinder. Despite the results observed in table 1, that show that improvement still has to be done, the final results for BTEX/N_2 certified reference materials were satisfactory since the verification and the stability criteria were attended. The final results for the uncertainty are shown on table 2, for gravimetry, verification and stability. Afterwards, the uncertainties are combined and will be presented on the certificate along with the corresponding gravimetric concentration for each component of the mixture. In the case of meta and para xylene, since the peaks cannot be separated with this type of column, they are shown together.

The certificate values of the BTEX/N_2 certified reference material are shown in table 3 and considering that the standard that originated this mixture has a 3% uncertainty the results show that the methodology applied is satisfactory and are in accordance with the gas standards [7,8].

Table 2. Uncertainty values of the BTEX/N_2 certified reference material.

Component	Uncertainty		
	Gravimetry (μmol/mol)	Verification (μmol/mol)	Stability (μmol/mol)
Benzene	0.089	0.065	0.009
Toluene	0.089	0.060	0.013
Ethylbenzene	0.089	0.058	0.007
Meta + para - xylene	0.130	0.158	0.008
Orto-xylene	0.089	0.058	0.008

Table 3. The certificate value of the BTEX/N_2 certified reference material.

Component	Certificate value	
	Gravimetric concentration (μmol/mol)	Combined uncertainty, k=2(μmol/mol)
Benzene	6.01	0.22
Toluene	5.74	0.22
Ethylbenzene	5.80	0.21
Meta + para - xylene	12.21	0.41
Orto-xylene	5.84	0.21
Nitrogen	Balance	

Additional studies were performed to check if the surface treatment of the cylinder could have any influence on the final concentration of the BTEX/N_2 gas mixture. Transferring mixtures from a cylinder to another is a test that is reported by the recent literature especially for low concentrations [6]. The tests performed in this present work show that the BTEX loss averaged 0.4% for all the components after two transferring steps were done in cylinders that have never been used previously (figure 1).

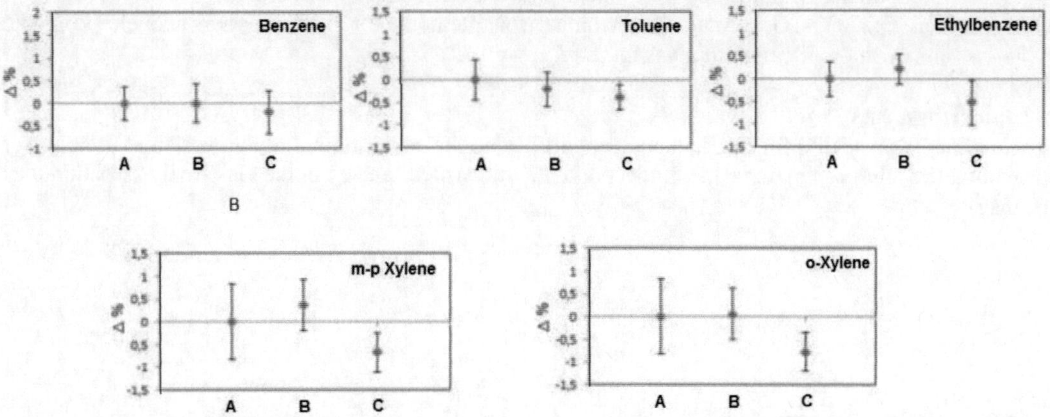

Figure 1.Plot of the delta obtained for the BTEX/N_2, 6 µmol/mol mixture after the transferring steps in the same type of aluminum cylinders, A, B and C (the bars are the relative standard deviation of the analysis).

Since the pressure value of the cylinders are decreasing after each transfer this could influence the results and future tests will be conducted to evaluate this influence before taking into account on the uncertainty budget.

4. Conclusions

The BTEX in nitrogen mixtures were produced and compared to the ones purchased from another NMI. The materials were certified obeying all the requirements.

Preliminary adsorption tests showed a 0.4%. loss of BTEX that can be influenced by the adsorption on the cylinder wall, on the tubing and valves during transferring and by pressure inside the cylinder. Future work is still needed to elucidate this effect and perhaps consider inserting it in the final uncertainty budget. Also studying other sources of loss and testing cylinders from gas vendors are needed.

References

[1] World Meteorological Organization (WMO) GAW Report No. 171: A WMO/GAW Expert Workshop on Global Long-Term Measurements of Volatile Organic Compounds, 2007.

[2] Słomin´ska M, Konieczka P, Namies´nik J 2010 Standard gas mixtures - indispensable reference materials in the analysis of gaseous media Trends in Analytical Chemistry 29 5 419-429.

[3] Grenfell R J P, Milton M J T, Harling A M, Vargha G M Brookes, C Quincey PG and Woods P T 2010 Standard mixtures of ambient volatile organic compounds in synthetic and whole air with stable reference values *Journal of Geophysical Research* **115** 1-16.

[4] Milton M J T, Vargha G M and Brown A S 2011 Gravimetric methods for the preparation of standard gas mixtures *Metrologia* **48** 121-129.

[5] van der Veen A M H *et al.* 2011, Supplementary comparison EURAMET.QM-S4: automotive gas mixtures *Metrologia* **48** Issue 1A, 08016.

[6] Miller W R, Rhoderick G C and Guenther F R 2015 Investigating Adsorption/Desorption of Carbon Dioxide in Aluminum Compressed Gas Cylinders *Anal Chem* **87** 1957-1962.

[7] ISO 6142:2001 - Gas analysis - Preparation of calibration gas mixtures - Gravimetric method,

International Organization for Standardization, Geneva, Switzerland.
[8] ISO 6143:2001 - Gas Analysis - Comparison methods for determination and checking the composition of calibration gas mixture.

Acknowledgments
The authors thank CNPq for the financial support during the development of the work and Inmetro for providing the reference materials, cylinders and the infrastructure used in the Gas Analysis Laboratory (Lanag).

Metrological performance of indoor-GPS in a simulated measurement assisted assembly process

Gustavo Heiden[1], Mauricio de Campos Porath[1]

[1]Federal University of Santa Catarina (UFSC), Department of Mobility Engineering, Rua Dr. João Colin 2700, 89218-035 Joinville/SC, Brazil

Email: mauricio.porath@ufsc.br

Abstract. This paper presents an experimental evaluation of the performance of an indoor-GPS (iGPS) system in a measurement assisted assembly process. A device was designed to simulate an assembly process with two rotational and one translational degree of freedom. The assembly of the device was performed with the assistance of iGPS and the result evaluated with a coordinate measuring machine. The results confirm the applicability of iGPS in assembly processes with production tolerances down to 1 mm.

1. Introduction

The assembly of large structures is a key challenge in the production of equipment such as wind- and water turbines, aircrafts, trains and ships. In modern shipyards, the construction of medium to large-sized vessels is performed block-wise. In one of the final assembly processes, called erection, previously manufactured blocks, often weighting hundreds of tons, are put into position by special vehicles, cranes and hydraulic jacks and welded to form the hull. Theodolites, laser levels, plumb-bobs and measuring tapes are used as measurement systems. The assembly is performed in an iterative fashion, intercalating positioning and measuring processes.

A paradigm shift in large-scale assembly is expected through the development of measurement assisted assembly (MAA). MAA consists in using advanced 3D measurement systems to enable direct, fast and sometimes even automatic part-to-part assembly without iteration. Quality and productivity enhancement and thus cost reduction is expected by implementing MAA processes [1]. The present work is part of a project that aims to evaluate the feasibility of MMA in ship construction. One of the measurement systems available for MAA is the Indoor GPS (iGPS).

Mosqueira et al. evaluated this system for the robotic alignment of fuselages, using a laser radar as reference measurement equipment [2]. Muelener et al. [3] and Schmitt et al. [4] assessed the overall performance of iGPS in 3D position measurement. They also identified the main sources of uncertainty of measurement processes using this system. However, to the best of our knowledge there are no results reported on the evaluation of the metrological performance of iGPS in assembly processes using an accurate calibrated reference measurement system.

The experiment we performed consists in assembling a special device with assistance of iGPS and checking the result with a coordinate measuring machine (CMM).

2. Materials and Methods

2.1. Indoor-GPS (iGPS)

The optical measurement system iGPS, also known as rotary-laser automatic theodolites (R-LAT) [3], operates according to the triangulation principle. On triangulation-based systems the position of a target point is determined based on azimuth and elevation angle measurements from at least two stationary measurement systems with known position and orientation.

The system consists of three basic components: transmitters, which act as measurement stations, detectors (or receivers) and position calculation engines (PCE). The transmitters are equipped with a rotary head that sweeps two fan-shaped laser beams and a LED ring that emits a strobe signal at the beginning of each second rotation. Photosensitive sensors in the receivers detect the light pulses and convert them to an electrical signal. This signal is amplified, digitized and processed in the PCE units that send receiver position and orientation data to a workstation. The rotation frequency of the rotors lies between 40 and 50 Hz. Each rotor operates with a slightly different frequency, enabling signal isolation [5].

The angle measurement principle is depicted in figure 1. Azimuth is calculated from the strobe time t_1 and the average of t_2 and t_3 and elevation is derived from the difference between t_2 and t_3 [6].

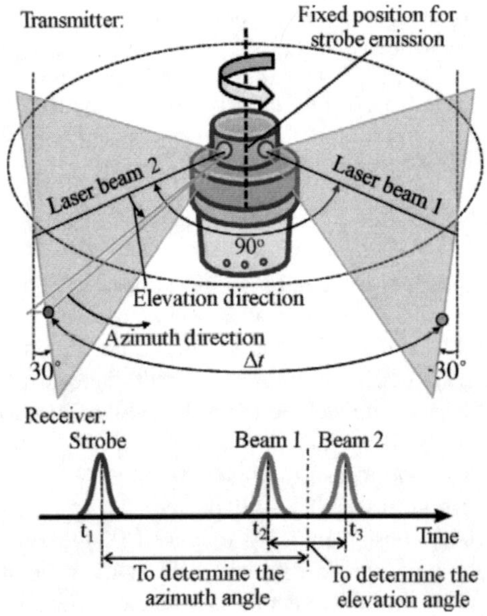

Figure 1. Azimuth and elevation angle determination of a transmitter-receiver vector.

The detectors are integrated in measurement tools. The measurement tools available in our system are the scale-bar, which is used in the system setup, the mini vector bars, used as static reference points (monuments) as well as in dynamic tracking tasks and the handheld vector bar probe (HHVB) which is the device used to capture single points. Each one of these tools is equipped with two receivers, enabling the determination of position and orientation in five degrees of freedom.

A six degrees of freedom tracking frame is created by fastening at least two mini vector bars to the object to be tracked. We used this kind of frame to assess the alignment and positioning movements during the assembly process.

The typical static length measurement uncertainty of the iGPS under good environmental conditions is 0.5 mm [5]. Our iGPS system is a Nikon Metrology (formerly Metris) model iSpace 6i, manufactured in 2008.

Dephental et al. state that the highest accuracy of an iGPS system with four transmitters can be obtained if the transmitters are placed on the corners of a rectangle and if the measurements are taken next to the centre of this rectangle [5]. The configuration we used in our experiments is depicted in figure 2. The transmitters where positioned at 1.8 m height in relation to the floor. We performed the assembly on a workbench of 1 m height located approximately at the centre of the rectangle formed by the four transmitters (T1 to T4).

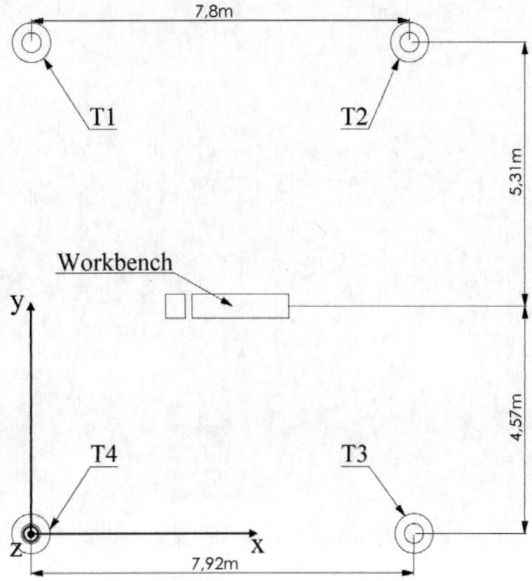

Figure 2. Transmitter configuration.

2.2. Experimental device

We designed a simple experimental device to simulate an assembly process with two degrees of freedom of rotation and one degree of freedom of translation. This device consists of two triangular parts that can be aligned and positioned with aid of iGPS and then rigidly fastened to a base plate. The size of the device was chosen to fit the measurement volume of the available CMM.

The two triangular parts are made of aluminium profiles and are fastened to a 1000 mm x 1000 mm plate by three sets of screw/nut mechanisms (figure 3). This configuration allows three degrees of freedom: translation in 'z' and rotation around 'x' and around 'y'.

Figure 3. Screw/bolt mechanism.

Two integrated detector units (IDU), including a mini vector bar and a PCE each, and three nests are placed on top of each part. A nest is a part with an inner cone (magnification in figure 4). The location of a nest can be measured with a probe (e.g. HHVB or CMM probe), featuring a spherical tip of appropriate size, by centering the probe tip in the cone. The triangular parts where designated 'A' and 'B'. The nests of each part are numbered 1 to 3 (figure 4). The coordinate system was defined as follows: xy-plane defined by A1, A2 and A3; x-axis pointing from A1 to A2 and origin at A1.

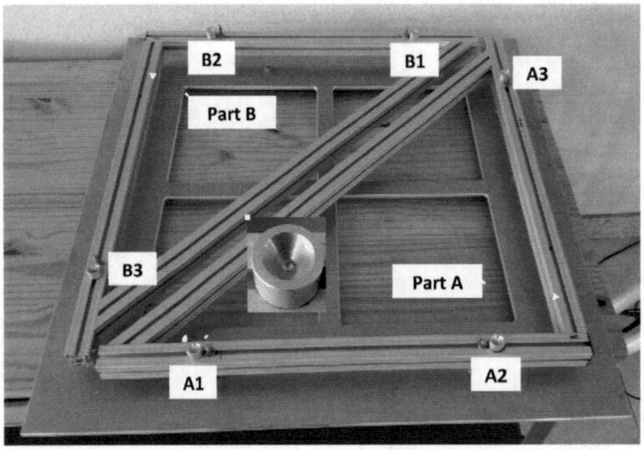

Figure 4. Experimental device.

2.3. Procedure

Part A was kept stationary and fastened to the base plate. Part B was moved with the screw/nut mechanisms until its nests where brought to the xy plane, i.e., until z-coordinates of all nest where close to zero (we admitted a tolerance of ±0.3 mm). After the positioning process, we also fastened part B to the base plate and evaluated the result by measuring the position of all nests 10 times with the HHVB probe.

After the assembly and measuring processes, we took the device to a metrology laboratory where the positions of the nests where measured using a CMM (figure 5). A second measurement cycle with iGPS was performed after the return of the device to our laboratory to detect accidental misalignments during transport.

Figure 5. CMM measurement.

3. Results

The results of the CMM measurement are listed in table 1 and the results of the iGPS measurements (average of 10 measurements) in table 2. The repeatability (two standard deviations) of the iGPS measurements is presented in table 3. The maximum expanded measurement uncertainty for the CMM measurements was estimated to be U = 0.004 mm, with 95,45% coverage probability.

Table 1. Results of CMM measurements.

	x [mm]	y [mm]	z [mm]
A1	0.0	0.0	0.0
A2	489.8	0.0	0.0
A3	624.8	664.2	0.0
B1	433.6	818.6	0.0
B2	-54.7	818.8	0.4
B3	-189.1	153.9	0.5

Table 2. Results of iGPS measurements.

	Before CMM measurements					
	Results			Difference to CMM results (iGPS-CMM)		
	x [mm]	y [mm]	z [mm]	x [mm]	y [mm]	z [mm]
A1	0.0	0.0	0.0	0.0	0.0	0.0
A2	489.8	0.0	0.0	0.0	0.0	0.0
A3	624.7	664.1	0.0	-0.1	-0.1	0.0
B1	433.6	818.6	0.2	0.0	0.0	0.2
B2	-54.9	818.8	0.1	-0.2	0.0	-0.3
B3	-189.2	153.9	0.2	-0.1	0.0	-0.3
	After CMM measurements					
	Results			Difference to CMM results (iGPS-CMM)		
	x [mm]	y [mm]	z [mm]	x [mm]	y [mm]	z [mm]
A1	0.0	0.0	0.0	0.0	0.0	0.0
A2	489.8	0.0	0.0	0.0	0.0	0.0
A3	624.5	664.2	0.0	-0.3	0.0	0.0
B1	433.3	818.5	0.1	-0.3	-0.1	0.1
B2	-55.0	818.7	0.3	-0.3	-0.1	0.1
B3	-189.1	153.8	0.4	0.0	-0.1	-0.1

Table 3. Repeatability of iGPS measurements.

	Before CMM measurements			After CMM measurements		
	x [mm]	y [mm]	z [mm]	x [mm]	y [mm]	z [mm]
A1	0.01	0.03	0.04	0.02	0.09	0.03
A2	0.03	0.05	0.05	0.11	0.09	0.02
A3	0.16	0.13	0.04	0.05	0.04	0.02
B1	0.09	0.04	0.05	0.08	0.08	0.08
B2	0.07	0.07	0.05	0.08	0.06	0.03
B3	0.06	0.06	0.06	0.02	0.09	0.02

4. Discussion

The CMM measurement results revealed a maximum residual difference of 0.5 mm between the z-coordinates of the nests of part A and part B. Although the device is much smaller than an industrial structure, we consider this result representative for the general performance of the system in an assembly process. Due to the measurement principle of iGPS, size-proportional error sources are expected to be less significant than effects such as triangulation condition and transmitter visibility.

The repeatability of the iGPS results ranges from 0.01 mm to 0.16 mm. This variation is probably related to the measurement condition (e.g. triangulation condition and transmitter visibility) at each specific point.

Although the length measurement performance evaluation was not within the scope of this experiment, it can be verified in table 3 that all the measured errors were smaller than 0.5mm, confirming the manufacturers specification for length measurement errors.

5. Conclusions

In this paper we reported the results of an experiment to evaluate the performance of an iGPS system in an assembly processes that includes alignment and positioning requirements. The quality of the assembly result was evaluated with a coordinate measuring machine.

The results show that it is possible to introduce iGPS as support to assembly processes. However, it is important to note that, due to metrological limitations of iGPS, the production tolerance should be in the order of one millimeter or larger.

A next project task is set to assess the system robustness in conditions typically found in a shipyard. The feasibility of outdoor measurements and the sensitivity to optical and electromagnetic interferences will be evaluated. Simultaneously, we're aiming to integrate iGPS in a self-developed robotic platform for automatic assembly purposes.

6. References

[1] Maropoulos P G, Muelaner J E, Summers M D and Martin O C 2014 A new paradigm in large scale assembly – research priorities in measurement assisted assembly Int. J. Adv. Manuf. Technol. 70 621-633 DOI: 10.1007/s00170-013-5283-4

[2] Mosqueira G, Apetz J, Santos K M, Villani E, Suterio R and Trabasso L G 2012 Analysis of the indoor GPS system as feedback for the robotic alignment of fuselages using laser radar measurements as comparison 2012 Robotics and Computer-Integrated Manufacturing, 28 DOI: 10.1016/j.rcim.2012.03.004

[3] Muelaner J E, Wang Z, Martin O, Jamshidi J and Maropoulos P G 2010 Estimation of uncertainty in three-dimensional coordinate measurement by comparison with calibrated points Meas. Sci. Technol. 21 DOI: 10.1088/0957-0233/21/2/025106

[4] Schmitt R, Nisch S, Schönberg A, Demeester F and Renders S 2010 Performance evaluation of iGPS for industrial applications International Conference on Indoor Positioning and Indoor Navigation (IPIN), 15-17 September, Zürich, Switzerland DOI: 10.1109/IPIN.2010.5647630

[5] Depenthal C and Schwendemann J 2009 IGPS – A New System for Static and Kinematic Measurements 9th Conference on Optical 3D Measurement Techniques, Vienna v. 1 131-140

[6] Gao W, Kim S W, Bosse H, Haitjema H, Chen Y L, Lu X D, Knapp W, Weckenmann A, Estler W T and Kunzmann H 2015 Measurement technologies for precision positioning, CIRP Annals - Manufacturing Technology DOI: 10.1016/j.cirp.2015.05.009.

Acknowledgements

The authors thank *Conselho Nacional de Desenvolvimento Científico e Tecnológico* (CNPq) for the financial support.

Characterization of spectral irradiance system based on a filter radiometer

M S Lima, T F Silva, I Duarte, J S Correa, D Viana, W A Sousa, G B Almeida, and I B Couceiro

Optical Metrology Division, National Institute of Metrology, Quality and Technology – Inmetro, Av. N. S. das Graças, 50, Xerém, Duque de Caxias, 25250-020, RJ, Brazil

E-mail: maufferreira@yahoo.com

Abstract. The spectral irradiance scale has been realized recently. It is based on a filter radiometer that was mounted and characterized. The optical system was assembled and the procedures of the methodology were defined, including the mounting of FEL lamp jig, alignment of the optical system, calibration of the instruments and optical devices used on the experimental system. The main uncertainty components were evaluated and the preliminary uncertainty budget of the spectral irradiance system is presented.

1. Introduction

At INMETRO, the realization of the spectral irradiance scale is based on detector. This methodology has been realized using filter radiometers. These devices are basically composed by an aperture, interference filters, and a trap detector [1, 2]. All these components should be characterized to give traceability in spectral irradiance measurements. For this the spectral responsivity of trap detector is linked to cryogenic radiometer [3]. The preliminary irradiance measurements were presented elsewhere [4, 5]. In this resume we describe the employed methodology and the uncertainty budget for the system.

2. Experimental Set-up

The experimental setup assembly consists of four main steps: definition of the optical axis with auxiliary lasers, alignment of the FEL lamp jig, of the FEL lamp, and of the filter radiometer. The Resin (Ceramacast 575-N powder) was used to fix the lamp jig and in its support, as depicted in Figure 1. The jig is a target used as a reference for the reproducible alignment of the lamp.

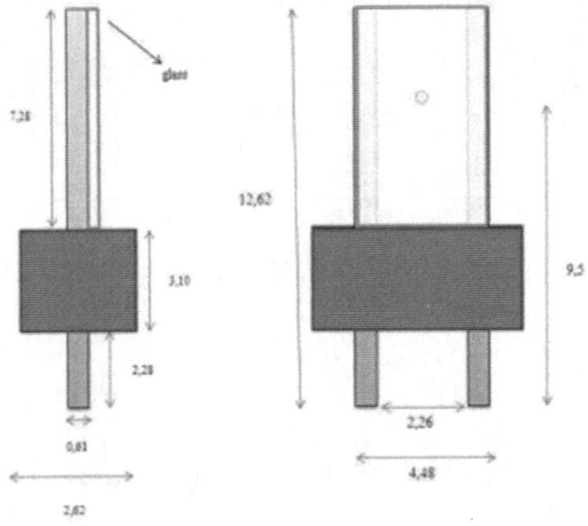

Figure 1 – FEL lamp Jig. A) Side view; B) Front view. All quotes are in centimeter.

The laser was fixed to a support at a height of about 28 cm, corresponding in about to the lamp middle height. First, a target was used to align the laser on the optical table. The target was moved along the trail defining the optical axis. It was then removed and two apertures were used to finish the alignment and as markers.

The lamp holder is fixed in an adjustable support on the optical rail with several degrees of freedom. The height of this structure relative to the optical table was measured (~ 28 cm). The center of the lamp filament was aligned in relation to the optical axis. The FEL lamp was replaced by the jig. Angular alignment was completed by returning the reflected beam back on the incident laser beam, as shown in Figure 2.

A **B**

Figure 2 – A) FEL lamp aligned with the He-Ne laser beam. B) FEL lamp jig aligned.

The two apertures are displaced close to the radiometer, once the optical axis has already been marked with them. Another laser was added to the system – similar to the other laser used before – with a plane mirror to align the filter radiometer on the optical axis. The laser is aligned to the apertures, reaching the filter radiometer, which has its center aligned. The back-reflected beam is used for angular alignment (see Figure 3).

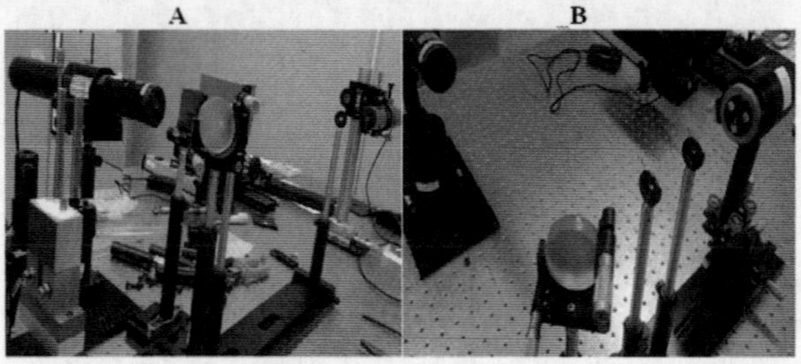

Figure 3 – A) Alignment of the second laser relative to the first. B) Filter radiometer alignment.

The Figure 4 shows the experimental system assembly. This system is composed by a filter radiometer with a thermally-controlled filter holder and a current-stabilized lamp.

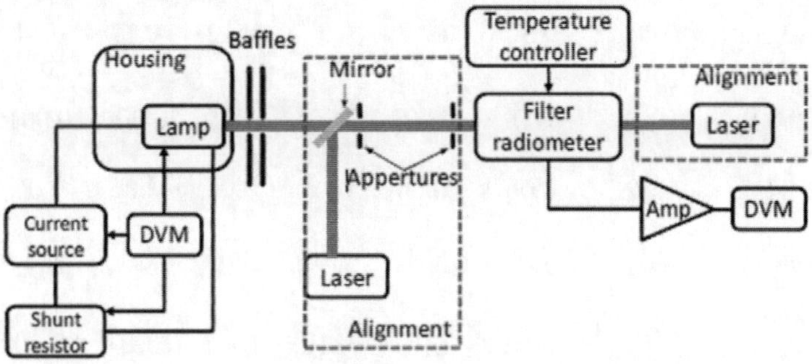

Figure 4 – Schematics for the experimental setup

A trans-impedance amplifier and a digital voltmeter (DVM) are used to acquire the electrical signal of the filter radiometer. Additional DVMs are used to monitor the voltage and current (through a shunt resistor) applied to the FEL lamp and also the electrical signal of the temperature controller. Figure 5 shows a picture of the setup.

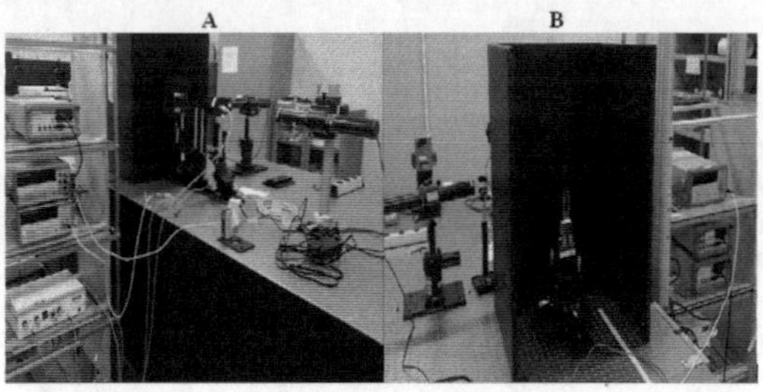

Figure 5 – The view of the irradiance system, A) by the side of the filter radiometer B) by the side of the FEL lamp jig.

3. Results and Discussions

The uncertainty components were evaluated and the uncertainty budget is presented in Table 1. In Table 1 are detailed some of the main uncertainty components related to spectral irradiance scale. The uncertainties analyzed are linked to the trap detector, the interferential filters, the apertures, and setup characterization. The dominant component comes from the calibration spectral responsivity of the trap detector. The second one comes from the calibration of interference filters.

Table 1. Uncertainty Budget

Sources of uncertainty										
Wavelenght (nm)	394	420	450	500	550	600	694	800	905	940
Spectral Responsivity (%)	0.83	0.77	0.72	0.65	0.59	0.54	0.46	0.40	0.35	0.34
Spectral transmittance (%)	0.40	0.40	0.41	0.40	0.40	0.40	0.40	0.40	0.23	0.43
Filter temperature (%)	0.26	0.00	0.31	0.25	0.38	0.12	0.17	0.14	0.23	0.26
Lamp current (%)	0.001	0.001	0.001	0.001	0.001	0.001	0.001	0.001	0.001	0.001
Distance (%)	0.06	0.06	0.06	0.06	0.06	0.06	0.06	0.06	0.06	0.06
Aperture area (%)	0.1	0.1	0.1	0.1	0.1	0.1	0.1	0.1	0.1	0.1
Interreflection in filter (%)	0.10	0.10	0.10	0.10	0.10	0.10	0.10	0.10	0.10	0.10
Combined uncertainty (%)	0.96	0.88	0.89	0.81	0.81	0.69	0.65	0.59	0.49	0.62
Expanded uncertainty (%)	1.9	1.8	1.8	1.6	1.6	1.3	1.3	1.2	1.0	1.2

The detector responsivity was measured using a cryogenic radiometer as a primary standard at wavelengths of 457.8 nm, 476.0 nm, 488.0 nm, 501.0 nm, 514.5 nm, and 632.8 nm. The data was interpolated and extrapolated (provided the linear behavior of this kind of detector) in the visible and near infra-red spectral ranges. The uncertainty component due to linear adjust of the data dominates the total uncertainty (see Table 2).

Table 2. Spectral responsivity uncertainties of Trap detector

λ (nm)	**s (A/W)**	**u (A/W)**	**u_c (A/W)**
400.0	0.3167	8.5E-05	2.6E-03
420.0	0.3331	8.5E-05	2.6E-03
450.0	0.3577	8.5E-05	2.6E-03
457.9	0.3642	8.5E-05	2.6E-03

472.7	0.3763	7.6E-05	2.6E-03
488.0	0.3889	9.7E-05	2.6E-03
500.0	0.3987	8.0E-05	2.6E-03
501.7	0.4001	8.0E-05	2.6E-03
514.5	0.4106	9.1E-05	2.6E-03
550.0	0.4397	9.1E-05	2.6E-03
600.0	0.4807	9.7E-05	2.6E-03
632.8	0.5075	9.7E-05	2.6E-03
694.0	0.5577	9.7E-05	2.6E-03
800.0	0.6446	9.7E-05	2.6E-03
905.0	0.7306	9.7E-05	2.6E-03
940.0	0.7593	9.7E-05	2.6E-03

The interferential filters are linked to the reference spectrophotometer; as shown in Table 3. The main characterization steps of the instrument include evaluation of the stray light, photometric noise, linearity, temporal drift of the standard, FWHM, scan rate, among others. The total uncertainties of the filters are determined measuring the peak transmittance, wavelength scale and wavelength setting repeatability. The temperature uncertainty is analyzed with the temperature controller attached to the interferential filters during the transmittance measurement.

Table 3. Uncertainties of Transmittance, $\Delta\lambda$ and Temperature of Interferential Filters

λ (nm)	T (%)	U (%)	u (%)	$\Delta\lambda$ (nm) (FWHM)	u (nm)	θ (°C)	u_c (°C)
340	89.61	0.74	0.40	9.7	1.7	24.97	0.01
365	92.00	0.74	0.41	8.9	1.5	24.98	0.28
394	95.31	0.76	0.40	7.8	1.4	24.99	0.26
420	96.78	0.77	0.40	9.5	1.6	24.93	0.00
450 (445)	88.32	0.73	0.41	8.6	1.5	24.94	0.31
500 (502)	96.42	0.74	0.40	8.9	1.5	24.94	0.25
550	95.68	0.76	0.40	8.5	1.5	24.94	0.38
600	95.42	0.76	0.40	8.2	1.4	24.94	0.12
694 (695)	96.29	0.76	0.40	8.5	1.5	24.95	0.17
800 (798.35)	95.98	0.75	0.40	8.7	1.5	24.95	0.14
905 (907)	89.79	0.38	0.23	10.9	1.9	24.95	0.23
940 (943)	90.32	0.73	0.43	10.1	1.8	24.96	0.26

The other components are connected to spectral irradiance measurement where the uncertainties of lamp current, distance between lamp and detector, and repeatability are analyzed. The interreflections of the filter radiometer have been evaluated too. The components related to electronics of the system are linked to electrical standards. The aperture area of the filter radiometer is determined at INMETRO with a non-contact method [6], which is based on the image analysis of the aperture area with a microscope. The final value is calculated based on a mean radius, so the circularity of the aperture is important. The calibrated area is 7.051 mm^2 with uncertainty 0.004 mm^2. Additionally, studies on the stability and detection limits of spectral irradiance system are being made to characterize it better. Figure 6 shows the electrical signal detected by the filter radiometer during a period of exposure to radiation emitted by a FEL lamp. It can be seen that the data oscillate around a mean straight line and the amplitude of this oscillation does not exceed too much 0.05%.

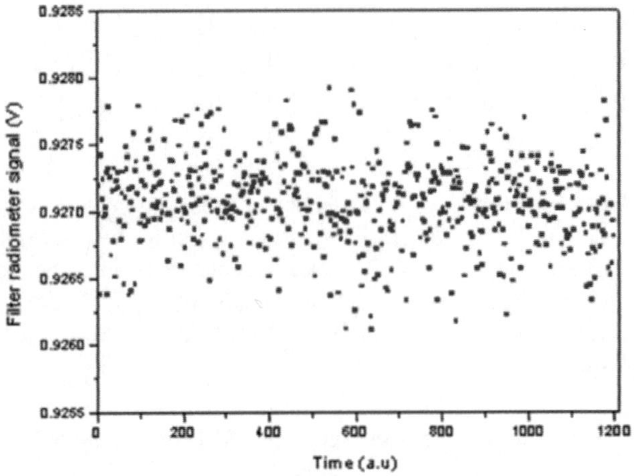

Figure 6 – Temporal stability of the irradiance system

4. Conclusions

This work described the definition of the alignment methodology and the uncertainty budget for the spectral irradiance scale developed in Brazil, which covers the spectral range of visible and near infra-red. The preliminary spectral irradiance measurements show results compatible with the literature. The spectral responsivity of trap detector was measured by direct comparison with the cryogenic radiometer using Ar+ and HeNe laser lines as radiation sources. The uncertainty component due to the linear adjust of this data dominates the total uncertainty. A characterized commercial spectrophotometer was used to measure the spectral transmittance of the interferential filters. The analysis was also performed as a function of the temperature of the filters. The measurement corresponds to the second dominant uncertainty contribution. Other uncertainty sources as detector nonlinearity, out-of-band filter peaks, and diffraction will be determined in the near future.

5. References

[1] Kärhä P, Toivanen P, Manoochehri F, and Ikonen E 1997 *Applied Optics* **36** 8909-8918.

[2] Yoon W H and Charles E G 2011 *NIST special publication* **250-89**.

[3] Menegotto T, Ferreira da Silva T, Simões M, Sousa W A T, and Borghi G 2015 *IEEE Trans. Instrum. Meas.* 64 1702-1708.

[4] Simões M, Menegotto T, Duarte I, Ferreira da Silva T, Alves L C, Alvarenga A D, Almeida G B, Teixeira R N, and Couceiro I B 2015 *J. Phys. Conf. Ser.* **575** 012023.

[5] Simões M, Menegotto T, Ferreira da Silva T, *et al.* 2014 *12th International Conference on New Developments and Applications in Optical Radiometry – NEWRAD 2014*, Espoo, Finland.

[6] Costa P B and Leta F R 2010 *17th International Conf. on Systems, Signals and Image Processing*, Niterói, Brazil.

Acknowledgments
This work has been supported by the National Research Council of Brazil (CNPq/ PROMETRO).

Determination of cut front position in laser cutting

M Pereira [1] and U Thombansen [2]

[1] Federal Institute of Santa Catarina (IFSC), Santa Catarina, Brazil
[2] RWTH Aachen University, Chair for Laser Technology (LLT), Aachen, Germany

E-mail: milton@ifsc.edu.br; ulrich.thombansen@ilt.fraunhofer.de

Abstract. Laser cutting has a huge importance to manufacturing industry. Laser cutting machines operate with fixed technological parameters and this does not guarantee the best productivity. The adjustment of the cutting parameters during operation can improve the machine performance. Based on a coaxial measuring device it is possible to identify the cut front position during the cutting process. This paper describes the data analysis approach used to determine the cut front position for different feed rates. The cut front position was determined with good resolution, but improvements are needed to make the whole process more stable.

1. Introduction

The Laser cutting is used in processing of flat metal sheets for construction, cutting of pre-formed parts or trimming of hardened 3D parts in several areas of manufacturing. Today laser cutting is the most common industrial application of the laser. In Japan, around 80% of industrial lasers are used in this way [1].

To reach a reliable cutting performance associated with the best productivity, the cutting process should operate close to its technological limits, but with good quality in face roughness and with dross free processing results [2]. For fusion cutting, the feed rate can be limited by the insufficient ejection of the melt out of the cut kerf, or by a power demand not covered by the absorbed laser beam power [3]. This lead to the possibility of monitoring the cutting phenomena "in process" (at real time) to make possible the self-optimization for the next generations of laser cutting machines.

One way to realize this monitoring in by acquiring information about geometric properties of the cut kerf during the process through a coaxial process observation [2].

Figure 1 shows a sketch of the cutting process observed by a coaxial perspective. The point of interest for monitoring is the melting point, named cut front position. This cut front position changes with different operational parameters as feed rate, laser power or laser focus position.

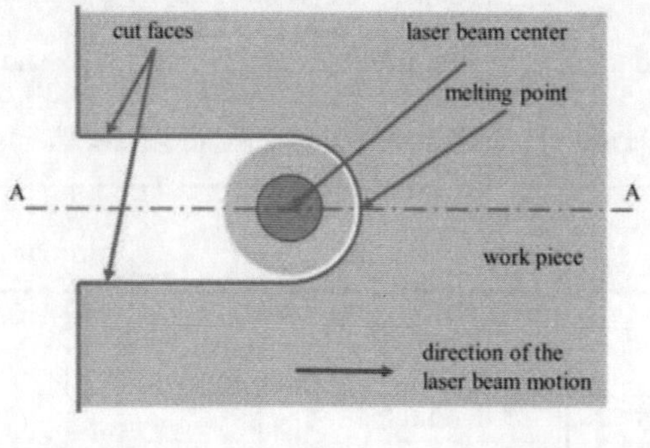

Figure 1. Sketch describing the cutting process [2].

2. Process Observation

The process observation with coaxial positioning is accomplished by combining and separating the optical path depending on the respective wavelength [4]. Figure 2 shows the coaxial setup. The processing radiation of the CO_2 laser is fully reflected by the beam splitter to the work piece, while the illumination laser beam with different wavelength crosses the beam splitter and reaches a high-speed camera mounted on the cutting head.

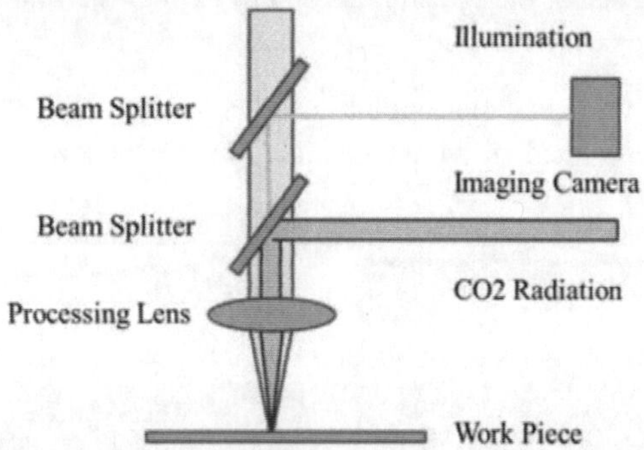

Figure 2. Coaxial setup for process observation [2].

The resulting image acquired by this system is shown in figure 3. This image has 280 x 300 pixels and represents only one frame of a set of 1000 frames acquired at a rate of 4 kHz for each test done. In the center of this image is easy to identify the cut kerf generated by the laser cutting process.

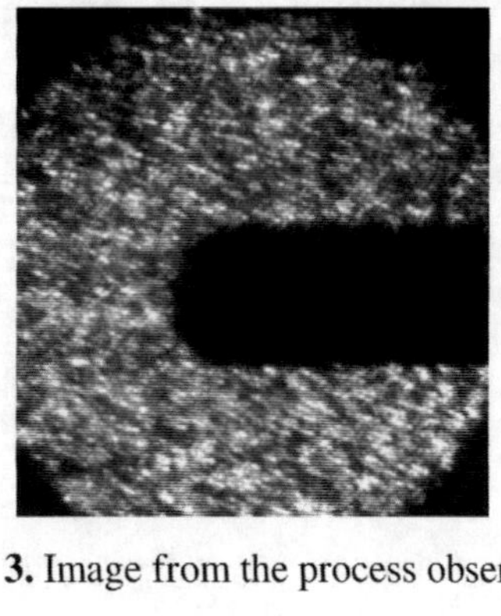

Figure 3. Image from the process observation.

3. Data Analysis

The experiments were conducted following pre-defined parameters, as shown in table 1.

Table 1. Setting parameters for the experiments.

Setting parameter	Value
Feed rate	900, 1200, 1500, 1800, 2100, 2400, 2700 mm/min
Laser Power	3.000 Watts
Focal length	250 mm
Focus positions	top surface level – 0 mm
Material	Stainless Steel 14301
Thickness	5 mm

For the focus position in 0 mm (position of the laser beam focus in relation to the metal sheet top surface), a test was done in each of the 7 feed rates and in each of the four cutting directions, resulting in a total of 28 cutting tests.

Figure 4 summarizes the investigations and the searched results to identify the cut front position for each test. Algorithms were created in Matlab® and Simulink® software to extract the geometric properties from the images.

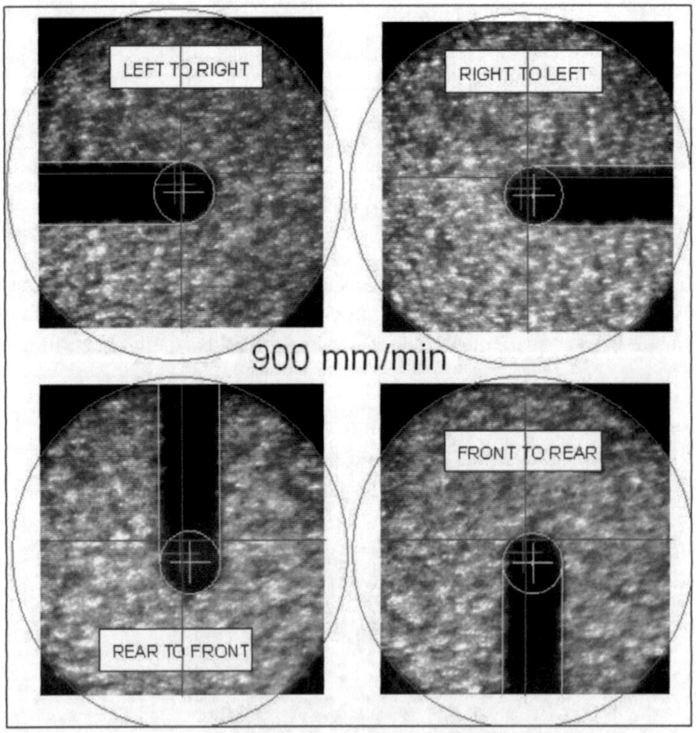

Figure 4. Analysis of results for cutting in four directions with the same process parameters.

In figure 4, the lines in red represent the centerlines for the image, identifying the camera center. The green circle represents the nozzle inner hole and the green cross represents the nozzle center. The lines in yellow represent the kerf geometry. Assuming that the laser beam has a circular profile, the yellow circle represents its position and the yellow cross represents the laser beam center.

Based on the behavior of the cut front position in relation to the nozzle center for each set of parameters used for the tests, it is possible to determine if the current set of parameters is near the critical condition or if the cutting process can be accelerated within a safe condition.

All the 1000 frames acquired for each test represents the same cutting phenomena with steady conditions.

The main task for the analysis consists in determining the cut kerf geometry to identify the cut width and the cut front position. As the use of only one frame can lead to an inaccurate determination of these properties, the analysis is based on a statistical superposition of the 1000 frames.

3.1. Analysis Algorithms for data extraction

The first step consists in converting the images in values that can be processed to enable the geometric analysis. Three processing algorithms were created for this purpose: one to process the kerf width, other for the cut front position, and the last one for determining the nozzle inner diameter.

The conversion is based on the determination of the intensity profile for each region of interest, as shown in Figure 5. In this case, the aim is to determine the kerf width, so the kerf was split in three regions (1, 2 and 3) and the intensity profile was determined for each half of these regions. The use of three regions is important to avoid errors caused by images aberrations and to verify if the kerf width keeps its uniformity. This algorithm automatically processes the 1000 frames and gives as result vectors with the pixel position and the related cumulative intensity of ten nearby lines, with each frame producing six vectors.

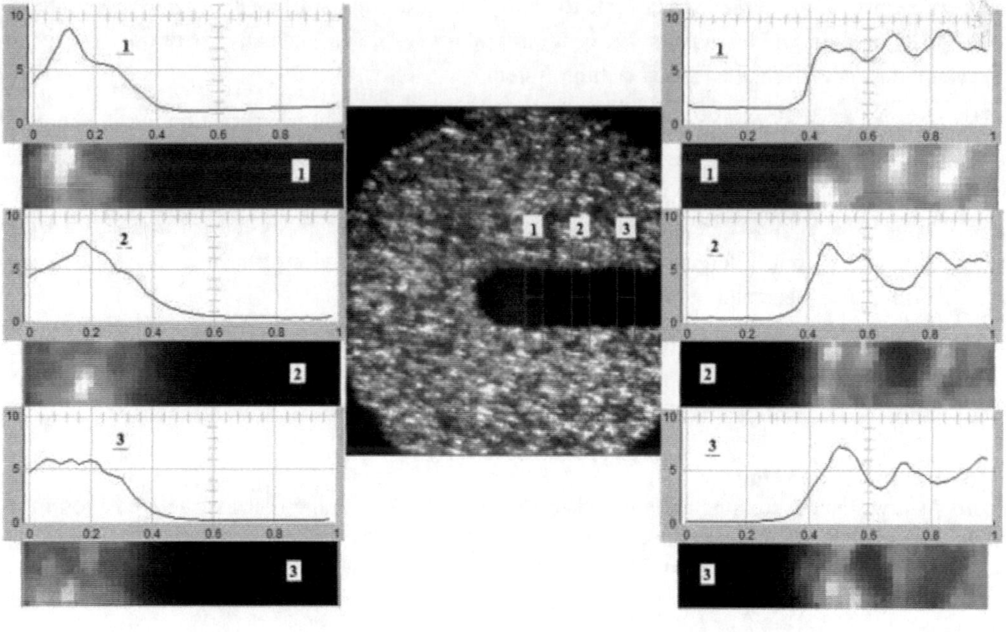

Figure 5. Intensity profile determination in kerf width analysis.

The same intensity profile extraction was also used to identify the cut front position.

To identify the nozzle inner diameter a contour filter was used and the resulting profile was converted in pixels positions, as shown in figure 6. This algorithm also automatically processes the 1000 frames and gives as result vectors with the pixel position of the contour profile for each frame.

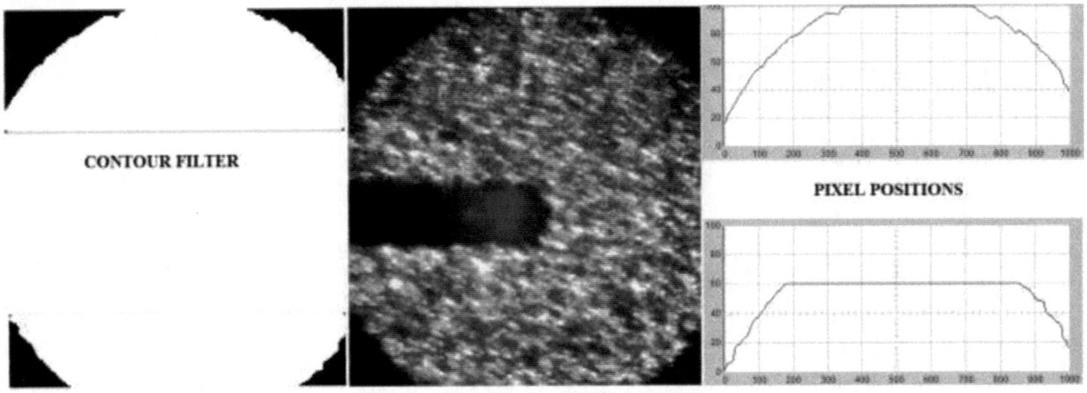

Figure 6. Nozzle inner hole determination.

3.2. Analysis Algorithms for geometric determination

The main concept for the analysis consists in making a statistical superposition of the 1000 vectors for each acquired attribute. This is done by fitting all individual intensity profile curves in one high degree polynomial curve using least squares method.

After finding the best curve that describes the nozzle inner diameter, the routine can calculate the circumference radius and its center coordinates (represented as the green cross in figure 4). This nozzle center is reference point to evaluate the cut front position displacement for different tests parameters. The nozzle center was checked for all tests to certify that the contrast parameters in the software were well adjusted resulting in the same nozzle center for all analysis. Mechanically, there was no change in the camera position in relation to the machine system.

From the best curve for each intensity profile during width and cut front position measurement, the criteria that was used to determine the geometrical point known as the edge between the region with material and without material was the determination of the minimum value of the polynomial derivate curve. This may not be the exact point in absolute terms, but it will be always the same point for all tests and can easily be used as reference to determinate the relative behavior of the width and the cut front position during all tested conditions.

As the laser beam is not perfectly centered in relation to the nozzle, the tests made in both sides for horizontal and vertical directions helped to determine the static laser beam incidence center. The cut front position can then be determined in relation to the laser beam center.

4. Results and Conclusion

Figure 7 shows the cut front position in μm for the tests made with the focus vertical position in zero (coincident with the sheet top surface) and showing the four directions of cut (horizontal, left to right and right to left and vertical, rear to front and front to rear) with different speeds.

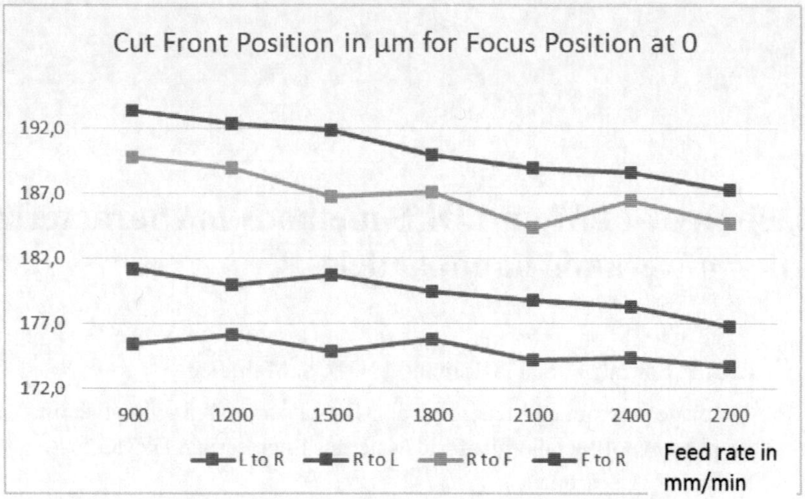

Figure 7. Cut front position for different feed rate.

The energy distribution of the laser is not uniform, which makes the cut front position different for different directions of cut. Also, the machine vibrations generated during the operation interferes significantly on the results. Despite this, it is clear that the cut front position changes with different feed rates and has potential to be used as monitored variable to adjust the feed rate during the machine operation to improve the process productivity.

5. References

[1] Steen, W. M., e Mazumder, J., "Laser Material Processing", Springer-Verlag London, 2010.
[2] Thombansen, U., Hermanns, T., Molitor, T., Pereira, M., Schulz, W., "Measurement of cut front properties in laser cutting". Physics Procedia, Volume 56, 2014, páginas 885-891.
[3] Poprawe, R., "Tailored Light 2: Laser application technology", Springer-Verlag Berlin Heidelberg, 257-259, 2011.
[4] Kratzsch, C., Abels, P., Kaierle, S., Poprawe, R., Schulz, W., 2000. Coaxial process control during laser beam welding of tailored blanks. pp. 472–482. doi:10.1117/12.377055.

A comparison of TEM and DLS methods to characterize size distribution of ceramic nanoparticles

T. G. F. Souza[1], V. S. T. Ciminelli[2], N. D. S. Mohallem[3*]

[1] Graduate program on Metallurgical, Materials and Mining Engineering, UFMG
[2] Department of Metallurgical and Materials Engineering, UFMG
[3] Chemistry Department,
Universidade Federal de Minas Gerais, Belo Horizonte, CEP 31270-901, Brazil

E-mail: nelcy@ufmg.br

Abstract. The accuracy of dynamic light scattering (DLS) measurements are compared with transmission electron microscopy (TEM) studies for characterization of size distributions of ceramic nanoparticles. It was found that measurements by DLS using number distribution presented accurate results when compared to TEM. The presence of dispersants and the enlargement of size distributions induce errors to DLS particle sizing measurements and shifts its results to higher values.

1. Introduction

Nanoparticles (NP) present unique properties related to their size. Therefore it is important to be able to accurately characterize NP size distributions. Different techniques may provide different results [1, 2]. TEM is an excellent tool for characterizing NP [3], since its resolution reaches about 0.07 nm depending on sample thickness and accelerating voltage [4], but have a high cost and is operationally complex. For nanoparticles dispersed in solvents, DLS is a suitable technique that may allow access not just to particle size but also to the presence of agglomerates and aggregates [5]. It is a non-invasive, non-destructive, and low cost technique, and its operation is relatively simple and rapid. The main disadvantage is the data interpretation for polydisperse samples [4].

The present work compares size distributions of two ceramic NP with different sizes and assesses the effect of mixing (polydisperse system) in the final results.

2. Material and methods

2.1. Materials

Titanium oxide (TiO_2) was obtained from Minérios Ouro Branco. The cobalt ferrite ($CoFe_2O_4$) powders were prepared with stoichiometric amounts of cobalt and iron nitrates, precipitated by sodium hydroxide as described in previous works [5]. Both powders were dispersed (0.1% w/v) into deionized water with citric acid from Cargil. The two materials were analysed separately and together by mixing equal volumes of dispersions (TiO_2- $CoFe_2O_4$).

2.2. Transmission Electron Microscopy (TEM)
The diluted solution was dropped onto carbon-coated copper grids and then dried at environmental conditions. Images were acquired using an transmission electron microscope Tecnai G2-12 – Spirit Biotwin - 120 kV, from FEI. At least 10 locations on the TEM grid were examined. The quantity of NPs necessary to obtain reliable measurements was evaluated similarly as described in NIST protocol [3]. Image J software, freely available on the internet, was used for image analysis. Area sizes enclosed by the oval selection tool previously calibrated to the scale bar imprinted on the TEM images were determined. The diameter was calculated considering perfectly spherical shape.

2.3. Dynamical Light Scattering (DLS)
The DLS methodology followed recommendations outlined in the NIST-protocol [9]. The DLS measurements were performed using a Microtrac Intruments-Zetatrac 173 and its software. Measurements were performed at 25°C. The results were presented at intensity and number-based distributions.

2.4. Statistical Analysis
The NPs were prepared and analyzed in triplicate from the same batch and for each technique, under condition of repeatability. The normality of each sample data was tested using the Shapiro-Wilk Test. A p-value of ≤ 0.05 was considered statistically significant for all analyses, and ran using the R free Software [6], with an interface Excel via Action Software [7].

3. Results
TEM images show a good sample preparation and dispersion of the titanium oxide nanoparticles (figure 1). After each particle size was manually measured, the set of data was submitted to a normality test, since a normal distribution is mandatory for several hypothesis tests. Besides, the curve distribution determines the standard deviation and confidence interval calculation [10].
The Shapiro-Wilk test showed that the particle sizes distribution from the material analyzed do not follow a normal distribution, as depicted in figure 2 for TiO_2 NP. Particle size distributions are usually fitted by log-normal, Weibull or log-hyperbolic probability distributions [11].

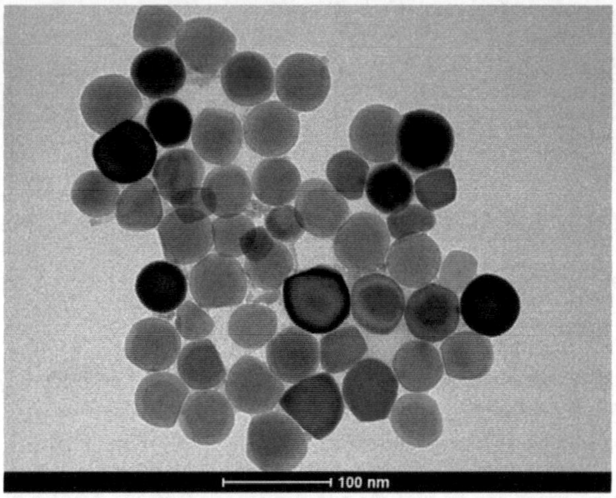

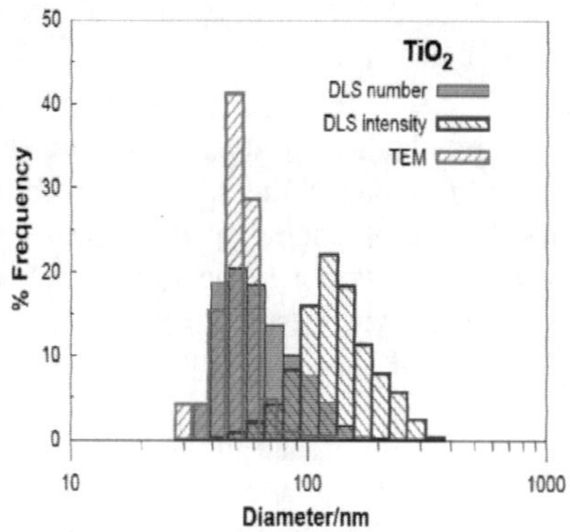

Figure 1. TEM images of TiO_2 NPs and the correspondent TEM and DLS particle size distribution.

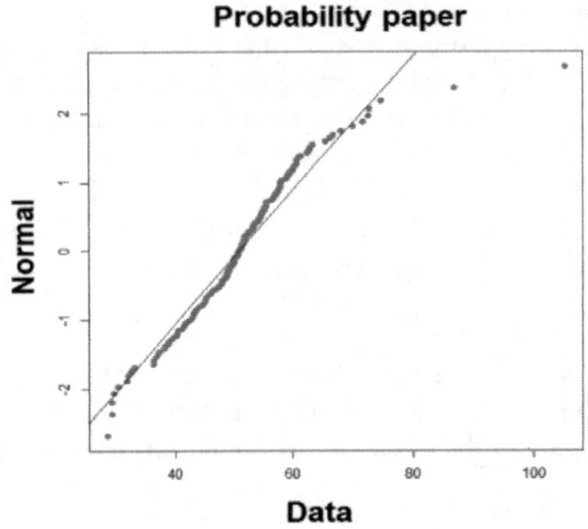

Figure 2. Probability paper of normality tests of TiO_2 nanoparticles sizes analyzed by TEM.

Table 1 summarizes the size measurements results of TiO_2 nanoparticles. It was expected that mean and median values from DLS would be a slightly higher than TEM due to the interference of the dispersant into the hydrodynamic diameter. However, DLS-numbers are close to the TEM results whereas DLS-intensity presents a large difference with TEM. Since the particle size distribution is not narrow, the presence of bigger particles may contribute to an increase light scattering, shifting the measured particles size towards larger values. The DLS measurements show a higher quantity of bigger particles (D95).

Table 1. Particle size measurements of TiO_2 in nm.

	TEM 190 NPs	DLS-Number	DLS-Intensity
D10	39.3	38.6	77.2
D95	66.2	108.3	224.8
Median	51.0	54.2	117.9
Mean	50.9	61.5	127.8

The results obtained from $CoFe_2O_4$ are summarized in figure 3 and table 2. The DLS technique was not able to identify the particles with sizes below 15nm. These results may be related to the agglomeration of smallest particles as evidenced by the red circle in figure 3. For $CoFe_2O_4$ the DLS-number results were also near to TEM values and the difference may be explained by the increased size in the latter by the presence of dispersant.

The mixture of TiO_2 and $CoFe_2O_4$ (TiO_2- $CoFe_2O_4$) dispersion was also analysed and the results are presented in figure 4 and table 3. The size distribution obtained by TEM is not bimodal as one would expect by the difference in the mean sizes. Besides, the dispersion used had the same concentration, and therefore, the number of $CoFe_2O_4$ NPs in a given mass was bigger than the number of TiO_2 NP, since the size of $CoFe_2O_3$ NP is smaller. Thus, the particle size distribution is centred at 20nm and TiO_2 has a minor contribution to the curve due to a smaller number of particles. The values from DSL are much higher than those from TEM since DLS-intensity shows values about five times larger than TEM. These results may be related to the broad distribution formed by mixing the two particles. These results corroborates the previous affirmation were the distribution is not narrow and the presence of

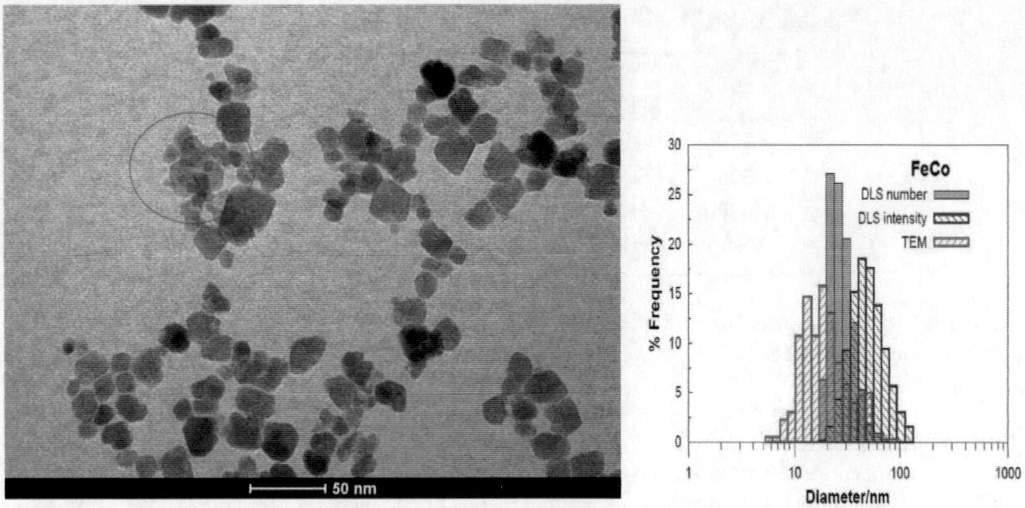

Figure 3. TEM images and DLS particle size distribution of $CoFe_2O_4$ NPs.

Table 2. Particle size measurements of $CoFe_2O_4$ in nm.

	TEM 360 NPs	DLS-Number	DLS-Intensity
D10	10.8	18.6	27.9
D95	45.8	38.8	84.5
Median	18.1	23.9	43.5
Mean	21.2	25.7	47.4

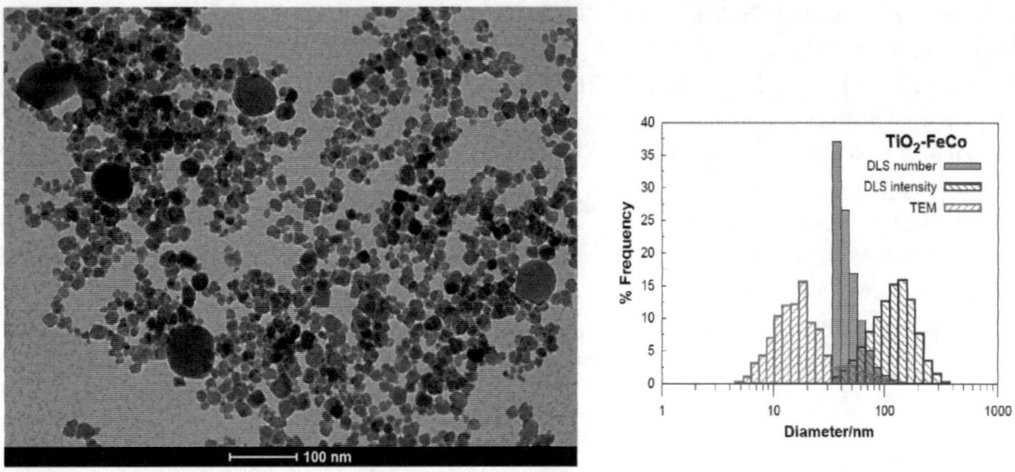

Figure 4. TEM Images of TiO_2-$CoFe_2O_4$ NPs and the correspondent TEM and DLS particle size distribution.

Table 3. Particle size measurements of TiO_2-$CoFe_2O_4$ in nm.

	TEM 800 NPs	DLS-Number	DLS-Intensity
D10	9	32.2	57.4
D95	50.5	71.4	207
Median	16.7	38.9	111
Mean	20.9	43.8	116.6

bigger particles shift the measured particle sizes towards larger values. Even the DLS - number shows large errors compared to TEM.

4. Conclusions

The DLS – number for a monomodal sample shows a good approximation to the size parameters obtained by TEM while DSL – intensity does not. The DLS mean is app. 20% higher than TEM mean size; the errors for the other parameters are larger. The difference is ascribed to the presence of dispersant. The enlargement of size distributions induces errors to DLS particle sizing measurements when compared to TEM analyses. These errors are more evident on the results of mixed samples with different sizes, where the DLS-intensity present values five times larger than TEM.

5. Acknowledgements

The authors acknowledge CNPq, CAPES and FAPEMIG for financial support, the Nanum Nanotecnologia Ltda and Microscopy Centre-UFMG for technical support and INCT-Acqua.

References

[1] Hoo C M; Starostin N; West P and Mecartney M L 2008 *Journal of Nanop. Research* **10** 89
[2]Anderson W *et al.* 2013 *Journal of Colloid and Interface Science* **405** 322
[3] Bonevich J E and Haller W K 2010 Measuring size of nanoparticles using TEM Gaithersburg:NIST 20p
[4]Kim H A; Seo J K; Kim T and Lee B T 2014 *Environmental Health and Toxicology* **29** 1
[5] Linkov P; Artemyev M; Efimov A E and Nabiev I 2013 *Nanoscale* **5** 8781
[6]Contadini J F, Souza T G F, Mohallem T D S 2014 US Patent nº 8815393 B2, p. 07
[7] http://www.R-project.org/
[8] http://www.portalaction.com.br/
[9]Hackley V A and Clogston D J 2007 Measuring the Size of Nanoparticles in Aqueous Media Using DLS Gaithersburg:NIST, 22p.
[10]Merkus, H G 2009 *Particle Size Measurements* e-book:Springer, 533p.
[11]Purkait, B 2002 *India Journal of Sedimentary Research* **72** 367.

Speed of sound as a function of temperature for ultrasonic propagation in soybean oil

P A Oliveira[1], R M B Silva[1], G C Morais[1], A V Alvarenga[1] and R P B Costa-Félix[1]

[1] Laboratory of Ultrasound, National Institute of Metrology, Quality and Technology (Inmetro), Av. Nossa Sra das Graças 50, Duque de Caxias, RJ, 25250-020, Brazil

Email: passuncaooliveira@gmail.com

Abstract. Ultrasound has been used for characterization of liquid in several productive sectors and research. This work presents the studied about the behavior of the speed of sound in soybean oil with increasing temperature. The pulse echo technique allowed observing that the speed of sound decreases linearly with increasing temperature in the range 20 to 50 °C at 1 MHz. As result, a characteristic function capable to reproduce the speed of sound behavior in soybean oil, as a function of temperature was established, with the respective measurement uncertainty.

1. Introduction

The production the use of biodiesel has attracted attention because it is a renewable source of energy. Thereby, numerous oleaginous have been studied with the aim of evaluating them as raw material for biodiesel production. In Brazil, soybean oil predominates as raw material for the production of biodiesel due its availability.

There is agreement about potential application of methods for characterization of liquid in several productive sectors and research. The acoustic techniques have non-destructive characteristics, with the possibility of online non-invasive applications 0.The ultrasonic parameters more frequently measured to characterize materials are the speed of sound, attenuation coefficient and acoustic impedance. Many studies have used the speed of sound for characterization of liquids.0-0.

Generally, the speed of sound is determined by measuring the arrival time of the ultrasonic pulse both with and without the test sample positioned in the ultrasonic beam. The difference in these times is the time-shift caused by the different speed of sound in the material0.

This article reports the measurement of the speed of sound in soybean oil along the temperature range 20 °C at 50 °C using longitudinal waves at 1 MHz, with its respective measurement uncertainty.

2. Methodology

In this study, the ultrasound technique known as pulse/echo was applied at low power. The measuring systemconsists of a cell containing the sample immersed in a thermal bath, a transducer with a nominal center frequency of 1 MHz and 12.7 mm diameter (A303 S, Panametrics - NDT Olympus Corporation, Japan) excited by an arbitrary function generator (model 33250A, Agilent Technologies, CA, USA) with a of 4-cycles burstand 20 volts peak-to-peak emission.

The emitted ultrasonic signal propagates through the sample, reaches the reflector target, and returns to the same transducer. The echo signals acquired are digitized by an oscilloscope (model DSO-X 3012A, Agilent Technologies, CA, EUA) and saved on a computer in a given time interval (20 s) with the aid of a software developed in LabView™ (National Instruments, Austin, TX, EUA).The bath temperature is set initially at 20 °C and increases in a steps of 5 °C up to 50 °C. The sample temperature is monitored during the measurement process with the aid of a thermocouple type K and a temperature measuring system (model 34970A,Agilent Technologies, CA). Figure 1 illustrates the experimental setup of the measuring system described above.

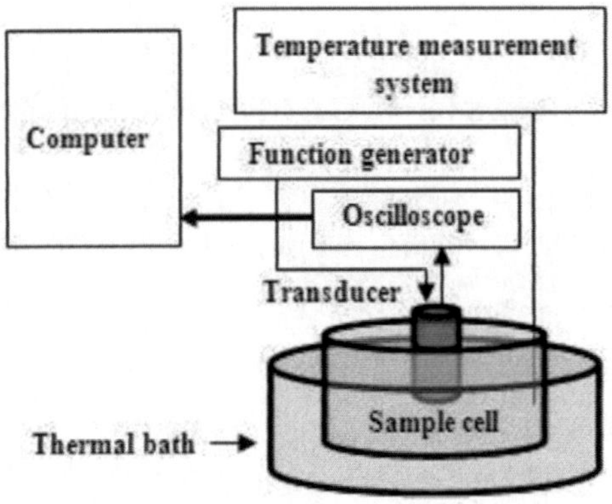

Figure 1. Speed of sound measuring system.

Five repetitions were performed under repeatability conditions for each temperature. The speed of sound of the oil is given by equation (1):

$$v_{oil} = \frac{c_{H_2O} \cdot t_{H_2O}}{t_{oil}} \tag{1}$$

wherec_{H_2O} is the speed of sound in the water [m.s^{-1}],t_{H_2O} is the time of flight [s] in the cell filled with water, and $t_{oil}$is the time of flight [s] in the cell filled withsoybean oil.

The speed of sound in the water was determined as a function of temperature, using equation (2), given by0:

$$C_{H_2O} = A + B.T - C.T^2 + D.T^3 - E.T^4 + F.T^5 \tag{2}$$

where: *A*, *B*, *C*,*D*, *E*and*F* are constants and *T* is the temperature of the water in °C.

The measurement uncertainty is a parameter that provides positive dispersion characteristics of a measured quantity. Here, the measurement uncertainty is calculated according to the Guide of the Expression of Uncertainty in Measurements0.

Standard uncertainties Type A and Type B were considered. The standard uncertainty Type A was obtained by calculating the standard deviation of the mean for five repetitions. The standard uncertainty Type B was estimated from the temperature uncertainty acquired from the digital thermometer calibration certificate (0.08 °C), time of flight uncertainty obtained from the oscilloscope calibration certificate (0.0018%), and the speed of sound uncertainty in water, obtained from the curve fit error (0.002 m.s^{-1}) declared in 0.

The combined standard uncertainty $uc(v_{oil})$is the positive square root of the combined variance, and is given by (3):

$$uc(v_{oil}) = \sqrt{\sum \left(\frac{\partial v_{oil}}{\partial x_i} . ux_i\right)^2} \quad (3)$$

The expanded uncertainty $U(v_{oil})$was obtained by multiplying the combined standard uncertainty $uc(v_{oil})$ by a coverage factor k that takes into account a *t*-distribution with coverage probability of 0.95.

3. Results and Discussion

The measurement results of the speed of sound in soybean oil, with its respective uncertainties are shown in table 1.

Table 1. Average values of speed of sound in soybean oil and their respective expanded (U_{exp}) and relative uncertainties (U_{rel}).

T(°C)	Speed of sound (m.s^{-1})	U_{exp}(m.s^{-1})	U_{rel}(%)
20	1484.1	4.3	0.3
25	1466.6	4.9	0.3
30	1452.8	6.4	0.4
35	1434.9	5.9	0.4
40	1419.9	5.2	0.4
45	1402.0	4.0	0.3
50	1383.6	2.0	0.1

The results from five repetitions are presented in Figure 2. Averaging those results, one can obtain a curve representing the relationship between the speed of sound and the temperature, which is shown in Figure 3,with its associated uncertainty.Analyzing Figure 3,one can observe a linear relation between the speed of sound and temperature, in the temperature range studied. Moreover, the speed of soundin soybean oil decreases with the increasing of temperature.The speed of sound can vary according several factors, though it regarded as a constant characteristic of each material, particularly for liquids.

Note that the contribution to uncertainty in the propagation velocity was barely significant. A possible explanation is given by thedocking of the transducer in the vessel lid used; thisfit must be performed carefully as a possible variation directly affects the distance.

From measurements, it was established a characteristic function to demonstrate the behavior of the speed ofsound in soybean oil asa function of temperature (4).

$$v_{oil} = a + b * T \quad (4)$$

where*a* and *b* are constants and *T* is the temperature in °C.

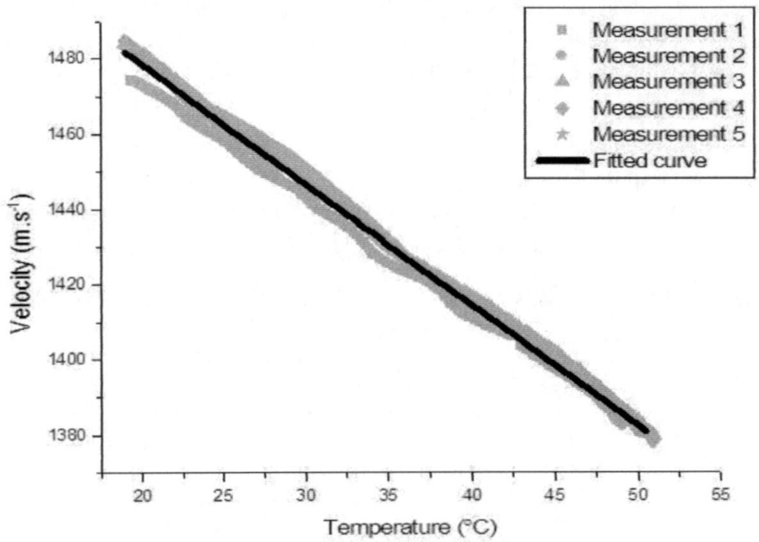

Figure 2. Curves from the five repetitions and the fitting curve.

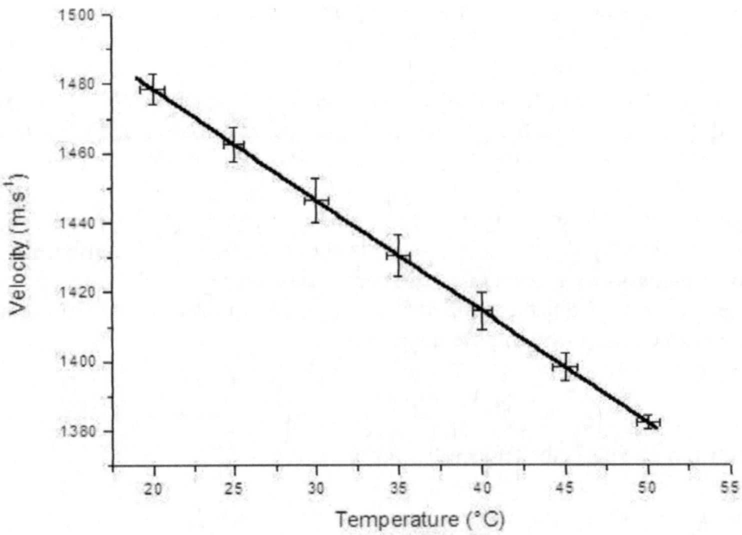

Figure 3. Relationship between the speed of sound and the temperature with its associated uncertainty.

The temperature measurement uncertainty was also calculated and depicted in Table 2. By analyzing the measurement uncertainty values it is possible to verify that the relative uncertainty of the temperature is higher than the relative uncertainty of the velocity of propagation. However, this uncertainty does not influence significantly the temperature measurement efficiency.

Table 2. Values of expanded (Uexp) and relative (Urel) uncertainties at different temperatures.

T (°C)	U_{exp}(°C))	U_{rel}(%)
20	0.74	3.7
25	0.62	2.5
30	0.75	2.5
35	0.70	2.0
40	0.57	1.4
45	0.77	1.7
50	0.68	1.4

Table 3 shows the linear regression parameters of the characteristic function that relates the speed of sound in soybean oil and temperature, with respective standard deviation. Assessing the variation coefficient values, one can suggest that the data set is homogeneous.

Table 3. Values of the constants a and b for Eq. (4).

Propagation velocity parameters	
a (m.s^{-1})	1542.60
σ_A	5.4122
c_{vA}(%)	0.35
b (m.s^{-1}.°C^{-1})	-3.2037
σ_B	0.1072
c_{vB}(%)	3.35
R^2	0.99748

4. Conclusion

Ultrasound, being a non-destructive technique and low-cost has great potential to contribute to the industrial development. Here, it was developed and a method for measuring the speed of sound in soybean oil. The results demonstrated that the propagation velocity varies linearly with temperature increasing in the range of 20 °C to 50 °C. A characteristic function representing the behavior of the speed of sound in soybean oil as a function of temperature was determined and a linear regression was estimated. With the help of linear regression function one can establish the speed of sound in soybean oil at different temperatures, with an expanded uncertainty of 7.0 m/s-1 (0.5%), and maximum error of 6.4 m/s-1.

5. References

[1]Carvalho R M, Pessoa F L P, Machado J C, Pereira F R and Adamowski J C 2014 Evaluation of Solid Content in Petroleum and Water in Oil Emulsion by Ultrasonic Spectroscopy. *Rev. Virtual Quím,* v. 6, p. 352-362.

[2]Higuti R T, Nazário S L S, Isepon J S, Buiochi F, Kitano C and Adamowski J C 2001 Characterization of Lubricating Oil Using Ultrasound. *J. Braz. Soc. Mech. Sci*, Vol.23.

[3]Koc B A 2009 Ultrasonic monitoring of glycerol settling during transesterification of soybean oil. *Bioresour. Technol*,Vol.10.

[4]Figueiredo M K K, Costa-Felix R P B, Maggi L E, Alvarenga A V and Romeiro G A 2012. Biofuel ethanol adulteration detection using an ultrasonic measurement method. *Fuel*, Vol.91, 209–12.

[5]Zeqiri B, Scholl W and Robinson S P 2010 Measurement and testing of the acoustic properties of materials: a review. *Metrologia*, Vol.47 (2), 156-71.

[6]Del Grosso V A 1998 Speed of Sound in Pure Water. *Ultrasound in Med. &Biol,*Vol. 24, 1065-68.

[7]JCGM, 2008, Evaluation of Measurement Data – *Guide of the Expression of Uncertainty in Measurements*, JCGM 100:2008.

Acknowledgements
The authors would like acknowledge the financial support from PROMETRO/CNPq and FAPERJ.

Journal of Physics: Conference Series **733** (2016) 012041 doi:10.1088/1742-6596/733/1/012041

Metrology in Pharmaceutical Industry – A Case Study

Priscila D. Yuvamoto[1], Ricardo K. S. Fermam[1], Elizabeth S. Nascimento[2]

[1] National Institute of Metrology, Quality and Technology (Inmetro). General Coordination for Accreditation (Cgcre). Rua Santa Alexandrina, 416, 7th floor, Rio de Janeiro, RJ 20261-232, Brazil.
[2] São Paulo University (USP). Av. Prf. Lineu Prestes, 580, Butantã, 05508-900 - São Paulo, SP – Brazil.

pdyuvamoto@inmetro.gov.br

Abstract. Metrology is recognized by improving production process, increasing the productivity, giving more reliability to the measurements and consequently, it impacts in the economy of a country. Pharmaceutical area developed GMP (Good Manufacture Practice) requeriments, with no introduction of metrological concepts. However, due to Nanomedicines, it is expected this approach and the consequent positive results. The aim of this work is to verify the level of metrology implementation in a Brazilian pharmaceutical industry, using a case study. The purpose is a better mutual comprehension by both areas, acting together and governmental support to robustness of Brazilian pharmaceutical area.

1. Introduction

The importance of Metrology is concerning to provide reliability in the measurements. This occurs in standardizing some practices, demonstration of measurements uncertainties and expression of trend results to take decisions. This information is used to process control and verification of quality requirements of a product [1]. Correct measurements also contributes to the economy (it is estimated around 2.7% of GNP – gross national product - in European Union), because they avoid to rework, and consequently, reduce costs and increase productivity, besides more equitable commercial relations [2].

In pharmaceutical industry, several requirements were developed to regulate this area, as the "Good Manufacture Practices" (GMP), searching for quality, security and effectiveness. However, metrological concepts like traceability, uncertainties, were not introduced at that time.

Metrology emerged firstly in calibration of physical elements, as mass, dimension, temperature, and others. Afterwards, this concept was applied in chemistry, originating the CRMs (certified reference materials). Its importance for several sectors of chemical industry (pharmaceutical, petroleum, foodstuff) is expanding, the investments of certification, human resources, equipment and others can reach 5 to 10% of the total production costs [3].

Besides the improvement of process (Productivity, more reliability in analytical results in Quality Control), metrology in Research and Development can also contribute for new drugs, as nanomedicines for example [4]. It is essential for robustness of a company, and Brazilian pharmaceutical industry is characterized by low international insertion, negative trade balance and development of generic drugs for the national market [5].

So, the purpose of this study is to know about the level of implementation of metrology in Brazilian pharmaceutical industries, using a case study. It is expected that this study contributes to public policies, helps to orient investments and joint actions by both areas, Metrology and Pharmaceutical industries.

2. Methods

Use of scientific references in database as SciELO, Google Scholar, "Periódicos Capes", in Portuguese and English languages during all the study.

A questionnaire was made using Surveymonkey® and was sent to a national pharmaceutical company. The identification is not mentioned because of confidenciality policy. The purpose of this case study was confirms the data from published articles and updates the level of implementation of metrology in this area.

3. Results and Discussion

A questionnaire was sent to a Brazilian pharmaceutical industry, whose content was about the implementation of metrology.

This is a national company and there is no affiliate in another country, confirming the low international insertion. It exports just some raw materials, mainly for European Union. According to them, the more difficult activity in exporting are the tariffs. There were options like importer requirements, transportation, reliability of Brazilian product and metrological subjects – all of them characterized by low level of difficulty. So, the REACH (Registration, Evaluation, Authorization and restriction of Chemicals) - an European Union regulation, published in 2006 and before considered a probable technical barrier to commerce, was complied by this company. But certainly it contributed to influence negatively in trade balance of this area [6].

Most part of the sale revenues of this company is from national market, constituted by similar and generic drugs, and 5 to 6% of this is invested in Research and Development. There is no nanomedicine produced by this company, justified by the difficulties to manufacture them in large scale and nowadays there is no regulation and the risks are unknown. The approach among international entities, like ISO, ASTM and FDA is happening to concentrate efforts [4].

One of the questions was about the rules or regulamentation used in Management System. There were options relating to ISO rules, as ISO 9001, ISO 17025, ISO 14001, GAMP (Good Practice Guide: Calibration Management), RDC ANVISA 17/2010 and there was the possibility to choose more than one answer. The result was just the national Good Manufacture Product or known as RDC 17/2010, regulated by ANVISA and required use. This question was searching to know some approach or familiarity with ISO rules that mention about metrology, but there is no contact with it.

This answer can be reinforced with the following question: Is there any human resources that have specific knowledge about metrology? The answer was no, and there is no qualification requirement in this area. It can be explained by historically, not only national but also international pharmaceutical industry, adopted the GMPs and not the metrological system.

And this absence of specific human resources in metrology can also justify the next answer. The company uses non accredited laboratories to calibrate the equipment, confirming that when there is no obligation, it is difficult to implement and invest on it. However, the consequences of non use or an inadequate use of metrological standards during a calibration can produce false results. A false disapproval can impact loss in productivity and a false approval can impact negatively in the quality of a medicine.

It was asked which was the level of difficulty in calibration in the following activities:

- For comprehension of calibration certified, the answer was low. It should consider the difficulties of a metrological standard and the amount of data of an accredited calibration certified. As reported before, this company do not use accredited calibration laboratories.

- For establishing the frequency of calibration, there is no difficulty. If there is no human resource to analyse it, maybe the frequency is established by the equipment producer.

- The difficulty to establish the frequency of intermediary verifications was considered intermediate level of difficulty, and it is explained by no published rule about it, even for accredited laboratories. A study evolving several influences has to be considered.

- And low level of difficulty to qualification of calibration service. And this answer is explained by no metrological requirements. For this company, it is not necessary that the laboratory be accredited.

The following question: During the root cause analyse, with is the frequency of the item Measurement in the Ishikawa diagram? The answer was Low. There are two possibilities to interpret it: one, that all the management system is functioning well or other, "Measurement" can not be analysed so deeply, with all the involved factors.

Among the non conformities there is more incidence on analytical measurement instruments than in process measurement instruments. The metrological traceability can be damaged in this case, without use of CRMs and calibrated equipment.

There was a question about the frequency of using CRM and PT schemes, and the answer was: The company uses reference materials, but no certified reference materials. They do not participate of PT schemes and have no intention to do it.

The last question was about the Nanomedicines, if the company has it in its portfolio. The answer was no. And when asked about the challenges, the answer was the production of nanomedicines, more than the quality control. And to produce an innovative product, it is necessary a constant improvement, invest in research and development and robustness in quality and productivity.

4. Conclusion

The national pharmaceutical industry focused on competitiveness based on quality and productivity, following the good manufacture practices, however, the calibration management and metrological insertion has not been accomplished. It is necessary to train human resources, disseminate the benefits of the insertion of metrological concepts.

Although some technical barriers can damage some markets, as REACH, Brazilian companies like this case study have robustness enough to continue to export to them its raw materials.

However, it is necessary to notice about the importance of metrology to innovation, quality and productivity. And these three elements are essentials to keep in the market and also increase the exportation, also with innovate goods, as Nanomedicines, for example. It is essential the governmental participation and public policies to promote joint actions.

References

[1] Weckenmann A, Kramer P, Akkasoglu G 2012 IV MESIC 2011. AIP Conf. Proc. 1431, 283-292; doi: 10.1063/1.4707576.

[2] Okrepilov V V, Krutikov V N, Elkin G I 2014 Measurement Techniques, 57, 2.

[3] Souza T L 2008 Tese de doutorado, UFRJ, Rio de Janeiro.

[4] Tinkle S, Mcneil S E, Mühlebach S, Bawa R, Borchard G, Barenholz Y, Tamarkin L, Desai N 2014 Ann. N. Y. Acad. Sci. 1313, 35-56.

[5] Pimentel V P, Gomes R P, Mitidieri T L, França F, Pieroni J P 2014 BNDES Setorial 40, 5-42.

[6] Bastos V D, Costa L M 2011 BNDES Setorial 33, p 163-206.

Use of ultrasound to monitor physical properties of soybean oil

R M Baêsso, P A Oliveira, G C Morais, A V Alvarenga, R P B Costa-Félix

Laboratory of Ultrasound (Labus), Directory of Scientific and Industrial Metrology (Dimci), National Institute of Metrology, Quality and Technology (Inmetro), Av. Nossa Sra. das Graças 50, Duque de Caxias, RJ, Brazil, ZIP 25250-020

raphaelabaesso@gmail.com

Abstract. The study of the monitoring physical properties of soybean oil was performed. The pulse-echo method allowed measuring the density and viscosity of the oil in real time and accurately. The physical property values were related to the acoustic time of flight ratio, dimensionless parameter that can be obtained from any reference. In our case, we used the time of flight at 20°C as reference and a fixed distance between the transducer and the reflector. Ultrasonic monitoring technique employed here has shown promising in the analysis of edible oils.

1. Introduction

Liquid viscosity and density are important properties for designing equipment for the fatty acids industries. One example is the use in estimating the efficiency of distillation columns aimed at the separation and purification of acids. Data fluid viscosity and density are also used in heat transfer designs, process piping and determining pressure drop[1][2].

Currently, there is considerable demand for measurement instruments that are able to characterize liquids with high sensitivity, robustness and accuracy in the chemical, food, petrochemical plants, among others. Furthermore, due to process automation, in-line measurements become necessary. Thus, the use of ultrasound techniques is interesting because it is a robust method, accurate, non-destructive and can be applied in the process line [3][4].

Literature shows the possibility of using ultrasound to measure the physical properties from the acoustic parameters of the liquid [5][6].

The aim of this study was to establish a relationship between the viscosity and density of the soybean oil, and the acoustic time of flight ratio.

The work purpose is to facilitate obtaining reliable data online and automated, allowing acquisition of immediate and accurate results. The calculation of uncertainties of measurements was carried throughout the study and is described here.

2. Methodology

In this project, we used low-power ultrasound and pulse/echo techniques, in which the ultrasonic pulse signals pass through the sample, reach the reflector and return to the same transducer. The study was carried out with the assistance of a transducer with nominal frequency of 1 MHz and diameter of 12.7 mm (NDT-Panametrics-Olympus Corporation, Japan), a waveform generator (33250A, Agilent Technologies, CA, USA), an oscilloscope (DSO-X 3012A, Agilent Technologies, CA, USA), a data acquisition unit (34970A, Agilent Technologies, CA, USA), and a computer with a program developed in LabView™, (National Instruments, Austin, TX, USA). In Figure 1, the experimental arrangement scheme is shown. It was possible to identified changes that occur in the density and viscosity of the soybean oil over the variations in temperature, through the signal of the ratio of acoustic time of flight.

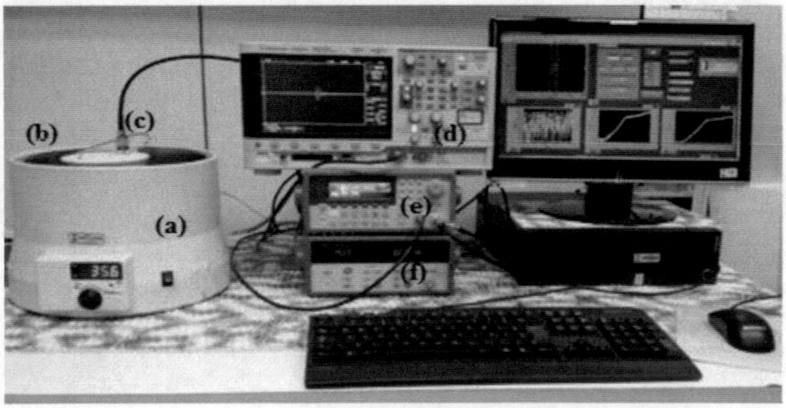

Figure 1. Experimental arrangement used for determining the viscosity and density of the soybean oil. (a) Heating bath, (b) glass recipient with oil, (c) transducer attached to the system, (d) osciloscope, (e) waveform generator and (f) data acquisition unit.

According to the literature [7][8], the dependency of the viscosity of neat vegetable oils with temperature can be satisfactorily described by the Andrade relationship (1)

$$\mu = A_1 \,.\, e^{\left(\frac{E}{RT}\right)} \tag{1}$$

where μ is the dynamic viscosity, T is the absolute temperature, A_1 is a constant, E is the activation energy and R is the universal gas constant.

The values of the parameters A_1 and E to soybean oil have been obtained from literature [7] (Table 1), and used to establish the relationship between the dynamic viscosity of the soybean oil and the time of flight ratio.

Table 1. Values of the constants A_1 and E for Eq.(1) and data fitting statistics for soybean oil.

Oil	Soybean
A_1^*(10^{-3}mPa.s)	1.616
E^* (J/mol)	25649
r²	0.9964

**Valid only in the range (293.15-353.15)K*

Source: *Adapted from Nguyen, Q.D. (2011).*

On the other hand, studies have shown [7][8] that the density of oil, varies linearly with increasing absolute temperature from the equation (2)

$$\rho = \alpha + \beta . T \tag{2}$$

where α and β are empirical constants. The values of the parametersα and β for the soybean oil were taken from the literature [7] (Table 2), and used to establish the relationship between the density of the soybean oil and the time of flight ratio.

Table 2. Values of the constants α and β for Eq.(2) for the pure components of the soybean oils.

Oil	Soybean
α(kg/m³)	1197.0
β(kg/m³ K)	-0.959
r^2	0.9978

Source: *adapted from Nguyen, Q.D. (2011).*

3. Results and Discussion

The data obtained by processing the ultrasonic signals from the 5 repetitions provided two curves capable to specify relationships of viscosity (Figure 2(a)) and density (Figure 2(b)) versus time of flight ratio, respectively.

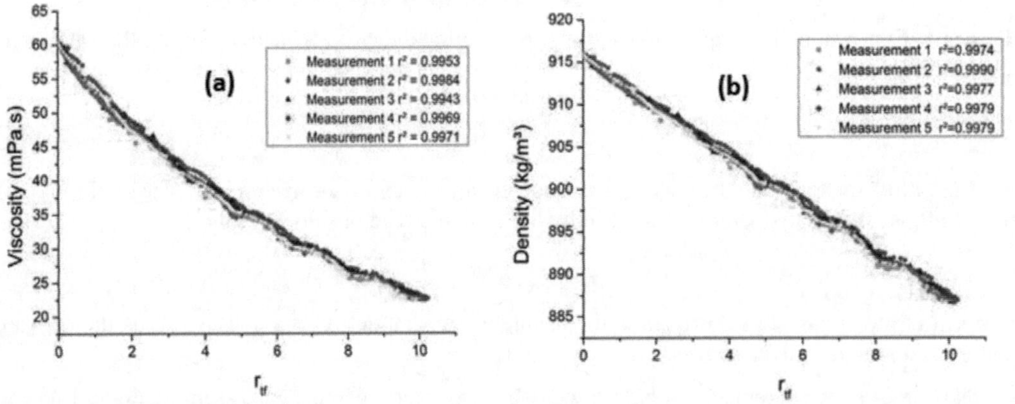

Figure 2. (a) Relation between dynamic viscosity and time of flight ratio and **(b) r**elation between density and time of flight ratio.

Characteristic functions were established, equations (3) and (4), able to describe the behavior of μ e ρ, respectively, as a function of time of flight ratio,

$$\mu = A . e^{(B.r_{tf})} \tag{3}$$

$$\rho = a + b . r_{tf} \tag{4}$$

where *A*, *B*, *a* and *b* are constants and r_{tf} is the ultrasonic time of flight ratio (dimensionless parameter). From the trials, it was calculated for each set of the four parameters, the average (*A*, *B*, *a* and *b*), the standard deviation (σ_A, σ_B, σ_a and σ_b) and the coefficient of variation (c_{vA}, c_{vB}, c_{va} and c_{vb}) (Table 3).

Table 3. Values of the constants *A*, *B*, *a* and *b* for Eq. (3) and (4) for the pure components of the soybean oil.

Oil	**Soybean**
Viscosity parameters	
***A*(mPa.s)**	58.75
σ_A	0.64
c_{vA}**(%)**	1.1
B	-0.09864
σ_B	0.000934
c_{vB}**(%)**	0.9
Density parameters	
***a* (kg/m³)**	914.14
σ_a	0.13
c_{va}**(%)**	0.01
B	-2.8914
σ_b	0.06611
c_{vb}**(%)**	1.7

It is important to note that as the data taken from the literature for soybean oil could only be applied in the range of 20-80 °C, the benchmark set for r_{tf}calculation was obtained at 20.0 °C.

For the calculation of uncertainties obtained from these models, we used the equations (5) and (6) from the literature [9].

$$U_\mu = \sqrt{\Sigma(\frac{\partial\mu}{\partial x_i}.U_i)^2} \quad (5)$$

$$U_\rho = \sqrt{\Sigma(\frac{\partial\rho}{\partial x_i}.U_i)^2} \quad (6)$$

Tables 4 and 5 show the expanded uncertainty values calculated for the established models.

Table 4. Values of the calculated uncertainties (p=0,95) for dynamic viscosity.

Oil	ToF -Rat	μ (mPa.s)	U_{exp} (mPa.s)	U_{rel} (%)
Soybean	0	58.7	0.8	1.3
	2	48.2	0.7	1.4
	4	39.6	0.6	1.4
	6	32.5	0.5	1.4
	8	26.7	0.4	1.4
	10	21.9	0.3	1.5

Table 5. Values of the calculated uncertainties (p = 0,95) for density.

Oil	ToF-Rat	p (kg/m³)	U_{exp} (kg/m³)	U_{rel} (%)
Soybean	0	915.3	0.4	0.04
	2	909.6	0.4	0.04
	4	903.9	0.4	0.04
	6	898.2	0.4	0.05
	8	892.4	0.5	0.06
	10	886.7	0.6	0.07

Based on the data obtained from the study developed, there is a great capacity to estimate the viscosity and density from ultrasonic data.

Is important to clarify that since it uses the time of flight ratio as ultrasonic parameter, the dependence of the results with the distance between the transducer and the reflector does not exist anymore, being able to use any spacing whereas it has a reference established.

The development of easy techniques, inexpensive and capable to provide reliable results, makes the ultrasound feasible to be implemented in various industry segments.

4. Conclusion

Preliminary results suggest that the proposed method is feasible to monitor the physical properties in real time of soybean oil, making it possible to implement in production lines.

Thus, with the aid of ultrasound, any unexpected results that may occur during the production processes can be perceived, avoiding the waste of raw materials and products that do not meet compliance.

5. References

[1] Noureddini H, Teoh B C and Clements L D, 1992, *J. Am. Oil Chem. Soc.*, 69, 1189-1191.
[2] Noureddini H, Teoh B C and Clements L D, 1992, *J. Am. Oil Chem. Soc.*, 69, 1184-1188.
[3] Buiochi F, Adamowski J C., and Furukawa C M, 1998 *Proc. - IEEE Ultrason. Symp.*, 1193-1196.

[4] Greenwood M S and Bamberger J A, 2002, *Ultrasonics*, 39, 623- 630.
[5] Figueiredo M KK, Silva C E R, Alvarenga A V, *Costa*-Félix, R P B, 2015, *J. Phys.: Conf. Ser.*, 575, 012001.
[6] Figueiredo M KK, Costa-Félix, R P B, Alvarenga A V, Maggi, L E, Romeiro, G A, 2012, *Fuel*,91, 209-212.
[7] Nguyen Q D, Franco Z, 2011, *Fuel*,90, 838–843.
[8] Krisnangkuraa K, Yimsuwanb T, Pairintraa R, 2006, *Fuel*, 85, 107–113.
[9] JCGM, 2008, *Evaluation of Measurement Data – Guide of the Expression of Uncertainty in Measurements,* JCGM 100:2008.

6. Acknowledgments

Research reported in this paper has been partially funded by Faperj (E-26/201.563/2014), and CNPq (310.392/2014-4).

Uncertainty evaluation of dead zone of diagnostic ultrasound equipment

R M Souza[1], A V Alvarenga[1], D S Braz[1], L I Petrella[2], R P B Costa-Felix[1]

[1] Laboratory of Ultrasound, Directory of Scientific and Industrial Metrology, National Institute of Metrology, Quality and Technology. Av. Nossa Sra das Graças 50, Duque de Caxias, RJ, 25250-020, Brazil

[2] Institute of Nuclear Sciences Applied to Health, University of Coimbra, Portugal

E-mail: raquel.ms.eng@gmail.com

Abstract. This paper presents a model for evaluating measurement uncertainty of a feature used in the assessment of ultrasound images: dead zone. The dead zone was measured by two technicians of the INMETRO's Laboratory of Ultrasound using a phantom and following the standard IEC/TS 61390. The uncertainty model was proposed based on the Guide to the Expression of Uncertainty in Measurement. For the tested equipment, results indicate a dead zone of 1.01 mm, and based on the proposed model, the expanded uncertainty was 0.17 mm. The proposed uncertainty model contributes as a novel way for metrological evaluation of diagnostic imaging by ultrasound.

1. Introduction

Medical ultrasound imaging equipments (USIE) are widely used in the most of Brazilian health centers. Although the emergence of new medical equipment with increasingly sophisticated technologies provides more confidence in medical activities, there is still need for quality control of such equipment [1].

It is worth pointing out that measurement procedures for evaluating USIE images are often subject of research, but, typically, no proper uncertainty evaluation is undertaken. These measurement procedures are based on standards and documents published by international organizations, such as the American Association of Physicists in Medicine (AAPM) [2] and the International Electrotechnical Commission (IEC) [3]. Specific indicators as contrast, axial and lateral resolutions, and dead zone, among others are quantified using phantoms and used to evaluate the ultrasound imaging quality [2][3].

Measurement uncertainty is a non-negative parameter that characterizes the dispersion of the quantity values being attributed to a measurand [4][5]. Considering the expression of uncertainty of the indicators used in assessing the quality of medical images, the literature does not present conclusive works. It is noteworthy that estimating the measurement uncertainty is essential to evaluate variations in the measurand, providing a metrological approach on quality control.

The dead zone occurs because the ultrasound system cannot send and receive data simultaneously [2]. A change in the system's dead zone is indicative of a problem with the transducer, the pulsing system or both. The IEC/TS 61390 defines the dead zone as the distance from the phantom surface to the nearest target that can be unequivocally imaged [3]. This standard also states that an estimate of the dead zone can be obtained by noting the most superficial target that can be imaged on a phantom having a set of targets at the top surface. Hence, if the first target observed is located at 1 mm away from the phantom top surface, the dead zone will be considered 1 mm.

In this paper, we propose also measuring the distance between the first target imaged and the top of the image, using the USIE caliper. This procedure gives a measured value of the dead zone, and allows expressing the measurement uncertainty.

2. Materials and Methods

2.1. Phantom

The phantom adopted for the tests is the Multi-purpouse, Multi-tissue, Ultrasound Phantom 040GSE (CIRS, Virginia, USA). Within the phantom, the targets group used to determine the dead zone consists of nylon monofilament wires (100 µm diameter), horizontally spaced 6 mm apart from center to center (figure 1).

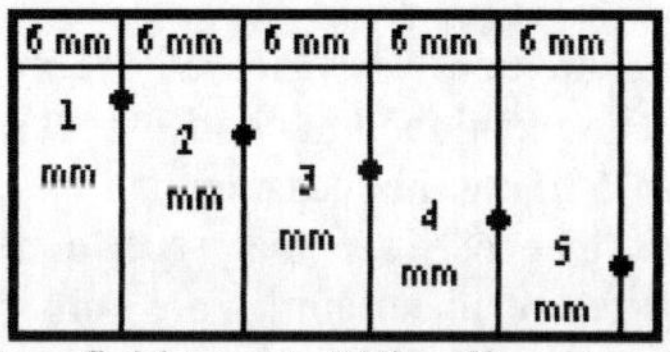

Figure 1. Near field target. (Wire diameter: 0.1 mm).

Vertical distance from the center of each wire to the top edge of the scanning surface ranges from 5 mm down to 1 mm in 1 mm increments. These targets allow assessing the distance from the front face of the transducer to the closest distinguishable echo.

2.2. Ultrasound Equipment

The USIE used for this evaluation was the CTS-5500V with a transducer C5F20, presenting convex shape, and operating at 5 MHz (SIUI, Guangdong, China).

2.3. Positioning system

To carry out the measurements, a positioning system was used to allow the stable position of the transducer on the phantom scanning surface. Pressure on the phantom surface was avoided to prevent any distortions in the image.

2.4. Environmental conditions

The environmental conditions were monitored and were always within the limits established in [3], namely: room temperature between 21°C and 25°C and relative humidity from 45% to 75%.

2.5. Instrument Settings

The instrument settings were performed in two steps. The first one, called "basic instrument setting" was carried out scanning the phantom as if it were a patient and the controls were adjusted to produce the best possible clinical image. Care was taken to not emphasize or exaggerate a particular image attribute [2]. The "specific instrument setting" was carried out for better viewing the targets. The focal zone was defined closest to scanning window and the gain was adjusted to reduce the background echoes [2].

2.6. Dead Zone Testing Procedure

The measurement procedure was defined based on references [2] [3] and it is described as follows:

- Fill the water on top of the phantom as the transducer coupling medium;

- Adjust the basic instrument setting;
- Position the transducer in a vertical plane;
- Apply the specific instrument setting;
- Scan the region in the phantom containing the dead zone test targets (figure 1);
- Freeze the image and determine the closest target that can be imaged;
- Use the electronic caliper of the ultrasound imaging equipment to measure the depth of this target and record dead zone.

2.7. *Technicians*

The inter-observer variability in assessing US images is well documented. In order to include the inter-observer variability in the uncertainty assessment, two trained technicians assessed the dead zone. Both technicians repeated the measurements thrice, under the same environmental conditions, using the same instrument settings and testing procedures.

2.8. *Expression of uncertainty*

To express the uncertainty of measurement, it was used the method described in the Guide to the Expression of Uncertainty in [4]. The mentioned method for evaluating and expressing the uncertainty of the result measurement starts by defining the measurand.

Here, the measurand is the distance between the target and the top of the image, which is the particular quantity itself. Two sources of uncertainties are considered: (i) the number of independent observations obtained under the same conditions of measurement, and (ii) the resolution (R) in millimeters of caliper.

For the estimate obtained from the statistical analysis of series of observations, it is carried out the Type A evaluation of standard uncertainty (u_s), which is calculated by (1).

$$u_s = \frac{s}{\sqrt{n}} \quad (1)$$

Where s is the experimental standard deviation and n is the number of independent observations obtained under the same conditions of measurement.

The resolution of the diagnostic equipment caliper (R = 0.18 mm) was used to calculate the Type B standard uncertainty (u_R). Assuming a rectangular distribution, all values have equal probability of occurring. Therefore, u_R is given by (2).

$$u_R = \frac{R}{\sqrt{12}} \quad (2)$$

Thus, u_R = 0.052 mm.

The combined standard uncertainty (u_C) can be obtained by combining the uncertainty components that characterize the dead zone (3).

$$u_C = \sqrt{(u_R{}^2 + u_s{}^2)} \quad (3)$$

Finally, the expanded uncertainty (U) is calculated by (4):

$$U = k \times u_C \quad (4)$$

Where k is the coverage factor determined by considering the effective degrees of freedom and the level of confidence of 95%.

2.9. *Difference test between means*

The difference test between means is applied in order to verify if the measurements carried out by two technicians can be considered statistically equal for a given coverage probability. If there is not enough statistical evidence to reject the equal variances hypothesis, the combined standard uncertainty (u_{cP}) can be determined according to (5).

$$u_{cP} = \sqrt{\frac{(n_1-1)u_{c1}{}^2+(n_2-1)u_{c2}{}^2}{n_1+n_2-2}} \tag{5}$$

Where u_{c1} and u_{c2} are the combined uncertainty from technicians 1 and 2, while n_1 and n_2 are the respective number of measurements.

Subsequently, it is considered the null hypothesis of equality between means and the alternative hypothesis of differences between means; thus, the statistical t_{calc} test is being given by (6).

$$t_{calc} = \frac{X_{tec.1}-X_{tec.2}}{u_{cP}\sqrt{\left(\frac{n_1+n_2}{n_1 \times n_2}\right)}} \tag{6}$$

Where $X_{tec.}$ is the mean value of the measurements of each technician.

The value obtained for t_{calc} is compared with t_{tab}, which is extracted from Student's t-distribution table, on the basis of the degrees of freedom $(v \mid v = n_1 + n_2 - 2)$. If t_{calc} is less than t_{tab}, the null hypothesis cannot be rejected, thus the result of each technician can be combined to determine the final result of the measurement. Finally, the expanded uncertainty arising from the measurement combination of both technicians is given by:

$$U = t_{tab.} \times u_{cP}\sqrt{\left(\frac{n_1+n_2}{n_1 \times n_2}\right)} \tag{7}$$

3. Results

Table 1 provides results obtained by technician 1, as well the coverage factor, expanded uncertainty and absolute error(E_{abs}). Similarly, Table 2 provides results obtained by technician 2.

Table 1. Result of measurements carried out by technician 1.

Reference value (mm)	X (mm)	s (mm)	u_s (mm)	u_R (mm)	k	U (mm)	E_{abs} (mm)
1.0	0.92	0.065	0.038	0.052	2.12	0.13	0.08

Table 2. Result of measurements carried out by technician 2.

Reference value (mm)	X (mm)	s (mm)	u_s (mm)	u_R (mm)	k	U (mm)	E_{abs} (mm)
1.0	1.09	0.035	0.020	0.052	1.98	0.11	0.09

The figure 2 presents the region in the phantom containing the dead zone test targets imaged by the USIE.

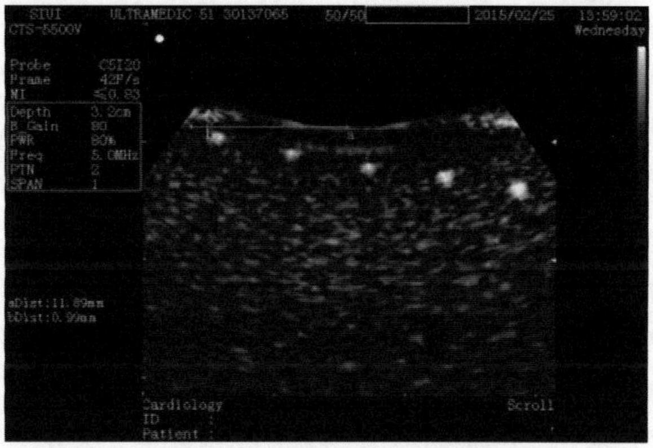

Figure 2. Image of dead zone.

The dead zone measured by the technician 1 resulted in 0.92 mm ± 0.13 mm, whereas for the technician 2, the result was 1.09 mm ± 0.11 mm. Applying the difference test between means for a coverage probability of 95 %, it was observed that the null hypothesis cannot be rejected. Thus, the results from both technicians may be combined by providing a dead zone of 1.01 mm ± 0.17 mm.

4. Discussions and conclusion

The result suggests that the dead zone of the equipment tested is 1.01 mm, with an expanded uncertainty of 0.17 mm, considering the proposed uncertainty model. Moreover, the procedure presented here can be useful to metrological evaluate the dead zone periodically. Hence, carrying out periodical tests and evaluating the uncertainty of dead zone, it is possible to estimate eventual variations in that value, which may indicate a possible change in image quality.

5. References

[1] A Oliveira L M, Maia J M, Gamba H R, Gewehr P M, Pereira W C A 2010 Avaliação da qualidade de imagens de equipamentos de ultrassom modo B; *Revista Brasileira de Engenharia Biomédica*, v. 26, n. 1, p. 11-24.

[2] Mitchell M, Goodsitt and Paul L. Carson 1998 Real-time B-mode ultrasound quality control test procedures Report of AAPM Ultrasound Task Group No. 1, Department of Radiology, University of Michigan.

[3] IEC/TS 61390 ed. 1.0 1996: Ultrasonics – Real-time pulse-echo systems – Test procedures to determine performance specifications, International Electrotechnical Commission, Genève, Suisse.

[4] JCGM 2008 Evaluation of Measurement Data – Guide of the Expression of Uncertainty in Measurements JCGM 100:2008.

[5] JCGM 2012, International Vocabulary of Metrology – Basic and general concepts and associated terms, JCGM 200:2012.

Acknowledgments

The authors would like to acknowledge the financial support from the CNPq and the Faperj.

Standard operating procedure to prepare agar phantoms

R M Souza[1], T Q Santos[1], D P Oliveira[1], R M Souza[1], A V Alvarenga[1], R P B Costa-Felix[1]

[1] Laboratory of Ultrasound (Labus), Directory of Scientific and Industrial Metrology (Dimci), National Institute of Metrology, Quality and Technology (Inmetro)

E-mail: rebecasouzam@gmail.com

Abstract. Agar phantoms are widely used as soft tissue mimics and some preparation techniques are described in the literature. There are also standards that describe the recipe of a soft tissue mimicking material (TMM). However some details of manufacture process are not clearly defined. The standardization of the phantom's preparation can produce a metrological impact on the results of the acoustic properties measured. In this direction, this paper presents a standard operating procedure (SOP) to prepare the agar TMM described on the IEC 60601-2-37.

1. Introduction

Ultrasonic phantoms are probe bodies used to mimic the ultrasound properties of biologic tissues such as acoustic impedance, ultrasound propagation velocity, and attenuation coefficient, allowing the study of its interactions with the ultrasound [1]. In general, the use of these tissues mimicking material (TMM) is intended to many applications, such as the calibrating ultrasound imaging diagnostic and therapeutic equipment, evaluating heating generated by ultrasonic transducers, among others.

The advantage of tissue phantoms is that tissue models can be constructed with defined acoustic properties, dimensions and internal features, thereby simplifying and standardizing the imaging environment. Phantoms are available commercially, mimicking many tissues organs and organ systems. Commercial phantoms costs hundreds to thousands of dollars and are often preferred for training and calibration of ultrasound systems [2].

It is worth pointing out agar phantoms are widely used as soft tissue mimics and some preparation techniques described in the literature [2][3].

The standards IEC 60601-2-5 [4] and IEC 60601-2-37 [5] present a recipe of agar TMM to evaluate safety on ultrasound therapeutic and diagnostic equipment, respectively. Although described in standards, some details of manufacture procedure are not clear, or some aspects that could influence the homogeneity of the TMM and some details on its manufacture process are not well defined, which lead each one to perform different procedures.

2. Materials and Methods

2.1. Phantom

Many materials have been studied looking for the optimum acoustical properties for the manufacture of phantoms such as gelatin, agar, polyacrylamide Zerdine®, urethane rubbers epoxy, among others [1]. This paper presents the agar TMM recipe.

The broad use of agar phantom is a result of their well characterized performance, the ease of fabrication and the flexibility that the process provides, allowing the incorporation of additional components to achieve a range of acoustic properties [2].

The accurate reporting of acoustic properties is highly dependent on preparation, and handling of TMM and the inherent dependence of acoustic properties, especially attenuation and backscatter. According to the standards [4][5] the acoustic properties of the TMM are listed in Table 1.

Table 1. Acoustic properties.

Material	**Velocity (m/s)**	**Density (kg/m³)**	**Attenuation Coefficiente (dB/cm MHz)**	**Acoustic Impedance (10^8 kg/m^2 s)**
TMM	1540	1050	0.5	1.6

2.2. Materials

The list of materials used to prepare the phantom is described as follows:

- Beaker of 1000 ml;
- Beaker of 25 ml;
- Glass rod;
- Erlenmeyer of 2000 ml;
- Mechanical Stirrer;
- Foil.

2.3. Components

The mixture is made from the materials provided in Table 2 [4][5]. The second column provides a percentage weight of each pure component. It should be calculated the weight of the component according to the necessary amount of water. Considering 660 g of water, the weights of components are shown in the third column of Table 2.

Table 2. Components.

Component	**Weight (%)**	**Weight (g)**
Glycerol	11.21	89.19
Deionized Water	82.95	660
Benzallkonium chloride	0.47	3.74
Silicon Carbide (Sic(-400 mesh)	0.53	4.22
Aluminium Oxide (Al2O3 (0,3μm))	0.88	7
Aluminium Oxide (Al2O3 (3μm))	0.94	7.48
Agar	3.02	24.03
Sum	100	795.66

2.4. *Stirring Technique*
There are three possible stirring methods: manual, mechanical or magnetic. None of them was mentioned in the standards [4][5]. The literature still does not present conclusive works about that. However, Culjat and colleagues show the difference among these methods [2]. In this paper, the method chosen was the mechanical stirring because of the large volume of mixture.

2.5. *Standard operating procedure*
The standard operating procedure to prepare agar phantoms was divided in seven steps, described as following.

1°) Initially, all the components must be weighed and the deionized water, to add into the mix, heated up to 90 °C (Figures 1 and 2). A calibrated balance shall be used. Thereafter, the components must be mixed together in the following sequence:

Add Al_2O_3, SiC, glycerin and benzalkonium chloride in a container and mix them using a glass rod until a homogeneous appearance is obtained (looks like a viscous paste light gray). Then, add the agar and the water homogenizing the mixture.

Figure 1. Component being weighed.

Figure 2. Heating of water.

2°) The mixture should be maintained in a water bath under continuous mechanical stirring. The water bath temperature can vary in the range of 90-100 °C. In the procedure described herein, the water bath temperature was around 97 °C. It is noteworthy that the water bath should be preheated while weighing the components of the mixture for 30 minutes with thermal bath at 120 °C. The rotational speed varies with the volume of the mixture in question. Initially, the mechanical stirring was set at 200 rpm. It is important that no vortex shall occur to avoid the bubbles formation in the mixture. As the viscosity of the mixture increases, the rotational speed must increase. During the phantom preparation, the mechanical stirring was increased in steps of 100 rpm, until it reaches 1000 rpm, according to the viscosity increasing.

3°) Heat the mixture for one hour in the temperature range of between 80-90 °C. In the present procedure, the mixture temperature was around 83 °C. To minimize evaporation to get out of the flask, it is desirable that the mixture container remains covered during the entire process. An Erlenmeyer was used due to its smaller nozzle diameter, in order to reduce evaporation.

4°) Therefore, the mixture should be cooled to 47 °C, under continuous mechanical stirring at 1000 rpm. Again, it is desirable that the mixture remains covered during this process in order to prevent evaporation.

5°) Dump the substance into a mold (Figure 3) and let it cool down. To avoid water evaporation, the mold is kept covered. After 24h the phantom can be demolded, cut (Figure 4), and it is ready to be stored.

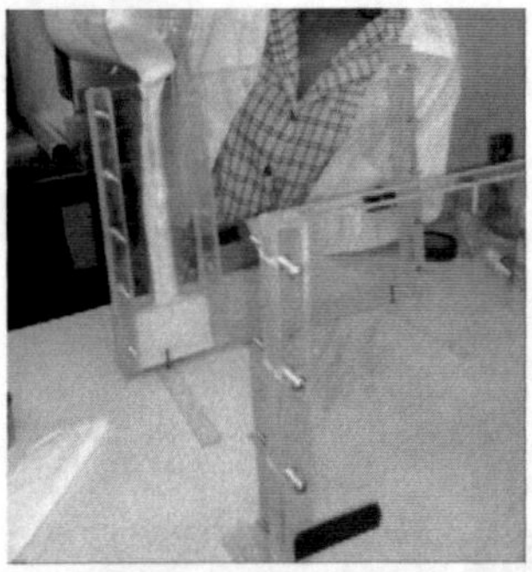

Figure 3. The mixture being poured into the mold.

Figure 4. The phantoms being cut.

6°) The material should be stored in a closed container under normal laboratory conditions (18°C – 25 °C). While stored, keep the material in a water/glycerol mixture to prevent it from drying out and to avoid air contact. This mixture contains 88.1% (weight) demineralized water and 11.9% (weight) glycerol (purity > 99%).

The shelf life of the material is at least one year, if it is preserved without air contact. The addition of a 0.5% (weight) solution of benzalkonium chlorid acts as an antifungal agent extending the life of the phantom. Produced samples with shelf lives over 2 years were found [4][5].

Two independent batches of phantoms where produced in different days, both applying the procedure described in this paper. The speed of sound of two phantoms, each one from one batch, is presented in Table 3. While its expanded uncertainties (U) were calculated based on the method described in [7] and are presented in the same table. The values of speed of sound were measured at Labus using the procedure described in [8]. One can observe that, based on the normalized error (En), the values of speed of sound cannot be considered statistically different. This result suggests that, using the proposed procedure, it will possible to produce phantoms with similar ultrasonic properties.

Journal of Physics: Conference Series **733** (2016) 012044 doi:10.1088/1742-6596/733/1/012044

Table 3. The values of speed of sound measured from phantoms obtained using the proposed standard operating procedure, but cooked in two different days.

Phantom	Speed of sound (m/s)	U (m/s)
Day 1	1551.8	5.5
Day 2	1550.3	5.3
E_n	0.2	

3. Discussions and conclusions

The lack of details in the standardizing document of agar phantom preparation could affect negatively on its ultrasonic properties. The aim of this paper is ultimately provide a step-by-step procedure to develop an agar phantom, according to the recipe disclosed at international standards. Furthermore, a metrologically based comparison between two phantoms produced in two dissimilar batches was performed. Ultrasonic quantity "speed of sound" was the comparison parameter. Results depicted from the statistical lead to the conclusion that the procedure to produce phantoms can be considered good enough, regarding its repeatability.

A worldwide international comparison is under agreement among many National Metrology Institutes and some ultrasonic devices manufactures. Agar phantoms will be the circulated among the participants, and the ultrasonic parameters "speed of sound" and "attenuation coefficient" will be the quantities to be assessed. The goal of such work is to stablish both a standardized agar phantom detailed procedure for manufacturing, as well as mutually accepted measurement procedure and uncertainties. The manufacturing procedure is the first step, and the work described within this paper will be used for such activity.

4. References

[1] PETRELLA, L. I. *et al.* 2014. "*Influence of subcutaneous fat in surface heating of ultrasonic diagnostic transducers*". Ultrasonics, v.54 (6), p.1476–1479.

[2] CULJAT, M. O. *et al.* 2010. "*A review of tissue substitutes for ultrasound imaging". Ultrasound in Medicine & Biology*, v.36, n.6, p.861–873.

[3] OLIVEIRA, D. P. *et al.* 2014. "*Ultrasound Propagation Velocity and Acoustic Attenuation on Agarose Phantoms' in Three Different Manufacture Techniques*". PAHCE.

[4] International Electrotechnical Commission. Part 2-5: Particular requirements for the basic safety and essential performance of ultrasonic physiotherapy equipment. 3rd Ed. Geneve: IEC, 2009, 66pp, IEC 60601-2-5:2009.

[5] International Electrotechnical Commission. Part 2-37: Particular requirements for the basic safety and essential performance of ultrasonic medical diagnostic and monitoring equipments. 2rd Ed. Geneve: IEC, 2007, 82pp, IEC 60601-2-37:2007.

[6] JCGM. *International Vocabulary of Metrology – Basic and general concepts and associated terms*. 3rd Ed. Paris: BIPM, 2012, 91pp, JCGM 200:2012.

[7] JCGM. Evaluation of Measurement Data – Guide of the Expression of Uncertainty in Measurements 1rd Ed. Paris: BIPM, 2008, 120pp, JCGM 100:2008.

[8] SANTOS, T. Q. et al. 2014 Validação do método de medição da velocidade ultrassônica de propagação longitudinal. In: Anais do XXIV CBEB 2014, Uberlândia. p.1644-1648.

Acknowledgments

The authors would like to acknowledge the financial support from the CNPq and the Faperj.

Journal of Physics: Conference Series **733** (2016) 012045 doi:10.1088/1742-6596/733/1/012045

Brief review of uncertainty quantification for particle image velocimetry

M H Farias[1], **R S Teixeira**[1], **J Koiller**[1] **and A M Santos**[1]

[1]National Institute of Metrology, Quality and Technology, Av.N. Sra. das Graças, 50, Prédio 06, Duque de Caxias - RJ, 25250-020, Brazil

E-mail: mhfarias@inmetro.gov.br

Abstract. Metrological studies for particle image velocimetry (PIV) are recent in literature. An attempt to evaluate the uncertainty quantifications (UQ) of the PIV velocity field are in evidence. Therefore, a short review on main sources of uncertainty in PIV and available methodologies for its quantification are presented. In addition, the potential of some mathematical techniques, coming from the area of geometric mechanics and control, that could interest the fluids UQ community are highlighted as good possibilities.

"We must measure what is measurable and make measurable what cannot be measured" (Galileo)

1. Introduction: 1984-1985

One could date the birth of particle image velocimetry (PIV) around 1984-1985, when several groups, notably in Germany (Oldenburg and Gottingen) began to photograph the scattering of light off particles seeded in the fluid, illuminated by a laser sheet. From a pair of pictures taken on a short time interval, the velocity components of particle motion in the plane of the sheet were obtained. Ronald Adrian, one of the most influential PIV researchers, actually mentions earlier work in 1917-1918 [1]. His bibliography up to 2009 is available in the web site of the meeting held in Gottingen to commemorate the 25th year of PIV[1].

1.1. Thirty years after

Particle Image Velocimetry is presently a mature experimental technique. It is a widely used tool for the study of gas and liquid flows due to its noninvasive, global nature. Applications of PIV range from high Reynolds number flows in aerodynamics to low and mid Reynolds in biofluidynamics; from multiphase to non-Newtonian; from turbomachinery and aeronautics flows to medical microfluidics. For recent trends and literature in we refer to the review by [2]. Symposia on PIV are held regularly, for instance the conference series which is now in its 11th edition[2] (September, 2015, two months after this assessment was submitted). A whole section will be devoted to PIV accuracy and uncertainty quantification. The purpose of this assessment is modest: we review some of the work on UQ since 2011 done by the PIV community. In the last section we outline, rather prospectively, a few mathematical techniques that could be potentially useful to address some of the uncertainty sources in PIV postprocessing.

[1] http://25-years-piv.dlr.de
[2] http://www.piv2015.org

1.2. Sources of Uncertainty

From the metrological perspective, PIV is still in its infancy. Uncertainty sources galore: in processing, they come from illumination, camera positioning and data recording. Seeding requires a mix of "art and science" and it is impossible to make it reproducible. Moreover, postprocessing has an "original sin": the estimated velocity at a image pixel comes from statistical correlations of a group of nearby particles motion. In addition, most commercial softwares extrapolate measurements to non-seeded areas in ways not detailed in the manuals. Seeds, being material particles, do not move precisely along the flow. There is a transversal component, due to several physical effects [3]. This is a serious issue specially for particle tracking velocimetry.

2. Review of Uncertainty Quantification effort in Particle Image Velocimetry

2.1. The 2011 meeting in Las Vegas

Probably the first concerted attempt to discuss uncertainty quantification in particle image velocimetry was the meeting held in May 11-13, 2011, Las Vegas, organized by Barton Smith (Utah State) and Pavlos Vlachos (Virginia Tech). The event website[3] contains slides/notes of fifteen presentations on uncertainty quantification for PIV and a short course by Coleman and Steele about their UQ book [4]. We now summarize some of the aspects discussed in that meeting, aimed to "establish the foundations for developing a comprehensive PIV uncertainty estimation methodology". The determination of the measurements accuracy with PIV was the main discussion topic in this meeting. The uncertainty sources analysis have been approached in numerous speeches. The Ronald J. Adrian presentation described an attempt to develop a systematic framework for the development and application of the analysis schemes for detections of the uncertainty sources on experiments with PIV [1]. The following presentations have described studies with different methods to detect the uncertainty sources. As such, experiments with complementary measurements, the use of CFD solutions for verification and evaluation of the auto-correlation used in postprocessing. Hence, this event shows the need of more investigations in this research field.

2.2. Measurement Science and Technology

MST recent issue 25:8 ([5], 2014) contains a whole section devoted to current developments in particle image velocimetry. Besides the focus on characterizing uncertainties sources, hardware and system reliability were also discussed. More questions were posed than answers: in fact, new challenges for UQ are coming from recent experimental techniques, such as tomographic PIV. For instance, an artifact that haunts experimentalists are so called "ghost particles". Curiously, these are easier to detect and eliminate in time resolved image sequences. As for hardware, the uncertainties of CCD and CMOS sensors (charge-coupled devices and complementary metal-oxide semiconductors) were addressed in the article by Abdelsalam et al, specially how individual pixels behave. From the quantitative perspective, an exciting development in this MST issue and other recent sources is the possibility to estimate dynamic quantities (accelerations, pressures) from the velocity measurement. These questions may give rise to interesting mathematical developments.

2.3. Towards an automatized UQ for PIV

As far as we know, only one research group, that leaded by Barton L. Smith in Utah has already proposed a thorough methodology for UQ in PIV [6]. Their work could serve as a departure point for further advances, leading to metrologically standardized procedures as advocated in GUM. The Timmins master's thesis describes a method to estimate the PIV uncertainty [6].

[3] `http://www.piv.usu.edu/PIV_Uncertainty_Workshop.html`

Initially, the relationship between error sources and their velocity measurements is investigated. Once choice the error source an "uncertainty surface" is provided for the PIV algorithm used. After PIV processing, it's possible to mesure the value of each of these parameters and estimate the uncertainty of each vector. This methodology was described to be general and adapted to any PIV analysis.

3. Mathematical Tools

3.1. Probabilistic Methods

We mentioned that seeding is not reproducible, but this does not prevent metrologists to think of a fluid flow measurement as a realization from a probabilistic ensemble of admissible initial conditions. In that sense, the practical need of doing repeated PIV recordings becomes an experimental feature. We refer to [7, 8] for data assimilation in fluids, where spectral methods are currently the state of the art. Such probabilistic methods fit well with the current tendency of GUM to move towards Bayesian statistics. We have looked into the literature of *gaussian processes*[4] and we believe there is an enormous potential there for uncertainty quantification, as an optimal distribution compatible with the available data comes also with error bars, albeit conditioned to a given probability of lying there.

We finish this brief report presenting three prospective mathematical techniques coming from geometric mechanics and control, and the planetary sciences community.

3.2. Observers

Observer theory gives life to Galileo's motto that opens our assessment. It is well now in geometric control [9]. It allows estimating parameters and dynamic variables not directly measurable in an experiment. The observer ODE or PDE adds artificial terms to the model equation so that it converges exponentially fast to a desired trajectory. Observed measurements y(t) are fed into the observer equation. The desired quantities appear in the output x(t), which is "best" fit in a Bayesian philosophy. In metrology the observer method it is not widely disseminated. The only reference we found is Ruhm [10], where he describes it as a "model-based measurement". Observer theory has been already used for fluids [11, 12]. Its use in PIV was advocated by Rouchon [13].

3.3. Data assimilation

In a series of papers, starting around 2010, Edriss Titi and collaborators have developed a research program which combines ideas from the geophysical and the control theory communities [14, 15, 16, 17, 18]. In a nutshell, consider a dynamical systems where partial information is known about its state, observed from an initial time to the present. This information is used to 'gently nudge' (say, by a fictitious forcing term in the right hand side the Navier Stokes equations) the system solution to the best full initial condition at current time. Evidently data assimilation has the same flavor as the observer method.

3.4. "Jetlets"

In a nutshell, a k-jetlet is a local kth degree polynomial that approximates the flow near a particle. Exact solutions of regularized Euler equations can be constructed from a finite set of particles, which follow the corresponding flow field. This flow is automatically volume preserving and can explicitly be numerically simulated [19, 20]. For metrology, jetlets could become a novel way to reconstruct the velocity field from a finite set of measurements of seed positions and velocities.

[4] `http://www.gaussianprocess.org`

4. Conclusion

The particle image velocimetry have been used as a powerful tool to investigate fluid dynamics behavior. Nevertheless, the PIV metrological approach began in the end of last decade. Therefore, many issues need to be investigated. The uncertainty sources and its quantification are a new scientific brand in this research area. In this paper, different types of mathematical techniques coming from geometric mechanics and control and planetary science community were briefly presented indicating new ways to approach uncertainty quantification for particle image velocimetry.

References

[1] Adrian R J 1996 *Bibliography of particle velocimetry using imaging methods: 1917-1995* (Department of theoretical and applied mechanics, University of Illinois at Urbana-Champaign)

[2] Westerweel J, Elsinga G E and Adrian R J 2013 *Annual Review of Fluid Mechanics* **45** 409–436 (*Preprint* `http://dx.doi.org/10.1146/annurev-fluid-120710-101204`) URL `http://dx.doi.org/10.1146/annurev-fluid-120710-101204`

[3] Maxey M R and Riley J J 1983 *Physics of Fluids* **26** 883–889 URL `http://scitation.aip.org/content/aip/journal/pof1/26/4/10.1063/1.864230`

[4] Coleman H W and Steele W G 2009 *Experimentation, validation, and uncertainty analysis for engineers* (John Wiley & Sons)

[5] Elsinga G E and Scarano F 2014 *Measurement Science and Technology* **25** 080301 URL `http://stacks.iop.org/0957-0233/25/i=8/a=080301`

[6] Timmins B H 2011 *Automatic Particle Image Velocimetry Uncertainty Quantification* Master's thesis Utah State University

[7] Le Maître O P and Knio O M 2010 *Introduction: Uncertainty Quantification and Propagation* (Springer)

[8] Iaccarino G 2009 Quantification of uncertainty in flow simulations using probabilistic methods Tech. rep. DTIC Document

[9] Besançon G 2007 *Nonlinear observers and applications* vol 363 (Springer)

[10] Ruhm K H 2015 Measurement plus observation - a modern structure of metrology (Prague, Czech Republic: XXI IMEKO World Congress "Measurement in Research and Industry")

[11] Auroux D and Bonnabel S 2011 *Automatic Control, IEEE Transactions on* **56** 1046–1058

[12] Jahanshahi E, Skogestad S and Grotli E I 2013 Anti-slug control experiments using nonlinear observers *American Control Conference (ACC), 2013* (IEEE) pp 1056–1062

[13] Aghannan N and Rouchon P 2003 *Automatic Control, IEEE Transactions on* **48** 936–945

[14] Hayden K, Olson E and Titi E S 2011 *Physica D: Nonlinear Phenomena* **240** 1416 – 1425 ISSN 0167-2789 URL `http://www.sciencedirect.com/science/article/pii/S016727891100114X`

[15] Azouani A, Olson E and Titi E S 2013 *Journal of Nonlinear Science* **24** 277–304 ISSN 1432-1467 URL `http://dx.doi.org/10.1007/s00332-013-9189-y`

[16] Farhat A, Lunasin E and Titi E S 2015 *Journal of Mathematical Fluid Mechanics* **18** 1–23 ISSN 1422-6952 URL `http://dx.doi.org/10.1007/s00021-015-0225-6`

[17] Albanez D, Nussenzveig Lopes H and Titi E 2016 *Asymptotic Analysis* **97** 139–164

[18] Foias C, Mondaini C F and Titi E S A discrete data assimilation scheme for the solutions of the 2d navier-stokes equations and their statistics, http://arxiv.org/abs/1602.05995

[19] Cotter C, Holm D, Jacobs H and Meier D 2014 *Journal of Physics A: Mathematical and Theoretical* **47** 352001

[20] Cotter C, Eldering J, Holm D, Jacobs H and Meier D 2015 *arXiv preprint arXiv:1503.07650*

Quality management system for application of the analytical quality assurance cycle in a research project

R S Camargo and I R B Olivares

Instituto de Química de São Carlos, Universidade de São Paulo, Av. Trabalhador São Carlense, 400 CP 780, São Carlos, SP CEP 13560-970, Brazil

igorolivares@iqsc.usp.br

Abstract. The lack of quality assurance and quality control in academic activities have been recognized by the inability to demonstrate reproducibility. This paper aim to apply a quality tool called Analytical Quality Assurance Cycle on a specific research project, supported by a Verification Programme of equipment and an adapted Quality Management System based on international standards, to provide traceability to the data generated.

1. Introduction

The Research and Development (R&D) Laboratories have great importance in the construction of pure or applied know-how. In Brazil, the majority of these kind of laboratories are in public universities. These laboratories generate data that could interest Academia or enterprises however, the problem arises when these data could not be reproducible and historically, the preoccupation with the quality assurance in these institutions is negligible [1].

To assure quality, the R&D laboratory could seek the formal accreditation of a quality management system (QMS) fulfilling the requirements of ISO/IEC 17025 or Good Laboratories Practices (GLP). It will discover that accreditation is not expensive but the maintenance of the QMS, it is.

For R&D laboratories that could perform tests and calibrations is technically easy to develop a QMS and extend it to the other activities (research and teaching). It is possible by the profit achieved with the tests or calibrations. However, there is an infinity of other kind of research projects that do not support parallel activities.

Another option is the development of a QMS without formal accreditation although based on the principles of international standards. The Eurachem Guide was developed to give guidance to make this possible [2]. Unfortunately, this guide was too much prolix.

Olivares and Lopes created a quality tool more concise and simple: the Analytical Quality Assurance Cycle (AQAC). It is based in the three most important technical requirements of ISO/IEC 17025:2005 standard: 5.4.5 Validation of Methods; 5.4.6 Estimation of uncertainty of measurement and 5.9 Assuring the quality of test and calibration results, resulting in three major concepts: I) Validation; II) Uncertainty and III) Quality Control. Two important details link those major concepts: the use of calibrated equipment and the use of certified standards (and certified reference materials). It is impossible to apply the AQAC without these items because there isn't traceability to the chain of verification of the conformity [3].

Nonetheless, it is necessary to make some adaptations in the AQAC because it is focused in the metrological traceability but there are the necessity to provide traceability of the information (reports, documents, records). Regarding these points, this paper aim to apply the Analytical Quality Assurance

Cycle in a specific research project to provide metrological traceability and the development of a QMS in the Research in Quality Assurance for Laboratories (RQA Lab), situated in São Carlos.

2. Methodology

2.1. The Analytical Method

The method developed was based on Stir Bar Sorptive Extraction (SBSE), but improved by cooling the sorptive phase then it had been called Refrigerated Sorptive Extraction (RSE) for extraction of polyaromatic hydrocarbons (PAHs) and alkylphenols (n-octylphenol and n-nonylphenol), in surface water. After the sample preparation, the chromatographic analysis was processed in a GC/FID.

2.2. Analytical Quality Assurance Cycle

The use of calibrated equipment and certified standards is very important and it is a key point to implement AQAC. However, the funding was end for the RQA lab and it was impossible to implement a calibration programme and the acquisition of certified reference standards. The solution was implement a verification programme for balances, based on Euramet Guidelines [4] and micropipettes based on ISO 8655-2 and 8655-6 [5,6] and the estimation of the uncertainty of the standard solution used in fortified blanks to did the Validation and Quality Control.

The Validation was performed contemplating: selectivity (matrix effect), linearity, precision, robustness and limits of detection. The estimation of uncertainty was processed using the guide Eurachem [7]. Quality Control was performed following the principles of control charts [8].

2.3. Quality Management System

The QMS developed was based on ISO/IEC 17025:2005 and Good Laboratory Practices. The Chemistry bachelor's program (IQSC-USP) have in its grade, the discipline "NBR ISO/IEC 17025 System" given by the professor Igor Renato Bertoni Olivares. The Standard Operational Procedures (SOPs) created by the students are harnessed and applied in the QMS.

3. Results and Discussion

Observing the figure 1, perhaps the common sense shows that the order to create the system was: i) QMS; ii) Verification Programme; iii) AQAC. Nonetheless, it was by the application of the AQAC that the necessity of the adaptations was noted.

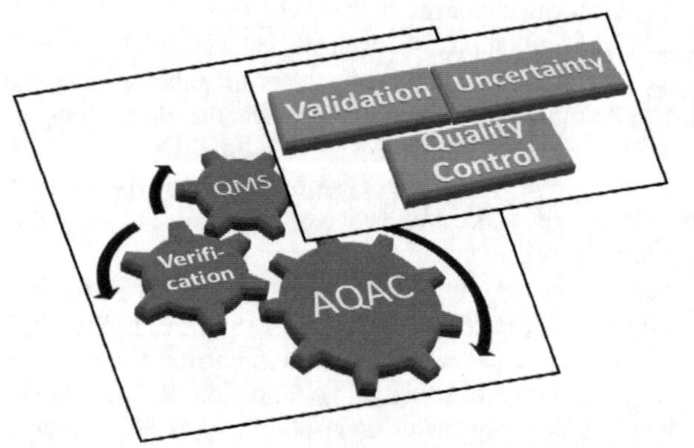

Figure 1: Schematic representation of the quality system idealized for the RQA Lab. In the first plan, the three major concepts: Validation, Uncertainty and Quality Control. In the second plan, the Quality Management System and the Verification Programme. The gearing represent that anything works solely.

The intermediate verification of balances is used to "know and control the drift of the balances in the period of two calibrations" and it could be performed by three tests: i) "repeatability test"; ii) "Test for errors of indication" and iii) "Eccentricity test" [4]. The verification of repeatability aim to know if the response of the balance changes with the weighing routine. However, in the Verification Programme, the verification of repeatability was used to generate an uncertainty (standard deviation) for the balance, similarly for the micropipettes.

The metrological traceability was guaranteed by verification of the equipment using standard weights calibrated by laboratories accredited in ISO/IEC 17025. The uncertainty of the verification was used to estimation of the analytical uncertainty (see figure 2).

Is the uncertainty of the verification overrated or underestimated concerning the uncertainty of a calibration? It could be both: for micropipettes, probably, the uncertainty of the verification is bigger than the calibration because in the first one, the variability probably is less controlled. For balances, the uncertainty of the verification could be smaller than the calibration because in the verification some uncertainty sources are suppressed.

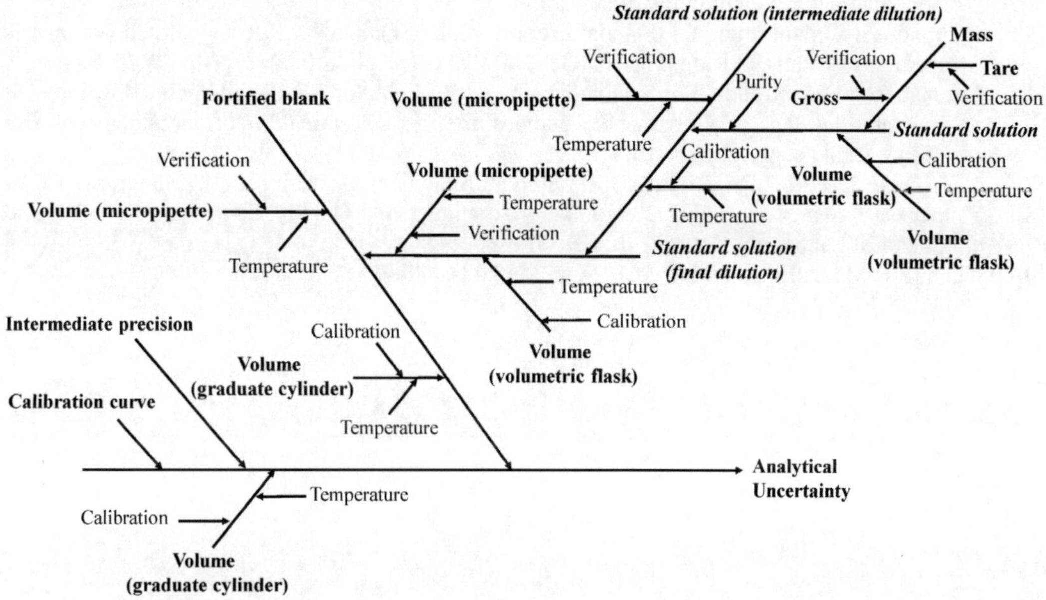

Figure 2: The Ishikawa diagram for analytical uncertainty. In the fortified blank, arise the uncertainty source of the dilutions of standard solution

Despite the doubts concerning the estimation of the uncertainty by the verification of equipment, the analytical uncertainty had as the major uncertainty sources, the precision of the measurement and the calibration curve. Both of them were performed using fortified blanks to obtain similar conditions to the sample analysis preventing that in the major sources there was an underestimate of uncertainty [9].

After the implementation of the Verification Programme, the Analytical Quality Assurance Cycle could be implemented. Although, there was the necessity to control documents and data. The Quality Management System was implemented using requirements of ISO/IEC 17025 and GLP. SOPs created by students in the audit simulation of the discipline "NBR ISO/IEC 17025 System" are adapted to be used in the laboratory. "Control of documents and records", "Corrective and Preventive Actions", "Personnel", "Accommodation and environmental conditions", "Measurement traceability" and "Reports" are examples of SOPs harnessed.

Moreover, the Coordination of Post-graduate Programme of the Institute of Chemistry of São Carlos showed the initiative to standardize the lab book for all students. The main motivation is to decide for the authors of patents in case of litigation, however it meet the GLP requirement that it should have a book for the research test. Moreover, these lab books could be the gateway to quality assurance for others research projects in the future.

4. Conclusion

The application of the analytical quality assurance cycle in a research project, supported by a quality management system with verification programme, could assure quality to the data generated, even that the lab operates without resources.

References

[1] Grochau I H Ferreira C A Ferreira J Z and Caten C S 2010 Accredit Qual Assur 15 681
[2] Eurachem/Citac 1998 Quality Assurance for Research and Development and Non-routine Analysis,Teddington: LGC)
[3] Olivares I R B and Lopes F A 2012 TrAC Trends Anal Chem 35 109
[4] European Association of National Metrology Institutes 2009 Guidelines on the Calibration of Non-Automatic Weighing Instruments
[5] International Organization for Standardization 2002 ISO 8655-2 Piston-operated volumetric apparatus — Part 2: Piston pipettes (Geneva: ISO copyright office)
[6] International Organization for Standardization 2002 ISO 8655-6 Piston-operated volumetric apparatus — Part 6: Gravimetric methods for the determination of measurement error (Geneva: ISO copyright office)
[7] Eurachem/Citac 2012 Quantifying Uncertainty in Analytical Measurement (Teddington: LGC)
[8] Konieczka P and Namiesnik J 2009 Quality Assurance and Quality Control in the Analytical Chemical Laboratory (New York: CRC Press)
[9] Thompson M and Ellison S L R 2012 Anal Methods 4 2609

Development and Implementation of Cgcre Accreditation Program for Greenhouse Gas Verification Bodies

Ricardo Kropf Santos Fermam, Andrea Barroso Melo Monteiro de Queiroz

National Institute of Metrology, Quality and Technology (Inmetro). General Coordination for Accreditation (Cgcre). Division for the Development of Accreditation Programs (Didac). Rua Santa Alexandrina, 416, 8th floor, Rio de Janeiro, RJ 20261-232, Brazil.

E-mail: rkfermam@inmetro.gov.br

Abstract. An organizational innovation is defined as the implementation of a new organizational method in the firm's business practices, organization of your workplace or in its external relations. This work illustrates a Cgcre innovation, by presentation of the development process of greenhouse gases verification body in Brazil according to the Brazilian accreditation body, the General Coordination for Accreditation (Cgcre).

1. Introduction

The official Brazilian accreditation body is the General Coordination for Accreditation (Cgcre), recognized by the Brazilian Government. Cgcre is signatory to the multilateral agreement (MLA) of the International Accreditation Forum (IAF), of the International Laboratory Accreditation Cooperation (ILAC) and of the InterAmerican Accreditation Cooperation (IAAC) and is part of the National Institute of Metrology, Quality and Technology (Inmetro) [1], according to the Act MDIC 82/2008, Federal legislation that approved Inmetro's bylaws [2].

The Development Division of Accreditation Programs of Cgcre (Didac/Cgcre) is responsible for establishing policy and systems for new accreditation programs, managing the activity of developing accreditation programs and disseminating knowledge generated in the accreditation activity, being responsible for organizational innovations in Cgcre, defined as the implementation of a new organizational method in the firm's business practices, organization of your workplace or in its external relations [3].

In the development process, based on internal procedure of CGCRE, NIE-CGCRE-015 [4], Didac/Cgcre has established two approaches based on the level of complexity that the new programme demands (levels 1 and 2). This work illustrates a Cgcre innovation, by presentation of the development and implementation process of the level 1 program: accreditation of greenhouse gas (GHG) validation and verification bodies, according to the international standard ISO 14065.

2. Accreditation of Greenhouse Gas Verification Body under ISO 14065

Greenhouse gases, as defined by ISO 14065 Clause 3.1.1, are gaseous constituent of the atmosphere, both natural and anthropogenic, that absorbs and emits radiation at specific wavelengths within the spectrum of infrared radiation emitted by Earth's surface, the atmosphere, and clouds. Include carbon dioxide (CO2), methane (CH4), nitrous oxide (N2O), hydrofluorocarbons (HFCs), perfluorocarbons (PFCs) and sulfur hexafluoride (SF6).

Verification is the process through which the verification body assesses a GHG assertion against defined verification criteria and the standard that is laid out by the GHG program [5]. The verification body is responsible for completing an objective assessment and providing a verification statement concerning the responsible party's GHG assertion based on evidence [5]. Greenhouse Gas Verification Body accredited under ISO 14065 must demonstrate that they are technically competent to undertake verification of greenhouse assertions in accordance with a greenhouse gas programme, such as Brazilian GHG Protocol Program, ISO 14064, Voluntary Carbon Standard (VCS) and others.

To perform this activity, a verification body must comply with ISO 14064 Part 3 [6], which specifies principles and requirements and provides guidance for those conducting or managing the verification of greenhouse gas assertions, established for the first time a process for conducting a verification of a GHG assertion, such as an inventory report of GHG of an organization.

ISO 14065 is a basis for assessing and recognizing the competence of verification bodies. This standard details a series of requirements that verification bodies must meet to become accredited. Another relevant international standard, ISO 14066:2011 [7], specifies competence requirements for verification teams of verification body.

According to ISO 14065, clause 6.1.a, the validation or verification body shall establish and maintain a procedure to determine required competencies for each sector in which it operates. In addition, ISO 14066:2011 clause 6 establishes that "a validation team or verification team collectively shall have applicable sector knowledge and skills". For each sector, the collective technical competence of validation team or verification team shall include (as applicable) the capability to identify greenhouse gas sources, sinks and reservoirs (SSRs) concerning to the sector, and others. Different sectors have different SSRs and require a different set of competencies for the verification team. This implies that to define the scope can be a very difficult task, since for each economy sector there are several different SSRs that should be considered to define an inventory of GHG. To solve this challenge, Didac/Cgcre defined the scope regarding the sector of economy associated with the main activity of an organization, interested to have its inventory of GHG verified by an accredited verification body.

3. Development of Greenhouse Gases Verification Body Accreditation by Didac/Cgcre

According to Didac/Cgcre's requirements, the development process requires consideration of the existence of the accreditation programme in other countries; update or creation of documents; competency criteria for the accreditation team and selection, training, evaluation and qualification of assessors and technical experts. To implement an accreditation program, identified as level 1, a pilot project must be performed before becoming the new programme available in the portfolio of accreditation. Figure 1 shows the approved accreditation programme implemented in Cgcre, which a body requesting accreditation must indicate (a) a GHG verification programme and (b) an organizational level for GHG inventories verification.

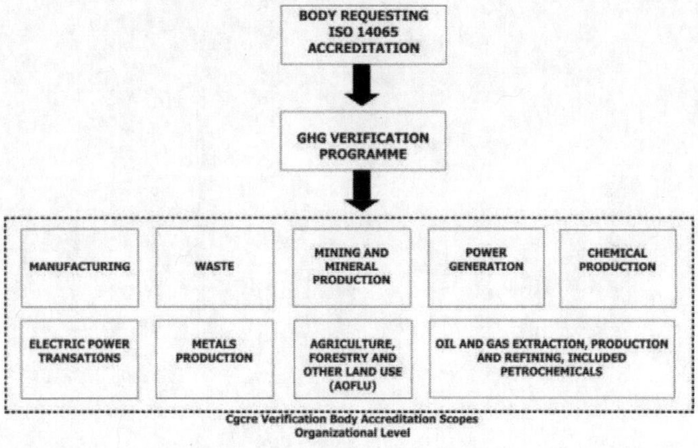

Figure 1. Approved ISO 14065 accreditation program by Cgcre.

4. Implementation of Greenhouse Gases Verification Body Accreditation in Cgcre

In implementation step, Didac/Cgcre determines that technical experts for this accreditation program must have experience with greenhouse gas verification processes and technical knowledge in one or more GHG methodologies (programmes) and must have Bachelor degree in technical field (for example Chemical Engineer). Experience also should include, at the minimum, areas as emissions monitoring and calculations, design of GHG emissions reduction systems, testing and calibration of emissions monitoring equipment, working with GHG quantification methodologies (as mathematical modeling, measurement, and others).

In order to qualify internal Cgcre's accreditation assessors and other personnel involved, for example, accreditation manager, a training program was selected. Starting with the three training programs offered, around 30 professionals were prepared, among them permanent staff of the Cgcre and external technical experts (from Brazilian Universities).

5. Pilot Project: Main Results

The main objective of pilot project is prepare infrastructure to become Cgcre competent to initiate a new accreditation program. At the end of accreditation in pilot project, a critical analysis is necessary to correct and improve the new program which will be included in the routine of accreditation´s Cgcre Division responsible for it.

During the pilot Cgcre assessed the competence, impartiality and performance capability of applicant organizations using the internationally recognized standard ISO 14065:2012. As result of Pilot Project, 09 verification bodies (of 15 initial applicants) were accredited by Cgcre (The list of Brazilian accredited verification bodies under ISO 14065:2012 can be found at http://www.inmetro.gov.br/organismos/index.asp and include several technical sectors).

The methodology of critical analysis carried on Didac/Cgcre consisted in a survey with organizations and to assessors and technical experts that participated in the pilot project. All participants evaluated the pilot as good. Regarding the competence of assessors, 89% of verification body evaluated it as good, while 44% evaluated the competence of technical experts as excellent.

Were also analyzed all documents in main accreditation process steps, namely initial assessment, office assessment and witness assessment. Main problems areas for applicants refer to impartiality (clause 5.4), liability and financing (clause 5.5), deployment of personnel (clause 6.3), information provided to a client or responsible party (clause 7.1), approach (clause 8.3), appeals (clause 9) and complaints (clause 10), as show in Figure 2.

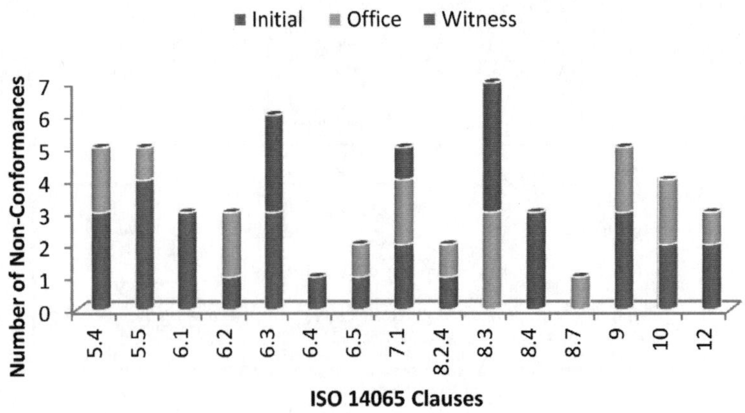

Figure 2. Non-conformances in ISO 14065 accreditation in Cgcre's pilot project.

6. Conclusions

According to ISO/IEC 17011:2005 clause 6.1.2, Accreditation Body shall have access to a sufficient number of assessors, including lead assessors, and experts to cover all of its activities. Especially for ISO 14065 accreditation, this is extremely important while remain a difficult task.

For internal assessors, training in ISO international standards and GHG Programme may not be sufficient to operate the accreditation program; in turn, technical experts in greenhouse gas emissions for all organizational sectors are not available yet. In this situation, improvements of training program with case studies, for internal assessors, remain a best option. For technical experts, may be necessary seeks qualified personnel in another countries for organizational sectors that isn't available in country of accreditation body, in a view of accreditation scoping extension.

Didac/Cgcre development step-by-step may be reproduce by others Accreditation Bodies (ABs), which seek launch this accreditation. This ABs will be founded this work very useful. Finally, suggests to international and/or regional accreditation forums offer training courses and technical cooperation in this new accreditation program for accreditation bodies.

References

[1] Nogueira, R., and Soares, M. A. (2013). Accreditation and recognition programs in Brazil: current situation and perspectives. Accreditation and Quality Assurance, 18(3), 217-223

[2] Inmetro (Instituto Nacional de Metrologia, Qualidade e Tecnologia) (2008). Rio de Janeiro: Inmetro http://www.inmetro.gov.br/legislacao/laf/pdf/LAF000408.pdf. Accessed 2 Out 2014

[3] STATISTICAL OFFICE OF THE EUROPEAN COMMUNITIES. Oslo Manual: Proposed Guidelines for Collecting and Interpreting Technological Innovation Data. OECD Publishing, 1997

[4] Inmetro (Instituto Nacional de Metrologia, Qualidade e Tecnologia) (2012) NIE-CGCRE-015: Estudo de Viabilidade e Desenvolvimento de Novos Programas de Acreditação. Rio de Janeiro: Inmetro http://www.inmetro.gov.br/sidoq/arquivos/Cgcre/NIE/NIE-Cgcre-15_07.pdf. Accessed 2 Out 2014

[5] ISO 14065 (2012) Greenhouse gases - Requirements for greenhouse gas validation and verification bodies for use in accreditation or other forms of recognition. ISO, Geneva

[6] ISO 14064 (2006) Greenhouse gases - Part 3: Specification with guidance for the validation and verification of greenhouse gas assertions. ISO, Geneva

[7] ISO 14066 (2011) Greenhouse gases - Competence requirements for greenhouse gas validation teams and verification teams. ISO, Geneva

Optimization of method a load cell calibration for the measurement of coefficient of friction

R M Castro[1], M Pereira[2], A R Sousa[2], E I M. Curi[1], C L Izidoro[1] and L C Correa[1]

[1]Faculty SATC - Department of Mechanical Engineering and Mechatronics
[2]Federal Institute of Santa Catarina (IFSC) - Department of Mechatronics Engineering

E mail: richard.castro@satc.edu.br

Abstract. The instrumentation of equipment for mechanical testing is used to optimize the time to deliver a result, besides minimizing errors associated with manual measurements. Given this context, this work aims to present a calibration method for a load cell to determine the measurement results of force and friction coefficient, developed from on rotary pin-on-disk tribometer. The results indicate that the procedure provides measurements reliable for the tribological phenomena, resulting in with proximity the values provided by the ASTM G99-04.

1. Introduction

The friction and gravity forces are the variables with which engineer mechanic finds more often throughout the development of projects, however, these are still the least understood forces of nature. Besides energy, friction also performs a vital role in systems that requiring motion. The science and technology that focuses on the study of the interaction between surfaces in relative motion and related practices, is referred tribology, including wear and lubrication phenomena [1]. Although the friction is extremely useful in most cases, much of the in tribology study, is concentrated in order to minimize its effects.

As a way to study the phenomena resulting from surface interactions, the American Society for Testing and Materials (ASTM) presents the method designated "Pin on Disk". It uses two test bodies, a pin or ball, and the other in the form of a thin disk (figure 1) [2]. The machine that performs the test is called tribometer, which is able to measure the friction coefficient between the two materials what will be in sliding.

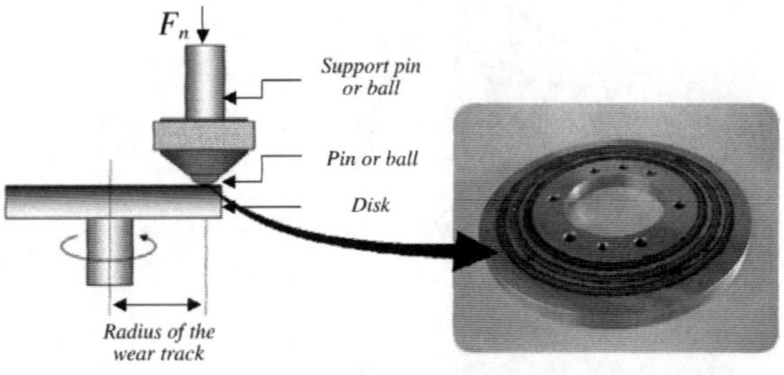

Figure 1. Schematic drawing pin on disk [2].

The fundamental laws of the friction in solid bodies are quite simple and are based on principle of Coulomb (1785) where the friction forces (F_a) is given by the following equation 1 [3]:

$$F_a = \mu . F_n \tag{1}$$

Where:
μ : Friction coefficient [---]
F_n : Normal force [N]

The standard guide this study it is ASTM G99-10, which differs from the other on Tribology, not determining which instrumentation technologies or even automation to be implemented for performing the test and determination of the friction results.
Basically there are two principles of devices for measuring the frictional force, the piezoelectric dynamometer, and the load cell designated strain gauge [4]. In this project was used a resistive load type cell, because makes the dynamic measurement and having a lower cost.

2. Experimental procedure

2.1 Definition and load cell calibration

For the selection of the load cell capacity was used the largest value of the normal force (F_n) of 30 N, suitable to design mechanic of tribometer and in accordance with the highest expected value for the coefficient of friction between the no lubricated materials, that is 0.8 [2]. The selected load cell was of 33 N.

The selected load cell is of the type single point model PW4MC3 manufactured by HBM. The sensitivity is 2.0±0.2 mV/V and the maximum load allowed is 30+3 N. In order to position the load cell in the tribometer, necessary was fabricating an articulation system, called arm of the lifting. To meet the proportional amounts of voltage to each load, the load cell calibration was performed using standard weights (figure 2). This procedure was done to verify the values specified by the load cell manufacturer and thus, performing the friction tests.

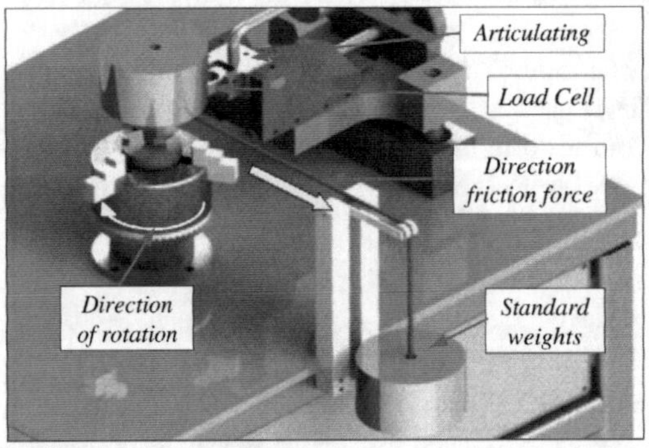

Figure 2. Methodology used for calibration.

Through the standard weights, the voltage was related. The voltage was collected by a data acquisition system and with the voltmeter, allowing the values can be stored and compared. Table 1

Journal of Physics: Conference Series **733** (2016) 012048 doi:10.1088/1742-6596/733/1/012048

presents the basic characteristics for each measurements instrument, where the accuracy data supplied by manufacturers refer to full scale (FS).

Table 1. Characteristics of the instruments (mV)

Instrument	Model	Accuracy	Resolution
Voltmeter	F106	±0.3% +3	1.0
Acquisition board	USB 6008	±7.7	9.7

For the loading of the load cell was used mass of 0.01, 0.02, 0.05, 0.1, 0.2, 0.5 and 1 kg, with uncertainty of 1×10^{-5} kg, employing a USB6008 data acquisition system and software LabVIEW®. The acquisition rate for obtaining measurements was set to 1 kHz, performing a time of 10 s, giving an average value that was compared to the voltmeter.

The excitation of the load cell was made with a voltage of 10 V, allowing an output proportional to the deformation of 0 to 10 V. After measuring and analyzing the data, the voltage (V) was converted in force ($F = m.g$), generated by the product of mass (m) with acceleration of gravity (g = 9.81 m/s^2). From the data, we performed the calculation of the average and standard deviation for each force obtained with accumulative loads by standard weights, ending with the curve trend shown in figure 3. With the calibration data, introduced the trend line and equation in order to demonstrate the coefficient of determination R^2, which indicates the quality of the regression. A value of 1 indicates a total correlation between the measurements and the linear behavior of the regression. In figure 3 the equation y = 0.2988x + 0.0175 represents data regarding USB-6008 and y = 0.2987x + 0.0192 corresponds to the voltmeter. In figure 4 the error curve related to the load cell calibration is displayed.

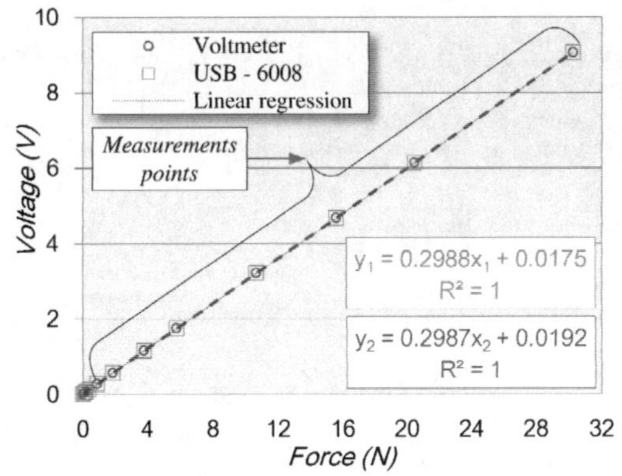

Figure 3. Curve de calibration load cell.

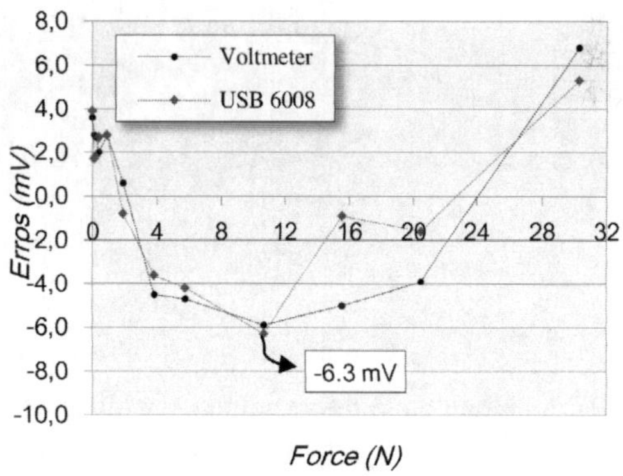

Figure 4. Errors curve.

According to the manufacturer of load cell, the sums of the errors are in the range of ±0.03% of the value of the sensitivity, meaning a ±0.045 N error for a full scale value (FS). In figure 4 it is still possible to identify the difference between the theoretical and experimental values, determining the how much the equations in figure 3 approaches the true value. Through the data, the biggest found was error of 6.3 mV, corresponding in a 0.021 N force. According to the data shown to maximum error of ±0.045 N of the cell, the maximum values determined by calibration are within its characteristic scale.

3. Results and discussion

The results of the calibration will be presented for the friction determination in the tribometer.

3.1 Procedures for friction testing

After calibration of the load cell, were performed the tribometer dynamic tests, where was established a number of six tests. In order to make a comparison between the experimental tests with standard ASTM G99, we used a sphere of aluminum oxide (Al2O3) (pin) and a steel disc AISI 52100 and a load by dead weight to a normal force (F_n). To perform tests under the same conditions and compare with the standard was followed by the data as Table 2.

Table 2. Input data for tests

v_d [m/s]	R_m [m]	F_n [N]	H_r [%]	Δs [m]
0.27	17.10^{-3}	10	44 a 65	1000

Where:

v_d is the sliding speed, R_m is the average radius, F_n is the normal force applied, H_r is the relative humidity and Δs refers to the total sliding distance.

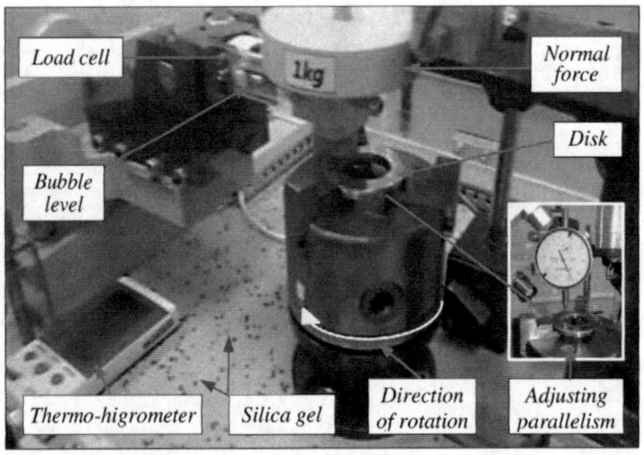

Figure 5. Friction testing procedures.

To maintain the relative humidity within the standards set by the standard (range of 12 to 70%), it was added a small amount of silica gel with a particle size of 1 to 4 mm. During testing, was observed through a thermo-hygrometer (figure 5), the decrease in humidity of 65 to 43%, justified also by the enclosure caused by acrylic protection and the natural warming of the atmosphere due to the testing.

Each test was also checked disk parallelism through measurement dial indicator to what the displacement of the load cell occur in a single direction. Furthermore, each test was adjusted arm of the level of articulation, made with the aid of a bubble level as shown in figure 5.

3.2 Friction measurement results

Although the values were recorded by the voltmeter, only the equation obtained by the data recorded by the USB 6008 was used, because this system would be fixed in the tribometer. From these data we determined the coefficient of friction for each test. The figure 6 shows the results of coefficient of friction versus distance for the six tests in the tribometer.

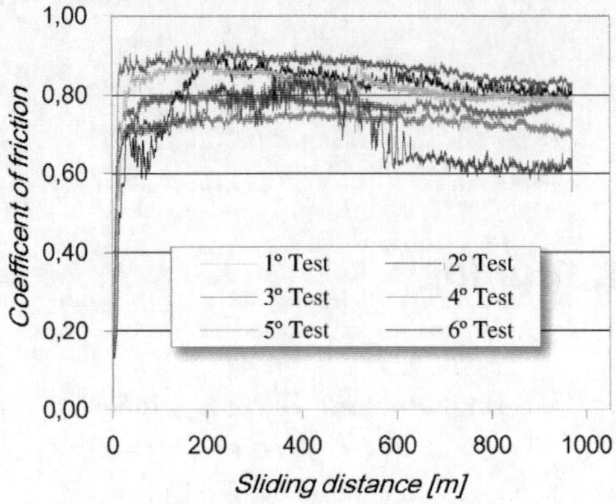

Figure 6. Friction results tribometer.

Table 3. Six tests of the coefficient friction

	Experimental tests						
	01	02	03	04	05	06	**Mean**
Average	0.76	0.81	0.86	0.73	0.79	0.71	**0.78**
Standard deviation	0.06	0.09	0.06	0.05	0.08	0.09	**0.07**

Through the values shown in figure 6, are observed with greater clarity the individual values of each test for the input conditions (Table 2) and at the end the final mean value and standard deviation for the friction coefficient for the developed test in rotary tribometer (Table 3).

The mean final value (six measurements) for the friction coefficient was 0.78 with a standard deviation of 0.07. According to Table 3, we can see a variation between the mean of each test, typically influenced by factors characteristic to the tests, as oxide formation, roughness changes, vibration, and others. However, making a comparison with the standard (ASTM G99) what indicating a value of 0.76±0.14 for 60 tests, it is clear that the calibration process and the methodology defined for the load cell measurement were consistent, validating measurement of the friction.

For all tests, the temperature and speed were also monitored. These were within the permissible variation and no significant influence to all results. The tests started with the average temperature of 19 °C and ended at about 28 °C. According to the results reported by Kim, 2014 [5], this difference of 7 °C is not sufficient to modification considerable in friction.

4. Conclusion

The application of the methodology used for the load cell calibration showed good results, since results were within the uncertainty values determined by the load cell manufacturer. Moreover, the procedure adopted for comparing the voltage values, using the acquisition system (USB-6008) and voltmeter, brought more reliable calibration results.

The measurements conducted in order to determine the strength and average coefficient of friction for the tribometer using equation 1, was satisfactory on the results supplied by the ASTM G99, with a mean small difference of 3%, relevant to a destructive type of test.

References

[1] POPOV, Valentin. Contact Mechanics and Friction - Physical Principles and Applications. 1ª ed. Springer, 2010.

[2] ASTM G99. Standard Test Method for Wear Testing With a Pin on Disc Apparatus. 2010.

[3] ASM International. ASM Handbook**.** Friction, Lubrification and Wear Technology. 4ª ed., 1992.

[4] WEBSTER, John. The Measurement Instrumentation and Sensors Handbook. 1ªed. Springer, 1999.

[5] KIM, D. W.; KIM, K. W. Effects of Sliding Velocity and Ambient Temperature on the Friction and Wear of a Boundary-Lubricated, Multi-Layered DLC Coating. Wear. p.95-102, 2014.

Portable compact cold atoms clock topology

R D Pechoneri[1], S T Müller[1], C Bueno[1], V S Bagnato[2] and D V Magalhães[1]

[1] Departamento de Engenharia Mecânica, Escola de Engenharia de São Carlos, Universidade de São Paulo, São Carlos, Brazil
[2] Departamento de Física e Ciência dos Materiais, Instituto de Física de São Carlos, Universidade de São Paulo, São Carlos, Brazil

E-mail: rodrigo.pechoneri@usp.br

Abstract. The compact frequency standard under development at USP São Carlos is a cold atoms system that works with a distributed hardware system principle and temporal configuration of the interrogation method of the atomic sample, in which the different operation steps happen in one place: inside the microwave cavity. This type of operation allows us to design a standard much more compact than a conventional one, where different interactions occur in the same region of the apparatus. In this sense, it is necessary to redefine all the instrumentation associated with the experiment. This work gives an overview of the topology we are adopting for the new system.

1. Introduction

The use of cold atoms results in a significant improvement in the stability and accuracy of atomic frequency standards by reducing the effects inherent to the atoms velocity [1-3]. However, the size of the atomic fountains is incompatible with some applications, like industry and space, which require a compromise between size and performance. Because of this demand, we decided, in the context of embedded systems, to develop a compact clock with Cs cold atoms.

Our research group has been working intensively to establish various experiments of frequency standards and to consolidate partnerships for the development of national references in Brazil [4].This is crucial for a country aiming stronger influence worldwide and that needs to reinforce strategic areas, such as defense and national industry. The project of the compact frequency standard was divided in three main stages. The first two have already been completed and in this work we will describe towards the third stage.

2. New apparatus – reducing instrumentation volume

The compactness of this kind of frequency standard is achieved by an operational mode purely temporal, instead of having different regions like atomic fountains (preparation, interrogation, detection). Indeed, during each cycle Cs manipulated in a single area: the microwave cavity.

A portable atomic standard is being developed with an instrumentation designed specifically for this role. Each part should play with minimal complexity and use hardware resources in order to allow compactness. It is then necessary to reduce the volume of sub-systems, such as laser drivers, microwave synthesizer, small-sized vacuum system and data acquisition.

3. General system topology

Our standard works with a cyclic process, where the atoms are trapped, cooled, interrogated and detected, successively. All the sequence process is realized using digital and analog signals that control acousto-optical modulators, microwave pulses, fluorescence detection and microwave signal corrections. A computational platform is used to process data, setup the sequential steps and user interface, working directly on the control of the closed loop operation.

In order to achieve robustness we decided to operate independent subsystems that have individual processing and are monitored by a supervisory interface. This defines a distributed architecture for control, monitoring and interface subsystems as diode lasers and microwave synthesizers.

Figure 1 is a general illustration of the topology for the portable standard in which the components operate independently, supervised by an embedded computer platform and connected using a data communication.

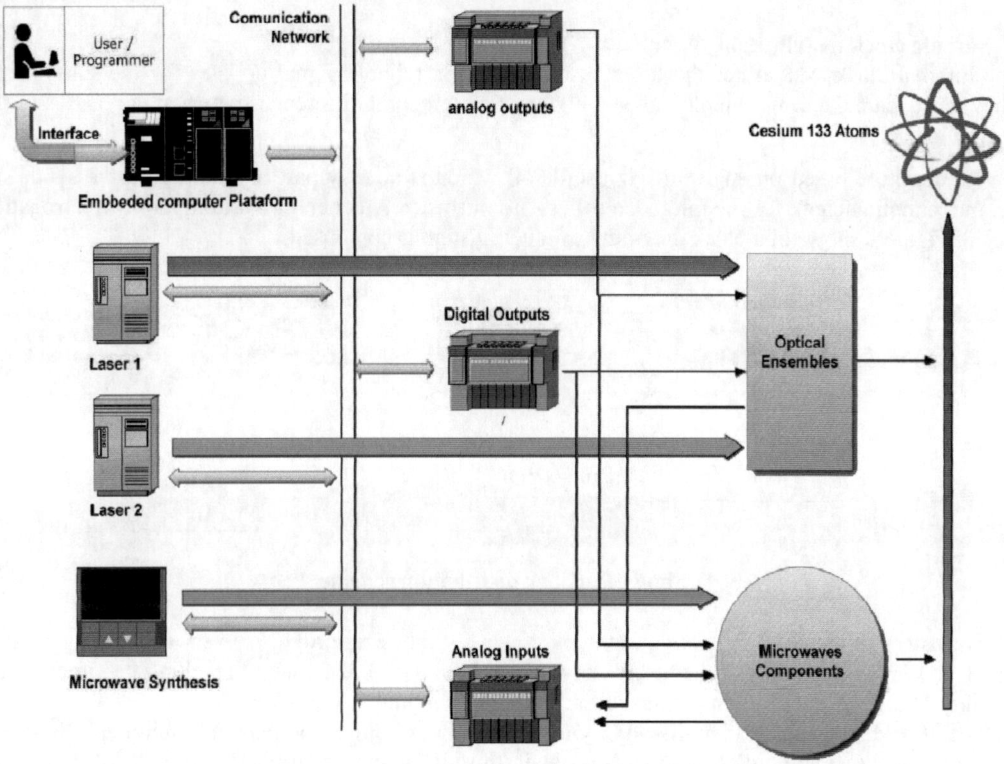

Figure 1. Overview of the system topology, with the supervisory action performed by the embedded computer platform and the subsystems operating with some degree of independence.

We adopted the LPC1768 microcontroller (NXP) [5] as a base processing unit because of some of this features. It has CANBUS native communication port, that allows the implementation of a robust communication network [6]. This microcontroller is being used in the laser systems, generation of control signals and acquisition.

The supervisory system is responsible for graphical user interface, setting parameters and microprocessor systems management, recording and analysis of the data obtained. Consists of an embedded computing platform able to communicate with a remote computer by universal serial bus (USB), also communicates with the microcontrollers in each module using CANBUS network.

The choice of an industrial communication protocol (CANBUS) is due to the fact that it has reduced cabling, real time support for data transmission rates, possibility of use of various network points and multi-master approach.

CANBUS network has been tested and validated for LPC1768 microcontrollers using the MBED development platform [7]. We are now developing the proprietary message protocol, specifically for use in our experiment.

The acquisition of analog data makes the voltage reading of the atomic fluorescence and sends to the processing unit the data obtained in synchronism with the remaining clock operating procedure. This data is set by the supervisory system and assigns the frequency correction for the microwave synthesizer.

4. Sample clock distribution

The internal clock system has the function of generating reliable signal updates for the subsystems. They must share the same sample rate signal, avoiding temporal differences between the phases of the operating cycle.

We therefore based on an 100 MHz oscillator, which is already part of the microwave synthesis, i.e. our sampling clock for the microcontrollers are intrinsically coherent with the atoms interrogation signal. Figure 2 shows in a block diagram how such system is composed.

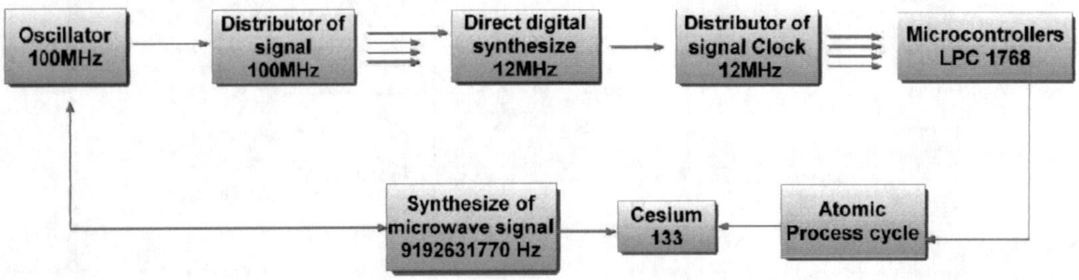

Figure 2. Clock distribution scheme.

The use of an oscillator belonging to the microwave synthesizer provides a very reliable signal. It is an OCXO (oven controlled crystal oscillator - Wenzel) with +13 dBm output level at 50 Ω impedance, stability of 0.1 parts per billion per day after 30 days of continuous operation [8].

The 100 MHz signal is then divided by a power splitter with 10 outputs (Minicircuits PSC-10-1) and amplifiers (Minicircuits MAV11). We obtain thus 10 outputs at 100MHz with +10dBm. One of these outputs is derived for the input sampling signal of a direct digital synthesizer (DDS - AD9912 Analog Devices) which provides 12 MHz signal which is also split and amplified, feeding the sampling clock of the LCP1768 microcontrollers.

We use a 12 MHz clock for the LPC1768, since all libraries are prepared for this rate, which is internally multiplied and provides a frequency of 96 MHz. Figure 3 shows the preliminary phase noise measurement of the splitter and amplifiers, working at 100 MHz.

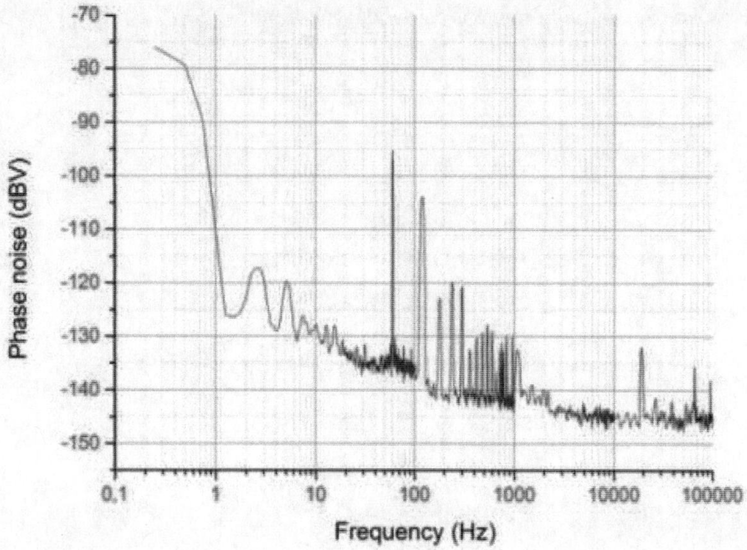

Figure 3. Phase noise level measurement of splitter and amplifiers.

5. Conclusion

We have been developing a compact system to be used as a portable reference for frequency signals. The core idea is the use of cold atoms as a high performance and reliable clock source. We are developing various embedded distributed systems, aiming that the apparatus has the flexibility of modular systems, allowing to make changes in hardware configurations.

Acknowledgments
The authors acknowledge financial support from INMETRO (Programa PRONAMETRO), FAPESP, CNPq and CAPES.

References

[1] Esnault F, Perrin S, Tremine S, Guerandel S, Holleville D, Dimarcq N, et al 2007 Stability of the compact cold atom clock HORACE. Proc. Int Freq Con. Symp. Expo.1342–5
[2] Müller S T, Magalhaes D V, Alves RF, Bagnato V S. 2011Compact Frequency Standard Based on an Intra-cavity Sample Cold Cesium Atoms,15. Available from: http://arxiv.org/abs/1108.4407
[3] Scherer DR, Lutwak R, Mescher MJ, Stoner R, Timmons B, Rogomentich F, et al 2015 Progress on a Miniature Cold-Atom Frequency Standard. arXiv Prepr arXiv14115006.
[4] Ahmed M, Magalhães D V., Bebeachibuli A, Müller ST, Alves RF, Ortega T a., et al 2008 The Brazilian time and frequency atomic standards program. An Acad Bras Cienc. 2008;80(2):217–52
[5] UM10360 LPC176x/5x User manual - UM10360.pdf [Internet]. [cited 2015 Aug 31]. Available from: http://www.nxp.com/documents/user_manual/UM10360.pdf
[6] Online Browsing Platform (OBP) [Internet]. [cited 2015 Aug 31]. Available from: https://www.iso.org/obp/ui/#iso:std:iso:11898:-1:ed-1:v1:en
[7] CAN - Handbook | mbed [Internet]. [cited 2015 Aug 31]. Available from:

https://developer.mbed.org/handbook/CAN
[8] Microsoft Word - 501-16843 - 501-16843.pdf [Internet]. [cited 2015 Aug 31]. Available from: http://www.wenzel.com/wp-content/parts/501-16843.pdf

Journal of Physics: Conference Series **733** (2016) 012050 doi:10.1088/1742-6596/733/1/012050

Review of the Technical Metrological Regulation for flow computers and volume converters

R O Almeida[1], E A Aguiar Júnior[1], R P B Costa-Felix[1]

[1] National Institute of Metrology, Quality and Technology (Inmetro), Av. Nossa Sra. das Graças 50, Duque de Caxias, RJ, Brazil, ZIP 25250-020

E-mail: roalmeida@inmetro.gov.br

Abstract. With the publication of Inmetro's regulation n° 373/2014, Inmetro presented the proposal to review the regulation of flow computers and volume converters, whose scope now includes fiscal measurement, ownership, custody transfer, among others, of hydrocarbons liquids and natural gas. This new proposal provides improvements to the legal metrological control of these instruments in Brazil, while also broaden the scope from the previous regulation, the Inmetro's regulation n° 64/2003. The purpose of this paper is to present these changes from a metrological point of view, and also clarify the transitional rules for the process.

1. Introduction

Flow computers are calculating devices used within fluid measurement systems. They functionality is to perform volume conversion, as well as total value summation. Temperature and pressure typically varies in a pipeline flow in many different situations, such as exploitation, transportation and storage. The use of flow computers allow a proper mathematical assessment of the actual volume, regardless of the atmospheric condition the fluid is at a moment. The assessment is based on the law, in the case Brazilian, item 3.8 of the Joint Resolution ANP / Inmetro No. 01/2013 [1]. The legal atmospheric conditions are temperature of 20 ° C and pressure of 0.101325 MPa. The standard measurement condition may vary according to a specific trade contract, and in this case a flow computer plays a key role as well. Therefore, flow computers are electronic devices similar to cash registers for oil and gas trade [2]. Figure 1 illustrates a typical setup of such equipment at a measuring point.

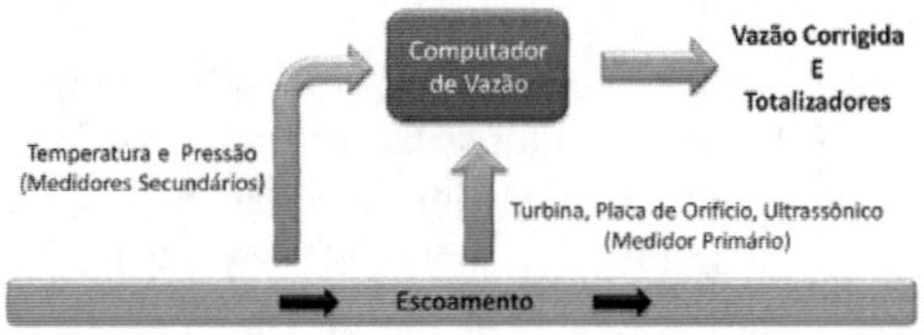

Figura 1. Typical flow computer connection to a pipeline and a flow measurement system.
Source: drawn by the author.

As illustrated, flow computers are fed with pressure and temperature transduced signals, and compile that information with the flow measurement input. A chromatographer (in the case of gas flow) or even a densimeter (for petroleum flow, for instance) could be included in the measurement system.

2. Legal Metrological Control

According to the International Vocabulary on Legal Metrology (VIML, [3]), the legal metrological control comprises a pool of activities aiming to assure the metrology quality of a device or service.

Presently in Brazil, the legal metrological control of flow computers is based on the rules established in Inmetro's normative n° 64 [4]. The mains metrological tests to be performed are the following:

- Pulse totalizing; dissimilar and pre-defined waveform are inserted into the flow computer, and the correctness of their computation is evaluated;
- Temperature and pressure input signals; analogic waveform type 4 (@ 20 mA) and type 1 (@ 5 V) are used to generate sensors output, and its performance and reliability is evaluated afterward;
- Simulation of a complete measurement arrangement, using standardized equations according to [5][6].

3. Regulatory Review

As aforementioned, flow computers are regulated with regard to its metrological aspects according to [4]. The legal reference to that normative was the Joint Normative Inmetro/ANP no /2000 [7]. The scope of the regulation was proposed to cover the main technical aspects of such kind of devices and related measurement systems.

A further step toward the technical enhancement of the mandatory regulation was settled within the Join Resolution ANP/Inmetro no 01/2013 [8]. That revised regulation defines many mandatory activities to Inmetro, which includes the necessity to renew the flow measurements metrological regulation. The importance of such revision is that flow computers, and flow measurement system as a whole, are fundamental for custody transference for the industry of oil and gas.

Aiming to adapt to the new legislation [8], Inmetro started a technical discussion with the goal to revise the Technical Metrological Regulation (RTM, acronym in Portuguese) for flow computers and volume correction devices. The Inmetro's normative no 373/2014 was issued in August 13th [1]. It includes a RTM draft, which represents the result of the hard work developed within the Technical Commission on Legal Metrology (TC/LM) for flow computers. That TC/LM was organized by Inmetro, and their members were many important stakeholders from oil and gas industry. That normative will supersede the presently valid regulation ([4]) according to a scheduled discussed further in this paper. Inmetro's normative nº 373 diverges in its scope from the former as it coves exclusively flow computers and volume correction devices for measuring oil and gas. Previously [4], the scope embraces all measurement devices applied to petroleum and their liquid products, as well as anhydrous and hydrated alcohols used as fuel.

Other paragraphs are indented (BodytextIndented style).

3.1. Main Technical Changes

Metrological requirements present in [1] were not significantly changed as compared to what was previously determined in [4]. The maximum admissible errors remain the same. However, inspecting the new RTM in deep, one can observe that the testing structure changed in such level that it cannot be considered similar to what is presented in the International Organization for Legal Metrology (OIML), as disclosed in OIML R117-9 [9]. That international document was formerly used as the technical base for Inmetro's normative no 64/2003 [4].

On the other hand, RTM's scope and metrological control for outdoor measurement were remarkably changed with the new regulation. The former document comprises petroleum measurement, what was preserved in the new document. However, metrological control shall now cover measurement system for gas flow in particular productive arrangements. Those productive arrangements are exploitation and production of gas and oil according to what is defined in [8]. It is worth mentioning that this document

([8]) does not cover natural gas, thus flow computers do not demand type evaluation as defined by OIML and Inmetro.

Finally, the new RTM establishes a mandatory the type (pattern) evaluation, i.e., conformity assessment procedure on one or more specimens of an identified type (pattern) of measuring instruments which results in an evaluation report and / or an evaluation certificate [10]. An initial verification is also mandatory, and it shall be undertaken by state institutes delegated by Inmetro (IPEM) for every device prior to its commercialization. However, subsequent verifications were not defined as mandatory to the governmental metrological system (Inmetro and IPEM). The final user, or the instrument owner, is in charge with the periodic verification. In that sense, the legal metrology control is finished after the device has been sold.

3.2. Transitional Rules

In order to minimize the impact due the new regulation in the industrial sector of oil and gas, Inmetro has introduced transitional rules in the normative itself. Measurement systems which type approval had already been issued will have their certificate valid for additional 12 months, as soon as the new normative is issued. Further on, their certificates will be withdrawn.

Due this systematic, dealers, manufacturers or commercial representatives are legally authorized to sell the measurement system with a previously issued certification. Meanwhile, a new type approval for their devices shall be demanded to Inmetro. No infraction is due to any final user that had bought a measurement system with a type approval issued during the transitional schedule. The validation of data is assured within the normative.

4. Conclusions

With the technical revision of the RTM related to the Joint Normative ANP/Inmetro no 1/2000 [7], the regulation regarding flow measurement system was changed. It affects mainly flow computers and volume converters.

Within the present paper, the main technical aspects, which concern the subject, were disclosed. Summarizing:

- Now the regulation explicitly defines the demand for type approval of flow computers, differently from what was formerly the case; just complete flow measurement system were under the metrological control;
- The scope of fluids possible to be measured was expanded, once natural gas had been included;
- Presently, the metrological control is restricted to type (pattern) evaluation and an initial verification, whilst the subsequent verification is of responsibility of the final user.

It is an intrinsic objective with the new RTM issue to let the metrological control, as realized in Brazil, updated. That document is coherent with the vision of minimum interference on industrial and trade affairs, but still acting close to the main technical and metrological aspects of a certification and type approval of measurement systems.

References

[1] Brasil. Portaria INMETRO, nº 373, de 13/08/2014. Aprovar o Regulamento Técnico Metrológico estabelecendo os requisitos técnicos e metrológicos aplicáveis aos computadores de vazão e conversores de volume, utilizados na medição de petróleo e gás natural

[2] Lazari R F, Aguiar Júnior E A de, Almeida R O de, Hartmann V N. Flow computers regulation in Brazil: state of the art and next steps. In: CONGRESSO BRASILEIRO DE METROLOGIA, 2011, Natal

[3] Brasil. Portaria INMETRO, nº 163 de 06/09/2005. Adota, no Brasil, o Vocabulário Internacional de Termos de Metrologia Legal, em anexo, baseado no documento elaborado pela Organização Internacional de Metrologia Legal, com a devida adaptação ao nosso idioma, às reais condições existentes no País e às já consagradas pelo uso

[4] Brasil. Portaria INMETRO, n° 64, de 11/04/2003. Aprovar o Regulamento Técnico Metrológico,

que com esta baixa, estabelecendo os requisitos técnicos e metrológicos aplicáveis aos sistemas de medição equipados com medidores de fluido, utilizados na medição de petróleo, seus derivados líquidos, álcool anidro e álcool hidratado carburante

[5] American Gas Association. Compressibility Factors of Natural Gas and Other Related Hydrocarbon Gases: AGA 8. Estados Unidos, 2003. Terceira Edição

[6] American Petroleum Institute. Manual of Petroleum Measurement Standards Chapter 11-Physical Properties Data Section 1-Temperature and Pressure Volume Correction Factors for Generalized Crude Oils, Refined Products, and Lubricating Oils: API 11.1. Estados Unidos, 2004. Primeira Edição

[7] Brasil. Portaria Conjunta ANP/INMETRO, nº 1, de 19/06/2000. Estabelece as condições e requisitos mínimos que os sistemas de medição de petróleo e gás natural devem observar, com vistas a garantir resultados acurados e completos

[8] Brasil. Resolução Conjunta ANP/INMETRO, nº 1, de 10/06/2013. Estabelece as condições e os requisitos técnicos, construtivos e metrológicos mínimos que os sistemas de medição de petróleo e gás natural deverão observar, com vistas a garantir a credibilidade dos resultados de medição

[9] Organização Internacional de Metrologia Legal. R117-en: Measuring systems for liquids other than water. Disponível em: <https://www.oiml.org/en/files/pdf_r/r117-e95.pdf>. 27/07/2015

[10] OIML, International Vocabulary on Legal Metrology on-line, http://viml.oiml.info/en/index.html. Accessed on Sep 11th, 2015

Metrological evaluation of a flaw detector used in non-destructive testing by ultrasound

R C Mayworm[1], A V Alvarenga[1] and R P B Costa-Felix[1]

[1] Laboratory of Ultrasound (Labus), Directory of Scientific and Industrial Metrology (Dimci), National Institute of Metrology, Quality and Technology (Inmetro), Av. Nossa Sra. das Graças 50, Duque de Caxias, RJ, Brazil, ZIP 25250-020
E-mail: rcmayworm@gmail.com

Abstract. Nondestructive tests are widely used in various sectors of industry, and its main techniques are based on the use of ultrasound. The flaw detector is a device used in ultrasonic testing. In this article, it is studied the influence of the flaw detection settings on the measurement result. It was observed that when the speed of sound and the thickness of the standard block are known, the equipment gives the best results, with a maximum error of 0.08%. The results show the importance of having a standard block calibrated in order to minimize errors.

1. Introduction

Nondestructive testing (NDT) are widely used in industry, particularly in the evaluation of mechanical parts. Among the NDT techniques, the most widely used are based on the use of ultrasound, for example, for measuring the thickness or location of flaws and discontinuities [1]. Many of these measurements are performed using specific equipment, as a flaw detector. This equipment is used to detect defects in materials, can be used in external or internal environments, and plays an important role in the reliability of nondestructive tests carried out [2].

The purpose of this paper is to obtain a metrological evaluation of measurements carried out by a flaw detector in order to show the importance of knowing the different features present in an ultrasound test.

2. Case study

In order to evaluate the results given by the flaw detector and to show the importance of understanding the fundamentals involved in the test, different measurements were performed. Three transducers with different nominal frequencies were used, and tests were conducted using different equipment settings. Tests were performed using a standard block type 1, with speed of sound and dimensions previously calibrated. The development and implementation of this research were performed at Inmetro's Laboratory of Ultrasound (Labus), which has the entire infrastructure to carry out the proposed work.

3. Materials and methods

3.1. Flaw detector

The flaw detector used in this work was the model portable analyzer EPOCH 600 (Olympus NDT, USA) – see figure 1. That allows changing various settings of the tests [2].

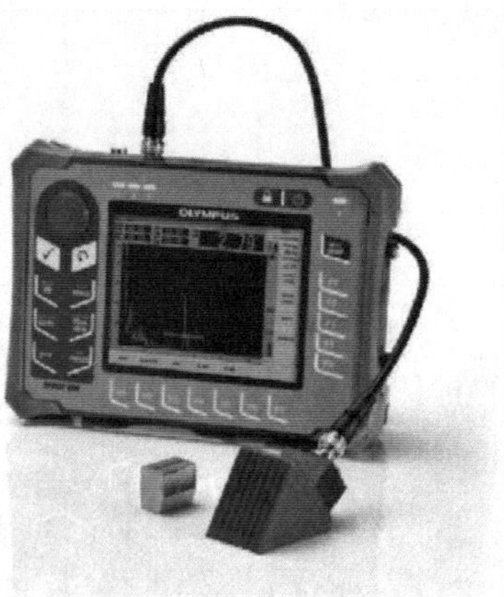

Figure 1. Flaw detector model Epoch 600 (Olympus NDT, USA).

It is noteworthy that the equipment measures the time of flight [s] of the pulse, however, presents the results in length [m]. Therefore, the speed of sound in material under test is an input on the equipment, which is used for converting the time of flight measured by the equipment, in length.

3.2. Test body

A standard block type 1 was used for the tests [3]. The block has a thickness of 25.036 mm with expanded uncertainty of 0.003 mm (k = 2.0 p = 0.95) and a height of 100.002 mm and expanded uncertainty of 0.003 mm (k = 2.0, p = 0.95). Those dimensions were calibrated in the Dimensional Metrology Laboratory of Inmetro (Lamed). The block has an ultrasonic propagation speed of 5910.6 $m \cdot s^{-1}$ with expanded uncertainty of 6.3 $m \cdot s^{-1}$ (k = 2.36, p = 0.95), as calibrated at Labus.

3.3. Experimental procedure

Tests were performed in different ways, all using transducers with different nominal frequencies and changing the configuration for calculating the time of flight considering the first peak or the maximum peak. The transducers used in the tests were the following: V306 2.25 MHz, V309 5 MHz, and A320s 7.5 MHz, all with a diameter of 0.5" (Panametrics NDT, USA). For the first, second and third tests, the calibrated speed of sound of the block (5.910 $m \cdot s^{-1}$) was used as input on the Epoch 600.

In the first, the block height measurements were carried out using as reference the time of flight between the start signal and the first reflection (first echo).

In a second moment, the height of the block was measured, with reference to the time of flight between the first reflection (first echo) and second reflection (second echo).

In the third test the equipment was set before the measurements. The adjustment was performed by measuring the block height and fixing it by means of the "zero setting" in the signal received by the equipment, in order to presented the same value calibrated by Lamed. Then, the third reflection of the block thickness was measured, equivalent to a size of approximately 75 mm (three times the block thickness). We chose to measure the third reflection, because according to the ABNT NBR 15824 - Nondestructive testing - Ultrasound - Thickness measurement, the measuring instrument shall be deemed fit to measure thickness in a range of ± 25% of the thickness of the block used in the default setting. In other words, the size of 75 mm is the limit for the measurement set in a block of 100 mm [4].

Finally, we used a feature of the Epoch 600, called "auto-calibration". By means of this tool, it is possible to estimate the value of the propagation velocity for the material tested by a known dimension value. Thus, the transducer was positioned on top of the block and it was informed to the equipment their height (100 mm). By doing that, the speed of sound of the block was estimated. Thereafter, it was measured the third reflection block thickness using the speed of sound estimated by the equipment.

3.4. Measurement uncertainty

In order to assess the reliability of measurement results, showing a quantitative indication of the quality of results, the measurement uncertainty was estimated. For a general relation $h = f(x_i)$, the general formula for the model uncertainty is obtained by the equation 1:

$$u_c^2 = \sum_{j=1}^{N} \left(\frac{dh}{dx_j}\right)^2 \cdot u_j^2 \tag{1}$$

Where u_c is the combined standard uncertainty associated with the outcome of the measurement (or calculation) h, and u_j the standard uncertainty, evaluated as Type A or Type B, associated with each variable parameter x_j used to express the value of h. [5].

Therefore, the initial step is to identify the sources of uncertainty. The following sources of uncertainty were identified: repetition (Type A), the speed of sound in block 1 (Type B) and the equipment resolution (Type B).

The uncertainty of repetition comes from ten measurements; the uncertainty of the speed of sound in block 1 comes from of the calibration certificate emitted by Labus ($c = 5910.3\ m \cdot s^{-1}$; $U = 6.3\ m \cdot s^{-1}$; $k = 2.36$); and, finally, the resolution of the equipment (0.01 mm).

4. Results and discussion

Table 1 presents the mean values and their respective uncertainties, for the test conducted using the time of flight between the excitation signal and the first echo as reference. Table 2 depicts the results for the test conducted using time of flight between the first and second echo as reference. In table 3, one can see the results for the test using the "zero setting", while table 4 shows the results achieved using the auto-calibration procedure.

Assessing table 1, it is noted that as the nominal frequency of the transducer increases, the values of the results tend to decrease. As the wavelength decreases when the frequency increases a higher resolution can be achieved. Moreover, transducers with higher nominal frequency have thinner coupling layer, than the total time of flight is shorter and, consequently, the thickness values are closer to the reference value. Besides, the values measured using as references the first peak are smaller to those measured in relation to the maximum peak. This is due to the maximum peak always appears after the first peak (bigger total time of flight). Here, it is important to mention that the previous analysis is valid only for results presented on table 1. In the other result, the differences in time of flight are mathematically annulled (table 2) or corrected using the NDT equipment settings (tables 3 and 4).

Table 1. Test 1 - References: Excitation signal and first echo. (Reference value = 100.002 mm)

		Mean [mm]	Uncertainty [mm]	Error [mm]
V306 2.25 MHz	First Peak	102.005	0.090	2.005
	Maximum Peak	103.092	0.091	3.092
V309 5 MHz	First Peak	101.613	0.090	1.613
	Maximum Peak	102.073	0.090	2.073
V320 7.5 MHz	First Peak	101.764	0.090	1.764
	Maximum Peak	101.765	0.090	1.765

The results achieved using the excitation signal and first echo as reference showed an error greater than the other tests, with the maximum error of 3.09% (table 1). On the other hand, determining the time of flight between the first and second echo, the maximum error falls to 1.12%. A possible explanation for this difference is that the time of flight between the excitation signal and the first echo contains the time of flight taken in the coupling layer of the transducer. Hence, that time of flight is bigger than the one measured between the first and second echo, where the time of flight in the coupling layer is annulled.

Knowing the speed of sound in the material and its thickness, it is possible to perform the zero adjustment. This setting gave the best result, with a maximum error of 0.08% (table 3). Here it is important to highlight that all errors were smaller than the estimated measurement uncertainty (table 3). Finally, knowing only the thickness, it is possible to use the "auto calibration". This setting showed a maximum error of 1.03% (table 4).

Table 2. Test 2 - References: first and second echo. (Reference value = 100.002 mm).

		Mean [mm]	Uncertainty [mm]	Error [mm]
V306 2.25 MHz	First Peak	101.000	0.090	1.00
	Maximum Peak	99.907	0.090	0.093
V309 5 MHz	First Peak	100.496	0.089	0.496
	Maximum Peak	101.122	0.091	1.122
V320 7.5 MHz	First Peak	99.926	0.088	0.074
	Maximum Peak	100.972	0.089	0.972

Table 3. Test 3 - Test carried out with zero adjustment and excitation signal as reference. (Reference value = 75.108 mm).

		Mean [mm]	Uncertainty [mm]	Error [mm]
V306 2.25 MHz	First Peak	74.951	0.067	0.049
	Maximum Peak	74.940	0.066	0.060
V309 5 MHz	First Peak	74.955	0.066	0.045
	Maximum Peak	74.964	0.066	0.035
V320 7.5 MHz	First Peak	74.964	0.066	0.036
	Maximum Peak	74.946	0.066	0.054

Table 4. Test 4 - Test performed using auto-calibration, and excitation signal as reference. (Reference value = 75.108 mm).

		Mean [mm]	Uncertainty [mm]	Error [mm]
V306 2.25 MHz	First Peak	75.516	0.071	0.516
	Maximum Peak	75.775	0.067	0.775
V309 5 MHz	First Peak	75.551	0.067	0.551
	Maximum Peak	75.526	0.067	0.526
V320 7.5 MHz	First Peak	75.347	0.067	0.347
	Maximum Peak	75.155	0.067	0.155

5. Conclusion

In general, the flaw detectors allow having control of various setting parameter. However, despite the advantage of having greater control over the testing carried out, it is clear that the operator must hold theoretical knowledge and experience about the performed test. Just one configuration incompatible with the test can generate inconsistent results compromising the inspection. Besides, the transducer frequency, as well as the reference used for measuring the time of flight, are factors that may influence on the final outcome.

Considering the different possible settings in the NDT equipment, the operator training has major importance in order to mitigate errors by choosing the appropriate parameters settings. Moreover, it is evident the importance of having a standard block with the speed of sound, as well as their dimensions calibrated, in order to minimize errors.

Acknowledgments
The authors thank the Program "PRONAMETRO" of the National Institute of Metrology, Quality and Technology (Inmetro), for financial support.

References

[1] ABENDI: Associação Brasileira de Ensaios Não Destrutivos. Available in: <www.abende.org.br> Jul. 24, 2014.
[2] Epoch 600: User manual. DMTA-10006-01PT –. December 2013.
[3] ISO 2400:2012 - Non-destructive testing Ultrasonic testing Specification for calibration block No. 1
[4] ABNT NBR 15824 – Nondestructive testing - Ultrasound - Thickness measurement.
[5] JCGM 100:2008 - Evaluation of measurement data - Guide to the expression of uncertainty in measurement. 1st Ed.
[6] Maggi L E, Silva C E R, Alvarenga A V and Costa-Felix R P B 2011 Ultrasonic calibration and certification of V1 and V2 type reference standard blocks for use in Non-Destructive Testing, Journal of Physics: Conference Series.

On the importance of measurement system calibration for underwater passive monitoring

S A Miqueleti[1], R P B Costa-Félix[1]

[1] Laboratory of Ultrasound (Labus), Directory of Scientific and Industrial Metrology (Dimci), National Institute of Metrology, Quality and Technology (Inmetro).

E-mail: samiqueleti@inmetro.gov.br

Abstract. The underwater passive acoustic monitoring of sound in oceans is growing in recent years and has served as a source of information on marine life and the interference of human activities on the environment. The recordings are used for species identification and prevention of potential adverse effects of vessel traffic, sonar and offshore activities as a whole. However, not much attention is given to the calibration of the hydrophone used to ensure the validity of the information collected. The resulting sound depends on the input audio, and the transfer function of the intensity of the input signal. This paper presents an assessment of how the lack of calibration of hydroacoustic systems might compromise the evaluation of the marine environment.

1. Introduction

Hydroacoustic, also known as underwater acoustics, studies the sound behavior in water and the influence of the environment on its propagation. A high quality hydroacoustic system is essential for a sound recording, and it can be used for evaluation purposes. Many echo sounders and depth probes are commercially available, but few have the desired quality for scientific and quantitative collection of acoustic data.

Currently there are few standards published in the field of underwater acoustics, such as [1,2,3], and some are being developed within the ISO (International Organization for Standardization) and IEC (International Electrotechnical Commission). This demonstrates how there is a growing concern about the quality of the performed measurements and the requirement of compliance with rules for possible use in regulatory measurement purposes.

The hydroacoustic systems should have stable electronic characteristics and are designed to maintain constant its inherent sensitivity properties, acoustic beam pattern, and transmission power, through a variety of conditions and for a prolonged period. This stability allows the system, once calibrated, to collect quantitative observations, including the repeatability within acceptable error limits. This paper presents and discusses some implications of metrology in the quality of the data used to assess the marine environment through the acoustic monitoring.

2. Acoustic monitoring

The acoustic monitoring is increasing by the scientific community to research marine mammals, especially cetaceans (whales, dolphins, etc.), many of which are easier to hear than to see. The acoustic monitoring is also used to support efforts to mitigate potential negative effects of human activities such as ship traffic, sonar and offshore. Most cetaceans produce sounds regularly, making the use of hydrophones (underwater microphones) to detect and monitor the presence of marine mammals by the sound they emit, or its "vocalization".

An important step in acoustic monitoring is the classification or the process of determining which species are making sounds. For its realization, it is necessary to know the vocalizations characteristics of the target species. Although the vocalizations of some cetaceans such as sperm whales and humpback whales are already well described, this is not the case for many other species of marine mammals. Many researchers are actively involved in the development of new techniques to improve the classification of animal sounds, plus the development of a marine audio library of sounds.

Currently there are three approaches to hydroacoustic detection system [4]:

2.1. Passive acoustic monitoring
In this approach a hydrophone monitors vocalizations or signals emitted by marine wildlife. This system does not allow the precise determination of the distance from the signal source and also has ambiguity in determining the direction of the source, i.e., you cannot determine which side comes the sign.

2.2. Active acoustic monitoring
In this case a focused pulse sonar "ping" is transmitted in the water and the system waits for a return signal when it finds a target. This system is used to locate schools, and military case, underwater targets. As it depends on a sound beam area that is covered at one time is small. In addition, the probe itself can cause environmental effects, albeit at a low level.

2.3. Daylight acoustic monitoring
This sonar class is new and based on the detection of existing background noise being scattered from an object. Have significant potential advantages, including wide area coverage, does not require acoustic emission and the ability to also act as passive sonar. However, it is the "cutting edge" technology and the limits of feasibility.

3. Electric transduction
The function of the hydrophone acoustic monitoring is transduction of cetacean's vocalization to electrical signals that can be recorded and analyzed. The main differences between a general measurement hydrophone and other special purpose are the characteristics of sensitivity and frequency response. Having a high sensitivity over a wide frequency range is desirable in transducers.

The hydrophone calibration consists in determining the system output (Volt) into a constant and known beep (Pascal). This ratio usually employed in the analysis of analog electronic circuits single input and single output is known as the transfer function. The audible signal when passing through the hydrophone undergoes changes and the recorded sound depends on the transfer function characteristics.

The transfer function modifies the original waveform modifying its characteristics, such as their pitch. As a result, waveform, duration, spectrum, and vocalizations level source cataloged animals, especially those with high-frequency components cannot be reliably determined from the recordings [5,6] if not well known to its transfer function. A system without calibration does not allow the correct identification of the species, and can generate a false classification [7].

4. Methods
An experiment was designed to demonstrate qualitatively as the calibration of a hydrophone and the environment can influence the recording of hydroacoustic signals. In this experiment, an acrylic tank filled with water was used, inside which were placed a sound source and a microphone. The sound source used was a speaker encased in a latex sheath to prevent water from entering the paper cone.

The hidroacustic transducer used was an electret microphone mounted on a nylon carrier and insulated by PVC film as the scheme Figure 1, below:

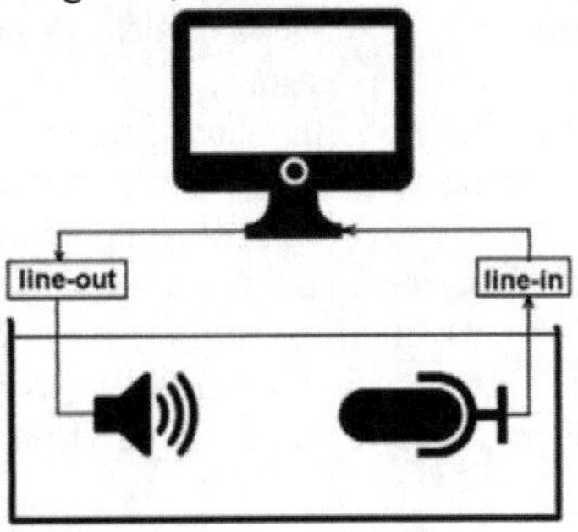

Figure 1 - simulated hydroacoustic system

The experiment was conducted in the audible frequency range (~100 Hz up to ~16 kHz). For playback and recording of the sounds we used the line-in and line-out interfaces to a PC and the software Cool Edit Pro aid (Syntrillium Software Corporation, 2003).

The acoustic signal used was the vocalization of a whale available in audio gallery of http://www.dosits.org/ website.

The experiment is to determine the difference in the frequency responses of the signals measured by the propagation on the water and in the air. This difference highlights the care that must be taken when using recorded signals to assign them to the vocalization of different sea creatures or other applications in passive monitoring.

The signals were recorded in 3 cases, namely:

a) Original signal, $s(t)$;
b) Propagation in water, $s_{ag}(t)$;
c) Propagation in air, $s_{ar}(t)$;
d) Propagation in water with interposition of a barrier (absorbent material), $s_{bar}(t)$.

To determine the frequency responses of various situations, it was applied to Fast Fourier Transform (FFT) on each sign up, define, respectively, $RF_s(t)$, $RF_{ag}(t)$, $RF_{ar}(t)$ e $RF_{bar}(t)$. The influence of water on the transmission compared with the transmission in air, is defined as the ratio:

$$RatRFag_{ar} = \frac{RF_{ag}\, f}{RF_{ar}\, f} \quad (1)$$

The same rationale applies to the determination of other frequency responses relations.

5. Methods

A Figure 2 shows the results obtained for the recording of the same sign in different conditions. The signal is reproduced in the time domain and frequency (spectral representation).

Figure 3 shows the ratio between the frequencies responses of the situations described below:

$RatRFag_{s}$ ⇒ ratio of the RF signal transmitted in the water compared to the original signal;
$RatRFar_{s}$ ⇒ ratio of the RF signal transmitted in the air in relation to the original signal;
$RatRFag_{ar}$ ⇒ ratio of the RF signal transmitted in the water in relation to the spread in the air;
$RatRFag_{bar}$ ⇒ ratio between the RF signals transmitted on water in relation to the signal with the interposition of the barrier.

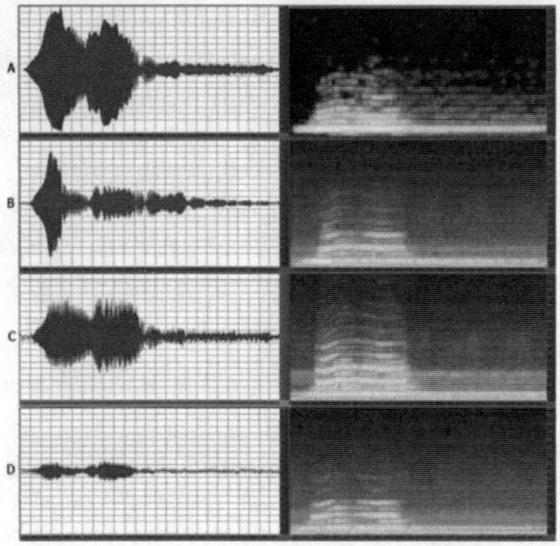

Figure 2 - (a) original electrical signal; (b) signal being transmitted through water; (c) A signal transmitted through the air; and (d) sign in water with acoustic barrier.

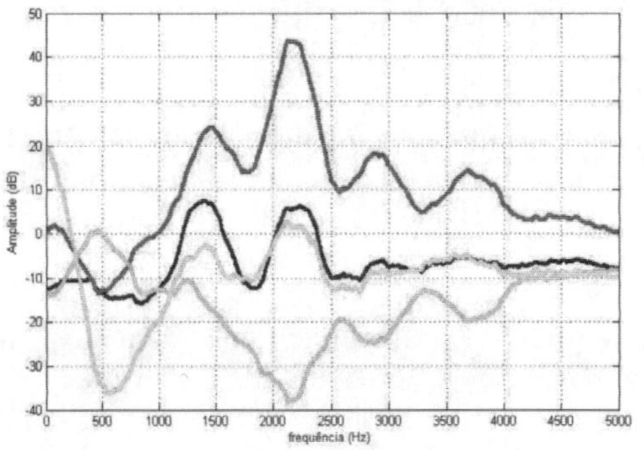

Figure 3 - (a) ratio of the FR signal transmitted into the water from the original signal: blue line; (b) ratio of the FR signal transmitted in the air compared to the original signal: red line; (c) ratio of the FR signal transmitted in the water in relation to the propagation in the air: green line; (d) the ratio between the FR signal transmitted on water in relation to the signal with the interposition of the barrier: cyan line.

6. Methods

Using a graphical analysis of Figure 2 is not possible to conclude that it is the same signal in all cases. The signal suffers changes due to propagation medium, air or water, as well as changes in the propagation path (sound barrier). The hydroacoustic system used can input harmonics and reverberations that are not present in the original signal.

The use of signal processing techniques and compensation hydroacoustic system, knowing the transfer function (hydrophone calibration) would make it possible to recover the original signal, but it should be an uncertainty associated with this conversion. Identification of software and classification of automatic signals that do not use information about system calibration may incur false positives or negatives to previously recorded patterns.

Analyzing Figure 3 in details, one could observe that the water influence on the frequency response of the recorded signal (green line) has its most remarkable effect in the frequency range between 1.5 kHz and 2.5 kHz. This frequency range corresponds to the less efficient for this particular system to propagate in water regarding the propagation in air.

Another remark to be outlined can be depicted observing the cyan line in Figure 3. Regarding line corresponds to the relation between the propagation in water with and without the insulation barrier, it is easy to find that the insertion loss is more effective in the frequency range around 500 Hz.

Finally, it is worthy emphasizing that a proper frequency response characterization is fundamental for a complete understanding of how a signal recorded can be analyzed. Moreover, it is clear that different propagation media could led to different frequency response behavior of the system, once the propagation medium itself is an important part of the propagation system. It is particular true for underwater long path propagation, such as those expected in wild life monitoring.

References

[1] IEC60500:1974 *Standard hydrophone*, International Electrotechnical Commission, Geneva, Switzerland

[2] IEC60565:2006 *Underwater acoustics -Hydrophones - Calibration in the frequency range 0.01 Hz to 1 MHz*, International Electrotechnical Commission, Geneva, Switzerland

[3] ISO/PAS 17208-1:2012 *Acoustics- Quantities and procedures for description and measurement of underwater sound from ships - Part 1: General requirements for measurements in deep water*, International Organization for Standardisation (ISO), Geneva

[4] Parvin S J, Nedwell J R and Harland E 2007 Lethal and physical injury of marine mammals, and requirements for Passive Acoustic Monitoring, *Subacoustech Repor*t No. **565R0212**

[5] Johnson M, Madsen PT, Zimmer WMX, Aguilar de Soto N, Tyack PL 2006 Foraging Blainville's beaked whales (Mesoplodon densirostris) produce distinct click types matched to different phases of echolocation. *J Exp Biol* **209**:5038–5050

[6] Madsen PT, Payne R, Kristiansen NU, Wahlberg M, Kerr I and Mohl B 2002 Sperm whale sound production studied with ultrasound-time-depth-recording tags. *J Exp Biol* **205**:1899–1906

[7] Harris EA 2012 Detection and Identification of Marine Mammals in Passive Acoustic Recordings from SCORE using a Visual Processing Approach Established for HARP Data, Thesis Naval Postgraduate School, California

8th Brazilian Congress on Metrology (Metrologia 2015) IOP Publishing
Journal of Physics: Conference Series **733** (2016) 012053 doi:10.1088/1742-6596/733/1/012053

Experimental findings on the underwater measurements uncertainty of speed of sound and the alignment system

T Q Santos, A V Alvarenga, D P Oliveira, R C Mayworm, R M Souza, R P B Costa-Félix

Laboratory of Ultrasound (Labus), Directory of Scientific and Industrial Metrology (Dimci), National Institute of Metrology, Quality and Technology (Inmetro), Av. Nossa Sra. das Graças 50, Duque de Caxias, RJ, Brazil, ZIP 25250-020

E-mail: taynara.qs@gmail.com

Abstract: Speed of sound is an important quantity to characterize reference materials for ultrasonic applications, for instance. The alignment between the transducer and the test body is an key activity in order to perform reliable and consistent measurement. The aim of this work is to evaluate the influence of the alignment system to the expanded uncertainty of such measurement. A stainless steel cylinder was previously calibrated on an out of water system typically used for calibration of non-destructive blocks. Afterwards, the cylinder was calibrated underwater with two distinct alignment system: fixed and mobile. The values were statistically compared to the out-of-water measurement, considered the golden standard for such application. For both alignment systems, the normalized error was less than 0.8, leading to conclude that the both measurement system (under- and out-of-water) do not diverge significantly. The gold standard uncertainty was 2.7 m·s^{-1}, whilst the fixed underwater system resulted in 13 m·s^{-1}, and the mobile alignment system achieved 6.6 m·s^{-1}. After the validation of the underwater system for speed of sound measurement, it will be applied to certify Encapsulated Tissue Mimicking Material as a reference material for biotechnology application.

1. Introduction

Speed of sound is an important quantity to characterize materials in ultrasonic applications. The accurate measurement of such property is a key step to certify reference materials used for calibrations in ultrasound field. The Inmetro's Laboratory of Ultrasound (Labus) has developed systems to calibrate the speed of sound of different materials (solids and liquids) [1][2][3], which are continually evaluated on metrological bases, and upgraded to improve the measurement uncertainty.

The implementation of new measurement procedure depends on their validation, and the guidelines for validating are presented in DOQ-CGCRE-8 [4]. Those guidelines are applicable to non-standard methods, standardized methods used outside of its original scope, methods modified to meet specific requirements, and new methods developed by the laboratory.

This article reports the improvement of the speed of sound measurement system developed at Labus/Inmetro. The new setup has more degrees of freedom in the mounting used to align the sample under test. The model to estimate the measurement uncertainty is presented and the obtained results are compared with a standardized method. The validate measurement system will be applied to certify the speed of sound of Encapsulated Tissue Mimicking Material (ETMM) for biotechnology application.

2. Methodology

The test body used in this work was an AISI 304 stainless steel cylinder (Figure 1), machined and polished to a surface roughness finishing better than 0.8 μm (Ra). Its thickness was calibrated using a caliper rule, and the result used as reference value (101.4 mm; u = 0.1 mm; k = 2.02; p = 0.95). Its diameter is 100 mm, and no calibration is needed for this dimension in the present work. The longitudinal speed of sound was measured in the central region of the test body, using the system described in [1], developed according to the standard ABNT-NBR 15824:2010 [5]. The calibrated speed of sound was defined as reference value ($c = 5724.7\ \mathrm{m \cdot s^{-1}}$; $u = 2.7\ \mathrm{m \cdot s^{-1}}$; $k = 1.96$; $p = 0.95$).

Figure 1. Stainless steel AISI 304 test body.

The new measurement system is composed of two ultrasonic transducers (Tx1 and Tx2), acting as a transmitter or receiver at different stages of the method, both with nominal central frequency of 5 MHz (Model v303, Panametrics, Olympus, MA, USA). The transducers and the test body were mounted on a mechanical system constructed to ensure their fixation and alignment. The transmitter transducer is mounted in a positioning with five degrees of freedom, while the receiver transducer and the test body are mounted on two degrees of freedom positioning system (Figure 2). All measurement system is immersed in a water tank filled with deionized water, and at least 30 min are necessary to achieve the thermal equilibrium before start the measurements. Here, it is important to highlight that the previous system set at Labus/Inmetro had only two degrees of freedom at receiver transduces, while the transmitter transducer and the test body were kept fixed (details in [3]).

An arbitrary function generator (33250A Agilent Technologies, CA, USA) is used to drive the transmitter transducer with a 20-cycles sine burst at 5 MHz, and 20 V peak to peak. The repetition period is defined as 10 μs. The signals acquired by the receiver transducer are digitized using the oscilloscope DSO 5012A (Agilent Technologies, CA, USA) with a sampling frequency of 500 $\mathrm{MSa \cdot s^{-1}}$.

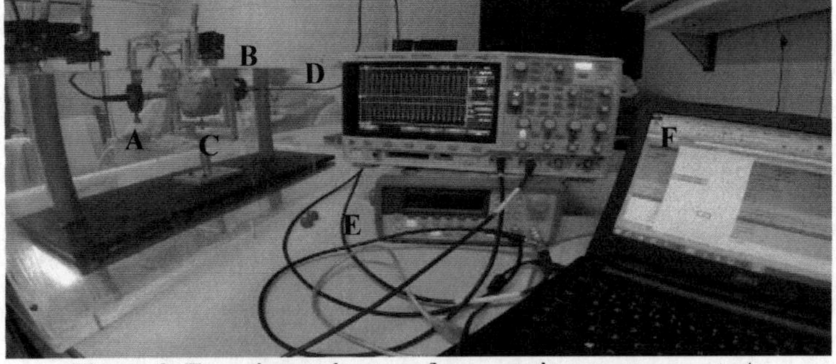

Figure 2. Experimental set-up for acoustic measurement. A, transmitting transducer with five degrees of freedom; B, receiver transducer with two degrees of freedom; C, specimen; D, oscilloscope; E, signal generator; F, computer.

The temperature was monitored throughout the measurements with a thermometer (Thermoscheneider, Baden-Württemberg, Germany). The whole procedure was repeated five times by two distinct operators in intermediate precision condition of measurement.
Different flight times are determined and are called here: t_{ref1} (reflection at the test body face nearest to the transmitter transducer), t_{ref2} (reflection at the test body face far of the transmitter transducer), t_{CP} (transmission between transmitter and receiver transducers with the test body in the acoustic path) and t_w (transmission between transmitter and receiver transducers without the test body).
The longitudinal speed of sound *c* [m/s] was calculated by (1).

$$c = \frac{x}{t} \tag{1}$$

where, *x* [m] is the thickness of the test body in calculated from (2), and *t* [s] is the time of flight of ultrasound in the test body calculated from (3).

$$x = \left[\left(2t_w - t_{ref1} - t_{ref2}\right) \cdot c_w\right] \tag{2}$$

$$t = t_{CP} - \frac{(t_{ref1} + t_{ref2})}{2} \tag{3}$$

The ultrasound propagation velocity in water c_w [m/s] is calculated according to (4) [6].

$$c_w = 1404,3 + 4,7T - 0,04T^2 \tag{4}$$

Where, *T* [°C] is the temperature in water.
The measurement uncertainty is estimated according to the guide to the expression of uncertainty in measurement [7]. The following sources of uncertainty were considered in measuring the speed of sound, as well the test body thickness: type A evaluation determined as the standard deviation between the five repetitions of the test; type B evaluation of the temperature obtained from the thermometer calibration certificate (0.03°C); type B evaluation of the ultrasound propagation speed in water (0.018 $m \cdot s^{-1}$) [6]; type B evaluation of the time base used to determine different times of flight from the oscilloscope calibration certificate (0.06%). The expanded uncertainty was calculated for a coverage probability of 0.95.
In this work, the benchmark is the normalized error (E_n) between the reference values, determined using the standardised measurement method, and the results achieved using the previous measurement system [3] and the improved one presented here. The results are considered satisfactory if $E_n \leq 1$.

3. Results

The results of speed of sound and thickness determined using the measurement system presented here (mobile system) and the previous system described in [3] (fixed system) are presented on Table 1 and Table 2, for operator 1 and 2 respectively.

One can observe that the results of thickness and speed of sound achieved using both systems when compared with the respective reference values presented $E_n \leq 1$. This result suggests that both methods can be considered validate for measuring the speed of sound, as well as the thickness of test bodies.

Table 1. Results of thickness (x) and speed of sound (c) achieved by operator 1 (Op. 1) using fixed and mobile systems.

Op. 1	Fixed System		Mobile System	
	x [m]	c [m·s^{-1}]	x [m]	c [m·s^{-1}]
Average Value	0.10130	5738	0.10130	5730
k	1.97	2.04	1.96	1.96
U	0.00011	13	0.00010	6.6
En	0.7	1	0.7	0.8

Table 2. Results of thickness (x) and speed of sound (c) achieved by operator 2 (Op. 2) using fixed and mobile systems.

Op. 2	Fixed System		Mobile System	
	x [m]	c [m·s^{-1}]	x [m]	c [m·s^{-1}]
Average Value	0.10134	5736	0.10134	5731
k	2.26	2.03	1.96	1.96
U	0.00020	12	0.00010	7.9
En	0.3	0.9	0.7	0.8

However, one can perceive that the speed of sound expanded uncertainties, achieved using the mobile system, are lower than the one obtained with the fixed system, for both operators. This result indicates that increasing the degrees of freedom in the system allowed a better alignment between the two ultrasonic transducers (transmitter and receiver), as well as between them and the test body, directly affecting the reduction of uncertainty.

4. Discussion and Conclusion

In literature, there are references mentioning speed of sound results for different materials [8][9]. However, rarely authors give attention to metrological rigors in their studies, or do not properly report their uncertainty results. Sometimes uncertainty is mistakenly confused with error or standard deviation and, when presented, the validation of measurement method is not detailed, neither it is explained how the uncertainties were calculated.

This work presented the validation of a new measuring system to determine the speed of sound of test bodies. The measuring system was first validated at 5 MHz, and new validation will be carried out considering other frequencies. The presented measuring system will be applied for certifying an Encapsulated Tissue Mimicking Material (ETMM) as a reference material for biotechnological applications.

References

[1] Maggi L E, Silva C E R, Alvarenga A V, Costa-Felix RPB Ultrasonic calibration and certification of V1 and V2 type reference standard blocks for use in Non-Destructive Testing 2011. *Journal of Physics: Conference Series 279 012029* doi: 10.1088/1742-6596/279/1/012029.

[2] Figueiredo M K K, Costa-Félix R P B, Alvarenga A V, Maggi L E, Romeiro G A 2012 *Fuel* **91** 209-212

[3] Santos T Q, Oliveira D P, Costa-Félix R P B, Alvarenga A V 2014 Validação do método de medição da velocidade ultrassônica de propagação longitudinal *XXIV Congresso Brasileiro de Engenharia Biomédica – CBEB* 1644-1648

[4] INMETRO. DOQ-CGCRE-008: *Orientações sobre validação de métodos analíticos* 2011. 4ª revisão, INMETRO: Duque de Caxias

[5] Associação Brasileira de Normas Técnicas *Ensaios não-destrutivos - Ultrassom - Medição de espessura* 2010 *ABNT NBR 15824*

[6] Lubbers J and Graaff R 1998 A simple and accurate formula for the sound velocity in water *Ultrasound in Med. and Biol.* **24** (7) 1065-1068

[7] JCGM 2008 Evaluation of Measurement Data – *Guide of the Expression of Uncertainty in Measurements*

[8] Cannon L M, Fagan A J and Browne J E 2011 Novel tissue mimicking materials for high frequency breast ultrasound phantoms *Ultrasound in Medicine and Biology*, **37**(1) 122-135

[9] Brewin M P, Pike L C, Rowland D E and Birch M J 2008 The acoustic properties centered on 20 MHz, of IEC agar-based tissue mimicking materials and its temperature, frequency and age dependence *Ultrasound in Medicine and Biology* **34** (8) 1292-1306

Acknowledgments

Research reported in this paper has been partially funded by the Carlos Chagas Filho Research Support Foundation (FAPERJ), Grant Numbers E-26/102.124/2013 and E-26/201.563/2014, and from the National Council for Scientific and Technological Development (CNPq), Grant Number 310.392/2014-4.

Primary calibration of measurement microphones in the world: state of art

T A B Milhomem and Z M D Soares

Electroacoustics Laboratory; Acoustics and Vibration Metrology Division; Scientific and Industrial Metrology Directory; National Institute of Metrology, Quality and Technology; Avenida Nossa Senhora das Graças, 50, Prédio 1, Xerém, Duque de Caxias, RJ, 25.250-020, Brazil

E-mail: tbmilhomem@inmetro.gov.br

Abstract. This paper presents an overview of state of art of measurement microphones primary calibration in the world with emphasis on Brazil practices. Initially, pressure field calibration is summarized being discussed mainly the couplers used to create pressure field conditions. After that, free-field calibration is presented being commented especially the anechoic chambers used to create free-field conditions. Concluding, it is showed diffuse-field calibration that is being investigated. It is presented, in particular, the reverberant chambers used to create diffuse-field conditions.

1. Introduction

Calibration of a measurement microphone commonly consists in determining its sensitivity (modulus and phase) as a function of frequency, i.e., determining the ratio between the incident sound pressure and the emerging voltage as a function of frequency.

Primary calibration of measurement microphones usually is performed by the reciprocity method using three microphones. Following this method, the microphones are acoustically coupled in pairs and using one of them as sound source (transmitter microphone) and other as sound receiver (receiver microphone), the electrical and acoustic transfer impedances are determined (figure 1). Thus, from the determined impedances for each pair of microphones, the sensitivity of each microphone is calculated [1-3].

Such as sensitivity changes according to the sound field, three theoretical sound fields were established for calibration purposes: pressure field, free-field and diffuse-field [1]. Therefore, a microphone must be calibrated in a sound field similar to that where it is expected to operate.

The motivation of this paper is to present an overview of state of art of measurement microphones primary calibrations in the world with emphasis on Brazil practices.

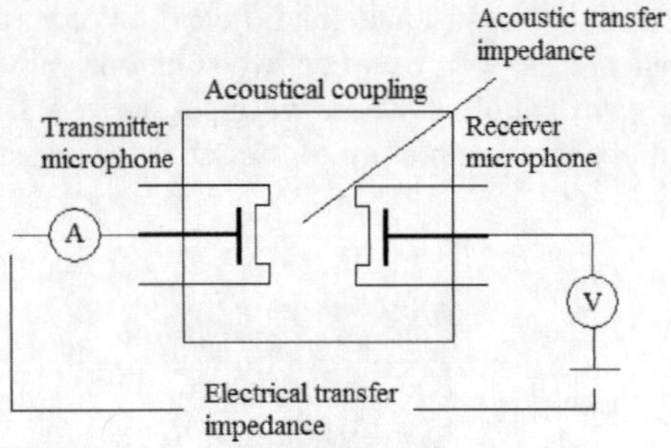

Figure 1. Primary calibration of measurement microphones.

2. Pressure field primary calibration

Pressure field reciprocity calibration is performed as stated by the international standard IEC 61094-2 [2]. It is performed in acoustic couplers and, according to International Bureau of Weights and Measures (BIPM), thirteen National Metrology Institutes (NMIs) are recognized for their accomplishment, being eleven along with the 1992 edition and two along with the 2009 edition [4]. The 2009 edition presents a significant change compared to the earlier edition: it presents a solution to heat conduction effect at low frequencies ("low frequency solution") and another solution that includes heat conduction effect and viscous losses at high frequencies ("broad-band solution") while the 1992 edition only presents one solution for heat conduction effect that applies to all frequency range [2,5]. The 2009 edition was tested for the first time in the CCAUV.A-K5 key comparison [6], whose results were presented in 2014 [7]. Brazil, through Inmetro, is recognized for its accomplishment to perform this calibration as stated by the 1992 edition and took part in the CCAUV.A-K5.

Pressure field calibration was tested in four key comparisons conducted by the Consultative Committee for Acoustics, Ultrasound and Vibration (CCAUV) of the International Committee for Weights and Measures (CIPM): CCAUV.A-K1, CCAUV.A-K2, CCAUV.A-K3 and CCAUV.A-K5, being CCAUV.A-K5 the most recent. Inmetro also took part in the CCAUV.A-K1 and CCAUV.A-K3 [7].

Twelve NMIs took part in the CCAUV.A-K5: NPL from United Kingdom, DPLA from Denmark, GUM from Poland, NIM from China, Inmetro, Cenam from Mexico, Inrim from Italy, Nmisa from South Africa, Kriss from Republic of Korea, NRC from Canada, Vniiftri from Russia Federation and NMIJ from Japan. The standards circulated among the institutes were two 1-inch laboratory standard microphones designed for pressure field. Their sensitivity (modulus and phase) were calibrated in the frequency range from 2 to 10000 Hz (being phase optional and range 2-20 Hz for modulus optional as well). The measurements took place between January 2011 - July 2012 and the final report was presented in 2014 [6].

In this comparison, NPL and GUM used a single coupler for the acoustical coupling. NPL used a 2 cc coupler and GUM used a 1.5 cc coupler without capillary tubes. NMIJ used two plane wave couplers, but it did not specify their volumes. Kriss and Viniiftri used two plane wave couplers, one 2 cc and other 4 cc. DPLA and NRC used four different plane wave couplers, but they did not specify their volumes being the couplers used by NRC did not have capillary tubes. NIM, Inrim and Nmisa also used four plane wave couplers. NIM used couplers with nominal volumes of 1.5, 2, 2.7 and 4 cc. The couplers used by Inrim were 1.2, 1.5, 2 and 2.7 cc and the ones used by Nmisa were 1, 1.5, 2 and 2.7 cc without capillary tubes. Cenam used five plane wave couplers with nominal volumes of 1.2, 1.5, 2, 2.7 and 4 cc being the three largest used in measurements below 40 Hz and the three smallest used in measurements from 20 Hz [6].

Regarding the effect on sensitivity due to the heat conduction, NPL, Inrim, Nmisa and NRC used the broad-band solution specified in the standard. GUM used the low frequency solution. DPLA combined the low frequency and the broad-band solution and, in order to minimize fluctuations on the microphones responses, a gradual transition between two solutions was made [6].

Inmetro used four plane wave couplers whose volumes were 1.2, 1.6, 3 and 5 cc and used the broad-band solution [6]. Figure 2 shows a picture of the setup for pressure field reciprocity calibration in Inmetro.

Figure 2. Setup for pressure field reciprocity calibration in Inmetro.

Results, showed in the final report [6], do not indicate a relationship between the number of couplers and a good performance or between the heat conduction solution used and a good performance.

3. Free-field primary calibration

Free-field reciprocity calibration is performed in anechoic chambers. It was performed as stated by the international standard IEC 61094-3 [3] and, according to BIPM, three NMIs are recognized for their accomplishment [4]. It was tested in only one key comparison conducted by CCAUV, the CCAUV.A-K4, and only the sensitivity modulus was measured [7]. Brazil also is recognized for its accomplishment to perform the free-field calibration.

Seven NMIs took part in the CCAUV.A-K4: DFM from Denmark, Cenam, Kriss, NMIJ, LNE from France, PTB from Germany and Inmetro. Two ½-inch laboratory standard microphones designed for pressure field were calibrated in the frequency range from 1 to 40 kHz (40 kHz optional). The measurements took place from February 2007 - February 2008 and the results were presented in 2010 [8].

In this comparison, DFM used a small anechoic chamber of 3.6 m^3 volume. Kriss and PTB, in turn, used a little higher anechoic chamber: 12 m^3 and 15 m^3 volume respectively. NMIJ used a large anechoic chamber of 547 m^3 volume, however, to compensate low signal-to-noise ratio at frequencies below 5 kHz it used, as transmitter microphone, microphones with higher sensitivity than the microphones of key comparison, which were used only as receiver microphone [8].

LNE determined the electrical and acoustic transfer impedances using three distances between transmitter and receiver microphones. The distances were 250, 300 and 350 mm. Cenam and NMIJ used four distances between the microphones. Cenam used distances of 160, 189, 220 and 250 mm, while NMIJ used distances of 100, 150, 200 and 250 mm being the three lowest used for the range 1-5 kHz and the three highest for the range 6.3-40 kHz. Kriss used five distances between microphones: 100, 150, 200, 250 and 300 mm [8].

Inmetro used a small anechoic chamber of 2.8 m^3 volume and determined the electrical and acoustic transfer impedances using four distances between microphones: 170, 200, 240 and 300 mm [8]. At present, Inmetro measures at only the distance of 300 mm because it uses an innovative signal processing technique that eliminates the needs to calculate the average on space [9]. This technique was checked in the CCAUV.A-K4. Figure 3 shows a picture of the setup for free-field reciprocity calibration in Inmetro.

Results, showed in the final report [8], do not indicate a relationship between the anechoic chamber volume and a good performance or between the number of distances used and a good performance.

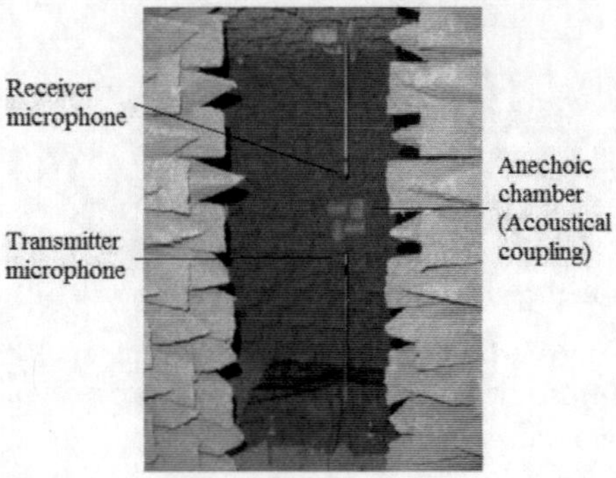

Figure 3. Setup for free-field reciprocity calibration in Inmetro.

4. Diffuse-field primary calibration

Diffuse-field reciprocity calibration is performed in reverberant chambers but it is not consolidated, i.e., there is no international standard that deals with this subject and no key comparison was conducted by the CCAUV [7]. Four NMIs are researching this calibration: DPLA/DFM [10], NPL [11], PTB [12] and Inmetro [13].

NPL measured in a reverberation chamber of 330 m^3 volume while DPLA/DFM measured in a short reverberation chamber of 2 m^3 and lastly PTB measured in a very short reverberation chamber of 0.23 m^3 [10-12].

NPL measured voltage, current and reverberation time in three positions, far enough away from all boundary surfaces, for determining the electrical and acoustic transfer impedances. In this context, "far enough away" means that the distance should be much greater than the diffuse-field distance which is the distance where the direct sound energy density equals the average energy density. DPLA/DFM, in turn, measured in sixteen positions representing spatially independent points, i.e., they should be spaced at least by half a wavelength. On the other, PTB measured in twenty-five spatial combinations result of a 5x5 used positions montage [10-12].

No research [10-13] presented satisfactory results yet, i.e., with adequate accuracy and precision (standard deviation).

Inmetro is measuring in a small reverberation chamber of 2 m^3 with different diffusers (hanging diffusers, panels, and/or boundary diffusers also known as volume diffusers, spherical caps) and measuring in sixteen and thirty-two spatial combinations (result of a 4x4 and 4x8 used positions montage). Figure 4 shows a picture of the experimental arrangement for diffuse-field reciprocity calibration in Inmetro.

Figure 4. Experimental arrangement for diffuse-field reciprocity calibration in Inmetro.

References

[1] Milhomem T A B and Soares Z M D 2015 Primary calibration in acoustics *Journal of Physics: Conference Series* **575** 012008

[2] IEC 2009 IEC 61094-2 Electroacoustics - Measurement microphones - Part 2: Primary method for pressure calibration of laboratory standard microphones by the reciprocity technique

[3] IEC 1995 IEC 61094-3 Measurement microphone - Part 3: Primary method for free-field calibration of laboratory standard microphones by the reciprocity technique

[4] BIPM Calibration and measurement capabilities Available in: www.bipm.org Access in: July 2015

[5] IEC 1992 IEC 61094-2 Measurement microphones - Part 2: Primary method for pressure calibration of laboratory standard microphones by the reciprocity technique

[6] Avison J 2014 *Report on key comparison CCAUV.A-K5* Available in: www.bipm.org Access in: July 2015

[7] BIPM Key comparison database Available in: www.bipm.org Access in: July 2015

[8] Barrera-Figueroa S, Nielsen L, Rasmussen K, Matzumoto A E P and Razo J N R *Final report on the key comparison CCAUV.A-K4* Available in: www.bipm.org Access in: July 2015

[9] Soares Z M D, Muller S, Milhomem T B and Massarani P 2008 Primary microphone calibration in free-field using time-selective technique *37th International Congress and Exposition on Noise Control Engineering – InterNoise 2008*

[10] Barrera-Figueroa S, Rasmussen K and Jacobsen F 2008 A note on determination of the diffuse-field sensitivity of microphones using the reciprocity technique *J. Acoust. Soc. Am.* **124 (3)** 1505

[11] Jackett R J 2012 Implementation of a diffuse-field microphone calibration system *41st International Congress and Exposition on Noise Control Engineering – InterNoise 2012*

[12] Weihe S Wittstock V and Blau M 2013 Diffuse-field reciprocity calibrations of measurement microphones *40th Italian Annual Conference on Acoustics (AIA) and 39th German Annual Conference on Acoustics (DAGA)*

[13] Milhomem T A B, Soares Z M D and Pinto C P 2013 Determination of the reciprocity factor for microphones primary calibration in a diffuse-field *42nd International Congress and Exposition on Noise Control Engineering – InterNoise 2013*

Calibration of sound calibrators: an overview

T A B Milhomem and Z M D Soares

Electroacoustics Laboratory; Acoustics and Vibration Metrology Division; Scientific and Industrial Metrology Directory; National Institute of Metrology, Quality and Technology; Avenida Nossa Senhora das Graças, 50, Prédio 1, Xerém, Duque de Caxias, RJ, 25.250-020, Brazil

E-mail: tbmilhomem@inmetro.gov.br

Abstract. This paper presents an overview of calibration of sound calibrators. Initially, traditional calibration methods are presented. Following, the international standard IEC 60942 is discussed emphasizing parameters, target measurement uncertainty and criteria for conformance to the requirements of the standard. Last, Regional Metrology Organizations comparisons are summarized.

1. Introduction

Sound calibrators are devices that generate a known sound pressure level in a known frequency. They are mainly used for adjustment of sound pressure level measurement systems, but they are also used for calibration of measurement microphones by the electrostatic actuator method. They are calibrated according to the international standard IEC 60942 [1-3]. This calibration comprises in measuring sound pressure level, frequency and total harmonic distortion (THD) or total distortion (TD) according to the edition of standard used in their project. This is an interesting point: sound calibrators are calibrated according to edition of standard used in their project no matter which is the current edition.

2. Calibration

Sound pressure level calibration can be performed by different methods being the more traditional the microphone method (using the insert voltage technique) and the sound calibrator comparison method.

In the microphone method, sound calibrator is coupled to a microphone, turned on and then it is measured the system output voltage. After that, sound calibrator is turned off and a voltage is inserted on preamplifier input (microphone output). This requires an insertion voltage unit and a preamplifier with insert voltage facility. When the inserted voltage is adjusted to give the same system output voltage as results from sound pressure level on microphone, this insert voltage is equal to the microphone output voltage (when the sound calibrator is turned on). Knowing the microphone open circuit sensitivity (determined from its calibration [4]), sound pressure and sound pressure level are calculated. Alternatively, from the inserted voltage and the system output voltage (measured when the voltage is inserted), system gain is calculated. From the system gain and system output voltage as results from sound pressure level on microphone, the microphone output voltage is calculated. Again, knowing the microphone open circuit sensitivity, sound pressure and sound pressure level are calculated.

In the sound calibration comparison method, sound calibrator under test is compared with a reference sound calibrator previously calibrated, usually by the microphone method. Sound calibrator under test is coupled to a microphone and it is measured the system output voltage. After that, reference sound calibrator is coupled to the same microphone and it is also measured the system output voltage. From measured voltages and knowing the sound pressure level of the reference sound calibrator, sound pressure level of the device under test is calculate. It should be paid attention to possible corrections for non-linearity of system if sound calibrators (reference and under test) generate sound pressure level at different frequencies i.e., for different microphone sensitivity or different preamplifier gain at each frequency.

Frequency and THD or TD (THD + noise) calibration is performed directly, measuring frequency in a frequency meter and THD or TD in a distortion meter. However, traceability for (low amplitude) distortion measurements is a global challenge [5].

3. IEC 60942

Standard that deals with calibration of sound calibrator is IEC 60942 [1-3]. It is in its third edition, being the first dated of 1988, the second one of 1997 and the newest of 2003. Nevertheless, a new edition is already under analyses by a technical committee of the International Electrotechnical Commission (IEC) [6]. Regards the measured parameters, 1988 edition indicates calibration of sound pressure level, frequency and THD while 1997 and 2003 editions replace THD by TD. Relative to the target measurement uncertainty, 1988 edition does not present anything about it, 1997 edition indicates different limits depending on microphone calibration and 2003 edition indicates an unique limit (independent on microphone calibration). Concerning to the criteria for conformance with the requirements of the standard, 1988 edition indicates that the measured deviations from designed goals do not exceed the applicable maximum permissible measurement error. The 1997 edition indicates that the measured deviations from designed goals extended by the expanded uncertainties of measurement do not exceed the applicable maximum permissible measurement error extended by the applicable target measurement uncertainties, and that the expanded uncertainties of measurement do not exceed the corresponding target measurement uncertainties. The 2003 edition indicates that the measured deviations from designed goals extended by the expanded uncertainties of measurement do not exceed the applicable maximum permissible measurement error that already includes the applicable target measurement uncertainties, and that the expanded uncertainties of measurement do not exceed the corresponding target measurement uncertainties [1-3].

The new edition under discussion uses the newest policy on conformance assessment of the technical committee number 29 of IEC. According to this policy, the conformance will be demonstrated when the measured deviations from design goals do not exceed the applicable maximum permissible measurement error not including the target measurement uncertainties, and that the uncertainties of measurement do not exceed the corresponding target measurement uncertainties.

4. Comparisons

According to International Bureau of Weights and Measures (BIPM), thirty four National Metrology Institutes (NMIs) are recognized for accomplishment of calibration of sound calibrator [7], but no comparison was performed under the International Committee for Weights and Measures (CIPM), just under the Regional Metrology Organizations (RMO): the APMP.AUV.A-S1 in Asia Pacific Metrology Program (APMP) and the SIM.AUV.A-S1 in Inter-American Metrology System (SIM) [8].

The regional supplementary comparison APMP.AUV.A-S1 has been carried out for the measurement of sound pressure level, frequency, THD and TD of a multi-frequency sound calibrator. The pilot laboratory was the NMI of Thailand and a multi-frequency sound calibrator was circulated through thirteen NMIs. The measurements took place between September 2008 - July 2010 and the final report was presented in March 2014. Deviations from supplementary comparison reference values (SCRVs) for sound pressure level and frequency are mostly within declared expanded uncertainties. For THD and TD it was not possible to calculate a SCRV because the absence of any consistency in the approach to calculating uncertainty which varied a lot, e.g. for 1 kHz at 94 dB, THD reported expanded uncertainties varied from 0.00938 to 0.5% [9].

SIM.AUV.A-S1 has been carried out for the measurement of sound pressure level, frequency and THD of two pistonphones. Cenam, NMI from Mexico, was the pilot laboratory and five NMIs participated on this comparison. The measurements took place between July 2001 - February 2002 and the preliminary report (Draft A) was finished in April 2004. Final report was not presented yet.

Brazil, through Inmetro, is recognized for its accomplishment to perform the calibration of sound calibrator and participated in the comparison SIM.AUV.A-S1 [7,8].

5. Conclusions

There is a request for traceability for (low amplitude) distortion measurements.

The new edition of IEC 60942 under discussion proposes a significant change with regard to the criteria for conformance to the requirements of the standard.

The absence of a comparison performed under CIPM and the existence of only two regional comparisons carried out for the measurement of sound pressure level, frequency and distortion (TD and/or THD) argument the need of more comparisons.

References

[1] IEC 1988 IEC 60942 Sound calibrators
[2] IEC 1997 IEC 60942 Electroacoustics: Sound calibrators
[3] IEC 2003 IEC 60942 Electroacoustics: Sound calibrators
[4] IEC 2009 IEC 61094-2 Electroacoustics - Measurement microphones - Part 2: Primary method for pressure calibration of laboratory standard microphones by the reciprocity technique
[5] Soares Z M D 2011 Electroacoustic devices: Traceability of distortion measurements *18th International Congress on Sound and Vibration – ICSV18*
[6] Webstore IEC Available in: www.iec.ch Access in: August 2015
[7] BIPM Calibration and measurement capabilities Available: www.bipm.org Access in: July 2015
[8] BIPM Key comparison database Available in: www.bipm.org Access in: July 2015
[9] Plangsangmas V, Leeudomwong S, Scott A, Zhong B and Huang Y 2014 *Report on regional supplementary comparison APMP.AUV.A-S1 - Final report* Available in: www.bipm.org Access in: July 2015

Experimental evaluation of the uncertainty associated with the result of feature-of-size measurements through computed tomography

T L Fernandes[1], G D Donatelli[1] and C R Baldo[2]

[1] Centre for Metrology and Instrumentation - CMI, Reference Centres for Innovative Technologies - CERTI, Florianópolis, SC, Brazil
[2] Centre for Engineering, Modelling and Applied Social Sciences - CECS, UFABC - Federal University of ABC, Santo André, SP, Brazil

E-mail: thf@certi.org.br

Abstract. Computed tomography for dimensional metrology has been introduced in quality control loop for about a decade. Due to the complex measurement-error cause system, generally no consistent measurement uncertainty reporting has been made. The ISO 15530-3 experimental approach, which makes use of calibrated parts, has been tested for estimating the uncertainty of CT-based measurements of features of size of a test object made of POM. Particular attention is given to the design of experiment and to the measurement uncertainty components. The most significant experimental findings are outlined and discussed in this paper.

1. Introduction

Computed tomography (CT) for industrial coordinate metrology has been part of dimensional quality control loop for about a decade. In general, complex-shaped parts with hundreds (even thousands) of features (hidden features as well) can be holistically inspected with great operational advantages over other existing coordinate metrology technologies, such as tactile and/or optical coordinate measuring machines.

CT measurement principle relies on the attenuation of X-rays when propagating through the test object, which depends on the object material and radiographic thickness. For a large number of beam directions, the intensity distribution of the remaining radiation is measured and digitally stored as a grey-value image. The resulting projections of the full object rotation are mathematically processed to create the 3D voxel matrix. Further processing steps over the voxel data allow performing dimensional measurements.

The CT principle and the metrological CT scanner setup give rise to influence factors that affect the performance of dimensional evaluations. They are related to the source (e.g. photon energy, focal spot size), to the detector (e.g. sensitivity, pixel size, exposure time, averaging), to the object (e.g. material, shape, size), to the CT kinematics (e.g. magnification axis and turntable repeatability and accuracy), and mathematical data processing (e.g. segmentation, measuring strategy) [1].

Due to that intricate measurement-error cause system, establishing traceable measurements with CT has been pointed as a key metrology issue. This paper outlines and discusses the use of calibrated workpieces, as specified in part 3 of ISO 15530, for estimating the task-specific uncertainty associated with feature-of-size measurement results. Section 2 shows the intrinsic characteristics of the test object under analysis. Section 3 summarizes the reference and CT measurements of the test object. Sections 4 and 5 present respectively the uncertainty assessment and the most relevant findings.

2. Case description

2.1. Test object intrinsic characteristics

The industrial part selected for this experimental uncertainty evaluation composes the drive system of a window lift mechanism, and is made of acetal photopolymer (POM). Three intrinsic characteristics were defined in accordance with the technical drawing of the test object: the diameter of the external circumferential line at surface mid-height, D_1; the diameter of the internal circumferential line at surface mid-height, D_2 and D_3. Figure 1 illustrates the test object and the intrinsic characteristics just described.

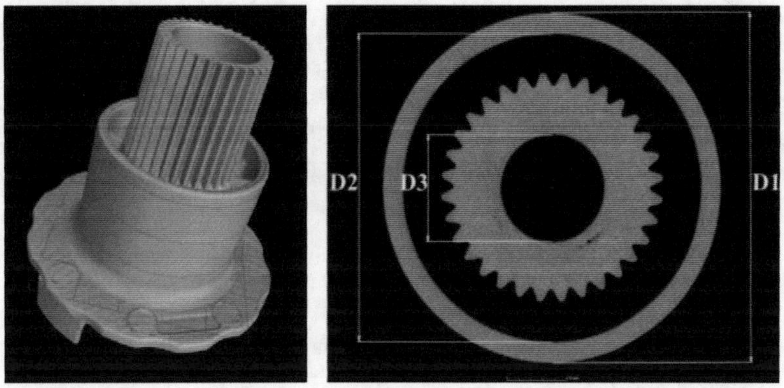

Figure 1. Illustrative image of the test object and the intrinsic characteristics (features of size).

2.2. Uncertainty evaluation background

In order to evaluate the uncertainty associated with CT measurements of the intrinsic characteristics, three uncertainty contributions were considered, which are represented by the following standard uncertainties: uncertainty of the parameter of the calibrated test object, u_{cal}; uncertainty associated with the measurement procedure, u_p; uncertainty associated with the systematic error, u_b. The expanded measurement uncertainty is obtained by combining the individual standard uncertainties in a root-sum-of-squares manner and multiplying by the coverage factor, k. Please refer to ISO 15330-3 for more details.

3. Measurement setup

3.1. Tactile reference measurements

The intrinsic characteristics of the test object were calibrated on a Carl Zeiss PRISMO ultra CMM, which is housed in a temperature-controlled room maintained at (20.0 ± 0.3) °C. All diameters were realized by associating ideal features of type circle to the sampled points (using least-squares method). Measurement uncertainties were estimated using the Virtual CMM software output as well as expert judgment. Table 1 lists the calibration results.

3.2. CT measurements

The test objected was measured on a Carl Zeiss METROTOM 1500 CT system, which is equipped with a 225 kV micro-focus tube and a 2048^2 pixels flat panel detector. The CT system is installed in a temperature-controlled room maintained at (20 ± 1) °C. The CT system manufacturer specifies a MPE for length measurements of (9 + L/50) μm, using a test piece consisting of 27 ruby spheres mounted on carbon fiber shafts, and then determining the sphere-center to sphere-center distances of several pairs of spheres [2]; which on the other hand is nearly insensitive to material influence [3].

Table 1. Calibration results for the object intrinsic characteristics (best estimates and expanded uncertainties in millimeters).

Feature	Calibration results
D_1	28.211 ± 0.002 (k = 2)
D_2	24.784 ± 0.002 (k = 2)
D_3	8.612 ± 0.002 (k = 2)

To scan the test object, the magnification axis was positioned to project the object using the maximum possible area of the detector (and thus reducing the resulting voxel size). The source voltage was set high enough to avoid beam extinction, and the detector integration time (and sensitivity) set to a convenient value. The source current was then tuned to enhance image contrast and brightness. The number of angular poses was selected as approximately the number of pixel covered by the resulting shadow of the test object in the projection. See table 2.

Table 2. Simplified list of the CT control parameters chosen for measuring the test object under analysis.

Parameter	Unit	Value
Source voltage	kV	110
Source current	μA	450
Focal spot size	μm	50
Integration time	s	1
Detector binning	--	2x2
Magnification	--	7.53
Voxel size	μm	53
No. of projections	--	800

In fact, the selected CT settings agree with some lessons learnt for reliable dimensional analysis of feature of size [4]: (a) the focal spot size should be closer or smaller than the voxel size; (b) the detector gain setting would not affect average size, albeit increasing the gain would result in noisier images; (c) the smaller the angular increment, the lesser the image background noise; (d) the higher the image resolution, the lesser the image background noise; (e) the higher the magnification factor, the lower the measured biases.

Regarding the surface definition from the voxel dataset, the standard 'iso-50%' threshold value was applied globally. From the material boundary thus defined, 3600 points evenly spaced around the circumferential line were extracted for each intrinsic characteristic, and the ideal feature of the type circle associated to the points using the least-squares fitting method.

4. Uncertainty evaluation

Measurements performed on the CT system were entirely repeated three times, and for each intrinsic characteristic, the mean value of the measurement result and the standard uncertainty associated with measurement procedure were determined. The test object temperature inside the CT system enclosure was measured in order to properly compensate the temperature effects and to estimate the standard uncertainty associated with the systematic error.

Table 3 shows the resulting individual standard uncertainties, expanded uncertainty and bias (after temperature correction) for each feature of size of the test object. Since the dominant factors were the calibration uncertainty and procedural uncertainty, which were nearly identical for all characteristics, the expanded uncertainty after correcting the bias would be virtually the same, $U_1 = 0.003$ mm.

From table 3 other inferences could be drawn, such as the dominant role played by the bias in CT measurement performance; bias eight times higher than the expanded uncertainty for characteristic D_3. Treating the empirically determined bias like other uncertainty component, U_2 [5], a ratio $|b| / U_2$ can be calculated, which shows that bias accounts for about 50% of the expanded uncertainty.

4.1. Edge detection method

Being the dominant contribution in the uncertainty estimate of all intrinsic characteristics investigated in this work, some factors that may introduce non-negligible bias between CT-based data and tactile CMM data are presented.

First, there is an explicit difference between the surface extraction principle related to tactile CMM and CT-based data. The first method identifies the surface by physically contacting it (thus subjected to mechanical filtering). The latter determines the surface by assigning a threshold grey value to the 3D voxel data, which is affected by magnification errors.

Image artefacts caused by beam hardening and scattering may disturb correct surface detection, as they change the grey value of the edge pixels, and thus affect the identification of threshold values. Image blurring and noise also introduce threshold and interpolation errors [6].

Surface determination based on the 'iso-50%' threshold value applied globally, which represents the average value between the background peak and the material peak on the grey value histogram, may result in an edge offset errors with respect to the actual material edge.

In fact, the effect of beam hardening artefacts combined with the global threshold value could be identified in this study, by comparing the bias for external and internal features: positive bias for D_1, negative bias for D_2 and D_3 (see table 3).

Table 3. Standard uncertainties and bias for each intrinsic characteristic (values in millimeters).

	Intrinsic characteristic		
Component	D_1	D_2	D_3
u_{cal}	0.0010	0.0010	0.0010
u_p	0.0010	0.0010	0.0010
u_b (1)	0.0002	0.0002	0.0001
U_1 (k = 2)	**0.003**	**0.003**	**0.003**
bias, b	0.021	-0.007	-0.023
U_2 (k = 2)	**0.043**	**0.014**	**0.046**
$\|b\| / U_2$	49.9%	48.9%	49.9%

(1) $u_b = D_i \,.\, (T - 20\ °C) \,.\, u_\alpha$
with $\alpha = 92 \,.\, 10^{-6}\ °C^{-1} \pm 15\%$ and $T = (20.8 \pm 0.2)\ °C$

5. Concluding remarks

The experimental uncertainty assessment of CT-based measurements using ISO 15530-3 approach has been shown effective for simple parts, and thus has been extended for other measurement cases in the laboratory routine. The good short-term repeatability observed in CT measurements and the relevant role played by the bias, are in consonance with results reported by other authors [6-7].

The estimated biases, clearly affected by edge-related determination errors, lie within the empirical error band proposed by the authors in a recent paper [4], and therefore reinforce their estimation. This error band has been proven fairly representative and acceptable for most dimensioning tasks performed on parts with similar shape and material.

References

[1] Kruth J P, Bartscher M, Carmignato S, Schmitt R, Weckenmann A and De Chiffre L 2011 Computed tomography for dimensional metrology *CIRP Annals* **60(2)** 821-42

[2] Lettenbauer H, Georgi B and Weiss D 2007 Means to verify the accuracy of CT systems for metrology applications *DIR* Lyon FR

[3] Bartscher M, Sato O, Neuschaefer-Rube U and Härtig F 2014 Current state of standardization in the field of dimensional computed tomography *Meas. Sci. Technol.* **25** 064013

[4] Baldo C R, Fernandes T L, Donatelli G D 2015 Experimental investigation of computed tomography dimensional capability using modular test parts *XXI IMEKO World Congress* Prague CZ

[5] Härtig F and Krystek M 2009 Correct treatment of systematic errors for evaluation of measurement uncertainty *IX ISMTII* **1** 16-9

[6] Hiller J, Maisl M and Reindl L M 2012 Physical characterization and performance evaluation of an x-ray micro-computed tomography system for dimensional metrology applications *Meas. Sci. Technol.* **23** 085404

[7] Müller P, Hiller J, Dai Y, Andreasen J, Hansen H N and De Chiffre L 2014 Estimation of measurement uncertainties in X-ray computed tomography metrology using the substitution method *CIRP Journal* **7** 222-32

Exploratory analysis of voxel size effects on CT measurements of situation characteristics of type point-to-point distance

T L Fernandes[1], G D Donatelli[1] and C R Baldo[2]

[1] Centre for Metrology and Instrumentation - CMI, Reference Centres for Innovative Technologies - CERTI, Florianópolis, SC, Brazil
[2] Centre for Engineering, Modelling and Applied Social Sciences - CECS, UFABC - Federal University of ABC, Santo André, SP, Brazil

E-mail: thf@certi.org.br

Abstract. In recent years, computed tomography (CT) has been applied as an industrial metrology tool for the dimensional evaluation of visible and even hidden features of production parts in a non-destructive manner. Considering the experimental findings of a recent work of the authors, this paper deals with the effect of voxel size on measuring distances between feature-of-size centers, which would be less sensitive to edge offset errors. Particular attention is given to the design of experiment and to the measurement uncertainty sources. The most significant experimental findings are outlined and discussed in this paper.

1. Introduction

In recent years, computed tomography (CT) has been applied as an industrial metrology tool for the dimensional evaluation of visible and even hidden features of production parts in a non-destructive manner. CT features great operational advantages over existing coordinate metrology techniques, as tactile and optical coordinate measuring machines.

However, CT working principle and scanner setup give rise to many influence factors that affect the performance of dimensional evaluations. They are related to the X-ray source (e.g. photon energy, focal spot size), the detector (e.g. response time, pixel size, exposure time, image averaging), the CT kinematics (e.g. magnification and rotation axis repeatability and accuracy), the object (e.g. size, shape, material), and the mathematical data processing (e.g. artefact correction, segmentation, measuring strategy, fitting algorithm) [1].

When choosing a CT scanner for composing the dimensional metrology infrastructure of an organization, VDI/VDE 2630 Part 1.3 can be very useful, since it defines specifications and methods for acceptance testing of coordinate measuring machines with sensors relying on the X-ray CT principle. The dimensional performance of the CT system is investigated by measuring the distance between two balls centers. Although this approach is considerably robust for comparison purposes, as it is nearly insensitive to threshold errors, it is not useful for measurement uncertainty estimation of CT-based data [2].

A recent technical paper of the authors brought some lessons learnt for more reliable dimensional analysis of feature of size, through an exploratory study using modular test artefacts [3]. Two of the lessons learnt are replicated here: (a) the focal spot size should be closer or smaller than the voxel size; (b) the higher the magnification factor, the lower the measured biases.

The first experimental lesson learnt agrees with the definition of geometric unsharpness (blurring), from which the focal spot size must be correlated with the voxel size prior to measurement [4]. The second lesson learnt is the subject matter of this paper, but focusing on distances between feature-of-size centers, which would be less sensitive to edge offset errors.

2. Case description

2.1. Test object situation characteristics

The test object selected for this exploratory study is made of acetal photopolymer (POM) and composed of a pattern of holes uniformly spaced. The distances between each circle center related to the reference circle center define the situation characteristics to be evaluated. This feature arrangement is suitable for investigating the voxel size effect on center-to-center distance measurement. This arrangement is also adequate for assessing the length-dependent error behavior, since it covers different distances.

Figure 1 shows the test object and the situation characteristics. All center-to-center distances are related to situation feature (hole center) 1, namely: L_{12} (distance between situation features 1 and 2). The same reasoning applies to the other situation characteristics: L_{13}, L_{14}, L_{15} and L_{16}.

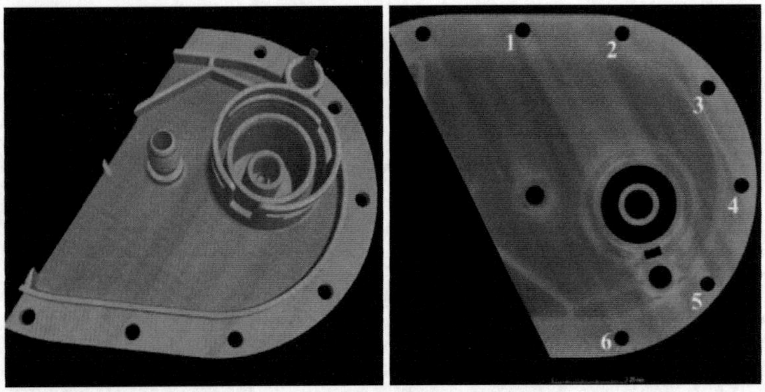

Figure 1. Illustrative image of the test object and the situation characteristics (distances between a given circle center and the reference circle center no. 1).

3. Measurement setup

3.1. Tactile reference measurements

The situation characteristics of the test object were calibrated on a Carl Zeiss PRISMO ultra CMM, which is housed in a temperature-controlled room kept at (20.0 ± 0.3) °C. All situation characteristics were realized by associating ideal features of type circle to the sampled points (using least-squares method), from which the situation characteristics (point-point distance) were derived. Measurement uncertainties were estimated by using the Virtual CMM software output as well as expert judgment. Table 1 lists the calibration results.

3.2. CT measurements

The test objected was measured on a Carl Zeiss METROTOM 1500 CT system, which is equipped with a 225 kV micro-focus tube and a 2048^2 pixels flat panel detector. The CT system is installed in a temperature-controlled room kept at (20 ± 1) °C. The CT system manufacturer specifies a MPE for length measurements of (9 + L/50) μm, using a test piece consisting of 27 ruby spheres mounted on carbon fiber shafts, and calculating the center-to-center distances of several pairs of spheres from the CT data [2]. This approach is on the other hand nearly insensitive to material influence [3].

Table 1. Calibration results for the object intrinsic characteristics (best estimates and expanded uncertainties in millimeters).

Feature	Calibration results
L_{12}	33.070 ± 0.002 (k = 2)
L_{13}	64.610 ± 0.002 (k = 2)
L_{14}	88.867 ± 0.002 (k = 2)
L_{15}	102.941 ± 0.002 (k = 2)
L_{16}	105.243 ± 0.002 (k = 2)

Due to the high aspect ratio (width-to-thickness ratio) of the test object, special attention to find the proper object orientation had to be taken, in order to avoid saturation or dying out of pixels during capture of the X-ray images. To scan the object, the source voltage was set high enough to avoid beam extinction, and detector integration time set to a convenient value. The source current was then tuned to enhance image contrast / brightness. The number of angular poses was chosen as about the number of pixels covered by the resulting shadow of the test object in the projection.

The magnification axis was, in the first attempt, positioned to project the part using the maximum possible area of the detector (and thus reducing the voxel size). For the second experimental trial, the magnification factor was halved (thus enlarging the voxel size by a factor of two). Table 2 shows the CT settings for each experimental trial.

Table 2. Simplified list of the CT control settings chosen for scanning the test object under analysis.

Parameter	Unit	Trial 1	Trial 2
Source voltage	kV	120	120
Source current	µA	450	450
Focal spot size	µm	54	54
Integration time	s	1	1
Detector binning	--	2x2	2x2
Magnification	--	2.50	1.25
Voxel size	µm	160	320
No. of projections	--	800	800

Regarding the surface definition from the voxel dataset, the standard 'iso-50%' threshold value was applied globally. From the material boundary thus defined, 3600 points evenly spaced around the circumferential line were extracted for each situation feature, and the ideal feature of the type circle associated to the extracted points using the least-squares fitting method.

4. Experimental results

Measurements performed on the CT system were repeated three times for each experimental trial. For each characteristic under analysis, the mean value of the measurement result and the standard uncertainty associated with measuring procedure were computed. The object temperature inside the CT system enclosure was monitored in order to properly compensate temperature effects and to assess the standard uncertainty associated with the systematic error. For more information regarding the experimental uncertainty evaluation, please refer to ISO 15530-3.

Table 3 shows the resulting individual standard uncertainties, expanded uncertainty and bias (after temperature correction) for each characteristic of the test object. Since the dominant factors were the calibration uncertainty and procedural uncertainty, which were nearly identical for all characteristics, the expanded uncertainty after correcting the bias would be virtually the same, $U = 0.005$ mm.

The metric normalized error (En-value) was calculated using the expanded uncertainty and bias absolute difference (table 3), which confirmed that the situation characteristic measuring difference related to each trial are not statistically identical, as the En-values are beyond the range [0,1]. That significant bias difference is a direct effect of the voxel size chosen for each experimental setup.

Table 3. Uncertainty components, bias difference and normalized error related to each situation characteristic (values in millimeters).

	Situation characteristic				
Component	L_{12}	L_{13}	L_{14}	L_{15}	L_{16}
u_{cal}	0.001	0.001	0.001	0.001	0.001
u_p	0.002	0.002	0.002	0.002	0.002
$u_b^{(1)}$	0.000	0.001	0.001	0.001	0.001
$U\ (k = 2)$	0.005	0.005	0.005	0.005	0.005
$\vert b_1 - b_2 \vert$	0.018	0.024	0.029	0.024	0.024
En	2.86	3.64	4.25	3.42	4.76

u_{cal} standard uncertainty of the parameter of the calibrated workpiece
u_p standard uncertainty of the measurement procedure
u_b standard uncertainty of the systematic error (bias)
$u_b = D_i \,.\, (T - 20\ °C) \,.\, u_\alpha :: \alpha = 92 \,.\, 10^{-6}\ °C^{-1} \pm 15\%$ and $T = (21.4 \pm 0.2)\ °C$
b systematic error observed during uncertainty evaluation

Figure 2 graphically shows the bias divergence between experimental trials 1 and 2. By not using the maximum possible area of the detector in trial 2, and therefore lowering the structural resolution, measurement results were adversely affected. In fact, the observed measurement bias exceeded the MPE for length measurements specified by the CT system manufacturer.

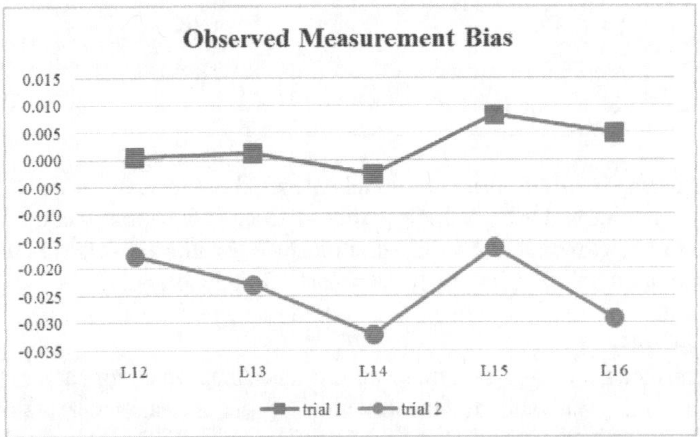

Figure 2. Chart of the observed bias for each experimental trial (point-to-point distance).

Although poorer structural resolution was a direct consequence of arbitrarily positioning the test object closer to the detector, in practice this effect can be very common, e.g., when scanning a test object that resembles a stepped cylinder [5].

5. Concluding remarks

In this paper, the effect of changing the voxel size (by changing the image magnification factor) was analyzed for some situation characteristics. By worsening the structural resolution, measurement accuracy was seriously undermined. This may be attributed to insufficient sampling at the image level (so-called aliasing artefacts) [1].

The proper location and orientation of the test object is an important duty of the CT metrologist. Finding the optimal CT setup may not be a trivial task, as the measurement error behavior depends on the workpiece material and geometry. The use of reference workpieces, either calibrated using a CMM or a CT system with multiple measuring sensors, makes possible improving CT measuring accuracy by correcting known systematic effects.

References

[1] Kruth J P, Bartscher M, Carmignato S, Schmitt R, Weckenmann A and De Chiffre L 2011 Computed tomography for dimensional metrology *CIRP Annals* **60(2)** 821-42

[2] Hiller J, Maisl M and Reindl L M 2012 Physical characterization and performance evaluation of an x-ray micro-computed tomography system for dimensional metrology applications *Meas. Sci. Technol.* **23** 085404

[3] Baldo C R, Fernandes T L and Donatelli G D 2015 Experimental investigation of computed tomography dimensional capability using modular test parts *XXI IMEKO World Congress* Prague CZ

[4] Weiss D, Shi Q and Kuhn C 2012 Measuring the 3D resolution of a micro-focus X-ray CT setup *CTC* Wels AT

[5] Krämer P and Weckenmann A 2010 A Multi-energy image stack fusion in computed tomography *Meas. Sci. Technol.* **21** 045105

8th Brazilian Congress on Metrology (Metrologia 2015) IOP Publishing
Journal of Physics: Conference Series **733** (2016) 012058 doi:10.1088/1742-6596/733/1/012058

Traceability of laser frequency/wavelength calibration through the frequency comb at Inmetro

I L M Silva, I B Couceiro, M A C Torres, P A Costa and H P H Grieneisen

Inmetro, Av. N Sra. das Graças 50 Xerem, Duque de Caxias, Rio de Janeiro, Brazil

E-mail: ilsilva@inmetro.gov.br

Abstract: The acquisition of a femtosecond laser comb by the Optical Metrology Division of Inmetro now allows for carrying out high precision calibrations of optical frequencies for lasers which are used as standards of the length unit with gauge block interferometers. The frequency comb is operated as an optical frequency synthesizer and is presently linked to the time unit by a 10 MHz oscillator which is disciplined by GPS. Laser frequencies are determined with accuracy in the range of few parts in 10^{12}. This measurement method now links the length unit, meter, to the SI-second attending the recommendation by the BIPM.

1. Introduction

The current SI definition of the meter adopted in 1983 specifies that the meter is the length of the path traveled by light in vacuum during a time interval of 1/299 792 458 of a second [1]. This new definition implies that the traceability of the length unit must be linked to the time unit. Following the new trends several National Institutes of Metrology started to use the laser frequency combs, linked to the primary cesium clocks, to maintain the traceability chain in the area of interferometric gauge block calibration.

At the Optical Metrology Division (Diopt) of Inmetro the length scale is maintained using a He-Ne laser stabilized to the a_{16} component of the iodine R(127) 11-5 hyperfine transition. The use of only one standard restricts the scope of the calibration for He-Ne lasers with a wavelength of 633 nm. In 2011, to broaden the measurement capabilities in this field with the use of new standard lasers operating at different wavelengths a commercial frequency comb synthesizer was acquired. In 2014, the characterization and validation of this new system was analyzed and approved by an international peer reviewer. The next step was constructing the length traceability chain with respect to the time unit. The operation and characterization of the comb frequency system at Inmetro is presented in this article.

2. Frequency comb synthesizer

The frequency comb has become a common tool in the area of high-resolution spectroscopy and optical frequency metrology in laboratories around the world [2,3]. The most important property of such system is that the repetition frequency of the femtosecond laser can be locked to an external standard RF frequency source in order to generate an output beam at optical frequencies consisting of millions of equally spaced modes of a monochromatic phase locked coherent radiations with frequencies given by

$$f_{comb} = n.f_r \pm f_0 \tag{1}$$

where n is an integer, f_r is the pulse repetition frequency of the femtosecond laser and f_0 is the carrier envelope offset frequency [4,5]. The latter depends critically on the laser power and dispersion properties of the laser cavity but f_0 can also be referenced and locked to the same RF-reference source.

2.1 Method of measuring a laser frequency

The optical frequency of a laser can be determined with the frequency comb using a heterodyning technique. A beat signal, f_b, is generated with a fast photodetector by superimposing the wavefront of the laser beam of frequency $f_{cw}$with the wavefront of a comb mode of similar frequency f_{comb} [6]. This frequency difference is the beat signal (f_b) [7] and is being measured with a frequency counter. The repetition frequency f_r is also measured straightforward through a photodetection of the output pulse train from the laser comb beam. To determine the offset frequency it is necessary that the frequency range of the comb spans a region larger than an optical octave. The offset frequency is determined by a self-referencing technique as proposed by Hansch [8-10].

The frequency of the laser under calibrations can be written:

$$f_{cw} = f_{comb} \pm f_b = n.f_r \pm f_0 \pm f_b \tag{2}$$

This equation requires knowledge of two RF-frequencies (f_r and f_0) and mode number n with absolute certainty. f_0 and f_r must be locked to a reference radio-frequency, which must be generated by frequency standard, traceable to the primary cesium standard [11].

If the laser frequency (f_{cw}) is known *a priori* with accuracy higher than half of the repetition frequency, the value of mode number n is determined by taking the integer part of the ratio given by:

$$n = \frac{f_{cw-apriori}}{f_r} \tag{3}$$

In this case only one measurement is sufficient to determine the absolute laser frequency. The *a priori* information of the laser frequency can be obtained through a previous calibration certificate or through the measurement performed with the aid of a wavemeter [12].

2.2 The Frequency Comb

Inmetro´s frequency comb is a commercial system based on a diode-pumped erbium doped fiber laser (M-comb), operating at 1560 nm wavelength, described in [13], schematically shown in figure 1. The M-Comb output is divided in four beams that are amplified in two erbium doped fiber amplifier (EDFA1 and EDFA2) units as shown in figure 2. Part of the amplified beam from EDFA1 is sent to a spectral broadening setup for generating the octave-spanning and then fed into a nonlinear interferometer for detection of the offset beat frequency.

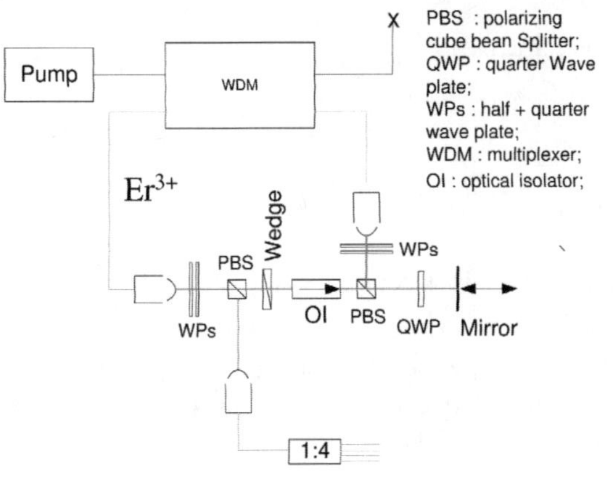

Figure 1. M-Comb, the green lines represent the fiber connection and the red lines the free space laser radiation.

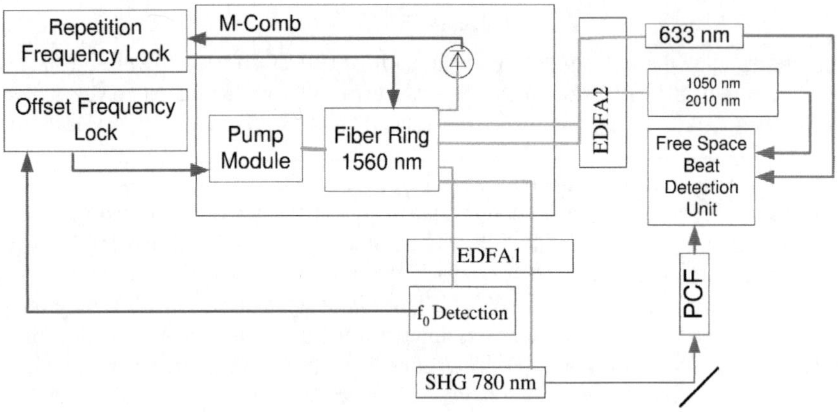

Figure 2 –FC 1500 setup, the green lines represent the fiber connection, blue lines the electrical/RF connection and the red lines the free space laser radiation.

The repetition frequency and the offset frequency are phase locked by controlling the cavity length and the pump power. A GPS controlled quartz clock calibrated at the National Observatory provides the traceability of the system. This unit provides the 10 MHz reference signal used to phase-lock the repetition frequency, the offset frequency, as well as the time base of the counter and the synthesizer, as shown in figures 3 and 4.

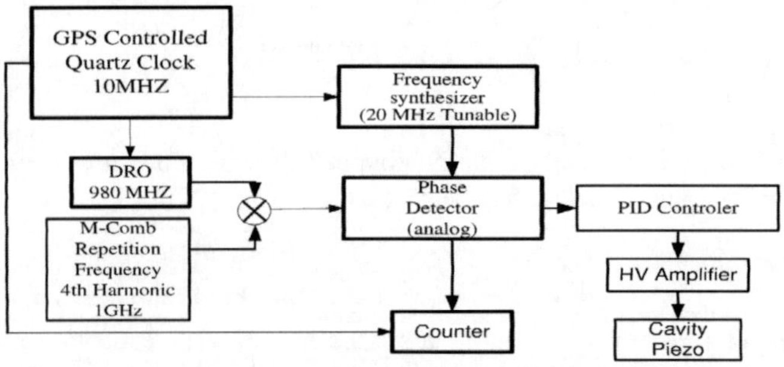

Figure 3. Repetition frequency phase lock setup.

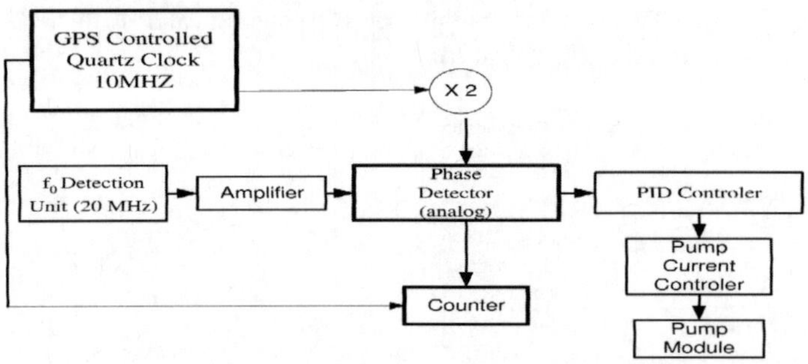

Figure 4. Offset frequency phase lock setup.

The nominal repetition frequency f_r of the M-Comb is 250 MHz, but its cavity is provided with elements for coarse and fine tuning about this value and f_r can be phase locked to a preset precise value linked to a RF-frequency synthesizer. Also the offset frequency f_0 can be tuned and controlled with a phase locked loop.

Part of the amplified beam from ADFA1 is sent to a frequency doubling stage generating a beam at 780 nm which is fed into a photonic crystal fiber in order to broaden the spectrum into the visible region. This beam is used to calibrate lasers with wavelength in the visible region of the spectrum. The EDFA2 is a high power fiber amplifier with internal spectral broadening. Part of the beam from the EDFA2 is filtered for generation of an output radiation at 1266 nm with subsequent frequency doubling stage for 633 nm generation used to calibrate the He-Ne laser.

3. Measurement and results

We measured the frequency of an I_2 stabilized He-Ne lasers, named PR020, which is the Inmetro´s primary standard. The cold finger temperature, the frequency modulation width and the one-way intracavity beam power were kept within the operational conditions required by the BIPM [14].

The reliability of the measurements is directly related to the occurrence of cycle slips in the phase-locked loops (PLL) or if miscounts occur in the measurement of the beat frequency between the comb and the laser under calibration.

To ensure that the offset and repetition frequency phase lock are operating properly, the offset and the repetition frequency signals are split and send to the locking electronics and to the counter. Any cycle slips are easily visible in the recorded counter readings. Any reading that differs by more than 0,002 Hz or 10 Hz from the repetition and offset reference frequency, respectively, is discarded.

The counting errors can be avoided if the beat signal has a good absolute power and a good signal to noise ratio (SNR). In our system the beat frequency, after detection, is low pass filtered and amplified and then band pass filtered at 30 MHz (filter center: 30 MHz, 3 dB bandwidth 25 MHz – 35 MHz). The signal is then fed to the Counter. Typical beat signal is shown in figure 5.

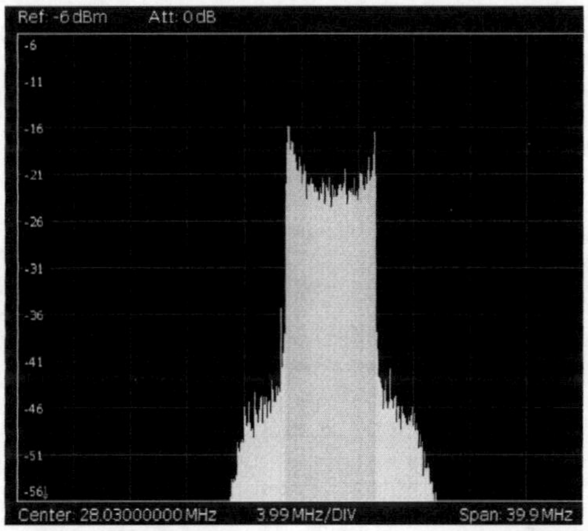

Figure 5. He-Ne (PR020) and comb beat note after filtering and amplification (100 kHz RBW).

Several tests to verify the counter´s sensitivity to the signal level as well as to the SNR have been performed. We have changed the absolute signal power and the SNR by changing the polarization of the beam that is coupled into the frequency doubling stage for 633 nm generation in the IDFA2 unit. To evaluate the reliability of this method we compared the standard deviation of the acquired data with the one predicted by the Allan deviation. We conclude that an absolute signal level above –25 dBm and with a SNR higher than 26 dB is appropriate to ensure a reliable measurement.

The figure 6 shows the stability of the beat frequency between the He-Ne laser and the comb. These measurements quantify the combined stability of the two systems demonstrating that good accuracy can be achieved for measuring periods larger than 100 s, when the Allan deviation falls below 10^{-12}.

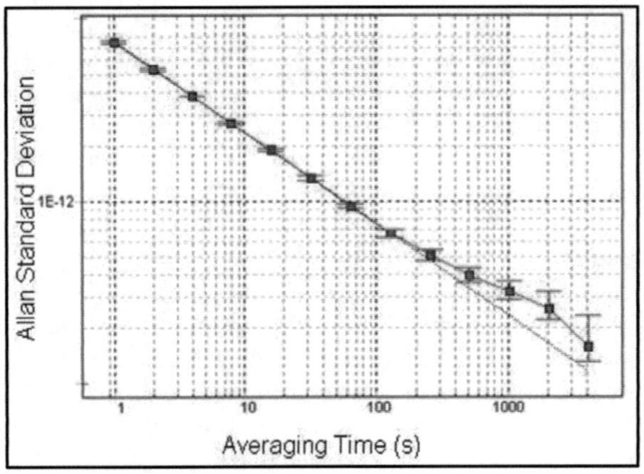

Figure 6. Stability of the beat frequency between the He-Ne laser (PR020) and comb.

The uncertainty budget for the calibration is composed by three parts, one from fluctuation of the beat signal, one related to the contribution of stability of quartz clock (4×10^{-12} , k =1), and one from the uncertainty in the settings of the laser´s working parameters.

The figure 7 shows the results of measurements with the Inmetro´s comb. The values calculated with (2) were correct to take into account the quartz clock offset from the National Observatory primary time standard.

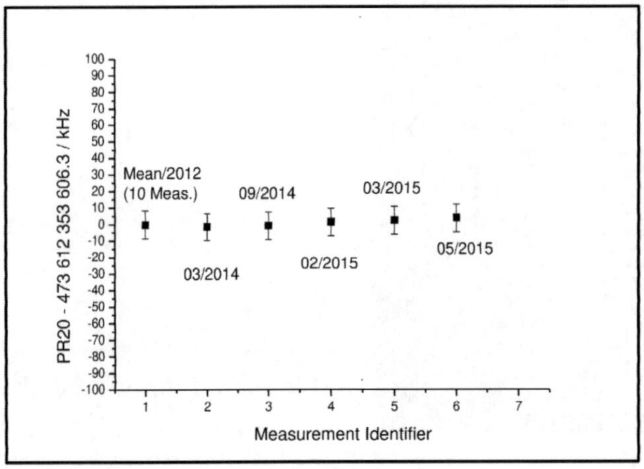

Figure 7. Measurements of the PR020 frequency compared to the CCL K11 value.

4. Conclusion

The frequency comb at Inmetro was successfully put into operation. The absolute frequency measurements of 633 nm standard are in good agreement with the reference value, which ensures the measuring capability of the system and the traceability for length metrology with minimum relative uncertainty of 4×10^{-12} (k=1).

References

[1] Giacomo P 1984 *Metrologia* **20** 25–30
[2] Hansch T W 2006 *Rev. of Mod. Phys.* **78** 1297-09
[3] Udem T, Holzwarth R and Hansh T W 2002 *Nature* 416 233-37

[4] Jones D J, Diddams S A, Ranka J K, Stentz A , Windeler R S, Hall j L and Cundiff S T 2000 *Science* **288** 635-639
[5] Hansch T W, Holzwarth R, Reichert J and Udem T 2001 *Proceedings of the International School of Physics* IOS Press, Amsterdam
[6] Fox R W, Diddams S A, Bartels A and Hollberg L 2005 *Appl. Opt.* **44** 113-20
[7] Stone J and Egan P 2010 *J. Res. Natl. Inst. Stand. Technol.* **115** 413-31
[8] Udem Th, Reichert J, Holzwarth R and Hänsch T W 1999 *Opt. Lett.* **24** 881-883
[9] Niering M et al. 2000 *Phys. Rev. Lett.* **84** 5496-5499
[10] Holzwarth R, Udem Th, Hänsch T W, Knight J C, Wadsworth W J and Russell P St J. 2000 *Appl. Opt.* **85** 2264-2267
[11] Smith P R, Roos P A, Wahlstrand J K, Pipis J A, Rivas M B and Cundiff S T 2007 *J. Res. Natl. Inst. Stand. Technol.* **112** 289-96
[12] Ma L S, Zucco M, Picard S, Robertsson L and Windeler R S 2003 *IEEE J. Sel. Top. Quantum Electron.* **9** 1066-71
[13] Menlo Systems GmbH, FC1500 Optical Frequency Synthesizer, User Manual, 2010
[14] Quinn T J 2003 *Metrologia* **40** 103-133

Modernization of Koesters interferometer and high accuracy calibration gauge blocks

R S França, I L M Silva, I B Couceiro, M A C Torres, M S Bessa, P A Costa, W Oliveira Jr, H P H Grieneisen

National Institute of Metrology, Quality and Technology – Inmetro
Optical Metrology Division-Diopt

E-mail: rsfranca@inmetro.gov.br

Abstract. The Optical Metrology Division (Diopt) of Inmetro is responsible for maintaining the national reference of the length unit according to International System of Units (SI) definitions. The length unit is realized by interferometric techniques and is disseminated to the dimensional community through calibrations of gauge blocks. Calibration of large gauge blocks from 100 mm to 1000 mm has been performed by Diopt with a Koesters interferometer with reference to spectral lines of a krypton discharge lamp. Replacement of this lamp by frequency stabilized lasers, traceable now to the time and frequency scale, is described and the first results are reported.

1. Introduction

In the early 1970s the team of John Hall of the NBS (renamed afterward as NIST) realized the first frequency chain connecting the microwave frequency of the cesium atomic clock to the frequency of a methane-stabilized helium-neon laser at 88.4 THz [1]. In the early 1980s it was possible to measure the frequency of some iodine stabilized He-Ne lasers allowing in 1983, during the 17th General Conference on Weights and Measures, a new definition of the meter, henceforth defined as "the path length traveled by light in vacuum during a time interval of 1/299 792 458 second" [2,3]. This definition fixes the speed of light to an exact value, thus requiring for the determination of the vacuum wavelength of a radiation through the measurement of its frequency [4]. Following this new definition of the meter, the International Committee of Weights and Measures (CIPM) published their *Mise en Pratique* with a list of radiations that can be used for its practical realization [5,6]. Among these there are a radiation of He-Ne laser, stabilized to the a_{16} component of the $^{2}I_{127}$ R(127) 11-5 hyperfine transition, which currently corresponds to the national primary length standard, and radiation transitions corresponding to $5d_5$-$2p_{10}$, $2p_8$-$5d_4$ and $1s_3$-$3p_{10}$ of the krypton spectral lamp (^{86}Kr), used as the working length standard in the interferometer Koesters. Since the 80s there were few spectral krypton lamps still in operation worldwide and with the advent of the stabilized lasers these lamps have been gradually replaced. In order to adapt the length measurements with traceability to current time and frequency standards the acquisition of a commercial frequency comb synthesizer [7] made it possible to equip the Koesters interferometer with three lasers stabilized at different frequencies, thus phasing out the now becoming obsolete spectral lamps.

With this substitution it is possible to further improve measurement capability, since the laser light provides sharper and more intense interference fringes for easier-viewing or image capture with digital cameras to carry out the reading of the fringe pattern. This also enables process automation, reducing measurement uncertainties and false operator judgments.

2. Materials and Methods

The materialization of the meter unit is carried out through artifacts known as gauge blocks, used by the industry for calibration of calipers, micrometers and performance verification of Coordinate Measuring Machines (CMMs). The traceability transfer between the length optical standards and the gauge blocks with nominal length between 100 mm and 1000 mm (long-blocks) is performed with a Koesters interferometer. The Koesters interferometer is a modified Michelson type interferometer, where the beam splitting and recombination is accomplished with a double prism [8] as shown in figure 1. In order to modernize length measurements with traceability to current time and frequency standards a femtosecond laser frequency comb was acquired and the interferometer equipped with three lasers, stabilized at different wavelengths, thus, phasing out now the becoming obsolete spectral discharge ^{86}Kr lamp.

For an unambiguous determination of the block deviation length value it is necessary to use at least three different wavelengths [9].

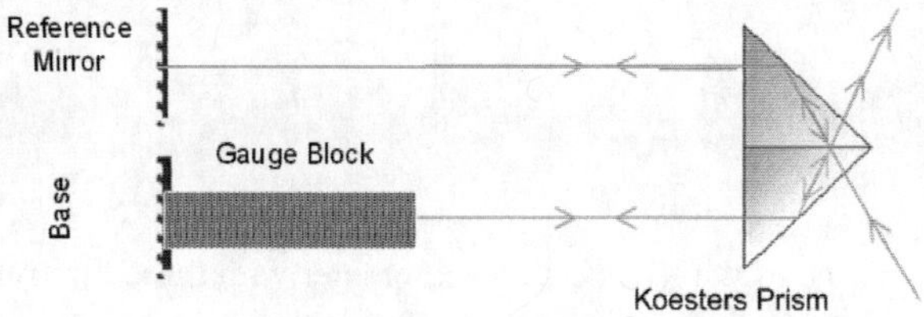

Figure 1. Simplified scheme of the optical layout of the Koesters interferometer.

2.1. Radiation sources used

Until 2012 the Koesters interferometer used a krypton (^{86}Kr) spectral lamp for measurements of gauge blocks with nominal length between 100 mm and 1000 mm. The relevant wavelengths of interest are listed in table 1.

Table 1. Radiation emitted by the krypton lamp used for the calibrations of large gauge block.

Transition	λ(nm)	U (nm, k=3)
$5d_5$-$2p_{10}$	605.780 210 3	$2.4x10^{-6}$
$2p_8$-$5d_4$	642.280 060	$1.3x10^{-5}$
$1s_3$-$2p_{10}$	565.112 860	$1.1x10^{-5}$

In 2012 a 633 nm stabilized He-Ne laser operating in a single mode TEM00 was tested with the interferometer system. Other two stabilized He-Ne lasers, operating at 543 nm and 594 nm, were incorporated in 2014. Their respective wavelength stabilities are shown in table 2.

Table 2. Lasers coupled to the interferometer system.

Laser	λ(nm)	U (nm)
Red	632.991 372	$1.3x10^{-5}$ (k = 2)
Yellow	594.096 550 9	$1.8x10^{-6}$ (k = 1.65)
Green	543.515 310 0	$1.1x10^{-6}$ (k = 1.65)

The lasers and krypton lamp are arranged according to the optical layout, as shown in figure 2.

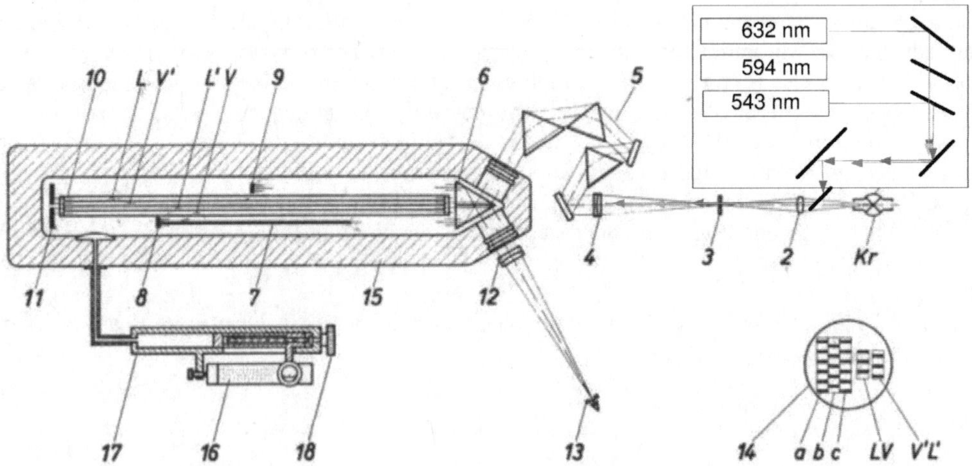

Figure 2. Graphic scheme of the Koesters interferometer with three different stabilized lasers.

In this figure also is shown the focusing optics (2), the collimating lens (4), the prism assembly used to select the wavelength (5), the Koesters prism (6), the reference mirror (9), the gauge block (7) and the wringing base (8). By means of the objective lens (12) and the circular diaphragm (13) the interference pattern is observed (14).

2.1.1. Traceability of Radiation Sources. The traceability of radiations emitted by the krypton spectral lamp as working length standard is ensured by an indirect and laborious calibration process, using a previously calibrated reference gauge block. With the use of lasers the traceability now is referenced to frequency measurements with an optical frequency comb synthesizer [10]. This equipment is located in a separated module of Optical Metrology Division, distant about 50 meters away from the Koesters interferometer. The laser beams are transmitted to the comb via optical fibers, as shown in figure 3. Thus, their calibration can be accomplished without removing the lasers from the system.

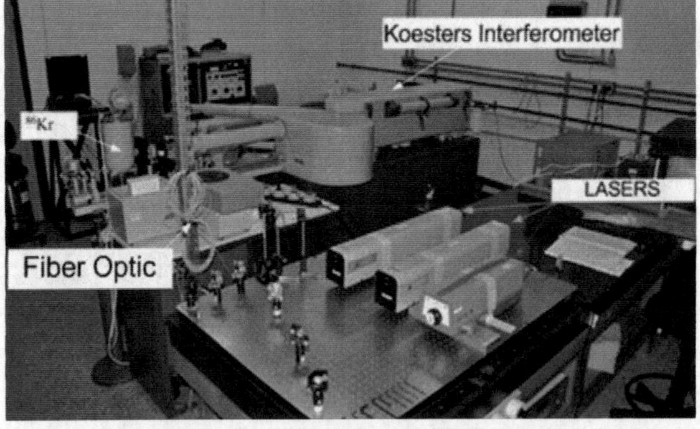

Figure 3. Koesters interferometer and lasers.

3. Results

The graph shown in figure 4 shows the monitoring measures of a gauge block used as a reference standard for over 17 years. Deviations from the nominal length were obtained until 2011 through measurements exclusively using three reference radiations emitted by the krypton spectral lamp.

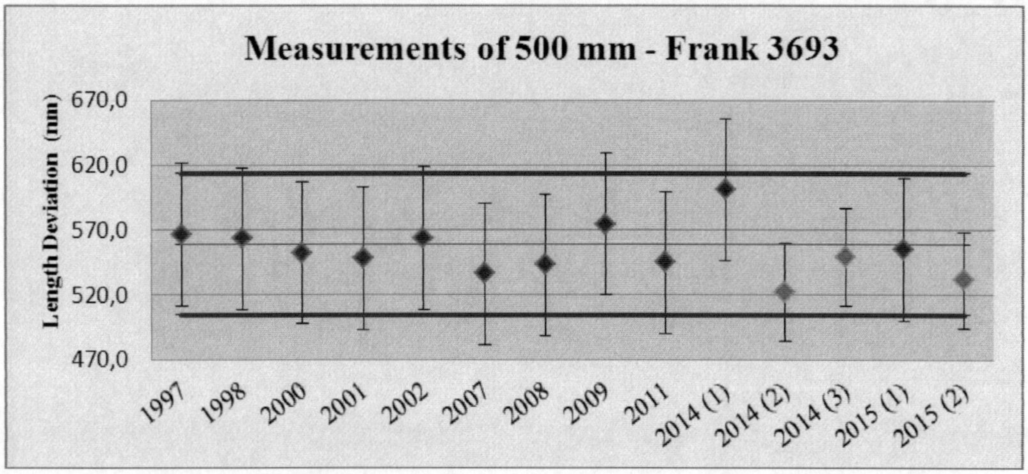

Figure 4. Evolution of length deviation measurements from a 500 mm reference block.

The blue indexed results 2014(1) and 2015(1) were performed using three spectral lamp radiations plus the radiation of a 632 nm red laser. The red indexed results 2014(2), 2014(3) and 2015(2), were obtained exclusively using the radiation emitted by three He-Ne stabilized lasers. Using the information presented in table 3 it can be concluded that the introduction of lasers in the system contributes to a global reduction of uncertainty of the final measurement of the gauge block. This reduction is due to the component associated with clearer interferometric fringe readings, and from the smaller contribution of the emitted laser radiation wavelengths as the emitted by the spectral lamp. The deviation values obtained with the introduction of lasers clearly converge in the graph with those obtained previously with the historical measurements made with the spectral lamp.

Table 3. Uncertainty components of interference fringe reading and radiation wavelengths (k=1).

L-independent terms		u_f	c_i	u_L (nm)
Fraction measurement	*^{86}Kr lamp*	3.3E-02	3.3E-02	2.9
	Laser	2.0E-02	2.0E-02	1.7
L-dependent terms (L=500 mm)		u_λ (nm)	c_i	u_L (nm)
Lamp wavelengths (^{86}Kr lamp)	*red*	4.5E-05	2.6E+05	11.7
	orange	4.0E-05	2.8E+05	11.0
	yellow-green	4.0E-05	3.0E+05	11.8
Laser wavelengths	*red*	6.5E-06	2.6E+05	1.7
	yellow	1.1E-06	2.7E+05	0.3
	green	6.6E-07	3.0E+05	0.2

The graph shown in figure 5 is the set of recorded values, over 12 hours, of the beat signal between the green laser and the laser comb. The average rate value calculated from the chart is used to calculate the laser frequency. Thus, it is shown that the maintenance of chain traceability of the Koesters interferometer can be performed more simply and directly, without intermediary standards and more laborious processes.

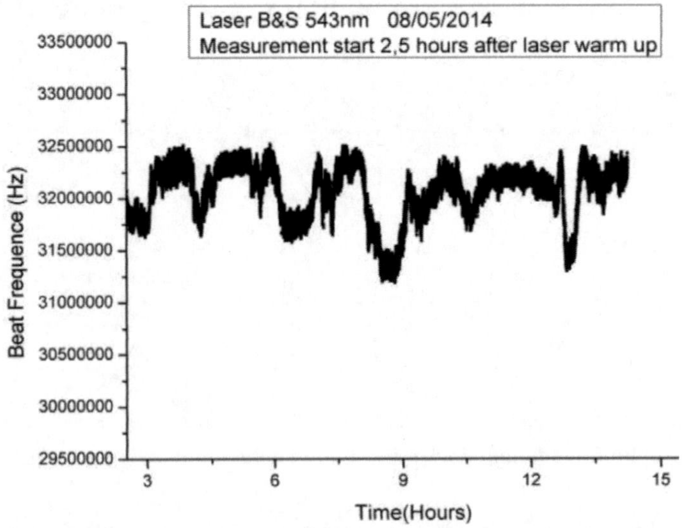

Figure 5. Beat frequency of the green He-Ne laser with the laser frequency comb.

4. Conclusions

Preliminary results indicate that using radiations emitted by stabilized He-Ne lasers has the following advantages compared to radiations emitted by the krypton spectral lamp.

Higher intensity and larger coherence length causes the formation of sharper and more intense fringes. Therefore they are easier to measure, and reducing in this way the final measurement uncertainty. Furthermore, a digital camera can be used to perform automatic fringe readings, which are currently still done by visual estimation.

Smaller wavelength uncertainties values with laser radiations are obtained in comparison with those emitted by the spectral ^{86}Kr lamp. This has a direct impact on reducing the final measurement uncertainty using the interferometer. The calibration of the lasers using an optical frequency synthesizer has been achieved.

As a next step for modernizing the Koesters interferometer one considers the incorporation of a CCD camera for image acquisition together with an automated fringe reading capability, thus, allowing for the automation of the entire calibration process and elimination of the need for visual observations.

References

[1] Evenson K M, Wells J S, Petersen F R, Danielson D L, Day G W, Barger R L, and Hall J L 1972 *Phys. Rev. Lett.* **29** 1346

[2] Hänsch T W 2006 *Rev. Mod. Phys.* **78** 1297-1309

[3] Giacomo P 1984 *Metrologia* **20** 25–30

[4] Udem T Holzwarth R and Hänsch T W 2002 Optical Frequency Metrology *Nature* **416** 223-237

[5] Quinn T J 1999 Practical realization of the definition of the metre (1997) *Metrologia* **36** 211-244

[6] Quinn T J 2003 Practical realization of the definition of the metre, including recommended radiations of other optical frequency standards (2001) *Metrologia* **40** 103-133
[7] Menlo Systems GmbH 2010 *FC1500 Optical Frequency Synthesizer User Manual*
[8] Darnedde H 1992 High-precision Calibration of Long Gauge Blocks Using the Vacuum Wavelength Comparator *Metrologia* **29** 349-359
[9] Decker J E and Pekelsky J R 1997 Gauge Block Calibration by Optical Interferometry at the National Research Council of Canada Measurement Science Conference NRC Internal Report No. 40002 (23–24 January 1997 Pasadena CA)
[10] Silva I L M Raymundo E V Couceiro I B and Oliveira A N 2013Aplicações de um Sintetizador de Freqüências Ópticas *Proc. Conf. 12ª Conferência sobre Tecnologia de Equipamentos* (18-21/jun/2013 Porto de Galinhas PE Brazil)

High-resolution interferometic microscope for traceable dimensional nanometrology in Brazil

I Malinovski, R S França, M S Lima, M S Bessa, C R Silva and I B Couceiro

Inmetro/Diopt, Av. N. S. das Graças, 50, Xerém, Duque de Caxias, 25250-020, RJ, Brazil

E-mail: rsfranca@inmetro.gov.br

Abstract. The double color interferometric microscope is developed for step height standards nanometrology traceable to meter definition via primary wavelength laser standards. The setup is based on two stabilized lasers to provide traceable measurements of highest possible resolution down to the physical limits of the optical instruments in sub-nanometer to micrometer range of the heights. The wavelength reference is He-Ne 633 nm stabilized laser, the secondary source is Blue-Green 488 nm grating laser diode. Accurate fringe portion is measured by modulated phase-shift technique combined with imaging interferometry and Fourier processing. Self calibrating methods are developed to correct systematic interferometric errors.

1. Introduction

Nanometrology is the science of measurement at the nanoscale level. It is considered as the basis of the nanotechnology in quality control of the products [1]. As a part of large international investments in nanoscience various instruments has been developed to support specific nanotechnology measurement needs [2]. We consider common approach of Co-ordination of Nanometrology in Europe (Co-Nanomet) [1], to be a good model for implementation in the near future. According to the Co-Nanomet strategy, measurements in the nanometer range should be traceable back to internationally accepted units, meter in our case of dimensional nanometrology. This requires creation of national validated measurement methods, calibrated scientific instrumentation as well as qualified reference samples. The traceability chain up to the primary nationally realized units for the required measurements must be maintained for any nanometrology quantity under consideration. So far nanometrology has been established internationally in only a few special cases. Principal problem of nanoscience instruments such as SPM, AFM, TEM, etc. is that they are not directly traceable to SI meter unit, but via additional laser interferometers or via secondary calibrated reference standard artifacts. Co-Nanomet delegates traceability maintenance to National Metrology Institutes (NMIs), serving as Centers of Excellence (CoE). The concrete nanotechnology measurements and instrumentation is delegated to Centers of Dissemination (CoD): i.e. R&D groups at technical accredited laboratories, institutions and companies involved in nano-tech products implementation. With those recommendations we at Interferometry Laboratory (Laint), Optical Metrology Division (Diopt) of brazilian NMI (Inmetro), as the CoE "par excellence" in Brazil, are working toward creation of a national micro-nano scale with target uncertainty smaller than 0.5 nm.

Here is reported an interference microscope (IM) for traceable nanometrology within most interesting range from micro to sub-nanometer. The IM is supposed to provide continuous transfer of the meter unit from traditional length metrology to nanometrology. Relatively easy from technical point of view and historically most studied approach is a height Z one point 1D calibration of the material artifacts using wavelength calibrated light sources. Classical interferometry and IM for many decades has been proved a reliable solution for this calibration needs. Secondary standards used are step height standards (SHS).

Previously at Inmetro/Diopt a nanometrology facility prototype was created, based on single He-Ne 633 nm stabilized laser and phase-shift interference microscopy (PSIM) principle [3]. Serious disadvantage of the system was large averaging time required for achieving target sub-nanometer resolution [4]. This makes low power system very inconvenient, expensive to run and virtually non-suitable for routine nanometrology. Here we report further development of IM with extra light source: 16 mW blue-green grating stabilized laser diode that can be "on-flight" calibrated via direct wavelength comparison to red reference standard. High power laser IM application permits implementation of various modern detection techniques such as sinusoidal modulated PSI and time-gated imaging interferometry. Those improvements are implemented in the new prototype reported here. Performance is evaluated by computer simulation and compared with direct measurements using several different types of SHS.

2. Setup and Samples

Optical setup consists of three main parts: (i) Light sources - stabilized lasers and beam conditioning anti-speckle unit with telescope; (ii) IM itself with reference and measuring arms on automated movable stages for sample manipulation and phase-shifting; (iii) detecting part with CCD and zooming telescopic unit.

PZT in figure 1 is the phase-shifter used both for PSI and calibrations of the blue laser relative to red reference laser in procedure similar to wavemeter operation. We move reference mirror certain distance measuring (digitizing) shape of the interferometer fringes. The phase difference gives the wavelength difference between the color lasers.

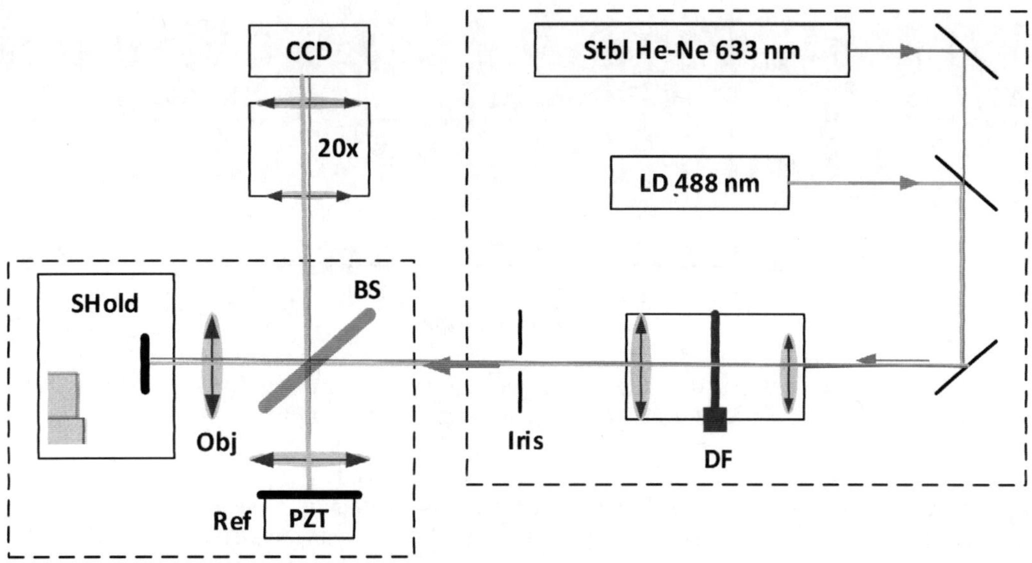

Figure 1. Optical setup of IM includes: **BS** - beam splitter; **SHold** - sample artifact on holder XYZ stage, **Ref** - reference mirror on PZT; **Obj** - two microscopic objectives; **DF** - rotating diffuser with beam collimator; **Iris** - variable diaphragm; **CCD** - imaging part with 20x zoom eye-piece.

To cover a broader target range of measurements we used two types of reference standards, macroscopic and microscopic artifacts: i) steel gauge block master set (GBS) from Mitutoyo, with two steps of 2 and 10 µm; ii) Al or SiO_2 coated steps on mono-crystal Si, so called step height standards (SHS).

3. Sub-nanometer Resolution

To reach highest possible resolutions required by modern nanometrology we improved several key techniques of data acquisition and processing. Detection part of hardware is based on HQ scientific grade 1.5 Mpix CCD, 14 effective bits intensity resolution with averaging. This imaging output is further improved up by pixel binning and 2D FFT Gaussian filtering processor. Final output is fit by multi-parameter sinusoidal function to extract phase values at point of interest. With our hardware we can use either fringes along the CCD image in frame (imaging interferometry) or PSI method. The above procedure is used for fringe fraction extraction in both imaging and phase-shifting recorder methods at resolution of about < 100 pm. Such high resolution can be fully benefited from correspondingly high accuracy of the PSI unit. PSI hardware is tilt-moved 3 degrees of freedom unit that is fed with 24 bit DAC output amplified by low noise HV unit. The most complete and informative technique we developed is multiple step phase shifting with 4D (XY-Z-A) imaging voxel cube hyper-data acquisition, where XY directions are pixels (SHS points), Z - phase scan and A - fringe intensity for phase extraction. Result of the single scan is 100-200 series of fringe images sampled at variable stepped Z length of the interferometer reference arm, so called *image stack* or voxel cube. In Z direction we have multiple point sinusoidal signal lines each corresponding to certain point on the sample (pixel). FFT procedure is applied and topography phase map is calculated. Processing of the 4D data sets requires substantial computational power so that parallel methods with either multicore or multi node architecture is desirable.

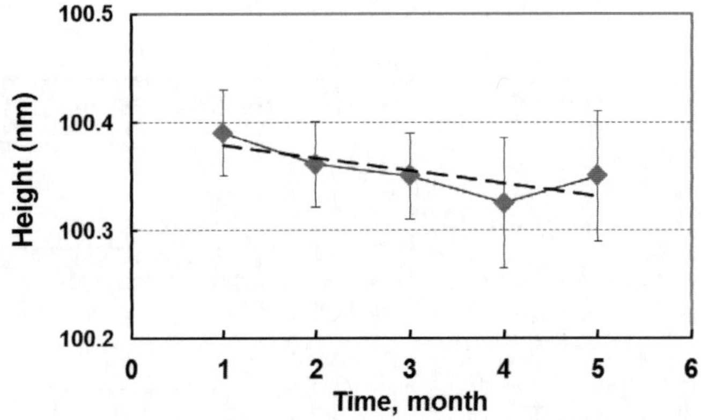

Figure 2. Drift of the measurement during several months of a 100 nm SHS made of Aluminum (manufactured at Inmetro).

Hardware and algorithm improvements resulted in accurate phase (length) determination down to sub-nanometer or sub-Ångström level. The resolution test is repeatability of real output (figure 2) that is good enough to detect drift of about 50 pm of the whole set-up or/and sample.

4. Noise and Systematic Effects

Detailed characterization and some comparisons of the IM were performed. Main systematic errors were evaluated, minimized and suitable corrections found.

4.1. Repeatability and statistics

The standards were measured using long series of automatic data acquisition with IM realignments between series. This involves significant amount of data processing and time. Experimental decrease of random noise is shown in figure 2 for He-Ne laser standard.

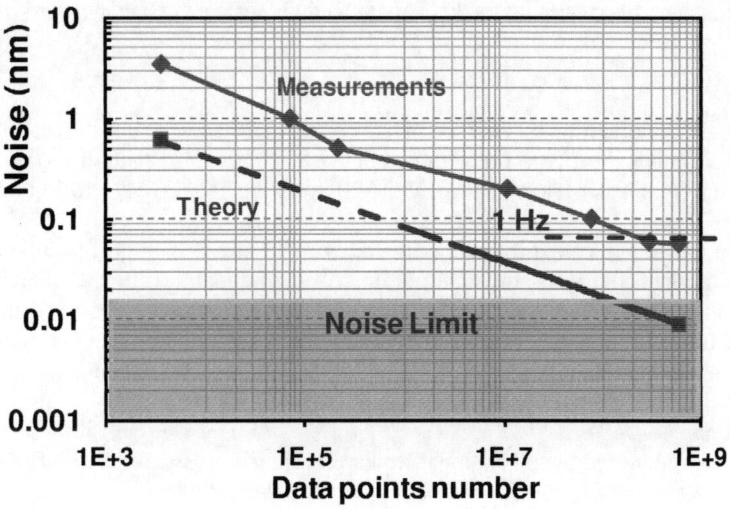

Figure 3. Stability of the measurement SHS 100 nm height deposited Aluminum.

The noise trend was not improving after certain periods of averaging, indicating that some residual noise is present, mostly related to vibrations as was confirmed with measurements of Inmetro Vibration Laboratory (Lavib). Basically we have reached the low frequency flicker-noise floor close to 1 Hz and the optical table resonant frequencies characteristic time. The solution is to move detection away from resonances and flicker-noise. Blue laser of larger power, modulation PSI and time-gated imaging, all those techniques were implemented to move data of interest detection to higher frequency with corresponding fast acquisition and noise reduction. Expected noise at same conditions is shown as horizontal dashed line (1 Hz) in figure 3.

4.2. Aperture corrections (obliquity error)

This error typically has the largest contribution to uncertainty budgets of IM instruments. Theoretical analysis and PC-simulation was made, resulting in numerically generated function: the value of correction vs. variable numerical aperture (NA). The function was compared with data measured at the same conditions to evaluate real influence of the effect. Variable iris diaphragm at the IM input was utilized. Similar effective NA decrease is produced by eye-piece, that is telescopic variable zoom unit (20x, in figure 1). The minimized correction in our geometry is relatively small: > 1.002 due to NA variation possibility down to > 0.1 or lesser values. For nominal 100 nm for SHS the correction was measured to be about 100 pm.

4.3. Phase-shifter nonlinearity

This effect results in phase differences at beginning and end of phase scan that can be detected analyzing full phase record waveform. This effect found to be significant for fringe displacements in four integer numbers or more. The effect was measured and adequate mathematical model function of modulation additional term is used to compensate this error within target 0.1 nm operation.

4.4. Wavefront distortion errors
These errors follow from unequal optics and alignment in IM sample and reference arms. The SHS height as defined by ISO 5436-1-2000 and measured by interferometry incorporates defects of optics associated with as on-off-axis path difference (considered as Abbe offset). Related wavefront correction is determined by height $\boldsymbol{Z_{corr} = 0}$ measurement on flat area of the same SHS. The correction is acquired during each measurement cycle. This procedure corrects associated error down to 100 pm target accuracy.

5. Conclusions
A new step height nanometrology CoE facility was developed and characterized. By principle of operation the facility is hybrid type IM with combined PSI and fringe pattern processing algorithms. Equipped with different color wavelength stabilized lasers the IM provides full automated traceable self-calibrated operation.

As compared to previous IM prototype with red He-Ne only [3], high resolution is achieved at shorter measuring time. Signal-to-noise ratio is improved of about x10 times. Improvements pushed system to physical limits of optical interferometry. This next step in optical nanometrology was possible due to implementation at Inmetro/Diopt of an optical frequency comb generator based on femtosecond laser facility that provides novel light source for precision dimensional metrology [5-6].

Acknowledgments
This work has been supported by the National Research Council of Brazil (CNPq/ PROMETRO) and also by FAPERJ (Research Support Foundation of Rio de Janeiro State).

References
[1] Burke T *et al* 2011 *European Nanometrology 2020* (European Comission Framework Programme 7: Co-Nanomet)
[2] Koenders L *et al* 2003 Comparison on nanometrology nano 2 step height *Metrologia* **40** Tech suppl 04001
[3] Malinovsky I, Couceiro I B, Franca R S, Mauricio S L, Azeredo C L S, Almeida C M and Von der Weid J P 2014 *ACTA IMEKO* **3** 28-32
[4] Titov A, Malinovsky I and Massone C 2000 *Metrologia* **37** 121
[5] Jin J, Kim Y-J, Kim Y, Kim S-W and Kang C-S 2006 *Optics Express* **14** 5968-74
[6] Silva I L M, Couceiro I B, Torres M A C, Costa P A and Grieneisen H P H 2015 Traceability of laser frequency/wavelength calibration through the frequency comb at Inmetro *Proc. Conf. Metrologia 2015* (26/nov–02/dec/2015 Bento Gonçalves RS Brazil)

A bilateral comparison on illuminance using a photometer between IPT and LABELO

A F G Ferreira Junior [1 3] and C J R Bindé Junior [2]

[1] Instituto de Pesquisas Tecnológicas do Estado de São Paulo, São Paulo, SP, 05508-901, Brazil

[2] Pontifícia Universidade Católica do Rio Grande do Sul, Porto Alegre, RS, 90619-900, Brazil

[3]Corresponding author's e-mail: agentil@ipt.br

Abstract. This work presents the result of bilateral illuminance comparison obtained from a photometer calibration. The bilateral comparison was performed comparing the calibration results from the same photometer at LABELO and IPT laboratories, which take part of Brazilian calibration network. Occasionally LABELO was chosen as a pilot laboratory and was responsible to calibrate the photometer at the beginning and end of comparison and define the reference illuminance value of photometer calibration. The illuminance calibration points ranged from 20 to 2000 lx and the comparison evaluation criterion was the normalized error (En numbers). The laboratory measurements are in agreement according to the evaluation criterion.

1. Introduction

One fundamental point of metrology is the measurement intercomparison. It is used to validate measurement equivalence among the National Metrology Institutes (NMIs) in the Mutual Recognition Arrangement of the General Conference of Weights and Measures (CIPM) [1]. It occurs in other levels of a metrological chain concerning the national metrological networks, test laboratories and factory laboratories, mostly based on laboratories accreditation standards [2].

Illuminance comparisons among NMIs are performed in ranges from 14 lx to 500 lx [3-5]. In the last key comparison a photometer was used as transference standard for the first time and it was observed that approximately 30 % of participant NMIs obtained a result out of reference values considering an uncertainty with k = 2 [3]. In another comparison in the Asia Pacific Metrology Program it was observed a drift of 30 % for one laboratory which could be explained by a drift on its primary standard [4]. These comparison results among the National Metrology Institutes plays an important task indicating that additional work should be done to assure a trustful measurement.

The purpose of this comparison is to confirm the metrological competence to perform the calibration of photometers by IPT and LABELO, within the scope of accreditation in Brazil, regulated by the General Coordination for Accreditation. The development of this comparison program and the results are reported in this paper.

2. Experimental method

2.1. The photometer

The photometer used as comparison equipment in this work was manufactured by Minolta, model T10 and is presented in figure 1. This equipment is a control standard at LABELO. It presents a very stable performance which was verified in this this work by a calibration before and after comparison measurements.

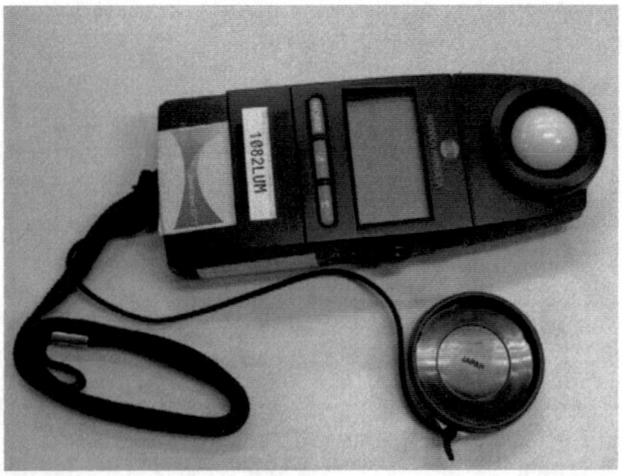

Figure 1. Photometer used in the comparison as comparison equipment.

2.2. Comparison procedure

The bilateral comparison was done comparing the calibration results of a photometer by both laboratories. The photometer circulated between the two laboratories and its drift was considered in reference values uncertainty estimation. LABELO was defined as pilot laboratory and evaluated the photometer at the begging and the end of the program.

Each laboratory performed the photometer calibration at environment temperature of (23 ± 3) °C. The photometer reference plane for the calibration was a plane tangent to its diffuser. The calibration points were defined as: 20; 110; 400; 1000; 1600 and 2000 lx. The calibration procedure was the same used by each laboratory to calibrate equipment on Brazilian metrological network (RBC).

The traceability of measurements from LABELO is a standard lamp type Wi 41/G calibrated in luminous intensity at Physikalisch-Technische Bundesanstalt (PTB) of Germany. LABELO uses the Inverse Squared Law as a calibration methodology to obtain reference illuminance.

The traceability of measurements from IPT is a standard lamp type FEL 1000 W calibrated in luminous intensity at Physikalisch-Technische Bundesanstalt (PTB) of Germany and a standard photometer calibrated in illuminance at National Institute Metrology, Quality and Technology (INMETRO) of Brazil. IPT uses illuminance measurement comparison as calibration methodology and the standard lamp is used to ensure high illuminance traceability, such as 2000 lx.

2.3. Comparison agreement criterion

The comparison agreement criterion was the normalized error (E_N) [6], which was obtained according to equation (1) using the participant laboratory's measurement (M_{Lab}), the pilot laboratory's measurement (M_{Ref}), the participant laboratory's uncertainty (U_{Lab}) and the pilot laboratory's uncertainty (U_{Ref}). If a measurement has absolute value of E_N is smaller than or equal to the unity ($|E_N| \leq 1$), this is a satisfactory result otherwise ($|E_N| > 1$) the result is doubtful.

$$E_N = \frac{M_{Lab} - M_{Ref}}{\sqrt{U_{Lab}^2 + U_{Ref}^2}} \quad (1)$$

3. Results

The results of photometer calibration obtained by each laboratory are presented at tables 1 and 2 including reference illuminance (E_{REF}), measured photometer illuminance (E_{MEA}) and the uncertainty with a coverage factor of 2 for a confidence level of approximately 95 %. The uncertainty results from LABELO, table 1, include the photometer drift observed during the comparison.

Table 1. LABELO photometer calibration results.

E_{REF} (lx)	E_{MEA} (lx)	Uncertainty (%)
19,74	20,00	3,7
106,4	110,0	3,1
398	400	3,7
1008	1000	3,9
1619	1600	4,0
1972	2000	4,1

The photometer drift obtained in the comparison is presented in table 3 where for each reference illuminance (E_{REF}) is presented the percentage of drift. The maximum drift value is small when compared to measurement uncertainty, though it is not negligible. Thus the photometer drift increased the pilot laboratory uncertainty in less than 20 % which do not invalidate this comparison.

Table 2. IPT photometer calibration results.

E_{REF} (lx)	E_{MEA} (lx)	Uncertainty (%)
19,80	20,00	3,3
108,4	110,0	3,4
395	400	3,3
987	1000	3,4
1581	1600	3,4
1981	2000	3,5

The normalized error (E_N) obtained is presented in table 4 and is related to each IPT reference illuminance (E_{REF}). For each calibration point of IPT result the E_N is calculated according to equation 1. The E_N result shows that all the calibrated points presented a satisfactory result in this comparison.

Table 3. Photometer drift in comparison.

E_{REF} (lx)	Photometer Drift (%)
19,74	0,6
106,4	0,1
398	0,5
1008	0,4
1619	0,3
1972	0,4

Table 4. Normalized error for IPT results.

E_{REF} (lx)	E_N
19,80	-0,06
108,4	-0,40
395	0,15
987	0,41
1581	0,45
1981	0,08

4. Conclusions

A bilateral comparison of illuminance in the range of 20 to 2000 lx was performed between IPT and LABELO and throughout this range all results were satisfactory considering the normalized error as evaluation criterion and confirming the convergence of results.

The photometer used in the comparison presented a drift which was included in pilot laboratory uncertainty. The obtained photometer drift corresponded to one fifth of calibration uncertainty.

Acknowledgments

Authors wishing to acknowledge the support from colleagues of both institutions, which made possible the accomplishment of this work.

References

[1] Bureau International des Poids et Mesures. Presents information about International equivalence of measurements. Available at <http://www.bipm.org/en/cipm-mra/> Access: 22 July 2015.

[2] ISO 2005 *ISO/IEC 17025: General requirements for the competence of testing and calibration laboratories* (Geneva: International Organization for Standardization)

[3] Köhler R, Stock M and Garreau C 2004 Final report on the international comparison of luminous responsivity CCPR-K3.b *Metrologia* **41** 02001.

[4] Gardner J L and Atkinson E 2004 Final report of the APMP comparison of luminous responsivity APMP.PR-K3.b *Metrologia* **41** 02002.

[5] Pons A, Campos J, Corróns A and Samedov F 2005 Key comparison EUROMET.PR-K3.b.1 bilateral comparison on illuminance responsivity between IFA-CSIC/Spain and UME/Turkey *Metrologia* **42** 02002.

[6] ISO 2010 *ISO/IEC 17043: Conformity assessment - General requirements for proficiency testing* (Geneva: International Organization for Standardization).

Journal of Physics: Conference Series **733** (2016) 012062 doi:10.1088/1742-6596/733/1/012062

Characterisation of optical filters for broadband UVA radiometer

Luciana C Alves[1], Carla T Coelho[1], Jaqueline S P M Corrêa[1], Thiago Menegotto[1], Thiago Ferreira da Silva[1], Muriel Aparecida de Souza[1], Elisama Melo da Silva[1], Maurício Simões de Lima[1] and Ana Paula Dornelles de Alvarenga[1]

[1] Instituto Nacional de Metrologia, Qualidade e Tecnologia – Inmetro; Divisão de Metrologia Óptica –Diopt;

E-mail: laraf@inmetro.gov.br

Abstract. Optical filters were characterized in order to know its suitability for use in broadband UVA radiometer head for spectral irradiance measurements. The spectral transmittance, the angular dependence and the spatial uniformity of the spectral transmittance of the UVA optical filters were investigated. The temperature dependence of the transmittance was also studied.

1. Introduction

Broadband ultraviolet (UV) radiometers usually consist of a radiometer head, a signal converter and a display unit. The detector head contains a photodiode, a spectral filter and an aperture. A diffuser is frequently used to improve the angular response and the spatial uniformity of the radiometer [1]. The spectral filter is used to shape the spectral response of the radiometer head to match the action spectrum of the actinic effect of interest [2]. The radiometer spectral responsivity is the product of the spectral responsivity of the detector and the spectral transmittance of the diffuser and optical filter. For UVA measurement, the required spectral response is a rectangular function from 315 nm to 400 nm and no response outside of this region. It is very difficult to find a commercial UVA meter with a perfect spectral match.

An UVA radiometer head is being assembled for the integrated irradiance measurements in the laboratory. In order to fully understand the response of such UVA meter head, some components properties are independently characterized. Considering the potential use in the device, some Hamamatsu G2119 GaAsP Schottky photodiodes and SUV-Detectors PtSi-*n*-Si Schottky photodiodes were previously investigated in terms of their spectral power responsivity, local responsivity and shunt resistance [3]. To define the effective area of the input flux for irradiance measurement, two precision apertures with cylindrical edge have been well characterized by using a non-contact technique used for area measurement [4].

In this paper, we present the optical characterization of a set of two UVA optical filters used to define the spectral bandpass measurement of the incident optical radiation in the radiometer head. The spectral transmittance of such filters was investigated inside and outside the specific region of UVA spectral region. The out-of-band transmittance effect was considered by using the spectral matching characteristic f_1' of the filter as an indicator of the closeness to the spectral function of interest. The angular dependence and the spatial uniformity of the spectral transmittance of the UVA optical filters were also analyzed. One UVA optical filter was thermalized and the temperature dependence of the transmittance was analyzed.

2. Characterisation of optical filters

2.1. *UVA optical filters*

Two XUVA315 optical filters (27SA-01 and 27SA-02), figure 1, manufactured by Asahi Spectra Co. Ltd., were measured. Both filters are 25 mm in diameter and 4.5 mm thick and are made of a thin film deposited on a fused silica substrate mounted in an aluminum ring. According to the manufacturer, these filters are designed for isolating a broad band of UVA, with a transmission band from 315 nm to 400 nm.

Figure 1. UVA optical filters

2.2. *Spectral regular transmittance measurements*

The spectral regular transmittance of the UVA optical filters was measured using the laboratory spectrophotometer Varian Cary 5000. The spectrophotometer repeatability in wavelength is better than 0.005 nm, and the expanded uncertainty is estimated to be equal to 0.13 nm (k = 2.02).

The out-of-band spectral transmittance of the UVA optical filters was investigated in order to determine its ability to block radiation from wavelengths outside the actinic band function. Spectral scans were carried out in the spectral range from 200 nm to 800 nm and 0.1 nm step. The spectrophotometer bandwidth was 1 nm. The angle of incidence was normal or near normal. Measurements were carried out under dark environment conditions. The spectral regular transmittance of the UVA optical filters was measured at the temperature of 21 °C. The ambient temperature and the relative humidity were monitored during measurements and varied no more than 0.4 °C and 1.5 % (1σ), respectively.

The temperature dependence of the UVA filter transmittance was investigated by measuring the spectral transmittance of the 27SA-02 optical filter at a set of different temperatures ranging from 21 °C to 25 °C. Spectral scans were carried out in the spectral range from 250 nm to 450 nm and 1 nm step. The UVA filter thermalization was performed with a PID temperature controller acting on an annular thermoelectric Peltier element. The accuracy of the whole temperature controller is about 0.3 °C.

2.3. *Angular dependence of the spectral transmittance*

Measurements of the angular dependence of the spectral transmittance of the UVA optical filters was realized in the UV dedicated system. The experimental set-up optical radiation source is a mercury arc lamp and measurements were performed at a fixed wavelength of 365 nm. The angular dependence of the spectral transmittance was analyzed from the measurements of the photocurrent as a function of angle of incidence starting with the stable optical beam at normal incidence and further rotating the optical filter around the vertical axis. An automated rotator was used to position the optical filter. The generated photocurrent by a silicon detector was amplified by a current-voltage converter and a digital multimeter was used to measure the voltage. The trans-impedance amplifier gain used for the test detector was typically 10^7 V/A.

2.4. Spatial uniformity measurements
Spatial uniformity measurements of spectral regular transmittance of UVA optical filters were performed in the spectral power responsivity facility adapted for this purpose. The experimental set-up optical radiation source is a mercury arc lamp. The measurement was performed in power mode. The generated photocurrents were amplified by a current-voltage converter and two digital multimeters were used to measure the voltage. The trans-impedance amplifier gain used for the test detector was typically 10^7 V/A. The measurement of spatial uniformity of spectral regular transmittance of UVA optical filters was performed at 365 nm. The diameter of the incident beam was equal to 1.0 mm. The spatial uniformity of optical filters was investigated by performing a scan of a (12 × 12) mm^2 area on the surface of each optical filter with a pitch of 0.5 mm.

3. Measurements results
3.1. Spectral regular transmittance
The UVA optical filters 27SA-01 and 27SA-02 showed spectral regular transmittance at 365 nm equal to 96.27 % and 96.30 %, respectively. In the pass band of interest, it was calculated a relative difference of 2.6 % between their spectral transmittances. The in-band transmittance does not match ideally to the UVA function and moreover both optical filters presented out-of-band transmittances below 315 nm, as can be seen in the figure 2. Both optical filters had undesirable leakages out-of-band transmittance with a peak at 760 nm. Up to 730 nm, the average out-of-band blocking of the optical filter 27SA-01 and 27SA-02 is 3.3×10^{-5} to 2.7×10^{-5} relative to the bandpass, respectively.

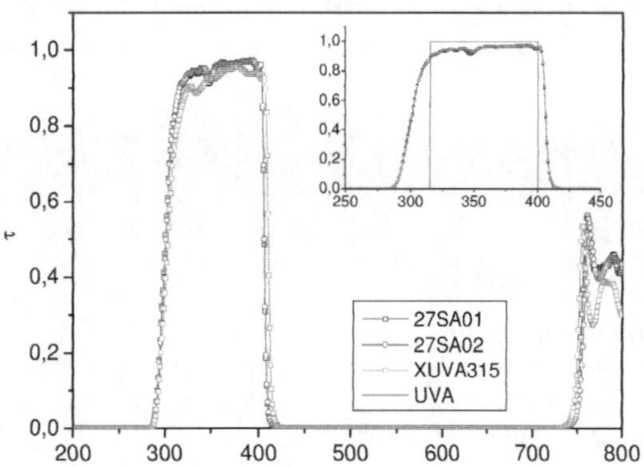

Figure 2. Spectral regular transmittance measured of the UVA optical filters inside and outside the passband.

Additionally, the relative differences between the measured spectral regular transmittance of the optical filters and the nominal spectral regular transmittance declared by the manufacturer Asahi Spectra were evaluated. In the pass band of interest, the optical filters showed a maximum relative difference equal to 9 %, approximately.

3.2. Temperature dependence

Within the passband, the spectral transmittance of the 27SA-02 optical filter was measured from 21 °C to 25 °C. The temperature stability was better than 0.1 °C. Figure 3 shows the temperature coefficient of the UVA filter transmittance as a function of wavelength and the region of the inflection points is the most sensitive to changes in temperature [5]. The typical temperature coefficient around the wavelength of 365 nm is about 1.2×10^{-2} °C^{-1}. Maximum temperature coefficients of transmittance were observed at the edges of the spectral band.

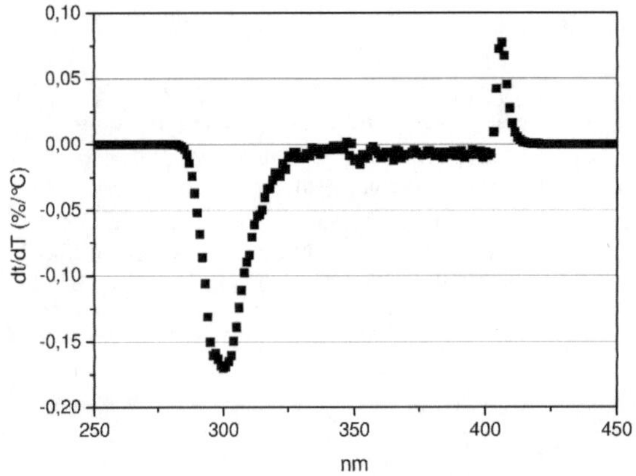

Figure 3. Temperature coefficient of the UVA filter transmittance.

3.3. Angular dependence of the spectral transmittance

The generated photocurrent was evaluated at different angles of incidence (0°, 5°, 10°, 15°, 20°, 25° e 30°) of the monochromatic beam at 365 nm. No significant variation of *T*% depending on the angle of incidence (ΔT %/° = 0.06 %) was observed.

3.4. Spatial uniformity

The spatial uniformity of spectral regular transmittance of the XUVA315 optical filter (27SA-01) was investigated. The spatial nonuniformity was calculated with a relative expanded uncertainty of 5×10^{-4} (k = 2). Figure 4 shows the measured spatial uniformity of the spectral regular transmittance of the UVA optical filter.

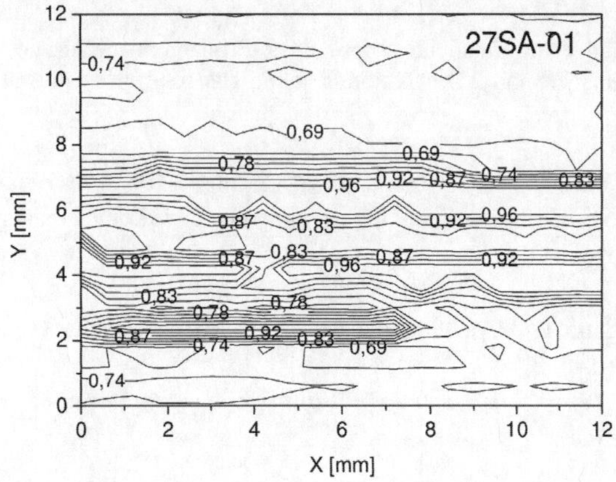

Figure 4. Spatial uniformity of the spectral regular transmittance of XUVA315 27SA-01 optical filter at 365 nm. The transmittances are normalized to the center of the active area.

4. Discussion

A set of UVA optical filters was evaluated by its ability to block radiation of wavelengths outside the band of interest. Both optical filters had undesirable out-of-band leakage in the near-infrared region, what may imply in a measurement overestimation by the assembled UVA meter head. The choice of photodiodes for use in the construction of the UVA meter head have already been made based on a previous study [3]. The best choice among the analyzed photodiodes, considering the out-of-band transmittance of the optical filters peaked at 760 nm, proved to be the GaAsP Schottky photodiodes due to the suppression of radiation above 650 nm and the high shunt resistance. The spectral responsivity of the UVA meter head based on a single GaAsP photodiode was preliminarily calculated as the product of its components characterized separately. The contribution of the out-of-band response of the UVA meter head in the short and long wavelength band was estimated to be only 5.8 % and 5.3 % of the total signal using a continuous spectral lamp as radiant source, respectively. Nevertheless, the presence of the transmittance outside the band of interest may imply the need to use blocking filter to reduce this effect. If the use of the blocking filter is not possible, the out-of-band responsivity must be considered in order to apply a correction, if significant. Assuming a UV radiant source with no flux out-of-band, the effect of the outer band response can be easily corrected by measuring the dark current.

5. Conclusion

The study of the optical characteristics of the UVA optical filters is an important step in the full characterization of the broadband UVA radiometers. Measurements were realized in order to characterize UVA optical filters for being used in the construction of a broadband ultraviolet radiometer for use as a transfer standard. The next step is to investigate some others optical characteristics functions of these UVA optical filters prior to use in radiometric measurements such as its spectral reflectance. As a further remark, filters and photodiodes should be individually selected to achieve a better realization of the UVA spectral function since the use of suitable components can contribute to lower the measurement uncertainty.

Acknowledgements

L. C. Alves acknowledges the financial support from Pronametro. Maurício S. Lima acknowledges support from CNPq/Prometro and A. D. Alvarenga acknowledges support from CNPq.

References

[1] Larason T. C. and Cromer C. L., J. Res. Natl. Inst. Technol., **106**, pp. 649-656, 2001.

[2] CIE 53:1982"Methods of characterizing the performance of radiometer and photometer"

[3] Alves L. C., Menegotto T., Silva T. F., Jaqueline S. P. M. Corrêa, and Duarte I. F., 2014, *UVNews* 10.

[4] Alves L. C., Menegotto T., Silva T. F., Corrêa J S. P. M., and Duarte I. F., 2014, *Proceedings of the NEWRAD 2014.*

[5] https://marketplace.idexop.com/Frontend/PDFs/interference filter coatings.pdf (acess: 10/12/2015)

An atomic beam of ^{6}Li – ^{7}Li for high resolution spectroscopy from matrix isolation sublimation

A N Oliveira[1], R L Sacramento[2], B A Silva[2], F O Uhlmann[3], W Wolff[2] and C L Cesar[2]

[1] INMETRO, Av. Nossa Senhora das Graças, 50 25250-020 Duque de Caxias, RJ, Brazil
[2] Instituto de Física, Universidade Federal do Rio de Janeiro, Caixa Postal 68528, 21941-972 Rio de Janeiro, RJ, Brazil
[3] Beuth University of Applied Science, Zip Code 13353, Berlin, Germmany

E-mail: anunes@inmetro.gov.br

Abstract. We propose the Matrix Isolation Sublimation (MISu) technique for generating cold lithium atoms for the measurement of the ^{6}Li – ^{7}Li isotope shift in D1 and D2 transitions. The technique is capable of generating cold ^{6}Li and ^{7}Li beams at 4 K with forward velocity of 125 m/s. Using this beam we offer a distinguished source of lithium atoms for transitions measurements, adding a new possibility to make high resolution spectroscopy towards improving the experimental checks of the theory.

1. Introduction

The energy structure of lithium can be solved from *ab initio* calculations and their predictions are compatible with experimental results. The fact that both results, experimental and theoretical, are within the same accuracy makes lithium atoms of great interest. In 2000 Yan and Drake [1, 2] proposed a combined theoretical and experimental approach to determine the relative nuclear charge radii based on the Splitting Isotopic Shift (SIS - difference for D1 or D2 line between different isotopes) (figure 1). Unfortunately, theoretical and experimental results reveals discrepancies as discussed in reference [3]. Since previous measurements where done using an effusive atomic beam at temperature around s 400 °C, including the recent one made in 2013 [4], we propose the use of a cold beam of Li atoms instead.

The present work shows that the MISu can produce simultaneous beams of ^{6}Li and ^{7}Li. Cryogenic beams of Cr, ^{7}Li, Ca, Cs and Li_2 molecules have been previously produced from the sublimation of Ne or H_2 matrices [5, 6, 7, 8, 9, 10].

2. Experimental setup

The basic experimental setup has been described in detail in [5] and a simplified diagram is shown in figure 2. The system consists of a closed-cycle cryostat that reaches 3 K at its cold plate with 1 W cooling power at 4 K. A sapphire substrate, with a deposited resistive NiCr film, is thermally connected, through a thermal link, to the cold plate. This configuration permits fast variations of the sapphire substrate temperature without affecting the cryostat temperature.

The gas to grow the matrix (Ne or H_2) is delivered through a 2 mm diameter tube whose exit points towards the sapphire substrate. The flow rate of the gas is controlled by the pressure in a

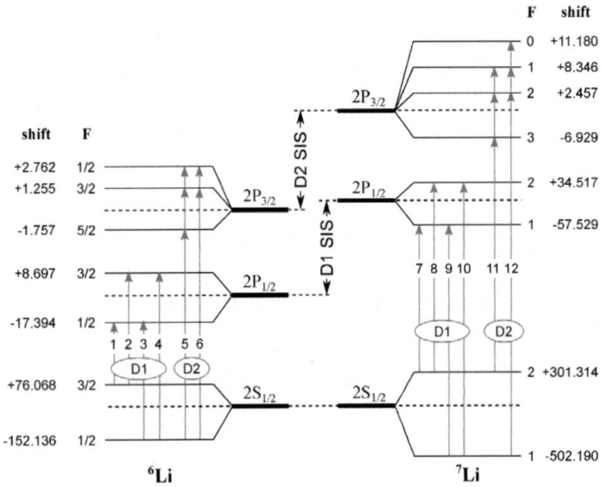

Figure 1. Lithium 6 and 7 isotopes energy levels, frequency shift relative to the state center of gravity energy ($E_{cg} = \sum_F(2F+1)E_F/\sum_F(2F+1)$) are shown in MHz [11].

reservoir followed by a high impedance line, achieving 1–10 mmol/h. The growth of the matrix film is monitored by a CW laser beam – designated "longitudinal" – through the interference fringes (with 10–12% visibility) on the etalon formed by the Ne film together with the $\sim$ 40% reflectivity of the NiCr resistive film on the sapphire substrate. This longitudinal beam comes from the bottom upwards, reflects off the Ne film and NiCr/sapphire mirror at near normal incidence, and is monitored by photodiode as indicated in figure 2. A typical time for growing a 330 nm thick Ne layer is around 40 s, but it can be reduced to less than 1 s. Once the cryostat starts cooling and cryopumping the vacuum line below 5×10^{-6} Torr the external vacuum pump line is closed.

After the matrix has been grown, or during its growth, a high-power pulsed laser ($\sim$ 10 mJ in 5 ns at 532 nm) promotes ablation of the solid precursor (Li or LiH), implanting the atoms into the matrix. Completed the matrix with the isolated atoms, we apply a heat pulse to the sapphire causing the matrix, together with the implanted atoms, to sublimate into vacuum. The atomic spectrum in flight is registered using the laser beam propagating in longitudinal and transverse direction, while scanning the laser frequency. The longitudinal beam is both co-propagating and counter-propagating with the sublimation plume. This leads to a splitting of a resonance line due to a positive and a negative Doppler effect, allowing a direct measure of the sublimation plume drift velocity (figure 5), as well as the sample's longitudinal temperature. The other component from the same laser designated transverse, propagates horizontally (transverse to the propagation direction of the sublimation plume). The transverse beam is insensitive to the Doppler shift from the drift velocity of the plume while allowing for the characterization of the transverse temperature of the sample. In the data presented here the CW laser is tuned to the ^{7}Li D1 and ^{6}Li D2 transitions at 670.795 nm. It is scanned at 2 kHz covering 4 GHz around the center frequency which is enough to detect both transitions. The frequency scan is monitored by a Fabry-Perot interferometer allowing corrections on the non-linear scan due to the PZT response of the laser.

3. Experimental results

The scan, around the central frequency 446 800 000 MHz, allows to cover the transitions labelled 5 – 10 in figure 1 since the frequency range of those transitions span around 1 GHz. We have used a LiH substrate and the expected natural abundance between ^{6}Li and ^{7}Li is respectively 7.6% and 92.4% [12].

Figure 3 shows a typical transverse absorption signal where the atoms transverse temperature

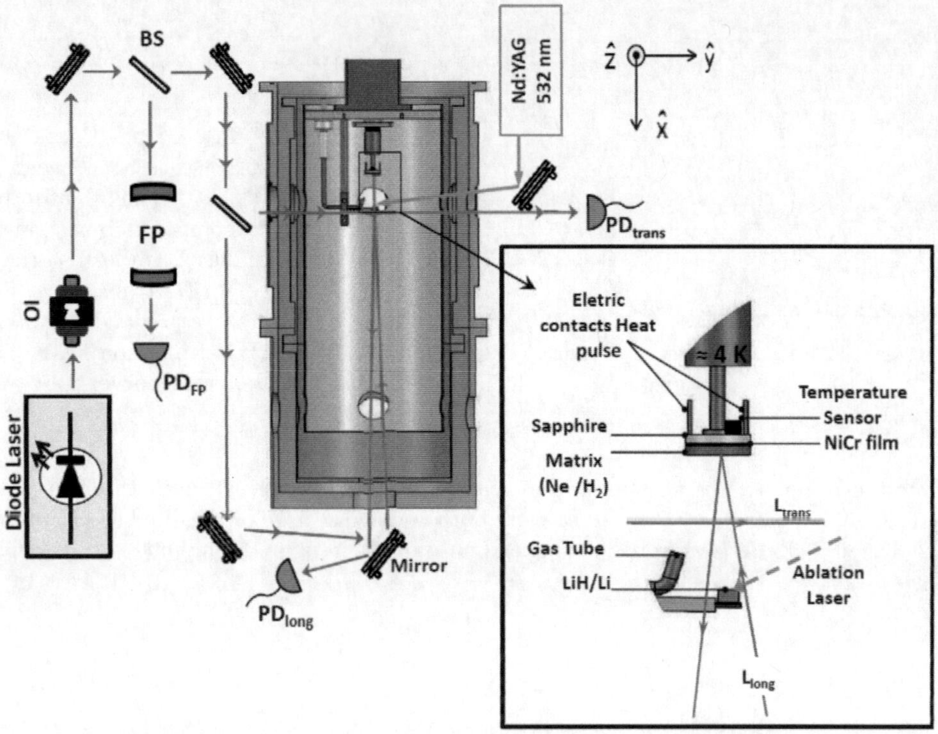

Figure 2. Schematics of a typical setup for MISu showing the cryostat with optical access, and the laser beams for absorption spectroscopy. The zoom in the central region shows the sapphire substrate with its NiCr film resistor and the deposited matrix of Ne. For the study of lithium, the atoms are implanted via laser ablation (dashed green line) on a solid Li or LiH precursor. The spectroscopy laser beams (solid red line and identified as L_{long} and L_{trans}), propagate along or perpendicular to the direction of the sublimation plume expansion. The longitudinal laser is also used to monitor the matrix film thickness. The lasers' transmissions through the system are monitored by photodiodes (PD_{long} and PD_{trans}).

was 2.5 K. Using the frequency difference between transition lines 5 to 10 from literature [13] and estimating the relative intensity by considering the isotopes proportion, it was possible to adjust the lineshape demonstrating a clear presence of ^{6}Li and ^{7}Li.

The advantage of the present technique is based on its versatility of operation on the choice of parameters governing the three steps of the method: matrix formation, ablation and sublimation. In order to achieve the best beam configuration for the experiment several sets of parameters can be chosen until exhausting the freedom of the method. The present set of parameters is not necessarily the best configuration for a high resolution experiment, the beam condition can be improved and made to last longer, achieving a quasi-cw condition.

In figure 4, a lithium beam is observed in the 5 to 15 ms range with a temperature between 3 – 5 K and longitudinal velocity of 120 – 130 m/s. A neon matrix of $\sim$ 170 nm thickness was grown in 210 s and 200 ablations pulses were applied during its formation. After matrix completion with implanted atoms we sublimated it in a slow regime [7], using a heat pulse of 4.0 W for 15 ms. In this regime the matrix takes awhile to start the sublimation ($\sim$ 5 ms). After

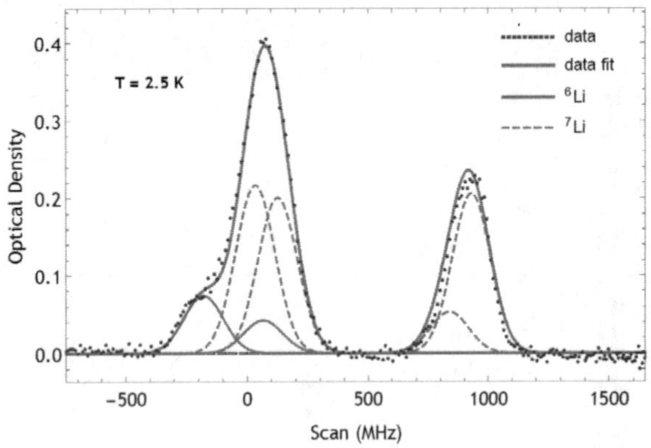

Figure 3. Single scan using the transverse laser converted to optical density. The full absorption signal is composed by 6 gaussian distributions: 2 gaussians originate from the ^{6}Li D2 transition – 5 and 6 (gray curves) – and 4 gaussians from ^{7}Li D1 transition – 7, 8, 9, and 10 (dashed gray curves).

this time the beam reaches an almost constant optical density regime, until the heat pulse is finished. The temperature was extracted from both transverse and longitudinal signal while the forward velocity was derived from the longitudinal signal. In figure 5 only one scan is depicted for illustration.

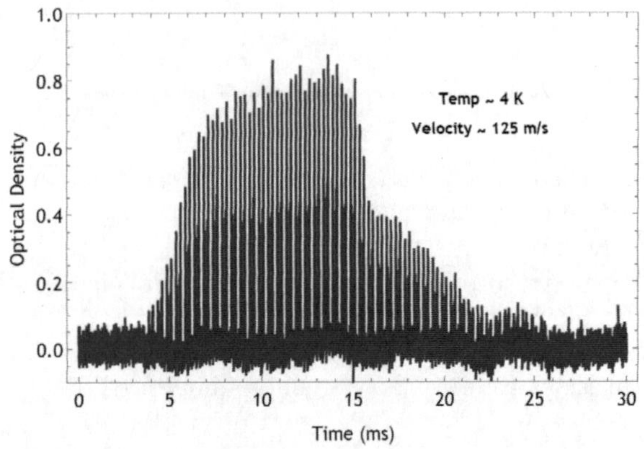

Figure 4. Processed data with absorptions converted to optical density. The peaks appear as the laser is continuously scanned through the resonances.

4. Conclusion and prospects

We observed the creation of a cold and slow atomic beam of ^{6}Li and ^{7}Li, suitable for high resolution spectroscopy. Due to the flexibility of MISu technique we can further improve the beam properties, in particular, its lasting time can be increased. Transverse temperature can be also reduced if we use collimating orifices in the beam path. Our purpose is to use an optical frequency comb available in Inmetro in combination with this "MISu-beam" to measure the lithium isotopic shift. This proposal will be the first step of a joint project based on MISu technique and high resolution spectroscopy in Inmetro Optical Division (Diopt). Another goal is placed on measuring the hydrogen 1S-2S transition.

Acknowledgments

This work is partially supported by the Brazilian institutions CNPq, CAPES, RENAFAE, and INCT-IQ.

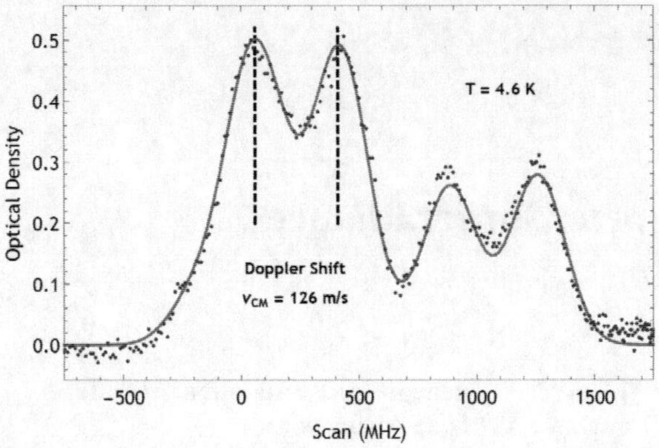

Figure 5. Processed data for a single scan of the longitudinal laser signal. Based on the Doppler shift the forward velocity is determined. Gaussian width is used to extract temperature. 12 gaussian distributions were used in this fit. The same parameters used in the analysis of figure 3 are used here to determine the two shifted distributions.

References

[1] Yan Z C and Drake G W F 2000 *Phys. Rev.* A **61** 022504
[2] Nörtershäuser N *et al* 2011 *Phys. Rev.* A **83** 012516
[3] Sansonetti C J, Simien C E, Gillaspy J D, Tan J N, Brewer S M, Brown R C, Wu S and Porto J V 2011 *Phys. Rev. Lett.* **107** 023001
[4] Brown R C, Wu S, Porto J V, Sansonetti C J, Simien C E, Brewer S M, Tan J N and Gillaspy J D 2013 *Phys. Rev.* A **87** 032504
[5] Sacramento R L, Oliveira A N, Alves B X, Silva B A, Li M S, Wolff W and Cesar C L 2015 *Rev. Scient. Inst.* **86** 073109
[6] Oliveira A N, Sacramento R L, Alves B X, Silva B A, Wolff W and Cesar C L2014 *J. Phys. B: At. Mol. Opt. Phys.* **47** 245302
[7] Sacramento R L, Alves B X, Almeida D T, Wolff W, Li M S and Cesar C L 2012 *J. Chem. Phys.* **136** 154202
[8] Sacramento R L, Scudeller L A, Lambo R, Crivelli P and Cesar C L 2011 *J. Chem. Phys.* **135** 134201
[9] Crivelli P, Cesar C L and Lambo R 2009 *Can. J. Phys.* **87** 799
[10] Lambo R, Rodegheri C C, Silveira D M and Cesar C L 2007 *Phys. Rev.* A **76** 061401(R)
[11] Das D and Natarajan V 2007 *Phys. Rev.* A **75** 052508
[12] Coplen T B *et al* 2002 *Pure Appl. Chem.* **74** 1987
[13] Noble G A, Schultz B E, Ming H and Wijngaarden W A 2006 *Phys. Rev.* A **74** 012502

Construction of UV-A radiometer for irradiance measurements

L C Alves, C T Coelho, J S P M Corrêa, T Menegotto, T F da Silva, M A de Souza, E M da Silva, M S de Lima and A P D de Alvarenga

Divisão de Metrologia Óptica, Instituto Nacional de Metrologia, Qualidade e Tecnologia, Av. Nossa Senhora das Graças 50, Duque de Caxias, RJ, Brasil

E-mail: laraf@inmetro.gov.br

Abstract. This work presents preliminary results aiming on providing Inmetro with a radiometric transfer standard in the UV-A spectral range. A broadband UV radiometer head was constructed using an UV photodiode, a commercially available UV-A optical filter and a precision aperture. These components have been characterized in the calibration and measurement facilities available at Inmetro. The preliminary characterization of the assembled UV broadband radiometer was carried out and the results are presented and discussed in this text.

1. Introduction

Ultraviolet (UV) light has many applications in materials processing, inspections of products and health care. These applications can require accurate monitoring of radiation levels [1-3].

Broadband UV radiation can be measured using filter radiometers whose spectral responsivities are matched to the desired spectral function. The UV-A spectral function is a rectangular-shape response function ranging from 315 nm to 400 nm which was standardized by the Commission Internationale de l'Eclairage (CIE) [4].

UV-A radiometers can be constructed using semiconductor photodiode, passband filters and a limiting aperture (figure 1). They are technically simple to construct and easy to use, however it is usually difficult to obtain good matching between its response function and the standardized spectral function [5].

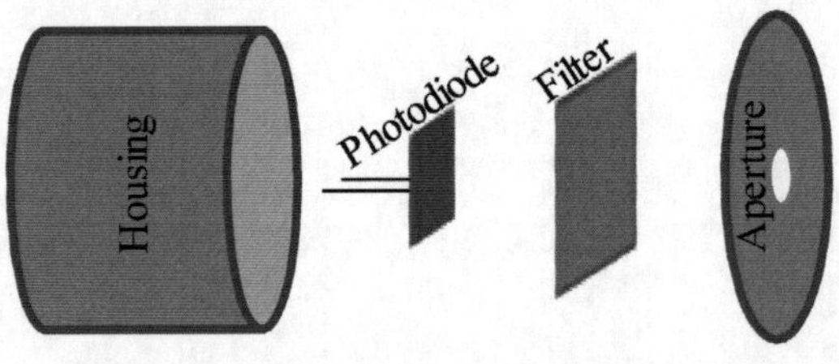

Figure 1. Exploded view of a typical radiometer.

This study reports introductory results aiming on providing Inmetro with a radiometric transfer standard in the UV-A spectral range. A broadband UV radiometer was assembled. The preliminary characterization of this broadband radiometer was carried out according to the guidelines elaborated within the Working Group of the Thematic Network for Ultraviolet Measurements [6].

2. Experimental procedure

Each element of the UV radiometer was characterized separately assuming that inter-reflections between filter and photodiode were negligible. The spectral transmittance of the optical filter was previously measured. The selected type of UV photodiode (GaAsP Schottky) used for the construction of the transfer standard was characterized with respect to spectral responsivity and spatial uniformity in the ultraviolet and visible spectral regions [7]. The power spectral responsivity values were measured with suitable transfer standards traceable to Inmetro´s cryogenic radiometer. A precision aperture with cylindrical edge and nominal area of 0.5 cm^2 was used in the UV broadband radiometer to define the area of the input flux for irradiance measurement. A non-contact technique was used for measuring the aperture area with an expanded uncertainty of 2.8×10^{-2} mm^2 ($k = 2$) [8].

The power spectral responsivity of the UV-A radiometer was then calculated from the product of the measured regular spectral transmittance of the UV-A optical filter and the power spectral responsivity of the UV photodiode.

The assembled radiometer was characterized in order to determine its performance according to reference [9]. A brief description of each characterized parameter as well as the measurement procedure is given under the following subsections. These parameters contribute to the uncertainty of measurement of the radiometer.

2.1. Directional response

The directional response describes the radiometer responsivity with respect to radiation incident at angles other than perpendicular. The ideal response would be a cosine function.

The measurement of directional response was performed under overfilled conditions using the output of a 1 kW FEL lamp as a radiation source. Lamp and radiometer were placed 1 m apart. The radiometer head was rotated about the axis passing by the center of the head aperture in the vertical direction (in the plane formed by the aperture surface) as shown in figure 2. The signal was acquired every 5° step. The characteristic function f_2 describing the directional response was calculated according reference [9].

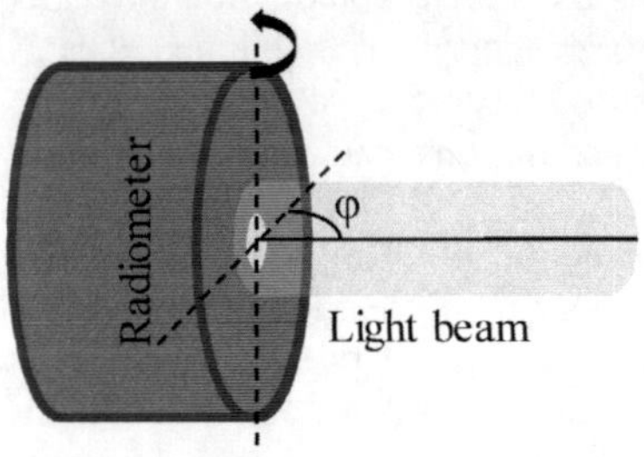

Figure 2. Setup for the measurement of the directional response.

2.2. Radiometer Linearity

The linearity describes the variation of the response of the radiometer with respect to different levels of radiation.

For measuring this property, a set of neutral density (ND) filters whose nominal transmittances were 12 %, 15 %, 20 % and 40 % were used. The radiation of a 1 kW FEL lamp was directly measured by the radiometer (100 % signal) followed by attenuation of the beam by each of the filters. At the end of the cycle, the stability of the lamp was checked by measuring again the non-attenuated beam. This procedure allowed testing the linearity in the range from 0.5 mW/cm^2 to 5.5 mW/cm^2.

The linearity is described by function f_3 [9], which express the maximum deviation of the linear behavior of the radiometer output when compared to the expected value calculated from the known transmittances of the ND filters.

2.3. Radiometer Fatigue
The fatigue describes a reversible change of the radiometer responsivity due to submission to radiation.

The fatigue of the UV-A radiometer was measured using the UV output of a 1 kW FEL lamp. The distance of the lamp to the radiometer reference plane was one meter. The output signal (photocurrent) was measured as a function of the irradiation period keeping the lamp operating conditions constant. Furthermore, in order to follow the recommended procedure [9], the radiometer head was not exposed to radiation for at least 24 hours before starting the test. The characterization was carried out in irradiance mode and the fatigue was tested over an elapsed time of 30 minutes.

2.4. Response uniformity
The response uniformity describes the influence of non-uniform irradiation of radiometers. This parameter was evaluated by clipping the beam to a size of 2.7 mm diameter. The signal of this beam was measured in five regions in the active area: at center, and more four positions located at 90° intervals around the center of the acceptance aperture as depicted in figure 3.

Figure 3. Radiometer front view displaying the active area (blue) delimited by the aperture (gray). The clipped beam was measured over the active area in the five indicated regions, one at a time.

2.5. Short and long wavelength responses
These parameters describe the responsivity of the radiometer outside the UV-A range.

The experiment was held using optical filters with known spectral transmittances, WG305 and GG420. The radiation from a 1 kW lamp was filtered and the short and long wavelength range responses were determined.

3. Results and discussion
The spectral responsivity of the radiometer was calculated from the measured spectral responsivity of the photodiode and the transmittance of the optical filter. The results are shown in figure 4.

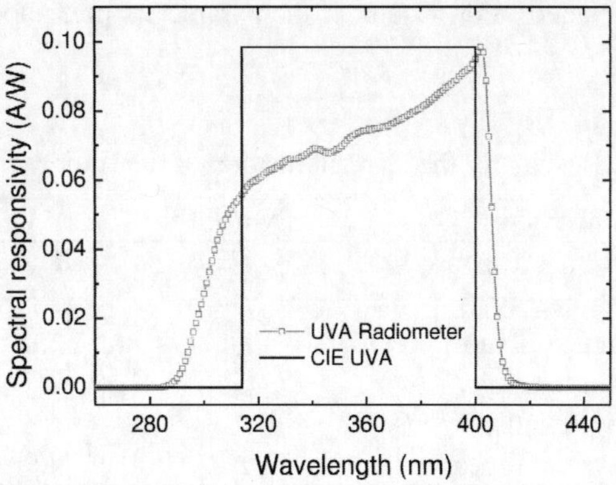

Figure 4. Calculated spectral responsivity for the broadband UV radiometer alongside the CIE UV-A spectral function.

Figure 4 also displays the CIE UV-A spectral function. It is clear that there is a large spectral mismatch of the radiometer responsivity to the proposed spectral function. Construction of UV radiometers with low spectral mismatch is challenging because of the rectangular shape of the response functions [5].

Figure 4 also indicates that there is a strong dependence of the power measured with the radiometer head on the spectral distribution of the source being measured. This poses a problem whenever the radiometer is calibrated with a kind of source and is used to measure radiation emitted by a source with distinct spectral distribution. The spectral mismatch correction factor F tries to account for and correct this problem. Table 1 shows this factor calculated for a set of commonly used UV lamps. The calculation was performed as described in [6] using the spectral responsivity of the radiometer and the nominal spectral distribution of the lamps [9]. The spectral mismatch correction factor can exceed 70 % when the radiometer is calibrated with a low pressure Hg lamp and used to measure a Xe lamp. Although table 1 displays rather high correction factors, those values are consistent with results found elsewhere [5,10].

Table 1. Spectral mismatch correction factor for a set of commonly used UV lamps: Xe – xenon; LPHg – low pressure Hg; MPHg – med-pressure Hg; QTH – quartz-tungsten halogen.

Test Lamp	Calibration lamp			
	Xe	**LPHg**	**MPHg**	**QTH**
Xe	1	1.7259	1.4063	1.0508
LPHg	0.5794	1	0.8149	0.6088
MPHg	0.7111	1.2272	1	0.7472
QTH	0.9517	1.6425	1.3384	1

Additionally, it is also possible to calculate the spectral mismatch factor, f_1', which accounts for the mismatch of the radiometer head responsivity to the desired spectral function [9]. This parameter cannot be used for correcting measurements. Considering the relative spectral distributions of the xenon arc lamp and the quartz-tungsten halogen lamp (QTH), the calculated f_1' for the assembled UV-A radiometer head were 0.3621 and 0.3925. The former result is consistent with f_1' values reported for commercial UV-A radiometers [10,11].

Table 2 presents the characterization results of the parameters presented in section 2 for the filter radiometer.

Table 2. Results for the characterization of the filter radiometer.

Parameter	Symbol	Value
Directional response	f_2	8.6×10^{-1}
Linearity	f_3	5.5×10^{-2}
Fatigue (30 min)	f_5	8.0×10^{-4}
Uniformity	f_9	1.7×10^{-3}
Short wavelength response	u	5.0×10^{-2}
Long wavelength response	r	4.9×10^{-2}

The directional response is high when compared to the other parameters. This indicates that the radiometer response is strongly dissimilar to the cosine response. Such high values for this parameter have also been reported for commercial UV radiometers [10]. A diffuser can be used to improve the directional response of this radiometer with the drawback that it usually deteriorates the uniformity, which is rather good for the current construction, as can be realized from the magnitude of f_9 in table 2.

4. Conclusion

A special UV-A radiometer head was assembled with characterized components. The spectral responsivity of the radiometer was calculated using the spectral transmittance of the filter and the responsivity of the photodiode. This allowed estimating the spectral mismatch correction factor, F, for a set of commonly used UV lamps, as well as the spectral mismatch factor, f_1'. Additionally, the assembled UV-A radiometer was experimentally characterized in order to evaluate the performance regarding to directional response, linearity, fatigue, uniformity, short and long wavelength responses.

The results indicate the difficulties of constructing this kind of radiometers with good spectral mismatch between its spectral responsivity and the standardized response function. This is essentially a consequence of the rectangular shape of the response function for the UV-A.

The constructed device display performance parameters similar to commercially available UV-A radiometers. However, the high value of the directional response indicates the need of using a cosine corrector (diffuser).

Acknowledgments

L. C. Alves acknowledges the financial support from Pronametro. Maurício S. Lima acknowledges support from CNPq/Prometro and A. D. Alvarenga acknowledges support from CNPq.

References

[1] Koutchma T, Forney L J and Moraru C I 2009 *Ultraviolet light in food technology: principles and applications* (Boca Raton: CRC Press)

[2] Saraiya M *et al.* 2004 Recommendations to Prevent Skin Cancer by Reducing exposure to Ultraviolet Radiation *Am J Prev Med* **27** 467

[3] Memarzadeh F, Olmsted R N and Bartley J M 2010 Applications of ultraviolet germicidal irradiation disinfection in health care facilities: effective adjunct, but not stand-alone technology *American Journal of Infection Control* **38** (5) S13

[4] CIE 1987 CIE 17.4 *International Lighting Vocabulary* (Vienna: Commission Internationale de l'Eclairage)

[5] Eppeldauer G P 2013 Standardization of broadband UV measurements *UVNEWS* **9** 34

[6] Working Group 1: Guidance for UV power meter classification for particular applications 2000 Thematic Network for Ultraviolet Measurements, Characterizing the Performance of Integral

Measuring UV-Meters – Final Report *UVNEWS*, **6**

[7] Alves L C, Menegotto T, Ferreira da Silva T, Corrêa J S P M and Duarte I F 2014 Characterization of UV photodiodes in the ultraviolet spectral region *UVNEWS* **10** 1456

[8] Alves L C, Menegotto T, Ferreira da Silva T, Corrêa J S P M and Duarte I F 2014 Development and Characterization of Broadband UV-A *Proceedings of NEWRAD* 255

[9] Working Group 1: Guidance for UV power meter classification for particular applications 2000 Thematic Network for Ultraviolet Measurements, Characterizing the Performance of Integral Measuring UV-Meters – Appendix A **6**

[10] Fusette S S, Facta S, Bonino A, Vaccarono M, Anglesio L and d'Amore G 2013 Characterization of instruments for the measurement of optical radiation *UVNEWS* **9** 28

[11] Xu G and Huang X 2000 Characterization and calibration of broadband ultraviolet radiometers *Metrologia* **37** 235

Eleven years of monitoring an ultra-stable 10 V zener-based voltage standard

H de Souza[1], **L Trigo**[1], **D Slomovitz**[1]

[1]UTE, Laboratory, Paraguay 2385, Montevideo-11800, Uruguay.

dslomovitz@ute.com.uy

Abstract. The long-term behaviour of a voltage standard based on 140 zeners is analyzed. This standard was developed in 2004 and was monitored up to now. Noise and drift are evaluated showing better results than commercial similar devices. This source is used as the National Voltage Standard in the Uruguayan Designated National Metrology Institute (UTE).

1. Introduction

Primary voltage standards are based on Josephson effects, but secondary standards on zener devices. There are some good commercial voltage standards using this technology [1], but their stability at long term is in the order of 1μV/V per year. Laboratories which their voltage standards are based on this type of sources must send them outside for calibration in periods around one or two years to limit the uncertainties to few μV/V. The problem to get low uncertainty is the drift these sources have. The drift has a predictable part and a random one. Although it is possible to correct for the first one, the second one can have an unpredictable behaviour. Then, the uncertainty increases with time.

All commercial sources are based on only one specially selected zener, so that any variation of this component cannot be detected. On the contrary, this project uses a large amount of zeners (140) for reducing the variation of the average voltage, and to detect variations between them [2, 3]. A prototype based on this principle has been continuously observed since 2004. Its deviations from their initial values and analysis of its noise are shown in the following sections.

2. General requirements

Figure 1 shows the schematic circuit. Each zener (REF102, 10 V) is connected to the output voltage (Va) through a 100 Ω resistor (Rj). Then, at that output, the voltage is the average of the individual ones. All zeners are divided in four independent groups of 35 units. Each group has its own regulated 15 V power supply. In this way, intercomparisons between groups are possible for evaluating random noises. Zeners and electronic devices are placed in a small oven at 44 °C with a standard deviation of 0.03 °C, if the external temperature is 23 °C ±1 °C (laboratory temperature). The oven has a thermistor to measure its internal temperature, which is continuously recorded.

The stability of the source depends on many factors. The first one is the stability of the type of zener. Its specifications states 5 μV/V per month, but this figure is very conservative. Complementary information on stability from the manufacturer seems to be contaminated by the measuring system. Our measurements show much lower values and not correlated drift between different samples. In this way, the 140 batch reduce the drift of the average voltage at very low level, as will be shown.

Temperature variations affect the output voltage. This type of zener has a thermal coefficient, according to its specification, of 25 μV/K, but in the batch there is some compensation. The measured temperature coefficient of a set of 35 zeners was 10 μV/K, which leads to an uncertainty due to the temperature factor of 0.03 μV/V, k=1.

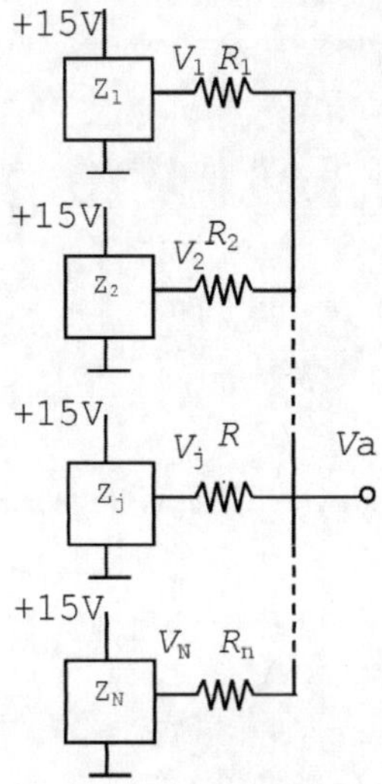

Figure 1. Schematic circuit of the developed device.

As a second uncertainty source, the variation in long time of the resistors is considered. They have 0.1% tolerance and under controlled temperature and practically zero power dissipation, the value is under that limit during years. A variation ΔR_j in the resistor R_j causes a change on the output voltage ΔV_a^j as

$$\Delta V_a^{\ j} = \frac{V_a - V_j}{N}\frac{\Delta R_j}{R_j} \tag{1}$$

The voltage differences between individual zeners are lower than 3 mV, then $|Va\text{-}Vj| < 3$ mV, and the value of ΔV_a^j will be lower than 2×10^{-8} V. This voltage represents a relative variation of 2 parts in 10^9.

Each group of 35 zeners is supplied by a regulated 15 V source. The influence of the stability of these sources was evaluated according to the zener specifications and with experimental corroboration. A variation of the source in 1 V produces a variation in V_a lower than 1 μV/V. To get an uncertainty contribution of 1 part in 10^8, the maximum variation of these power supplies must be 10 mV. Each source can be externally monitored to be within 15 V ±10 mV, but if any of them changes, there is an external adjust to set the voltage at its nominal value with a resolution of 1 mV.

Figure 2 shows the first prototype, which has been working since 2004. Another one was constructed in 2011, shown in figure 3. The average voltages of all independent groups of 35 zeners were continuously monitored separately, and differences between them were computed by a self-developed software. Each group has an independent output (four large binding posts in figure 2 and figure 3). They can be used as separate sources, or they that can be connected in parallel to get the analog average value. In this last case, the average value of all sections is automatically got. Two scanners and two multimeter, controlled by the software, measure the voltages and temperature of each source. The first scanner has 10 channels with very low contact emfs [4]. It selects a pair of voltages from the eight groups of the two prototypes and two commercial voltage standard sources. The output voltage of the scanner is the voltage difference between the two selected sources, which is connected to a 6½ digits multimeter [5] in the range of millivolt. Another less precision scanner with a similar multimeter measures the temperatures of the sources through the resistance of the thermistor each source has. Each measurement takes 20 s, which leads to 40 min for a complete test.

A conventional system of shielding and guarding was used.

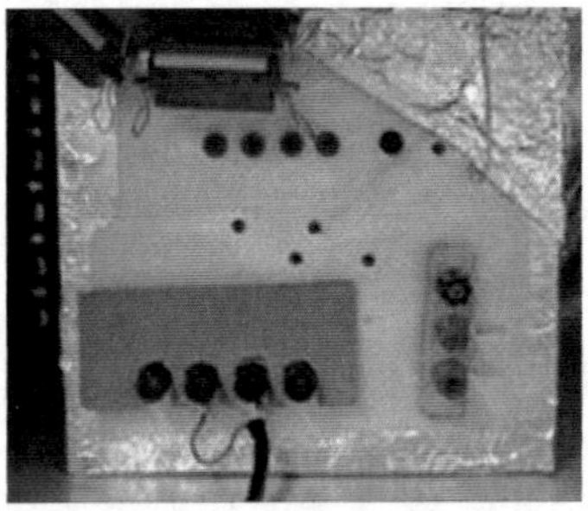

Figure 2. First prototype of the voltage standard source.

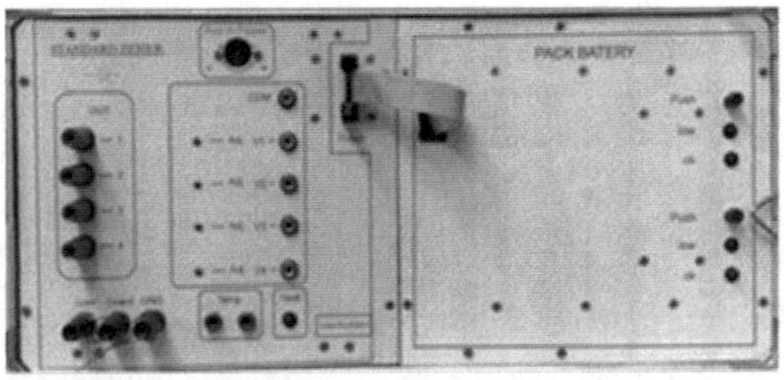

Figure 3. Second prototype of the voltage standard.

3. Stability

To study the long term stability, two types of comparisons were done: against Josephson standards and between different groups of zeners. Up to now, four calibrations against Josephson standards of external laboratories were done on the first prototype in 2004, 2008, 2010 and 2015. It was calibrated using as traveling standard a voltage source FLUKE 732B and an 8½ DMM (Agilent 3458) as comparator. Table 1 and figure 4 show the results. The second column of the table shows the value of the average voltage of the four groups, and the third column its uncertainty. It varies because these calibrations were done in different National Laboratories. The average value of the four groups was stated as the value of the prototype. The variation between 2004 and 2008 was -0.12 μV/V, and between 2010 and 2015 was 0.05 μV/V. However, between 2008 and 2010 a larger variation is observed, 0.64 μV/V. It was due to a failure in the temperature control.

Table 1. Calibrations against a Josephson standard.

Date	Average value (V)	Uncertainty (k=2) (μV)	Corrected value (V)
21/08/2004	10.0006014	2.9	
29/01/2008	10.0006002	2.5	
01/08/2010	10.0006066	1.3	10.0005999
07/05/2015	10.0006071	1.3	10.0006004

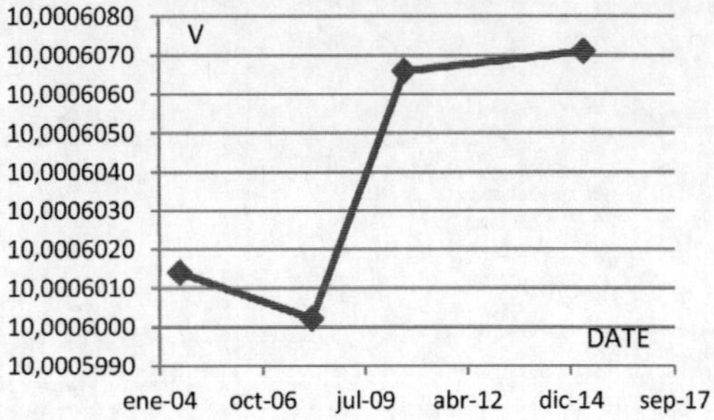

Figure 4. Calibration values against Josephson standards.

The source was out of service for nine months (10/2008 to 06/2009). Once repaired, it has been in operation until present. To correlate its voltage values previous and after the out-of-service period, the source was compared against a FLUKE 732B 10 V standard source taking into account its drift (-2.0 μV/year). The voltage variation due to repair was determined in 6.7 μV. The fourth column of Table 1 and figure 5 shows the corrected values, subtracting that voltage step. In this way, all values can be analyzed as if no break had occurred. The maximum variation during 11 years was 0.16 μV/V, with a drift of -0.009 μV/V per year which is well below the calibration uncertainties (0.13 μV/V in the best case).

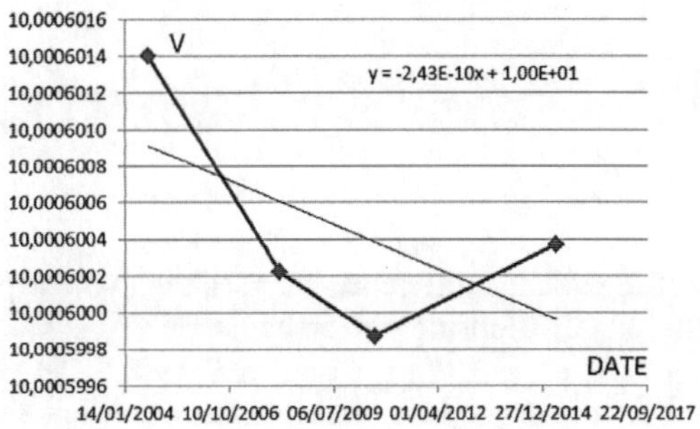

Figure 5. Corrected voltages, subtracting the variation due to repairs.

4. Noise

Other measurements were done to estimate the noise of the source. The voltage of one half of the source (70 diodes) was compared against the other half during 6 years. Allan deviation was computed and results are shown in figure 6 and figure 7. The general behavior at long term (figure 6) shows a non-white noise, similar to other zener based sources. In this case, the noise has the shape of $1/f^{1.5}$, approximately, lightly higher than flicker noise. The maximum value of Allan deviation, at one year, is 0.6 μV (6 parts in 10^8). In short term (figure 7), the noise is quite 1/f. The value of Allan deviation at 4 h, is 0.1 μV which is equivalent to 1 part in 10^8.

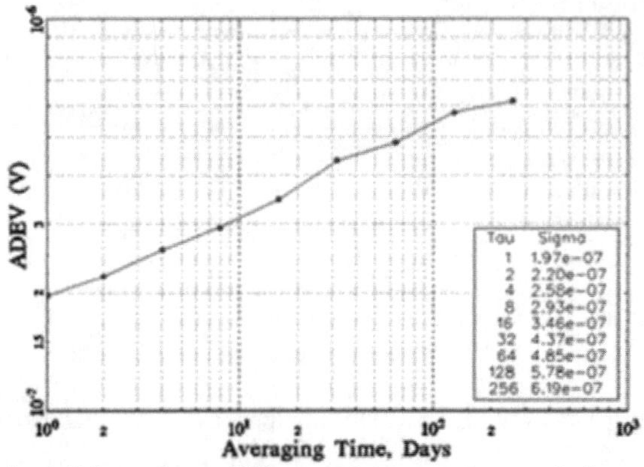

Figure 6. Allan deviation of the voltage difference between the two halves of the source, observed during six years.

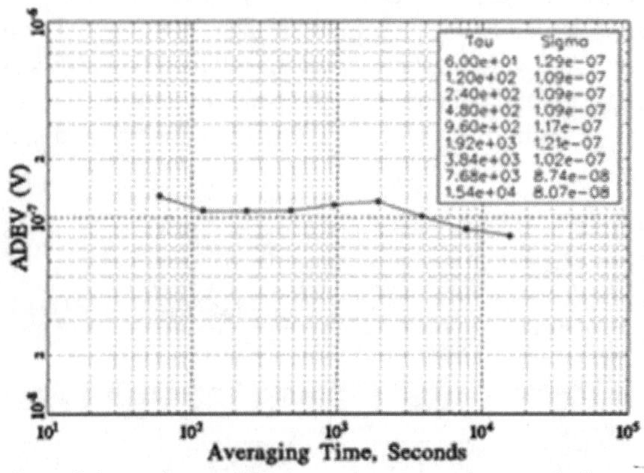

Figure 7. Allan deviation of the voltage difference between the two halves of the source, observed during sixteen hours.

Assuming that both halves have independent similar noise, the Allan deviation for the total source (140 diodes) is reduced by a factor of 2, leading to 3×10^{-8} V/V at long term and 5×10^{-9} V/V at short term. These values are smaller than most commercial zener voltage standards [1].

5. Conclusions

A proposed multi-zener based voltage-standard-source was analyzed during 11 years of operation. Its internal noise was evaluated in 0.03 µV/V at one year averaging time, and 0.005 µV/V at 4 h averaging time. The maximum variation during 11 years was 0.16 µV/V, with a drift of -0.009 µV/V per year which is well below the calibration uncertainty.

References

[1] http://us.flukecal.com/products/electrical-calibration/electrical-standards/732b-direct-voltage-standard-0?quicktabs_product_details=2

[2] Slomovitz D, Trigo L, De Souza H , "A Multi-Zener 10 V Voltage Standard – preliminary results", VI SEMETRO Digest, pp. 284-286, Rio de Janeiro, Brasil Set. 2005.

[3] Slomovitz D, Trigo L, De Souza H, "Long term behavior of a multi-zener 10 V voltage standard," Conference on Precision Electromagnetic Measurements (CPEM), pp. 211 – 212, 2010.
[4] http://mintl.com/DC/Products/Current/Scanners
[5] http://www.keysight.com/en/pd-1000001295:epsg:pro-pn-34401A/digital-multimeter-6-digit?&cc=UY&lc=eng

Comparison of two different methods to build micropotentiometers for low-voltage AC-DC transfer

G M Geronymo

Instituto Nacional de Metrologia, Qualidade e Tecnologia - Inmetro, Duque de Caxias, Brazil.

E-mail: gmgeronymo@inmetro.gov.br

Abstract. This paper shows a comparison between two different approaches to build micropotentiometers (μpots) for AC-DC transfer measurements at low voltages. An overview of μpots is presented, and two different approaches to build them are introduced. Two μpots built using the different approaches are compared using a calibrated thermal transfer standard, and the results are presented and discussed.

1. Introduction

In the past, AC-DC low voltage calibration (2 mV to 100 mV) was performed at Inmetro using a commercial thermal transfer standard traceable to another National Metrology Institute (NMI) [1]. In the last few years we started the implementation of a new system, using micropotentiometers (μpots). Two different approaches were adopted: a μpot using a Planar Multijunction Thermal Converter (PMJTC) [2] and a disk resistor (in-house built) [3, 4] and a "discrete μpot" [5], using commercial current shunts as output resistors. In this paper we present an overview of the two technics and compare the obtained results.

2. Micropotentiometers

The electrical circuit of a μpot, using a PMJTC as the thermal converter, is presented in figure 1.

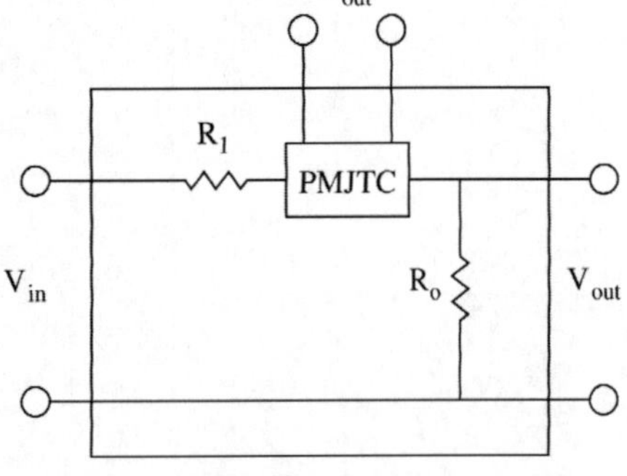

Figure 1. Micropotentiometer circuit. V_{in} is the input voltage, V_{out} is the output voltage, R_1 is the input resistor, R_o is the output resistor and PMJTC is the thermal converter, with heater resistance R_{pmjtc}. The output voltage of the PMJTC is E_{out} and the voltage applied to the PMJTC is V_{pmjtc}.

The device should work up to 1 MHz, so the output resistor should have a nearly flat frequency response up to that value. It is important, also, that its resistance value be small, to reduce the loading when connecting to a device with a finite input impedance [6]. Note that the input voltage (V_{in}) must be calculated in function of the μpot parameters to obtain the desired voltage at the output (V_{out}).

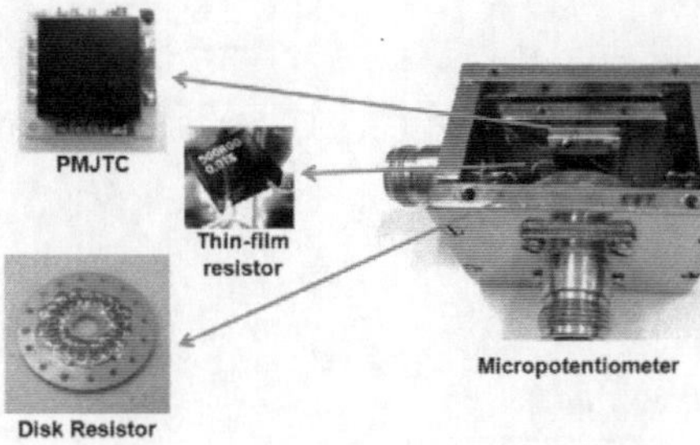

Figure 2. Disk resistor μpot.

Figure 3. Discrete μpot.

According to the circuit shown in figure 1, the output voltage of the μpot is given by equation 1 and the voltage applied to the PMJTC is given by equation 2.

$$V_{out} = \frac{V_{in} R_o}{R_1 + R_{pmjtc} + R_o} \tag{1}$$

$$V_{pmjtc} = \frac{V_{in} R_{pmjtc}}{R_1 + R_{pmjtc} + R_o} \tag{2}$$

The PMJTC used has a nominal heater resistance of 90 Ω, and is rated for a nominal input voltage of 1 V. The values of the input resistor (R_1) and the output resistor R_o need to be chosen to satisfy two main conditions: when the output voltage is the nominal voltage of the μpot, V_{pmjtc} should not exceed the maximum rating of the PMJTC; and when V_{out} is at the minimum working value, V_{pmjtc} need to be sufficient for a satisfying measurement of E_{out}.

For this work, a μpot for the range from 20 mV to 100 mV will be designed. Two different approaches to implement the output resistor (R_o) will be explored: (a) building a disk resistor by soldering SMD resistors on a disk-shaped PCB and (b) using a commercial current shunt.

The μpot using a disk resistor is shown in figure 2. R_1 is a thin-film resistor of 500 Ω, and the disk resistor was built using several SMD resistors in parallel. The nominal value of the disk resistor is 10 Ω. The housing is a brass rectangular box, with female N-type connectors for input and output.

The discrete μpot uses the same base circuit shown in figure 1, with only one difference: the output resistor, R_o, is a commercial current shunt. For the range from 20 mV to 100 mV, the current shunt used has a nominal resistance of 8 Ω. The housing of the discrete μpot was built using a brass cylinder, with N-type connectors. A thin-film resistor of 500 Ω was connected in series with a 90 Ω PMJTC. The discrete μpot can be seen on figure 3.

Table 1 shows the relations calculated by equations 1 and 2 and the output resistance for the built μpot.

Table 1. Electrical specifications for the μpots.

μpot	V_{out}/V_{in}	V_{pmjtc}/V_{in}	$R_o[\Omega]$
disk resistor	0.0166	0.150	10
discrete	0.0134	0.150	8

Table 2 shows the needed input voltage to obtain the maximum (100 mV) and minimum (20 mV) output voltages designed for the μpots. The voltage applied to the PMJTC (V_{pmjtc}) in both cases is shown, also.

Table 2. Input and output voltages.

μpot	V_{in} [V]	V_{out} [mV]	V_{pmjtc} [V]
disk resistor	6.02	100	0.90
	1.20	20	0.18
discrete	7.46	100	1.12
	1.49	20	0.22

3. Results

The μpots were measured against a calibrated commercial thermal standard. The measurement setup using the disk resistor μpot is shown on figure 4, and the measurement system using the discrete μpot can be seen on figure 5.

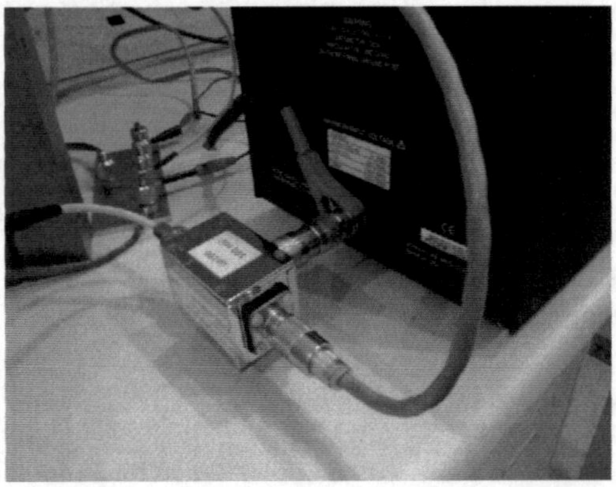

Figure 4. System with disk resistor μpot.

Figure 5. System with discrete μpot.

For each frequency, 12 measurements were made. The results for the disk resistor μpot and for the discrete μpot, corrected with the AC-DC difference of the standard, are presented in table 3.

Table 3. Measured AC-DC differences in μV/V.

	f [kHz]	0.01	0.065	1	10	100	1000
Disk resistor	60 mV	17	18	-3	-1	17	85
μpot	20 mV	2	17	-8	-1	14	82
Discrete	60 mV	20	20	-6	-7	-17	-261
μpot	20 mV	3	17	-4	-8	-18	-262

Table 4. Expanded uncertainties in μV/V.

f [kHz]	0.01	0.065	1	10	100	1000
60 mV	35	23	23	23	28	81
20 mV	51	26	25	26	32	85

The expanded uncertainties are presented in table 4.

4. Conclusions

The stability of the measurements using both μpot is similar. For 60 mV, the standard deviation of the 12 measurements for each frequency was around 1 ppm. For 20 mV, the standard deviation was around 5 ppm.

The disk resistor μpot has smaller AC-DC difference for high frequencies than the discrete μpot, as can be seen on tables 3 and 4. But, on the other hand, the discrete μpot is easier to be built, and a good connection between the output resistor and the rest of the system is guaranteed by the N-type connector. In the case of the disk resistor μpot, this connection is more difficult to be made, especially between the outer part of the disk and the housing (figure 2).

The results show that both approaches have similar results, with the advantage of easier construction for the discrete μpot.

References

[1] Ribeiro-Silva C I, Landim R and Silveira F 2011 Proposta de implantação de um sistema de padronização primária de transferência térmica ac-dc de baixa tensão no Inmetro *XVIII IMEKO TC4 Symposium and IX Semetro*

[2] Klonz M and Weimann T 1989 *Instrumentation and Measurement, IEEE Transactions on* **38** 335–337 ISSN 0018-9456

[3] Ribeiro-Silva C I and Geronymo G M 2013 Inmetro's micropotentiometer-based system for low-voltage calibration *X Semetro 10th International Congress of Electrical Metrology*

[4] Ribeiro-Silva C I and Geronymo G M 2013 Sistema de padronização primária de transferência térmica ac-dc de baixa tensão do Inmetro *7º Congresso Brasileiro de Metrologia*

[5] Geronymo G M, Ribeiro-Silva C I, Souza R M and Klonz M 2014 Ac-dc transfer measurements in the mv-range using discrete micropotentiometers *Conference on Precision Electromagnetic Measurements (CPEM 2014)*

[6] Klonz M and Funck T 2007 *IEEE Transactions on Instrumentation and Measurement* **56** 468–471 ISSN 0018-9456

Development of a system based on open source technology for DC resistor calibration

G M Geronymo and M C Silva

Instituto Nacional de Metrologia, Qualidade e Tecnologia - Inmetro, Duque de Caxias, Brazil.

E-mail: gmgeronymo@inmetro.gov.br

Abstract. This work present the development of a new system, based on open source technology, for the automation of DC resistor calibration. The new system is web-based, stores the measurement registers on a structured database and has new features that can increase the productivity of the laboratory. Some proposes of future development are presented, also.

1. Introduction

The Electrical Standardization Metrology Laboratory (Lampe), part of the National Metrology Institute of Brazil (Inmetro) is, among its tasks, responsible for providing traceability in DC resistance to all other laboratories and industries in Brazil. To accomplish this, the laboratory has automatic resistance bridges that cover a large range of resistance values, at different voltage and current levels.

The treatment of the results obtained by the resistance bridges, including the uncertainty calculation, was carried out by a simple software written in Visual Basic 6, and the registers was stored in simple text files. Due to the difficulty in the maintenance of that software and the incompatibility with newer versions of the Operating System (Microsoft Windows XP), it was decided to develop a new system.

The new system is multi-platform, stores the registers in a structured way, using an Open Source Data Base Management System (DBMS), and has new features that can increase the productivity of the laboratory. A comparison between the old system and the new system is summarized in table 1.

Table 1. New system versus old system.

Features	New System	Old System
Operating System	Multi-plataform	Windows 98/XP
Register storing	Open Source DBMS	Simple text-files
Read bridge report files	Yes	No
Modular Structure	Yes	No

The purpose of this paper is to present the progress obtained in the development of the new system, compare its features with the previous system and propose new improvements for future works.

2. System Description

2.1. System Architecture

The new system uses client-server architecture, illustrated in figure 1.

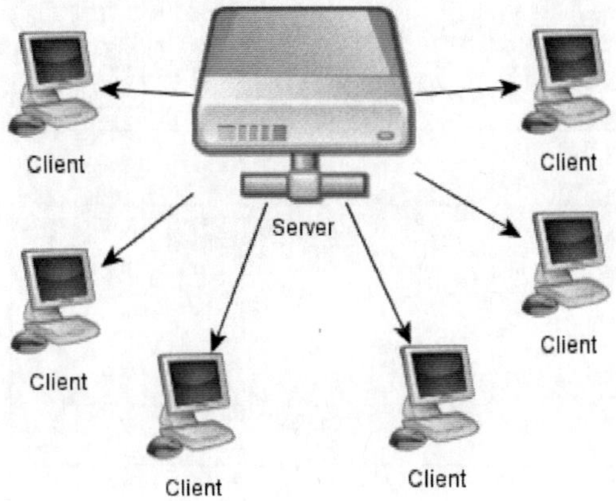

Figure 1. System model.

The clients can access the system using a Web Browser, and all the calculations are carried out in the server, in a centralized way. The server is based on a typical LAPP (Linux, Apache, PHP and PostgreSQL) stack, using Scilab for mathematical calculations, and runs on an Oracle VirtualBox virtual machine. As the system runs on a virtual machine, it is very easy to make backups of the full system and install it on other machines, if needed. This is very useful to create a production environment separated from the development environment, for example.

PHP is a programming language originally created for the implementation of dynamic contents in Web pages [1]. The language is provided with many extensions to interact with other applications, especially Web servers and databases (Apache and PostgreSQL, in our case).

Scilab is free and open source software for numerical computation [2]. It has a high level programming language and includes hundreds of mathematical functions, allowing faster development of scientific applications.

The concept of combine the efficient mathematical processing provided by Scilab with a web interface using PHP was already discussed [3]. The basic idea is to join the versatility of server-side scripting languages like PHP, Python and Ruby with the robustness of languages aimed at scientific computing, like Scilab, R and Matlab. This can be achieved in PHP using the function exec(), that permits the execution of a external program and capture its output. PHP can be used to control instruments via GPIB, also [4].

The following open source software packages are employed in this system: Oracle VirtualBox (virtual machine), Debian Linux (operating system), Apache (web server), PHP 5 (scripting language), PostgreSQL (DBMS), Scilab (scientific computation), Gnuplot (generation of graphics), ImageMagick (image manipulation) and Git (version control).

2.2. System modules

The system was built with independent modules (figure 2), in order to make the maintenance and addition of new features easier. All the mathematical processing is done in a small Scilab program. The interface between this program and the PostgreSQL database is managed by PHP scripts. In a similar way, the user interface is presented in HTML, with dynamic code generated by PHP.

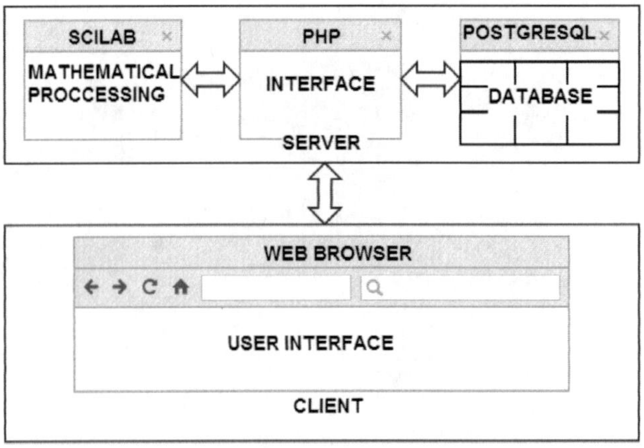

Figure 2. Integration of the system modules.

2.3. System Features

All the old system features were retained in the new system. Those features include the corrections of the resistance value due to environmental conditions, calculation of the drift of the standard and uncertainty calculations. All the calibration information of the instruments used for environmental conditions measurements and of the standard resistors was stored in text files in the old system. A tool was written to translate that data to the format of the new database.

One of the most useful new features included in this system is the ability to automatically read the measurement reports (text-files) generated by the automation system of the resistance bridges. As each resistance bridge uses a different format to present the measurement results, a specific reading function is required for each one. Up to now, this feature is compatible with the bridge Guildline 6675A and with the bridges 6000A, 6000B, 6010B and 6010D from Measurement International Limited (MIL). This feature not only speeds up the calibration process, avoiding the need to manually enter the values, but also stores in the database all the information provided by the measurement report, and not just the ratio Rx/Rs, as in the old system. Another useful feature is the possibility to store in the database observations and even pictures of the measurement setup. All this information can be easily retrieved later, when needed.

Storing all the calibration results in a structured database makes easy to plot the historical behavior of the calibrated resistors, for example. In order to add this feature to the system, it is being studied an efficient way to insert all the old calibration data into the database.

The user interface for the client computers is a dynamic web page, and can be accessed from a web browser. That interface is shown on figure 3.

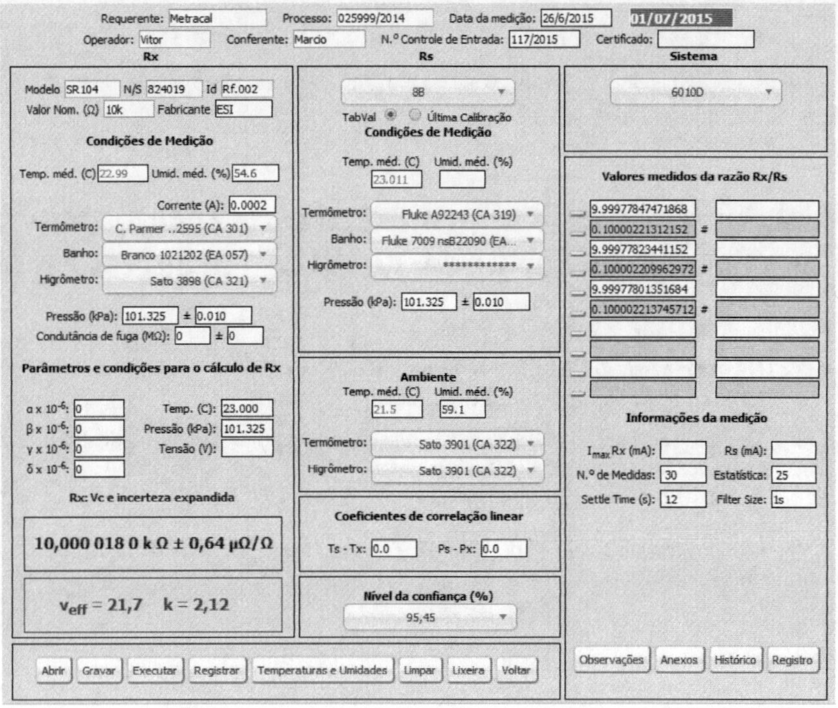

Figure 3. Client user interface.

3. Future Work

The next planned features to be added to the system include automatic generation of history plots and automatic generation of calibration certificates. For the history plots, the biggest challenge is to import all the old data of resistor calibration. This data is stored in different formats (in some cases, only in printed paper). For the automatic generation of calibration certificates, it is being studied an approach to use PHP functions that permit to interact directly with Microsoft Word files (the format used by the institution certificate templates).

4. Conclusions

The new system of automation of DC resistor calibration was presented. The systems new features can speed up the analysis of the results given by the resistance bridges, improving the productivity of the laboratory. Some other features have to be implemented yet, moving to an almost full automated process.

References

[1] *PHP: Hypertext Preprocessor.* Available at `http://php.net`. Accessed: 2015-07-20

[2] Scilab Enterprises 2015 *Scilab: Free and Open Source software for numerical computation.* Orsay, France. Available at `http://www.scilab.org`. Accessed: 2015-07-19.

[3] Lewis D 2010 *Efficient mathematical processing using Scilab through PHP.* IBM Developer Works. Available at `http://www.ibm.com/developerworks/library/os-php-scilab/os-php-scilab-pdf.pdf`. Accessed: 2015-07-19

[4] Santos J, Mendona J and Martins J 2008 Instrumentation remote controls through internet with PHP *IEEE International Conference on Virtual Environments, Human-Computer Interfaces, and Measurement Systems* (Istambul, Turkey)

Trilateral South-American project: a reference system for measuring electric power up to 100 kHz – progress report

G A Kyriazis[1], L Di Lillo[2], D Slomovitz[3], R Iuzzolino[2], E Yasuda[2], L Trigo[3], H Laiz[2], R M Debatin[1], A M R Franco[1], E Afonso[1]

[1] Instituto Nacional de Metrologia, Qualidade e Tecnologia, Av Nossa Senhora das Graças 50, Duque de Caxias, RJ 25250-020, Brazil
[2] Instituto Nacional de Tecnología Industrial, Av Gral Paz 5445, B1650KNA, San Martín, Buenos Aires, Argentina
[3] Administración Nacional de Usinas e Transmisiones Eléctricas, Paraguay 2385, Montevideo – 11800, Uruguay

gakyriazis@inmetro.gov.br

Abstract. Three countries in South America are jointly developing a reference system for measuring electric power up to 100 kHz. The objective is the construction of three similar measuring systems, one for each institute. The measuring system is described and its design requirements are presented. This project will contribute to provide calibration services in measuring ranges still not covered by the three institutes.

1. Introduction

In the last decades, the increased use of nonlinear loads in power distribution networks has caused a significant increase in the waveform harmonics degrading the quality of the power delivered to the customer. The reliable measurement of waveform harmonics requires refined techniques.

The nonlinear loads include the new technologies of compact lighting and the use of switching converters. The latter are used both in transmission and distribution networks in power electronics applications. Such loads also include switched-mode power supplies (SMPS) used in computers. SMPS switch voltages (in the order of 400 V) in frequencies in the range of kilohertz producing harmonics beyond 1 MHz.

Comparisons on harmonic power are being planned at the level of the Consultative Committee on Electricity and Magnetism (CCEM) and of the Inter-American System of Metrology (SIM).

To address this concern, the Instituto Nacional de Metrologia, Qualidade e Tecnologia (Inmetro), in Brazil, the Instituto Nacional de Tecnología Industrial (INTI), in Argentina, and the Administración Nacional de Usinas e Transmisiones Eléctricas (UTE), in Uruguay, are jointly developing a reference system for measuring electric power up to 100 kHz. The objective is the construction of three similar measuring systems, one for each institute. The project is supported in part by the Conselho Nacional de Desenvolvimento Científico e Tecnológico (CNPq), in Brazil.

This project will contribute to provide calibration services in measuring ranges still not covered by the three institutes. The project will also contribute to improve the traceability not only of electric power but also of related quantities like ac-dc transfer, voltage ratio, phase angle, ac voltage and ac current.

2. Measuring system

Calibration systems for harmonic power analyzers and power quality measuring instruments generally consist of a digital generator with two independent channels (one for voltage and one for current). Arbitrary voltage and current waveforms are programmed and applied to the instrument under calibration. The amplitude ratio and phase displacement of the two signals can be adjusted by changing the generator settings. Voltage dividers and current shunts with flat responses in all the frequency range of interest are used to convert with high accuracy the high voltages and currents to the relatively low voltages required by the digitizer inputs. The harmonic parameters of such waveforms and the electric power are estimated from the digitized data and the resulting estimates are compared with the readings of the device under calibration.

The reference measuring system comprises (see figure 1): an arbitrary waveform function generator with two synchronized channels, a power amplifier, a transconductance amplifier, a resistive voltage divider, a current shunt and a digitizer with two synchronized channels.

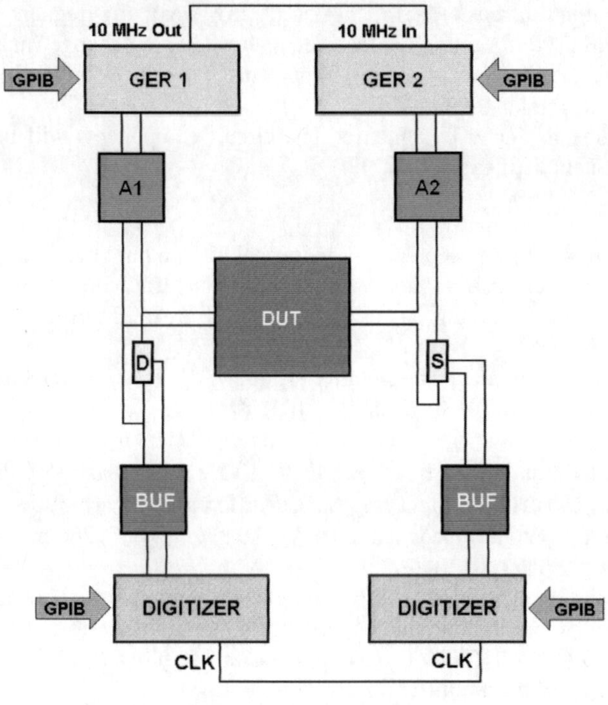

Figure 1. Simplified block diagram of the reference measuring system. (GER1 – channel 1 of the arbitrary waveform function generator, GER2 – channel 2 of the arbitrary waveform function generator, A1 – power amplifier, A2 – transconductance amplifier, D – resistive voltage divider, S – current shunt, BUF – buffer, DIGITIZER – two-channel digitizer, CLK – clock, and DUT – device under calibration)

2.1 Arbitrary waveform function generator

Arbitrary waveform function generators based on direct digital synthesis (DDS) present many advantages in comparison with other techniques. The DDS technique consists in the digital processing to generate signals at different frequencies and phases selectable by software from a reference clock. It shows a higher frequency resolution and a lower harmonic distortion than those obtained through other techniques (for instance, phase-locked loop). The frequency stability in the DDS technique depends only on the external reference oscillator employed, thus allowing multiple devices to be synchronized.

The design requirements are [1, 2, 3]: (a) two synchronized channels; (b) low total harmonic distortion: −80 dBc (@ 0.23 V, 1 kHz); (c) frequency resolution: 1.4 nHz; (d) frequency stability: 3.9 µHz (@ 0.23 V, 62.5 Hz, single-tone signal); (e) frequency stability: 25 µHz (@ 0.23 V, 1 kHz); (f) amplitude stability: 2.0 µV (@ 0.227902 V, 62.5 Hz, single-tone signal); (g) amplitude stability: 13 µV/V (@ 0.23 V, 1 kHz); (h) frequency range: DC to 150 kHz (with a slight increase in distortion); (i) internal reference clock: 400 kHz +/– 20 Hz; (j) output amplitude: 1 V, 2 V, 5 V and 10 V peak-to-peak; (k) output signal waveforms: single-tone sinusoidal, two-tone sinusoidal, distorted sinusoidal, squarewave, triangular and saw-tooth.

The generator circuits are currently being designed by INTI. The circuit components will be purchased by Inmetro as soon as the component listing is defined.

2.2 Power amplifier

The power amplifier is used to boost the output of a channel of the dual-channel generator to a nominal of 120 V rms or 240 V rms. In addition, the amplifier is designed to supply up to 100-mA rms current to accommodate the burden requirements without causing any significant error. A prime goal is to maintain the excellent short-term amplitude and phase stability inherent in the digital generator. In addition, there is a general need for a precision power amplifier that will provide an output swing of several hundred volts at moderate power levels while maintaining good dc characteristics, stable gain, and wide bandwidth.

The design requirements are [4]: (a) voltage gain: 40 (nominal); (b) maximum output swing: +/– 485 V, (340 V rms); (c) maximum output current: +/– 145 mA, (100 mA rms); (d) frequency response: DC to 100 kHz, 3 dB (240 V rms, 100 mA rms); (e) load regulation: 20 ppm change (null load to full load @ 120 V rms, 1 kHz); (f) output noise: less than 50 mV rms (100 kHz bandwidth); (g) input impedance: 10 kΩ; (h) maximum slew rate: 400 V/μs.

The amplifier circuits are currently being designed by Inmetro. The circuit components will be purchased by Inmetro as soon as the component listing is defined.

2.3 Tranconductance amplifier

A transconductance amplifier ideally produces a current in a load proportional to an input voltage and maintains that current independent of the load terminal voltage. A total of 4 (four) transconductance amplifiers are being designed each covering the following ranges (full scale): 20 mA, 200 mA, 5 A, and 20 A.

The design requirements are [5, 6, 7]: (a) transconductance stability (10 min): +/– (0.005% reading + 0.005% range) @ 10 Hz to 10 kHz and +/– (0.010% reading + 0.010% range) @ 10 kHz to 100 kHz; (b) input voltage range (full scale): 5 V; (c) frequency range: 10 Hz to 100 kHz; (d) compliance voltage: 7 V rms; (e) total harmonic distortion: –60 dB @ 10 Hz to 10 kHz, –50 dB @ 10 kHz to 40 kHz, –40 dB @ 10 kHz to 100 kHz; (f) input impedance (differential input): 500 kΩ from each input terminal to chassis ground; (g) output impedance: 20 mA – 10 kΩ // 1.4 nF, 200 mA – 10 kΩ // 1.4 nF, 5 A – 10 kΩ // 30 nF, 20 A – 2.5 kΩ // 120 nF.

The transconductance amplifiers will contain only analog circuits to reduce electrical noise. Linear power supplies are utilized in all amplifiers except the one that covers the 20-A range also to reduce electrical noise. Vishay Z-Foil resistors (VTA) with TCR of ±0.05 ppm/°C are employed in critical circuit locations to increase the output stability of the amplifier.

The amplifier circuits were designed by Inmetro. The printed circuit board (PCB) layout has been started by Inmetro. The circuit components are currently being purchased by Inmetro.

2.4 Resistive voltage divider

There are 9 (nine) resistive voltage dividers to be used one at a time whose (binary) nominal value (from 4 V to 1024 V) depends on the test voltage selected [8]. An additional 1024-V divider is being also designed with different power consumption for evaluation purposes. The (binary) input resistances range from 1 kΩ @ 4 V to 256 kΩ @ 1024 V. Their input capacitance is always 33 pF. Thus the power amplifier output current is always 0.004 A. The divider output resistance is always 200 ohms. Thus the nominal output voltage of the dividers is always 0.8 V.

The rated voltages were selected with binary ratios to ease the step-up calibration of the amplitude and phase errors with the dual-channel digitizer and a standard binary inductive voltage divider. The nominal ranges of the dividers are obtained from the voltage ratios 5, 10, 20, 40, 80, 160, 320, 640, and 1280, and the 0.8-V output voltage.

To get high accuracy in all ranges of voltage and frequency, it is necessary that the divider behaves as a linear device. The most important nonlinear effect in voltage dividers is dielectric losses, The voltage dividers employ Vishay Z-Foil audio resistors (VAR) with low dissipation factor whose dissipated power in each resistor does not exceed 100 mW. It was confirmed by UTE that the nonlinear effects are reduced considerably by soldering such resistors on PTFE PCBs [8]. Such boards have a very low dissipation factor. Once nonlinear effects have been minimized, stray capacitances are the most relevant linear parasitic effect. In this work a different shielding technique for nulling radial electric fields is explored [8].

The voltage divider circuits were designed by UTE. The circuit components were purchased by Inmetro and transported to UTE. The PCB layout has been started by UTE.

2.5 Current shunt

There are 12 current shunts to be used one at a time whose nominal value (20 mA, 50 mA, 100 mA, 200 mA, 0.5 A, 1 A, 2 A, 5 A, 10 A, 20 A, 50 A and 100 A) depends on the test current selected [9, 10]. Their input resistances range from 40 Ω @ 20 mA to 0.008 ohm @ 100 A. Thus the nominal output voltage of the shunts is always 0.8 V.

The current shunts employ Vishay Z-Foil resistors (Z201) with TCR of ±0.05 ppm/°C whose dissipated power in each resistor does not exceed 0.33 W.

The current shunt circuits were designed by INTI and Inmetro. The circuit components were purchased by Inmetro and are waiting an authorization from Argentinian customs to be embarked to INTI. The PCB layout was finished by INTI and the PCBs are currently being manufactured in Brazil.

2.6 Dual-channel digitizer

A high-resolution, digital sampling system based on a Sigma-Delta analog-to-digital converter (Σ–Δ ADC) has been developed and characterized [11]. The system has been modified to accommodate two synchronized channels.

The "equivalent sampling rate (ESR)" is the frequency with which the samples are outputted. As the Σ–Δ ADC works through oversampling, ESR = *fs*/OSR, where *fs* is the clock frequency and OSR is the oversampling ratio. The ADC allows a maximum clock frequency of 20 MHz and an OSR between 32 and 256, that is, 20 MHz / 32 = 625 kHz is the maximum ESR.

The design requirements are [11]: (a) two synchronized channels; (b) dynamic regime - total harmonic distortion (THD): < −80 dBc (@ −51 dBFS (4.382 mV), 1 kHz); signal-to-noise ratio (SNR): > 55 dBc (@ −51 dBFS (4.382 mV), 1 kHz); signal-to-noise-plus-distortion ratio (SINAD): > 55 dBc; effective number of bits (ENOB): 16 bits; intermodulation distortion (IMD): < −75 dBc (@ −57 dBFS (2.180 mV), f_1 = 1 kHz and f_2 = 1.25 kHz); settling time: 24 samples or 375 µs (@ OSR = 256, ESR = 64 kHz); (c) stationary regime – noise referred to input (rms): < 10.25 LSB or 1.2 µV (@ ESR = 64 kHz); noise-free code resolution: 18 bits; (d) overall bandwidth of the digital sampling system (ADC + digital filters): 250 kHz; (e) input impedance: 10 MΩ // 60 pF; (f) other important characteristics: flat frequency response, high linearity, minimum temperature dependence, minimum dc offset, and auto-calibration.

The digitizer circuits were designed by INTI. The PCB layout has been started by INTI. The circuit components were purchased by Inmetro and will have been transported to INTI by the time this article is published.

3. Project management

The project is coordinated by Inmetro, who is responsible for the purchase and transportation of the components and parts needed to the construction of the modules, and for the transportation of the modules needed to the assembly of the measuring systems, in each partner country.

Inmetro is also responsible for the development of the wideband power and transconductance amplifiers and of new digital sampling algorithms. INTI is responsible for the development of the arbitrary waveform function generator, the dual-channel digitizer and the current shunts (and their calibration). UTE is responsible for the development of the resistive voltage dividers (and their calibration).

The project activities include the design, layout and documentation of the printed circuit boards of the modules, the design and documentation of the electronic packaging of the modules, the calibration of resistive voltage dividers and current shunts, the design and documentation of the software and firmware, and the measuring system integration, testing and documentation.

Acknowledgments
This work is supported in part by the Conselho Nacional de Desenvolvimento Científico e Tecnológico (CNPq), of the Ministry of Science, Technology and Innovation of Brazil, under Grant CNPq/Prosul Processo Nº 490271/2011-1.

References

[1] Adad W 2010 *Investigación sobre la generación de señales arbitrarias por el método de síntesis digital directa para su aplicación a la metrología eléctrica*, Tese de Ingeniería, Facultad de Ingeniería, Universidad de Buenos Aires.

[2] Adad W and Iuzzolino R 2012 Arbitrary function generator using direct digital synthesis *CPEM Digest (Washington)* 622 ISBN 978-1-4673-0440-5

[3] Adad W and Iuzzolino R 2012 Low distortion signal generator based on direct digital synthesis for ADC characterization *ACTA IMEKO* **1** 59

[4] Laug O B 1987 A precision power amplifier for power/energy applications *IEEE Trans. Instrum. Meas.* **36** 994

[5] Laug O B 1985 A wide-band transconductance amplifier for current calibrations *IEEE Trans. Instrum. Meas.* **34** 639

[6] Laug O B 1996 A 100 A, 100 kHz transconductance amplier *IEEE Trans. Instrum. Meas.* **45** 440

[7] Hess D T and Clarke K K 1998 Evaluation of 100 Ampere, 100 kHz transconductance amplifiers *CPEM Digest (Washington)* 611

[8] Slomovitz D, Trigo L, Aristoy G, Brehm M and Kyriazis G A 2015 Wideband resistive voltage divider for a standard wattmeter *IOP Journal of Physics: Conference Series (JPCS) – this issue.*

[9] Yasuda E, Di Lillo L and Kyriazis G A 2015 Design and construction of new shunts for a wideband sampling wattmeter *Proc. Metrologia 2015 - XI SEMETRO (Bento Gonçalves)* ISBN 978-85-86768-10-1

[10] Rydler K -E and Tarasso V 2008 Extending ac-dc current transfer measurement to 100 A, 100 kHz *CPEM Digest (Boulder)* 28 ISBN 978-1-4244-2399-6

[11] Iuzzolino R 2011 *Josephson waveforms characterization of a Sigma-Delta Analog-to-digital converter for data acquisition in metrology*, Dr.-Ing. Dissertation, Technische Universität Braunschweig TU-BS Diss. ISBN 978-3-86387-095-9.

Long-term research in Inmetro on samples of quantum Hall resistance standards made by PTB

H R Carvalho[1], R E M Briones[1], K Pierz[2] and M Götz[2]

[1] Quantum Electrical Metrology Laboratory, Inmetro, Duque de Caxias, Brazil
[2] Physikalisch-Technische Bundesanstalt (PTB)

E-mail: hrcarvalho@inmetro.gov.br

Abstract. This paper shows up to date results of Inmetro's investigations on aging effects of quantum Hall samples fabricated by PTB in the frame of a mutual scientific agreement established between the Brazilian and German National Metrology Institutes.

1. Introduction

A mutual agreement on scientific cooperation in the frame of quantum Hall resistance measurements has been signed up between Inmetro and PTB on the 21st of December 2012 which will hold for the next five years. Related to this, there are two samples with appropriate semiconductor properties made by PTB team [1, 2]. These samples were sent to Inmetro in order to be periodically measured with its quantum Hall system (QHS). The resulting reports will be used by the PTB for improvement of quantum Hall samples manufacture, as well as by Inmetro for the fully characterization and knowledge of the behaviors of these two samples (which will stay at Inmetro after the end of the agreement). The two samples were labeled as P137-23 and P137-24 by PTB.

These samples were, prior to delivery, characterized by PTB at 2.2 K. This characterization and Inmetro's up to date results are presented in this paper.

2. Quantum Hall measurements

The Quantum Electrical Metrology Laboratory (Lameq) contains two modules. One of them is responsible for making measurements and research with quantum Josephson Voltage Standards (conventional and programmable ones). The other one is responsible for making measurements and research with quantum Hall system (QHS) to provide, in the near future, traceability to SI ohm [3-5]. The tasks reported here take place in this second module.

It is equipped with two cryostats. One of them, called "Quantum Hall Sample Cryostat" or simply "Sample Cryostat", is able to cool the sample down to 0.3 K (by pumping above liquid He-3) and enables magnetic field intensities of up to about 16 T. The equipment has a current source for the superconducting coil and a temperature control in three specific locations in cryostat.

The second Lameq's cryostat, in this second module, is equipped with a cryogenic current comparator (CCC), which is used to compare resistors (including the quantum Hall samples) with a relative measurement uncertainty of about 10^{-8}.

Additional instruments, common to QHS, were also used as an 8 ½ digit digital multimeter, a two-channel 7 ½ digit digital nanovoltmeter, a Zener voltage reference and a resistive decade.

A Labview application was developed in order to control the instruments and acquire measurements. It can change the field by controlling the coil current source and measures the Hall voltage (V_H) and the longitudinal voltage (V_{xx}) of the sample with the nanovoltmeter [6].

3. Measurements and results

3.1. Contact resistance of both samples

All contact resistances of both samples were measured by "three-terminal measurement technique" [6]. First, a "short-circuit sample" was put instead of the actual samples in order to find the resistance of each wire. The wire resistance measurements were processed under conditions identical to the samples (temperature was 0.4 K and the magnetic field was the same as used in plateau 2). Then, the samples were placed and the value of wire resistance was subtracted. All measurements were carried out with current through the sample in both directions to eliminate offsets of the measuring equipment. The current value used was 9.91 μA, because this value does not exceed the maximum allowed for all contacts [6].

The contacts of both samples show very low resistances yielding values around 0 Ω within a relative standard deviation of about 8×10^{-6} (measurement limited by sensitivity of the nanovoltmeter).

3.2. Results of the P137-23 plateaus scan

The graph of figure 1 shows three measurements (R_H and R_{xx}) while scanning the magnetic field up to near 14 T. They are related to two P137-23 sample measurements made in Inmetro-Lameq in December 2013 and one made in PTB before shipping.

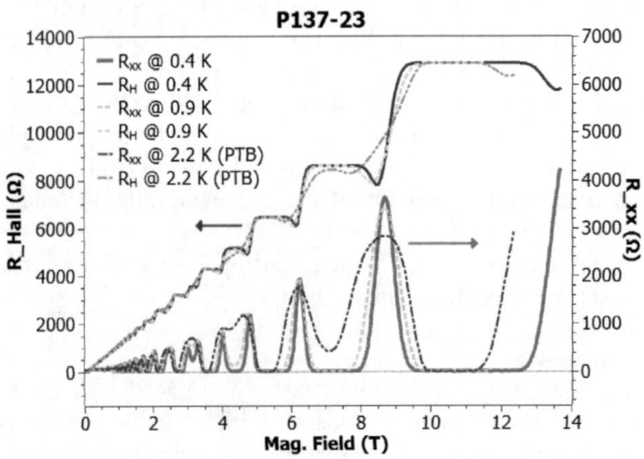

Figure 1. Two P137-23 measurements made in December 2013 and one made in PTB.

The graph in figure 2 shows P137-23 measurements made under temperature of 0.4 K in April 2015.

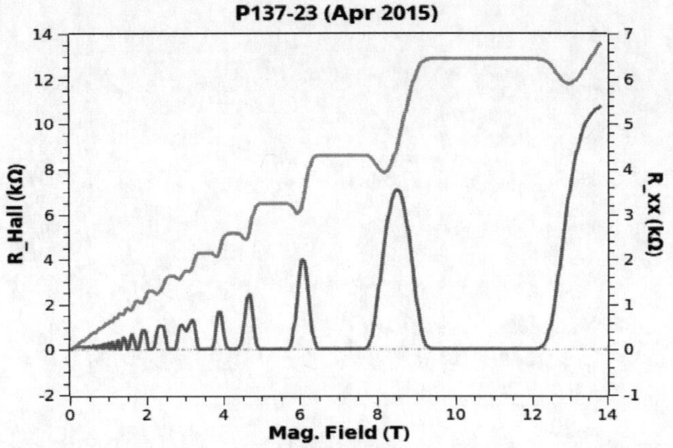

Figure 2. P137-23 measurements in April 2015 (R_H above and R_{xx} below).

Figure 3 shows detail of R_{xx} graph in figure 2 at plateau 2 region for both current directions.

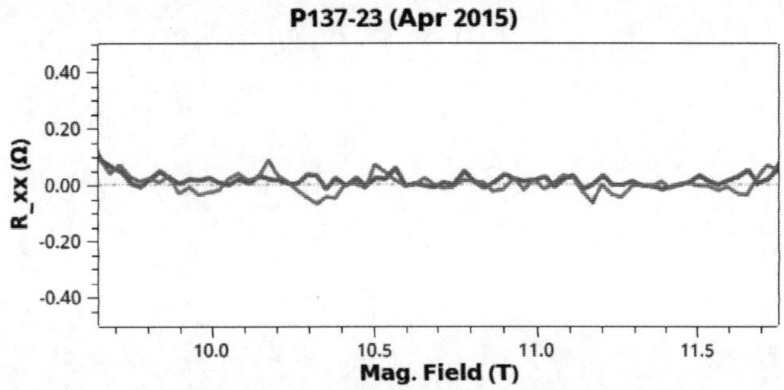

Figure 3. Graph of R_{xx} detail from P137-23 measurements (figure 2) at plateau 2 region in both current directions.

The measurement values between 10 T and 11.5 T are around zero and the standard deviation of these values is

$$\sigma = 2.5 \times 10^{-2}\ \Omega. \tag{1}$$

3.3. Results of the P137-24 Plateaus Scan

The same procedure was made with sample P137-24. First, figure 4 show the measurements made in December 2013 and one made in PTB before shipping.

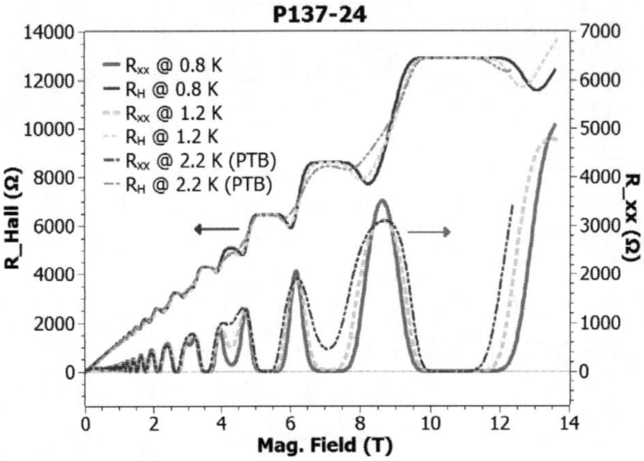

Figure 4. Two P137-24 measurements made in December 2013 and one made in PTB.

The next graph shows the measurements in April 2015 (figure 5).

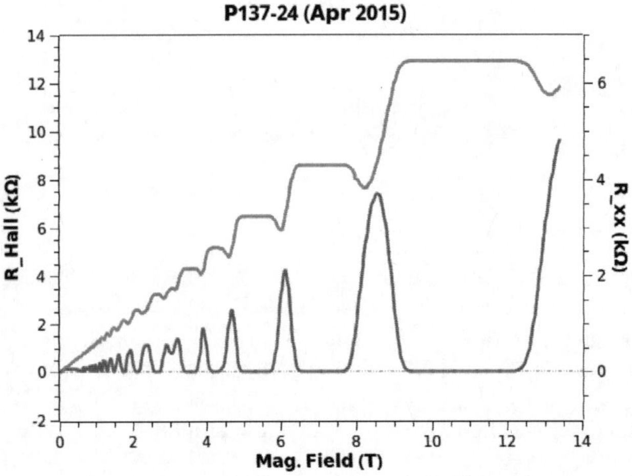

Figure 5. P137-24 measurements in April 2015 (R_H above and R_{xx} below).

The figure 6 shows detail of the R_{xx} graph at plateau 2 region from figure 5.

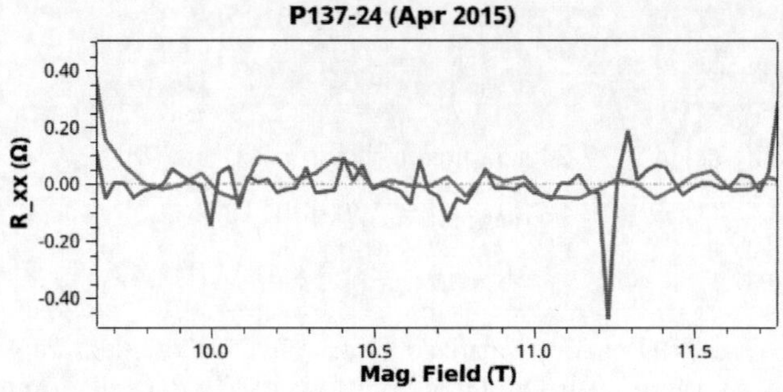

Figure 6. Graph of R_{xx} detail from P137-24 measurements (figure 5) at plateau 2 region in both current directions.

In this case, the measurement values between 10 T and 11.5 T are also around zero and the standard deviation is

$$\sigma = 7.3 \times 10^{-2}\ \Omega. \tag{2}$$

3.4. Comparison using CCC

In March 2015, these two samples were used to calibrate the standard resistor of 100 Ω using the CCC. Figure 7 shows these two measurements (the two points to the right) together with the historical evolution of this standard resistor (circle and square points and linear adjustment straight line) calibrated by using ordinary QHR samples.

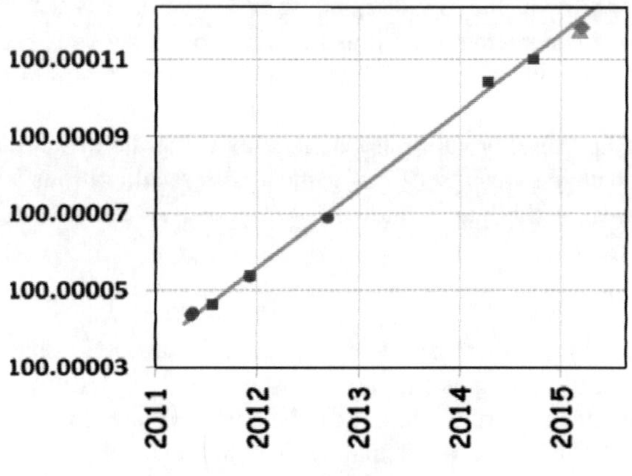

Figure 7. Historical evolution of 100 Ω standard resistor with two PTB sample calibrations (the two points to the right, diamond and triangle).

The value of standard resistor predicted by historical for March 2015 and its expanded uncertainty are shown in (3) and (4).

$$R_{100\Omega} = 100.0001211\ \Omega \tag{3}$$

$$U_{100\Omega} = 4.2 \times 10^{-6}\ \Omega \tag{4}$$

The value obtained by calibration, made in March 2015, of same standard resistor by using the P137-23 sample is shown in (5) and its expanded uncertainty is shown in (6).

$$R_{100\Omega(23)} = 100.00011764\ \Omega \quad (5)$$

$$U_{100\Omega(23)} = 6.3 \times 10^{-7}\ \Omega \quad (6)$$

The same as above by using P137-24 sample are shown in (7) and (8)

$$R_{100\Omega(24)} = 100.00011826\ \Omega \quad (7)$$

$$U_{100\Omega(24)} = 4.0 \times 10^{-7}\ \Omega \quad (8)$$

The relative difference (RD) and normalized error (NE) [5] of measurements made by using samples P137-23 are shown in (9) and (10), respectively. The results from similar procedure, concerning to P137-24 sample, are shown in (11) and (12).

$$RD_{100\Omega(23)} = \frac{R_{100\Omega(23)} - R_{100\Omega}}{R_{100\Omega}} = -3.5 \times 10^{-8} \quad (9)$$

$$EN_{100\Omega(23)} = \frac{|R_{100\Omega(23)} - R_{100\Omega}|}{\sqrt{U^2_{100\Omega(23)} + U^2_{100\Omega}}} = 0.81 \quad (10)$$

$$RD_{100\Omega(24)} = -2.8 \times 10^{-8} \quad (11)$$

$$EN_{100\Omega(24)} = 0.67 \quad (12)$$

4. Conclusion

These results show that the two samples, made by PTB and sent to Inmetro according to agreement between these two institutes, which were measured, have worked as expected. and Lameq contributed to improve calibration procedures for both laboratories, allowing not only to reduce uncertainties as well as to increase the reliability of the measurement results.

The contact resistance is undetected as well as the Vxx on the plateaus region. The plateaus width increase when the temperature goes down and have not had perceptive changing during these three years.

The results of the standard resistor calibrations made by using the official laboratory quantum Hall sample were compared with the results of the same resistor calibrations made by using the PTB samples. The results, shown in this paper, indicate that both PTB samples have worked as expected and have similar behavior.

Acknowledgments

The authors would like to thank JT Janssen from NPL and Thomas Holubar from Cryogenic Ltd. to assist in improving our understanding about Hall cryostat and CCC and to recover some problems of these Inmetro's QHS elements during technical visits in our laboratory in the course of 2013. Furthermore, we would like to thank Waldemar G. K. Ihlenfeld and Robson Gomes to assist us in reviewing the text.

References

[1] Pierz K and Schumacher B 1999 *IEEE Trans. Instrum. Meas. Phys*. **48-2** 293
[2] Pierz K, Götz M, Pesel E, Ahlers F and Schumacher H W 1999 *IEEE Trans. Instrum. Meas. Phys*. **60-7** 2455
[3] Carvalho Z V, Schechter H and Fernandes J B 2007 Análise e perspectivas dos processos de medição do laboratório de resistência elétrica do Inmetro *METROSUL, Conf. Digest*
[4] Carvalho H R, Silva J R B and Briones R E M 2013 Estabelecimento da base da escala do Ohm no Inmetro a partir da padronização quântica de resistência *X Semetro*
[5] Carvalho H R, Briones R E M and Silva J R B 2015 *J. Phys.: Conf. Ser.* **575** 012014
[6] Delahaye F and Jeckelmann B 2003 *Metrologia* **40** 217

Sigma phase detection in duplex stainless steel by induced magnetic field

E M Silva[1], W M L Fialho[1], H F B M Silva[2], J P Leite[2], J P Leite[3] and A V G Oliveira[4]

[1]Instituto Federal de Educação Ciência e Tecnologia da Paraíba, Av. Primeiro de Maio, 720 - Jaguaribe, João Pessoa - PB, 58015-435, Brazil. Email: edgard@ifpb.edu.br, wmlfialho@gmail.com

[2]Universidade Federal da Paraíba Jardim Universitário, s/n - Castelo Branco, João Pessoa - PB, 58051-900, Brazil. Email: josinaldo@ct.ufpb.edu.br, hicarofelipe@gmail.com

[3]Universidade Federal de Campina Grande, R. Aprígio Veloso, 882 - 58429-900, Brazil. Email: joao.leite100@yahoo.com.br;

[4]Instituto Federal de Educação Ciência e Tecnologia do Ceará, R. José Quintino, s/n - Prado, Cedro - CE, 63400-000. Email: venturagol14@gmail.com

Abstract. Duplex stainless steels when subjected to temperatures above 600 ° C have its tenacity decreased by the formation of sigma phase. This phase has high hardness and is rich in chromium and reduces the matrix of this element. In this study, field line density measurements, obtained in the reversibility region of magnetic domains, and application of artificial neural networks are used to monitor the formation of this undesirable phase. Samples of a stainless steel SAF 2205 were subjected to aging at temperatures of 800 ° C and 900 ° C, in order to obtain different amounts of sigma phase. The amount of this phase was obtained by image processing and the density of field lines through a Hall Effect sensor. Charpy impact tests were performed. The field lines densities were used for training of an artificial neural network and correlated with the presence of sigma phase and embrittlement of the material. The results showed that the method was able to correlate the parameters studied with the presence of the sigma phase and toughness of the material studied in both temperatures.

1. Introduction

Duplex stainless steels possess desirable properties of both the face-centered cubic (austenitic) and body-centered cubic (ferritic) phases within their microstructures. The ferrite and austenite phases are present in roughly equal volume fractions. They have excellent strength and toughness, improved corrosion resistance (especially to localized corrosion) and exceptional resistance to halide stress corrosion cracking. However, when these steels are heated above 600 ° C there is the formation of non-magnetic embrittlement σ phase. This has hardness around 900 Hv and is rich in chromium. An amount of 3% of this is able to compromise the material toughness [1-7].

The most important metallurgical process in duplex stainless steels is the eutectoid decomposition of the ferrite (δ) to sigma (σ) phase and secondary austenite (γ) through the reaction (δ - σ + γ 2) which happens due to thermal effects (Tavares, 2010).[3].

The ferrite phase is ferromagnetic in contrast to the paramagnetic austenite and sigma phases. Therefore, the magnetic methods seem to be adequate for testing duplex stainless steels, since they are sensitive to the amount and structure of the ferromagnetic ferrite phase. The intensive use of ferromagnetic materials within the engineering applications makes the loss of magnetic flux to be one of main techniques widely used to determine the presence of discontinuities [8, 9]. The disturbance of the magnetic flux near the discontinuities enables the contactless sensing of the vertical component of the magnetic field. Hall sensors have been used for detecting size and position of cracks in materials. These sensors are also sensitive detection of surface flaws in metallic materials, especially under ac excitation (Bi, 1998) [10].

Cavalcante, et al., 2008 [11], used Hall voltage measurement for training an artificial neural network, in order to make microstructures identification. The two techniques combined proved to be promising. The results were based on application of an external field, no saturation of the samples and therefore no measurement of the remaining field.

Some previous works have also proved the applicability of magnetic measurements on the detection of spinodal decomposition of the ferritic phase and the presence of σ in duplex stainless steels [12, 13]. The magnetic susceptibility was found to decrease during prolonged thermal aging due to precipitation of both phases. The non-magnetic σ phase reduce the permeability of the materials and so the magnetic susceptibility.

In this study, field line density measurements, obtained in the reversibility region of magnetic domains and application of artificial neural networks are used to monitor the formation of this undesirable σ phase. A duplex stainless steel 2205 with different amount of σ phase were used.

2. Experimental procedures

Samples of a SAF 2205 steel were subjected to treatments at temperatures of 800 ° C and 900 ° C for times of 0.25, 1 and 2 hours. Its surfaces were attacked using the Behara and 10% KOH reagents and analyzed through an optical microscopy (NIKON FX 35XD OM). The amount of sigma phase was determined by analysis of surfaces etched with the KOH reagent subjected to image processing by segmenting images and quantification by an image processing program developed in the laboratory. Forty images were captured and the volume fraction was determined with a 5% confidence interval. The microstructures were also analyzed by scanning electron microscopy (SEM) for microstructural characterization. Charpy impact test was performed on the samples subjected to the same treatments. A specimen of the as received material was also studied.

The measurements of induced fields in the material due to the interaction between an external field and the one were carried out as the configuration of the Fig. 1.

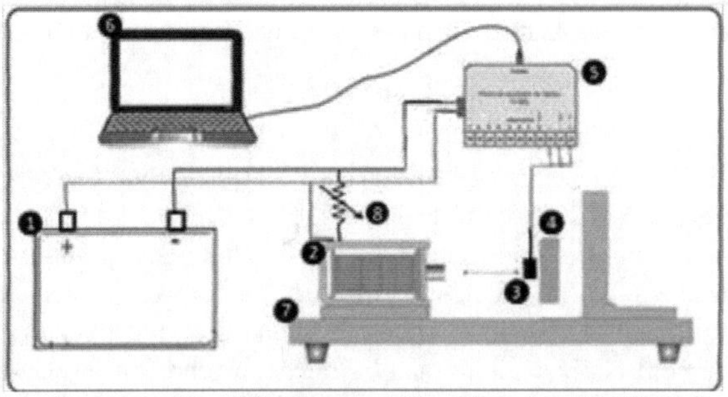

Figure 1. Experimental equipment of test: (1) Power supply; (2) Solenoid; (3) Hall sensor; (4) Sample; (5) Data acquisition board; (6) Computer; (7) Bench test; (8) Potentiometer.

It has a solenoid which is responsible to generate the external magnetic field. The magnetic flux density is determined by a Hall Effect sensor (SS495A model).

The neural network applied was a feed forward backpropagation type with feedback from output to input, with tansig activation function, supervised learning by error correction, standard mode, with the entrance with two neurons formed by induced fields and voltage values applied to the solenoid. Twenty neurons were used in the intermediate layers and an output with the absorbed energy values. The objective was to determine the impact energy of the samples analyzed from the applied voltages and field induced in the working region of the sensor.

3. Results and discussions

Optical and electron micrographs of the SAF 2205 steel for treatment condition of 800 ° C and 2 hours are shown in Fig. 2 and 3. Figure 2 shows an image of the material surface attacked with KOH and with Behara in figure 3.

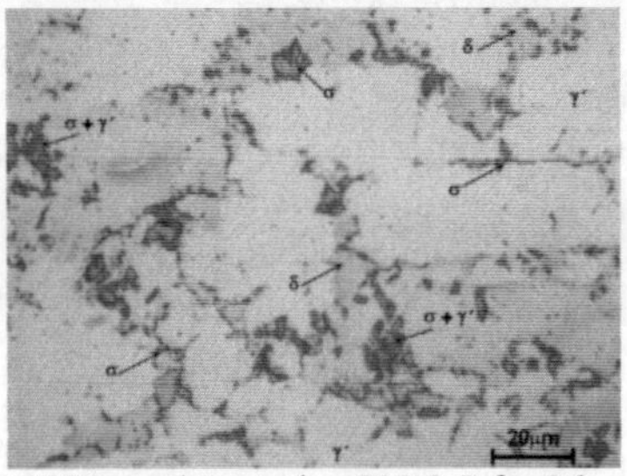

Figure 2. Micrograph of SAF 2205 steel treated at 800 ° C for 2 hours obtained by MO (Attack: KOH).

The precipitation of σ phase and the eutectoid reaction δ - > σ + γ' can be noted in both figures. Tavares, et al., 2010[3], studying the same steel found that σ phase can precipitate in the (δ / γ) grain boundary and also from the ferrite δ . The transformation in the interfaces (δ / δ) and (δ / γ) were observed in Fig. 3. Chen [1] and Tavares [3] observed that this phase precipitates is the only one revealed by electrolytic corrosion with KOH. In the regions of discontinuous precipitation the σ and γ'phases are formed side by side and as the σ grain boundary is attacked the γ' seems to revealed.

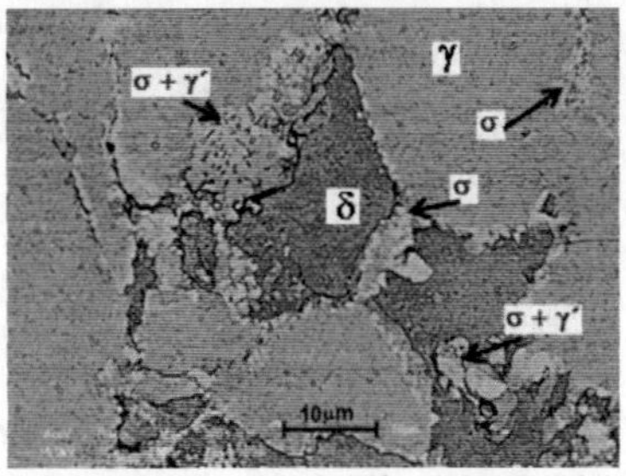

Figure 3. Micrograph of SAF 2205 steel treated at 800 ° C for 2 hours obtained by SEM (Attack: Behara).

By analyzing treated at 900 ° C for 2 hours in figure 4 it appears the same precipitation but in larger quantity due to the higher kinetic.

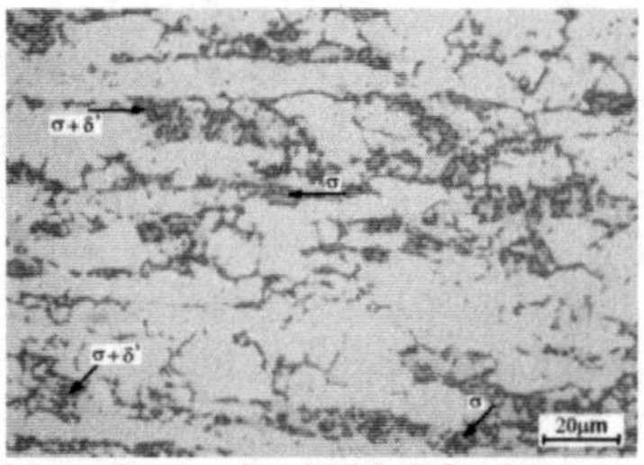

Figure 4. Micrograph of SAF 2205 steel treated at 900 ° C for 2 hours obtained by MO (Attack: KOH)

The embrittlement of duplex stainless steel in the temperature range from 600 oC to 1000 o C is due to the presence of sigma phase. This has a hardness about 1000 HV and a percentage of 4 % is sufficient to reduce the toughness around 90 % [14]. Figures 5 shows the change of the impact energy, induced magnetic field and the amount of σ phase, for a voltage of 10 V applied on the solenoid. It was plotted in figure 5 the values obtained at 800 oC and 900 oC temperature. From the analysis of Fig. 5 it can be seen that the absorbed energy and induced magnetic field values decreased with increasing of the amount of σ phase. Lo et al (2007)[12, 13] observed similar behavior studying the effect of formation of sigma phase by magnetic susceptibility measurements.

Figure 5 shows that there is a critical value of the induced field to the embrittlement of the material. A value of 690 Gauss was found to a 10 voltage applied in the solenoid. In order to correlate the different voltage values with the absorbed energy, a neural network was constructed, so any solenoid voltage could be used for future monitoring application.

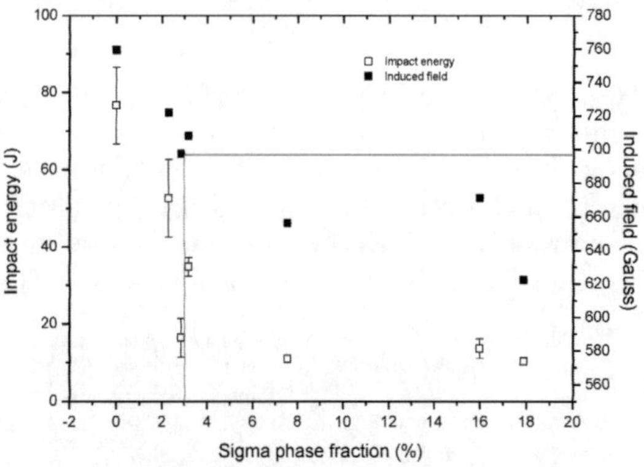

Figure 5 Change in absorbed energy and induced field due to the amount of sigma pass for a 10 V voltage.

Figure 6 shows the simulated results with the applied voltage, showing that there is a good correlation between them. However, some differences in simulated values were observed, but this can be improved by increasing the amount of data acquisition.

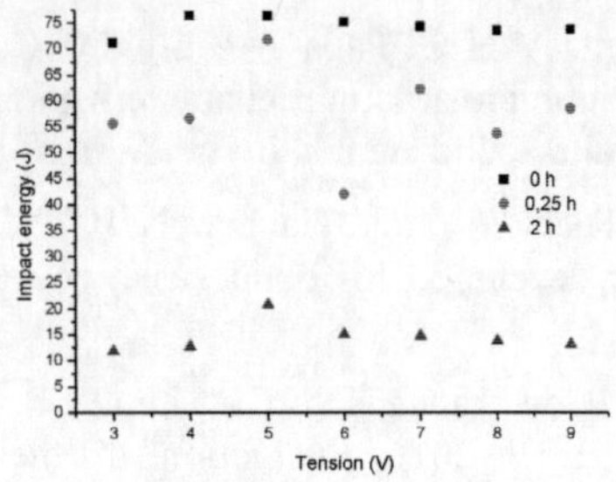

Figure 6. Simulated results of the neural network of impact energy.

4. Conclusions

A methodology formed by the use of induced magnetic field and neural network was applied to follow the σ phase transformation in an SAF 2205 duplex stainless steel, leading to the following conclusions:

The use of magnetic flux densities in the reversible region of magnetic domains is able to detect variations in the duplex stainless steel microstructure.

There is a critic induced field able to define the low impact energy of the material. A value of 690 gauss was obtain is the one to a voltage of 10V.

The magnetic field combined with a neural network can be used to follow the embrittlement of the studied stainless steel.

Acknowledgements

The authors thanks to the financial support of IFPB and CNPq.

References

[1] Chen T H., Weng K L and Yang J R. 2002 The effect of high-temperature exposure on the microstructural stability and toughness property in a 2205 duplex stainless steel *Mater. Sci. Eng. A* **338**, p. 259.

[2] Jiang Z L, Chen X Y, Huang H and Liu X Y 2003 Grain refinement of Cr25Ni5Mo1.5 duplex stainless steel by heat treatment *Mater. Sci. Eng. A* **363**, p. 263.

[3] Tavares S S M, Pardal J M , Guerreiro J L, Gomes A M and Da silva, M R 2010 Magnetic detection of sigma phase in duplex stainless steel UNS S31803 *J. Magn. Magn. Mater.* **322**, p. 129.

[4] Mohapatra J N, Kamada Y, Murakami T, Echigoya J, Kikuchi H and Kobayashi S 2013 Magnetic hysteresis loop technique as a tool for the evaluation of σ phase embrittlement in Fe–Cr alloys *J. Magn. Magn. Mater* **327**, p. 71–75.

[5] Normando P G, Moura E P, Souza J A, Tavares S S M and Padovese L R 2010 Ultrasound, eddy current and magnetic Barkhausen noise as tools for sigma phase detection on a UNS S31803 duplex stainless steel *Mater. Sci. Eng. A* **527** p. 2886–2891.

[6] Camerini C, Sacramento R, Areiza M C, Rocha A, Santos R, Rebello J M and Pereira G 2015 Eddy current techniques for super duplex stainless steel characterization *J. Magn. Magn.*

Mater. **388** p. 96–100.

[7] Nunes T M, De Albuquerque V H C, Papa J P, Silva C C, Normando P G, Moura E P and Tavares J M R S Automatic microstructural characterization and classification using artificial intelligence techniques on ultrasound signals *Expert Syst. Appl.* **40** p. 3096–3105.

[8] Enokizono M, Tokada T, Tuchida Y and Hachiki N 1999 Finite element analysis of moving magnetic flux type sensor developed for nondestructive testing *IEEE. Trans. Magn.* **35** p. 1853.

[9] de Macedo Silva E, Leite J P, de França Neto F A, Leite J P, Fialho W M L, de Albuquerque V H C and Tavares J M R S 2016 Evaluation of the Magnetic Permeability for the Microstructural Characterization of a Duplex Stainless Steel *J. Test. Eval.* **44** p. 20130313.

[10] Bi Y and JILES DC 1998 Dependance of magnetic properties on crack size in steels *IEEE, Trans. Magn.* **34** p. 2021.

[11] Cavalcante A F, Albuquerque V H C, Silva E M, Santos N C and Silva A A 2008 Um estudo da aplicação de um sensor Hall e uma rede neural artificial para reconhecimento dos diferentes tipos de tratamento térmico pelo qual um aço SAE 1045 foi submetido *Iberian Latin American Congresso on Computacion Methods in Engineering – XXIX CILANCE. Maceio, Alagoas, Brazil.*

[12] Lo K H, Lai J K L, Shek C H and Li D J 2007 Effects of pre-treatment on the magnetic susceptibility and ageing behaviour of duplex stainless steels *Mater. Sci. Eng. A* **452** p. 78.

[13] Lo K H , Lai J K L, Shek C H and Li D J, 2007 Magnetic and transformation behaviour of duplex stainless steels under non-isothermal conditions and temperature- fluctuation monitoring *Mater. Sci. Eng. A* **452** p. 149 .

[14] Glosh S K and Mondal S 2008 High temperature ageing behaviour of a duplex stainless steel *Materials Characterization* **59** p.1776.

Software validation applied to spreadsheets used in laboratories working under ISO/IEC 17025

J M Banegas [1,2] **and M W Orué** [1,2]

[1] Laboratorio de Metrología del Litoral (LabMET) – UTN Santa Fe / CCT Santa Fe, Predio CONICET SANTA FE, Colectora R.N. nº 168, Paraje "El Pozo", 3000 Santa Fe, Argentina

[2] Grupo Control y Seguridad Eléctrica (CYSE) – UTN Santa Fe, Lavaisse 610, S3004EWB Santa Fe, Argentina

E-mail: jmbanegas@frsf.utn.edu.ar

Abstract. Several documents deal with software validation. Nevertheless, more are too complex to be applied to validate spreadsheets - surely the most used software in laboratories working under ISO/IEC 17025. The method proposed in this work is intended to be directly applied to validate spreadsheets. It includes a systematic way to document requirements, operational aspects regarding to validation, and a simple method to keep records of validation results and modifications history. This method is actually being used in an accredited calibration laboratory, showing to be practical and efficient.

1. Introduction

Different kind of software is widely used in ISO/IEC 17025 accredited laboratories to perform a variety of tasks. Some of them are as simple and not critical as writing a document. In other cases, the use of software involves critical data processing and registering raw data; in such cases, some aspects like confidence, access security, and risk assessment, start to matter.

In order to cover this broad range of activities, laboratories use many different software, ranging from standard software packages, to custom made programs. Within the whole universe of available software, spreadsheets are of particular interest, since virtually every laboratory – metrology labs in special – uses them to process data, and calculate uncertainties.

The core requirements about the use of computers are in ISO/IEC 17025:2005 [1], clause 5.4.7.2. In particular, it states that "the laboratory shall ensure that… computer software developed by the user is documented in sufficient detail and suitably validated as being adequate for use". Regarding commercial off the shelf software (COTS) – including spreadsheets – the Note clarifies that the software itself "in general use within its designed application range may be considered sufficiently validated. However, laboratory software configuration/modifications should be validated…".

Validation is defined in [1] clause 5.4.5.1, as the confirmation by examination and the provision of objective evidence that requirements for a specific application are fulfilled [1]. Similar definitions are given in others standards like ISO 9000 [2] and the VIM [3]. Thus, validation implies three main aspects: documented software requirements for the specific application, evaluation process (namely, validation procedure), and records to provide evidence.

Although a recognized spreadsheet doesn't need validation itself, the formulas, logic, programming, etc., must be validated, since they are configurations/modifications performed by the laboratory. Even more, aspects like conditional formatting and cell protection should be taken into account, if they are necessary to guarantee the adequacy for the intended use of the spreadsheet. Usually, this is called a modified off the shelf software (MOTS) [i.e. 4, 5].

Many documents provide guidance on software validation and verification, but only a few are intended to be applicable to laboratories working under ISO/IEC 17025 [i.e. 4-7].

The aim of this work is to present a straight-forward validation method specifically applicable to spreadsheets used in laboratories, including software specifications documentation, validation itself, and associated records. This method is being successfully applied in an accredited laboratory, showing to be practical and simple.

2. Software life cycle and validation

Software life cycle is closely related to validation. The approach presented in the standard IEEE 1012 considers validation and verification processes "performed in parallel with all life cycle stages, not at their conclusion" [8]. This approach is shared in other publications [9].

On the other hand, validation in ISO/IEC 12207 is one of the "software specific processes", and it's mostly performed at the end of the development [10].

In any case, it's clear that a suitable model of the software life cycle has to be adopted. Most of the models depicted in the references are too complex to be used with a spreadsheet. Nevertheless, three typical stages can be clearly identified: development, operation and maintenance. This is probably the simplest software life cycle model, and it's the model adopted in this work.

First, validation takes part in the development stage. The earlier part of development is to establish the requisites and objectives for the spreadsheet, taking into account its intended use. It also provides the basis for the validation, namely, the software requirements to be evaluated. And finally, regarding of other verifications performed during the spreadsheet programming, the development stage must finish with the validation, before operation.

Later, in the maintenance stage, full or partial validation must be performed every time a new version of the spreadsheet is to be released.

3. Validation aspects and procedure

Usually a particular spreadsheet is modified and upgraded several times during its life cycle. Keeping track of the changes and the validation records of each version can be a hard task. Moreover, as the calibration procedures are updated, the requirements for the associated spreadsheets may change.

In order to simplify the process and minimize the paperwork, the proposed method uses the spreadsheet file itself for documentation. It is done by adding to the workbook a dedicated worksheet for validation. This worksheet is used for requirements documentation, validation records, and history tracking, as stated below.

A typical 'validation' worksheet would look as shown in figure 1.

ID	Requisite	Cply.	Notes

Figure 1. Example of 'validation' worksheet.

3.1. Software requirements

Spreadsheet requirements regarding behavior and expected results must be clearly established in order to perform validation.

Many requirements are derived from the applicable documented calibration procedures. These describe how data has to be processed (including formulas, coefficients, constants, etc.) and how the results are presented. Other requirements may be specific for each spreadsheet. In any case, they all must be documented. Requisites should be grouped in categories as follows:

- Behaviour (B): functional aspects of the spreadsheet, i.e. conditional formatting, date/time formatting, proper functioning of lists for selectable data (combos), etc.
- Formulas and calculations (F): calculations the spreadsheet is intended to perform, including factors, coefficients, etc.
- Logic and automation (L): automated actions the spreadsheet is intended to perform (i.e. allow/block data entry in specific cells, search in a matrix, select data), when particular conditions are fulfilled (i.e. data selected in a combo, data entered with certain values).
- Security and protection (S): access level to sheets and/or cells, in order to avoid undue modifications to the spreadsheet.

Requisites should be written in the 'validation' worksheet as short and as explicit as possible. If necessary, they may have references to the respective documented procedure. Each individual requisite should be coded (category and sequential number) in order to facilitate referencing. Here are some examples:

- B01. Combo 'Working standard' lists every possible instrument and works OK.
- B02. 'Emission date' cell painted in red if previous than 'Cal date'.
- F01. Error formula in cells ZBxx according to procedure formula (3).
- F02. Results as expected, with typical data set.
- L01. Correct selection of set of specifications according sensor type.
- L02. Multiplies reference value and uncertainties x50 when 'coil' used.
- S01. Every worksheet protected.
- S02. Cells for data entry left unprotected.

3.2. Examination procedure

Every single requirement must be evaluated to compliance, recording the results in the 'compliance' column of the 'validation' worksheet. Depending on the requisite, a black-box or a white-box testing [11] is used.

In a **black-box** (or functional) test, results are evaluated in response to inputs, ignoring the internal working of the software. It's very useful to evaluate compliance with specified functional requirements. For instance, requisites B01, B02, F02, L01, S01, and S02, are checked by black-box testing.

The procedure is checking the results for consistency and compliance, when individual actions – like selecting data from combos, introducing known set of values, etc. – are performed. It is important to notice that embedded spreadsheet formulas (i.e. LOG, SUM, STDEV) are considered sufficiently validated, thus no further validation is required.

The first validation of a particular spreadsheet should always include a black-box validation consisting of introducing a typical data set with known output results. Data should include boundary points, and even unexpected values.

In a **white-box** (or structural) test, internal working of the software is taken into account. In spreadsheets this can be done evaluating the algorithms and formulas used, checking for adequacy. For instance, requisites F01 and L02 should be checked by white-box testing, namely, analyzing if the formulas are written correctly and refer to the right cells.

Both types of testing are complementary, and sometimes one particular requirement may be evaluated by any of them. The laboratory must establish the method of evaluation for each requirement.

3.3. Evaluation records

The spreadsheet itself constitutes the evidence of the validation, since the 'validation' worksheet contains both specifications and results.

In case of a full validation – or in any case that the spreadsheet was loaded with data to perform black-box testing – it's necessary to save the filled spreadsheet as evidence. A blank copy is then used for everyday work.

3.4. Re-validation triggers

Every time the spreadsheet is modified, the changes have to be validated before use. If changes are small, only a partial validation can be done, covering the requirements that could have been affected by the modifications.

The laboratory must keep the versioning history, specifying changes between versions, and thus establishing the requisites to re-validate. It can be easily done in the same 'validation' worksheet, or using a dedicated one. History must include the last full-validation date, and the validations performed in each new version since then (i.e. the requisites that were re-evaluated).

4. Conclusions

Spreadsheets used for processing raw calibration or testing data must be validated, before use and after any modification.

The proposed method covers the 3 main aspects of validation: requirements, examination and records. Paperwork is kept to a minimum, since documentation of requirements and validation records are part of the spreadsheet itself, and no additional documents are required.

It also shows a simple way to re-validate modified spreadsheets, keeping track of history and performing only partial validations. Thus, the method is flexible and efficient.

This method is being successfully used in an accredited metrology lab. Several spreadsheets – covering calibrations in the field of temperature, time, electricity, pressure, and humidity – were validated and re-validated, showing the method is practical, simple and affective.

References

[1] ISO/IEC 17025:2005, General requirements for the competence of testing and calibration laboratories.
[2] ISO 9000:2005, Quality management systems — Fundamentals and vocabulary.
[3] JCGM 200:2012, International vocabulary of metrology – Basic and general concepts and associated terms (VIM).
[4] Eurolab Technical Report No. 2/2006, 2006, Guidance for the management of computers and software in laboratories with reference to ISO/IEC 17025/2005.
[5] Gogates G D, 2012, Software Validation in Accredited Laboratories A Practical Guide, Fasor Inc. (www.fasor.com).
[6] Wichmann B, Parkin G and Barker R, 2007, NPL Report DEM-ES 014 - Software Support for Metrology Best Practice Guide No. 1 - Validation of Software in Measurement Systems.
[7] Torp C E, 2003, Nordtest Report TR 535 - Method of Software Validation.
[8] IEEE Std 1012-2012, 2012, IEEE Standard for System and Software Verification and Validation.
[9] ESA PSS-05-10, 1995, Guide to software Verification and validation, European Space Agency.
[10] ISO/IEC 12207:2008, Systems and software engineering – Software life cycle processes.
[11] ISO/IEC/IEEE 24765:2010, Systems and software engineering – Vocabulary.

Journal of Physics: Conference Series **733** (2016) 012072 doi:10.1088/1742-6596/733/1/012072

Wideband resistive voltage divider for a standard wattmeter

D Slomovitz[1], L Trigo[1], G Aristoy[1], M Brehm[1], G A Kyriazis[2]

[1]UTE Laboratory, Uruguay.
[2]Instituto Nacional de Metrologia, Qualidade e Tecnologia, Brazil.

dslomovitz@ute.com.uy

Abstract. This work shows the state of development of resistive voltage dividers for use in a standard wattmeter that covers a frequency range from 50 Hz to 100 kHz. The voltage ranges go from 4 V to 1024 V with an expected accuracy of 20 µV/V and 60 µrad/kHz. Nonlinear influences are studied and different methods to compensate the effect of stray capacitances are analysed, as well as the measurement systems used to calibrate these devices.

1. Introduction
The amount of nonlinear loads in power networks has been increased during the last decades. This leads to distortion in currents and voltages which can disturb communication and control systems, as well as other electronic devices connected to the network. We are running a project [1] for designing and constructing a reference system for measuring electric voltage, current and power up to 100 kHz.

This project is being jointly developed by the National Metrology Institutes of Brazil, Argentina and Uruguay (Instituto Nacional de Metrologia, Qualidade e Tecnologia (Inmetro) in Brazil, Instituto Nacional de Tecnología Industrial (INTI) in Argentina, and Administración Nacional de Usinas y Transmisiones Eléctricas (UTE) in Uruguay). The objective is the construction of three measuring systems, one for each institute. This project will contribute to provide calibration services in measuring ranges still not covered by these institutes. The project will also contribute to improve the traceability not only of electric power but also of related quantities like ac-dc transfer, voltage ratio, phase angle, ac voltage and ac current.

For the voltage input, this standard system requires voltage dividers to scale the input values from 4 V to 1024 V, to the voltage value admitted by the analogue to digital converter (0.8 V). It is not easy to design such dividers with very low error at the highest frequency. There are parasitic capacitances and inductances that produce changes in the voltage ratio in amplitude and in phase, which have also been studied in other electrical fields [2]. Among this, the resistors and isolating materials used in the divider have nonlinear losses, which lead to nonlinear behaviours that cannot be compensated under distorted voltages. The following sections show the progress in the development of these dividers.

2. General requirements
To cover all ranges (from 4 V to 1024 V), nine resistive voltage dividers will be used, one at a time, with binary nominal ratios. Table 1 shows details. The input resistance and the dissipated power depend on the input voltage. For all cases, the output resistance is 200 Ω and the output voltage is 0.8 V.

Table 1. Characteristic details of the set of the dividers.

Input voltage (V)	Ratio	Input resistance (kΩ)	Power (W)
1024	1280	256	4.10
512	640	128	2.05
256	320	64	1.02
128	160	32	0.51
64	80	16	0.26
32	40	8	0.13
16	20	4	0.06
8	10	2	0.03
4	5	1	0.02

In this way, the nominal current is constant for all dividers (4 mA at nominal voltage). The ratios were selected with binary ratios to allow the calibration against a standard binary inductive voltage divider [3] at low frequencies, and for covering the whole voltage range with the same step-ratio between each adjacent pair. The input arms are constructed with multiple individual resistors in parallel-series connection to limit the power of each one at 100 mW. Nonlinear effects and stray capacitances will be discussed in the following sections.

3. Linearity

To get high accuracy in all ranges of voltage and frequency, it is necessary that the divider behaves as a linear device. If the voltage waveform is distorted, as can be in this project, nonlinearities mean that it is not possible to correct the response in a general way for an arbitrary harmonic content. On the opposite, in linear circuits a correction can be done according to the sweep frequency response analysis or the step response.

In voltage dividers the most important nonlinear effect is dielectric losses. High electric fields through isolation material can cause power losses that vary with frequency and field intensity in a nonlinear way. Other nonlinear effects, such as skin effect, can be neglected in this application. The main components of the divider that contribute to nonlinear effects are resistors, printed circuit board (PCB) and capacitors used for compensation and guarding [4]. In resistors, the larger contribution to nonlinearities is caused by their isolation covers due to dielectric losses. In some models it is possible to remove the cover, and for others, the manufacturer can deliver special units without cover (see figure 1). This last option was applied in this work, and tests confirm the behaviour improvement. A resistor (Vishay Z-Foil type) of 10 kΩ, 0.6 W, was tested with and without cover, mounted in FR-4 PCB, PTFE PCB or mounted in air, without any PCB.

Figure 1. Resistors without cover.

The variation of the modulus of the impedance against the frequency is shown in Table 2. The values are expressed in parts in 10^6. The test was done using an ac-dc standard thermal transfer [5].

Negative values of the second and third columns mean that the ac impedance is smaller than the dc resistance. Dielectric losses cause this behaviour because they can be modelled as a resistor in parallel with the resistor under test. As these types of losses increase with frequency, the ac-dc difference also increases. The absolute values of the third column are lower than those of the second one because the PCB losses were eliminated. For this test the resistor was mounted in air, directly fixed to the measuring structure, without any PCB.

Table 2. Variation of the modulus of the resistor impedance with resistors with and without cover using different types of PCBs. Values are expressed in parts in 10^6.

f (Hz)	With cover, with FR4 PCB	With cover, without PCB	Without cover, without PCB	Without cover, with PTFE PCB
53	2	-2	3	2
1000	2	-1	1	1
10000	-8	-7	5	2
50000	-39	-28	15	8
100000	-77	-50	22	23

The fourth column corresponds to a resistor without cover also mounted in air. In this case, results show that all main nonlinear effects were eliminated and the apparent impedance increased with frequency due to stray capacitances to ground. Part of the input current was shunted to ground without passing through the thermal converter. Fifth column corresponds to resistors without cover mounted in PTFE PCB which has low dissipation factor at high frequency. The ac-dc differences are similar to column four, which shows that the dielectric dissipation of the PTFE PCB can be neglected. As a conclusion, the dividers are constructed using PTFE PCBs and resistors without covers.

4. Capacitance compensating methods

Once nonlinear effects have been eliminated, stray capacitances are the most relevant linear parasitic effect. There are some methods to compensate for stray capacitances. In coaxial structures, the capacitance to ground is generally larger than the parallel capacitance that resistors have. For compensation, one simple technique consists in adding parallel capacitors to the input and output arms, in order to have the same time constant in both arms. Although this works well in low precision dividers, it is not easy to use in high precision dividers. The capacitors have dielectric losses and stray inductances that significantly affect the ratio at high frequencies.

Other proposal to compensate ground capacitances consist in connecting a capacitive guard arm [4] with the same partial potentials than the principal arm. This technique reduces radial electric fields, but it is difficult to get a shield shape to avoid edge effects. A third compensation method is shown in figure 2. It uses a partial shield connected to the high potential terminal (right), adding a capacitance between the resistor and that terminal. Adjusting the length of this shield, it is possible to compensate the response to get an almost flat response in all the frequency range.

Figure 2. Adjustable partial shield to compensate for ground stray capacitances.

Although this technique has worked well for thermal transfer applications, one drawback is that it does not eliminate radial electric fields at the surface of the resistor. So that, some nonlinear losses can remain due to these fields.

In this work we explore a different shielding technique with the aim of nulling radial electric fields. Two symmetric cone-shaped electrostatic shields are installed at each end of the resistor to get null radial electric field at the whole surface of the resistor (see figure 3). The dimensions of these shields were determined using an electric-field-simulation software.

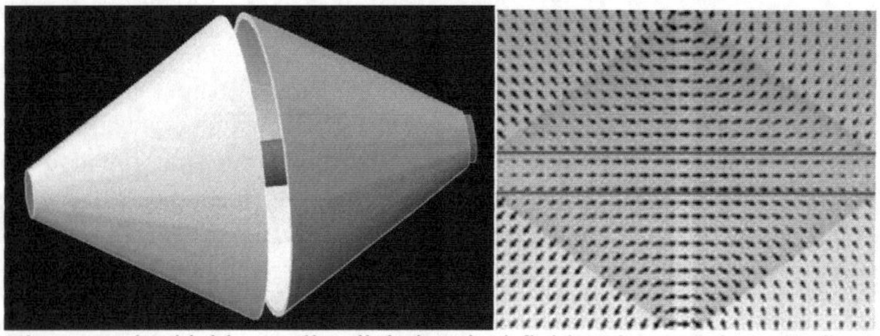

Figure 3. Electrostatic shield to null radial electrical fields. Shape and vector field diagrams.

5. Construction

A prototype of 120 kΩ was constructed. The cones are copper made and fixed to the external structure with an isolating ring between them. Inside, the PCB of the input resistor is fixed to the cones using two screws at the ends. These screws allow a small adjustment in the position of this resistor to compensate for asymmetries of the shields and for the time constant of the output resistor (200 Ω with a parallel capacitance of 15 pF), which is externally connected. Figure 4 shows a photo of the prototype without the external cylindrical cover.

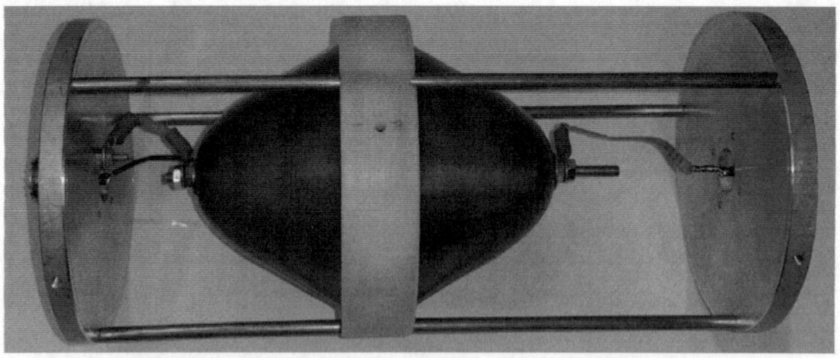

Figure 4. Construction details of the prototype.

6. Calibration

Two methods are used to calibrate the dividers. One of them compares the divider under test against a standard ac-dc thermal converter to get the modulus of the ratio error. The other method, based on the step response, gives the phase angle error.

6.1. AC-DC Method

To test the variation of the modulus of the voltage ratio when the frequency is varied between 50 Hz and 100 kHz, the dividers is compared against a standard voltage thermal converter. Figure 5 shows the ac-dc differences for the prototype. The uncertainty goes from 16 μV/V to 23 μV/V, k=2. The ac-dc variation was of 10 parts in 10^6 for the whole frequency range, so that these values are lower than the measurement uncertainty.

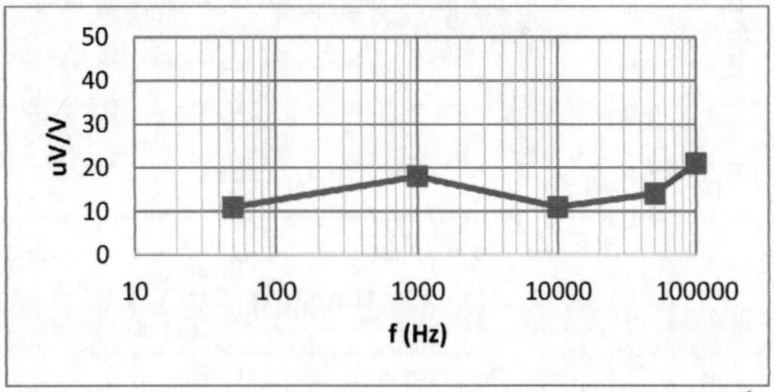

Figure 5. AC-DC differences of the prototype in parts in 10^6.

6.2. Step response

The step response method consists in applying a very fast step voltage at the input of the divider, measuring the output voltage waveform. This is a well-known technique for linear devices. As nonlinear effects were practically eliminated, this method can test modulus error as well as phase shift [6]. But, as the uncertainty of this method is generally greater than the ac-dc-method for modulus, it was only used for the calculation for the phase shift. The operational response H(p) of the divider can be expressed as

$$H(p) = \frac{1+a1.p+a2.p^2+a3.p^3+\cdots}{1+b1.p+b2.p^2+b3.p^3+\cdots} \tag{1}$$

If the excitation is sinusoidal, the complex variable p becomes $j\omega$, and at medium frequencies the higher order terms can be neglected. Then

$$H(j\omega) = \frac{1+j\omega.a1-\omega^2.a2}{1+j\omega.b1-\omega^2.b2} \tag{2}$$

For small phase displacement δ, its values is approximately

$$\delta \approx \omega(a1 - b1) \tag{3}$$

This value can be evaluated from the step response from the calculation of integral T defined as:

$$T = \int_0^\infty (1 - g(t))dt \tag{4}$$

were

$$\delta \approx -\omega T \tag{5}$$

To generate the step voltage, a dc current source is connected to the input of the divider and a mercury-wetted-relay is used to short-circuit this point to ground [7]. The transition to zero has a very short time, lower than 1 ns. A 10 bits 400 MHz digital oscilloscope records the output voltage which is processed by software that computes T, calculating δ in function of frequency [6].

Figure 6 shows the response of the tested prototype. In this case, T=-0.4 ns with an uncertainty of 0.8 ns, k=2. This leads to δ=250 µrad ±500 µrad, k=2 at 100 kHz.

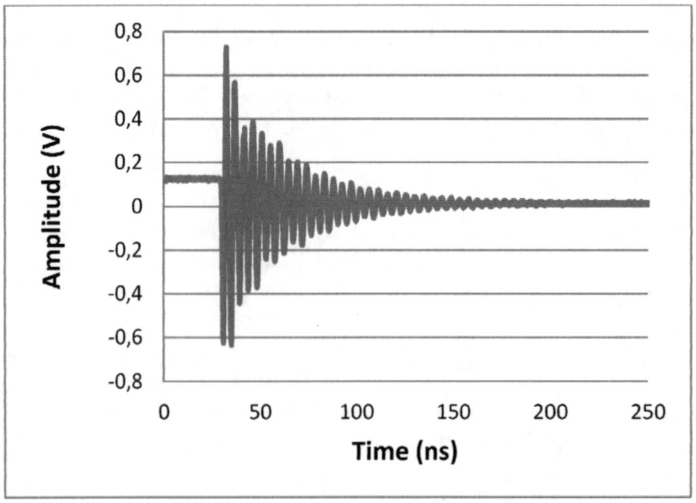

Figure 6. Step response of the prototype.

7. Conclusions

Methods to design and test resistive dividers have been proposed. They are intended for voltages up to 1024 V and frequencies up to 100 kHz. To get high accuracy, special resistors are used with very low non-linear effects and very high stability. Tests on a prototype shows an a ac-dc difference of (20±23) parts in 10^6 and a phase shift of (250±500) µrad at 100 kHz.

Acknowledgments

This work was partly supported by the Conselho Nacional de Desenvolvimento Científico e Tecnológico (CNPq), of the Ministry of Science, Technology and Innovation of Brazil, under Grant CNPq/Prosul Processo Nº 490271/2011-1.

References

[1] Kyriazis G A, Di Lillo L, Slomovitz D, Iuzzolino R, Yasuda E, Trigo L, Laiz H, Debatin R, Franco A M R, Afonso E, “Trilateral South American project: a reference system for measuring electric power up to 100 kHz – progress report,” IOP Journal of Physics: Conference Series (JPCS) – this issue.

[2] Rungis J, Schon K, "The evaluation of impulse divider response parameters," 1988, IEEE Trans. Pow. Del., **3**, No 1, 88.

[3] Slomovitz D, Faverio C, Izquierdo D, Trigo L, Castet C, “An electronically assisted binary inductive voltage divider,” 2006, Conference on Precision Electromagnetic Measurement, CPEM 2006 DIGEST, 274, Torino, Italy.

[4] Rydler K E, Svensson S, Tarasso V, “Voltage dividers with low phase angle errors for a wideband power measuring system,” 2002, Conference on Precision Electromagnetic Measurements, CPEM 2002 Digest, 382.

[5] Hermach F, Williams E, "Thermal voltage converters for accurate voltage measurements to 30 megacycles per second," 1960, AIEE Trans., pt. 1, **79**, 200.

[6] Slomovitz D, Bergalli P, Daher J F, "Evaluation of low-frequency voltage dividers, using the step response," 1998, Conference on Precision Electromagnetic Measurements, CPEM 1998 Digest, 319.
[7] Fernandez J, Slomovitz D, "Desarrollo de un generador de step rápido para pruebas de divisores de tensión de impulso," 1995, Electrotécnica, **81**, 233.

Effect of the overshoot level in an alternative method for processing of tail chopped lightning impulses

L S Ferreira[1, 2], C R H Barbosa[1], M T Silva[2] and L C Azevedo[2]

[1] Pontifical Catholic University of Rio de Janeiro, Posgraduate Programme in Metrology, Rio de Janeiro, Brazil;
[2] Centro de Pesquisas de Energia Elétrica (Cepel), Rio de Janeiro, Brazil

E-mail: lilianf@cepel.br, hall@puc-rio.br, thelio@cepel.br and luiz@cepel.br

Abstract. The construction of new Ultra High Voltage laboratories requires more studies about standards for equipment as well as for testing and measurement techniques. Currently, the analysis of tail chopped lightning impulse waveforms is possible only by prior application of a full reference lightning impulse, creating difficulties for testing and processing of waveforms. A new method for evaluating the relevant parameters of tail chopped waveforms has been previously developed and this paper studies the effect of the overshoot in this alternative method.

Keywords. UHV, testing, calibration, tail chopped lightning impulse.

1. Introduction

The energy transmission perspectives in Ultra High Voltage (UHV), with 1200 kV AC and ± 1000 kV DC, requires a new laboratory infrastructure for testing equipment used in UHV. The current facilities at high voltage equipment suppliers and at research labs, for the most part, are limited to 765 kV AC and ± 600 kV DC. The construction of new laboratories to cater to UHV systems requires more studies about standards for equipment, as well as for testing and measurement techniques. Consequently, with the increase in the voltage levels for AC and DC transmission systems, some adaptations and changes to existing standards might become necessary.

The current version of the standard IEC 60060-1: "High-voltage test techniques, Part 1: General definitions and test requirements" [1] issued in 2010 by the International Electrotechnical Commission (IEC) has as the main requirement, with regard to lightning impulse tests, the test voltage function, established based on experimental results of an European project. The test voltage function is frequency dependent and is the response of the equipment insulation to impulses with overshoot, being given by equation (1),

$$k(f)=\frac{1}{1+2.2f^{2}}, \tag{1}$$

f being the frequency in MHz.

The revision of IEC 60060-1 also includes a description of the assessment process to the test voltage, as currently all recording instruments are based on digital recorders.

In practice, the use of the test voltage function involves digitally filtering the original waveform and calculating the amplitude and time parameters from this filtered waveform. This processing, described in the standards for impulses, should be applied also to tail chopped lightning impulse waveforms.

The processing is briefly described with the help of figure 1, starting with the original data record $U_0(t)$, from which the base curve $U_m(t)$ is obtained by adjusting a double exponential. The difference between the base curve $U_m(t)$ and original curve $U_0(t)$ gives the superimposed oscillation $R(t)$ (basically the overshoot information), which is filtered by the test voltage function $k(f)$ according to equation 1. By summing the filtered oscillation to the base curve $U_m(t)$, one obtains the test voltage curve $U_t(t)$, from which the test voltage value U_t and the time parameters are determined. [2]

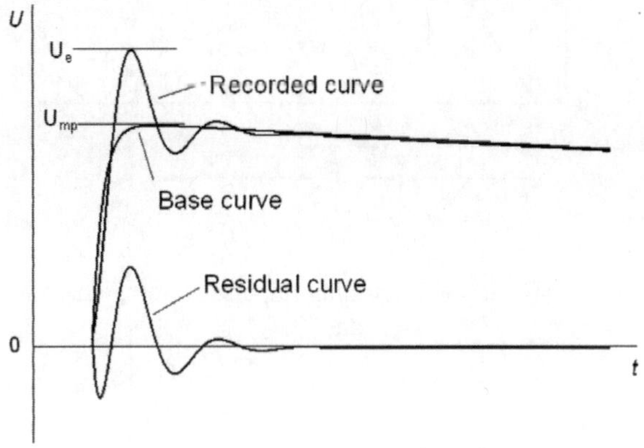

Figure 1. Recorded and base curve showing overshoot and residual curve.

Practically all tests involving high voltage equipment with large dimensions have the presence of overshoot, which is caused mainly by the inductance of the high voltage conductor, the capacitance of the object under test, as well as the intrinsic parasitic capacitances of the test arrangement. This LC circuit induces overshoot in the crest of the impulse waveform, being difficult to compensate it.

In the present paper, a study is performed on the effect of overshoot in tail chopped lightning impulses, based on experimental results with various waveforms, calculating the relevant parameters (value of test voltage U_t, front time T_1 and time to chopping T_c). These results are compared with the results of an alternative method which excludes the need for a full lightning impulse as a reference waveform.

2. Tail chopped lightning impulses

Tail chopped lightning impulses (LIC) are based on a standard full lightning impulse, which rises to its peak value in around 2 μs a 5 μs and slowly falls to zero. After the application of the full lightning impulse, using the same impulse generator settings, it is then inserted a chopping gap to cause disruption of the waveform.

Referring to figure 2, the parameters that need to be calculated to evaluate the tail chopped lightning impulse waveform, according to the IEC 60060-1 standard, are the time parameters T_1

Journal of Physics: Conference Series **733** (2016) 012073 doi:10.1088/1742-6596/733/1/012073

(front time) and T_c (time to chopping), which is the time interval between the virtual origin O_1 and the instant of chopping t_c (extrapolation of the line between the 70% and 10% points on the voltage collapse) and the amplitude parameter U_t (value of the test voltage), which is essentially the extreme value of the waveform. O_1 is the intersection with the time axis of the line through points A (30% of peak) and B (90% of peak).

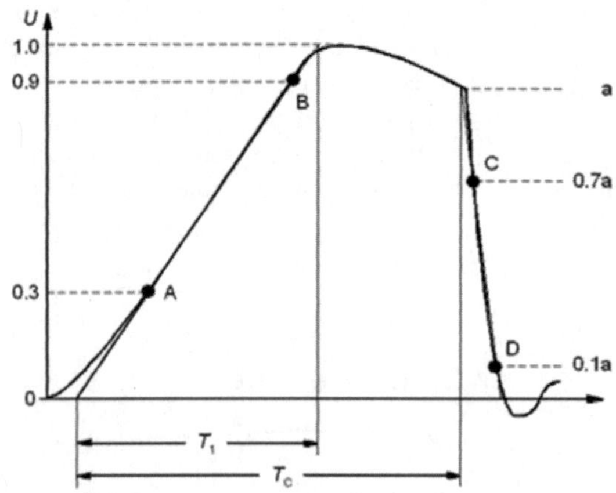

Figure 2. Typical normalized tail chopped lightning impulse showing the relevant time parameters T_l and T_c.

A lab bench recurrent surge generator RSG 482 from Haefely has been used to generate full reference lightning impulses and corresponding tail chopped lightning impulses. Tail chopped waveforms with front time T_l = 1.280 μs, time to chopping T_c = 6 μs and test voltage U_t = 49 V have been generated, with 3 levels of overshoot (0%, 4% and 20%). The data has been acquired by the system IMPREAL 2.0 (acronym for Impulse Response Analysis), developed in LabVIEW 2011, which acquires, analyses and calculates the relevant time parameters for the six types of high-voltage impulse waveforms [2].

3. Standard processing according to IEC 60060-1

This processing is used for the test voltage curve, from which the impulse parameters are calculated. This processing is dealt with more details in [3] with good results.

Table 1 presents the time parameters described in the previous section defined in the standard IEC 61083-2:2010, for the tail chopped lightning impulse using the standard processing with the aid of system IMPREAL 2.0 with 3 levels of overshoot (0%, 4% and 20%) and figure 3 shows a zoom on the tail chopped lightning impulse with 20% overshoot, together with the relevant points identified by the software.

Table 1. Results of the standard processing of the LIC impulse with 3 levels of overshoot.

Overshoot	Parameter	Value
0%	Front time T_1(μs)	1.280
	Time chopping T_c (μs)	6.144
	Test voltage U_t (V)	48.56
4%	Front time T_1(μs)	1.280
	Time chopping T_c (μs)	5.967
	Test voltage U_t (V)	49.74
20%	Front time T_1 (μs)	1.293
	Time chopping T_c (μs)	5.727
	Test voltage U_t (V)	60.76

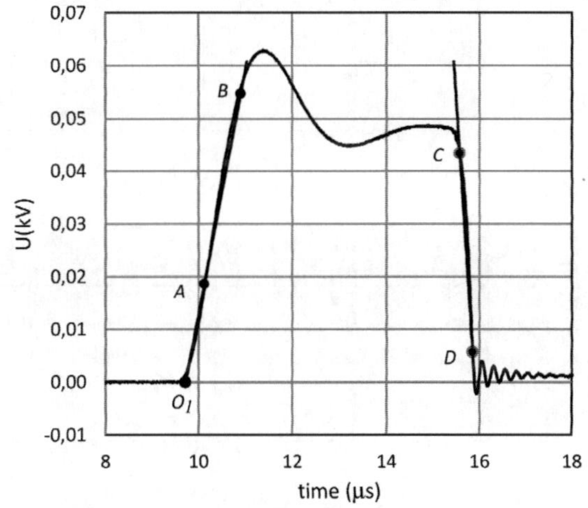

Figure 3. Results of the standard processing of the LIC with 20% overshoot. O_1 is the virtual origin, A is the 30% point, B is the 90% point, C is the 70% point and D is the 10% point.

4. Alternative processing

According with IEC standard, the need for a full reference lightning impulse is due to the fact that a tail chopped waveform does not follow the smooth double exponential shape, and this would preclude a proper fitting.

However, as the test voltage curve $U_t(t)$ needs to be calculated only up to the instant of chopping t_c, the fitting of the double exponential also needs to be done only for this period.

Thus, if the recorded curve $U_0(t)$ is truncated at t_c, it should be possible to fit a double exponential and proceed with the overshoot processing without the need of the full lightning impulse [4].

This alternative processing has been tested in this paper for the three different levels of overshoot shown in section 3 and the results are shown in table 2, together with a comparison with the standard method, by calculating the percent difference between the estimated parameters.

Figure 4 presents the result of the double exponential fitting described in [4], which yields an acceptable residual, with an RMSE (Root Mean Square Error) of 1.66%.

Table 2. Results of the alternative processing of LIC impulse with 3 levels of overshoot.

Overshoot	Parameter	Value	Difference
	Front time T_1 (μs)	1.293	1.02%
0%	Time to chopping T_c (μs)	6.166	0.36%
	test voltage U_t (kV)	48.60	0.08%
	front time T_1 (μs)	1.280	0.0%
4%	time to chopping T_c (μs)	6.005	0.64%
	test voltage U_t (kV)	49.86	0.24%
	front time T_1 (μs)	1.253	-3.09%
20%	time to chopping T_c (μs)	5.879	2.65%
	test voltage U_t (kV)	61.18	0.69%

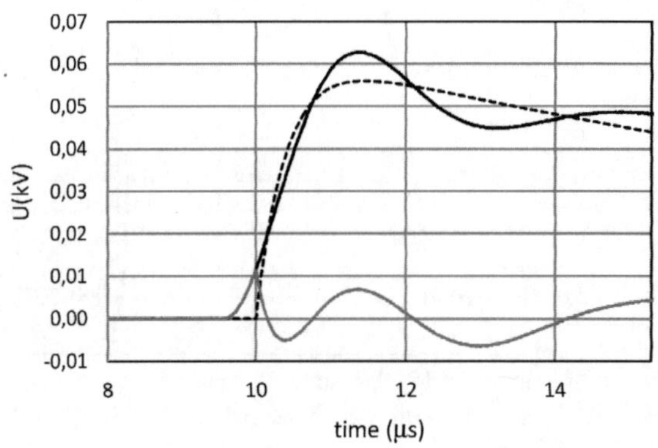

Figure 4. Fitting of a double exponential to the waveform shown in figure 4 truncated at the instant of voltage collapse. The solid line is the tail chopped waveform, the dashed line is the fitted double exponential, and the gray line is the residual.

5. Conclusions

This paper presented the effect of the level overshoot in an alternative method for processing of LIC impulse that has the potential to simplify the procedures for this type of high voltage testing. The method was tested with 3 levels of overshoot and compared with the standard method with good results.

For the largest overshoot there is a larger difference between the methods, especially regarding the front time T_1, and the alternative method is presently being refined in order to reduce this discrepancy. As expected, the difference in the time to chopping T_c increases as the overshoot level increases.

Another aspect that needs to be further studied is the effect of the chopping time on the accuracy of the alternative method.

Acknowledgements
The authors would like to thank Cepel and the Brazilian funding agencies CNPq (Conselho Nacional de Desenvolvimento Científico e Tecnológico) and CAPES (Coordenação de Aperfeiçoamento de Pessoal de Nível Superior) for their financial support.

References

[1] IEC 60060-1:2010 "High-voltage test techniques, Part 1: General definitions and test requirements".

[2] Barbosa C R H, Silva M T, Azevedo L C and Faria L C 2015 "Validation of a System for Evaluation of High-Voltage Impulses According to IEC 60060-1:2010", *IEEE Trans. Inst. Meas.*, vol. 64, nº. 6, pp.1378-1382.

[3] Barbosa C R H, Hällström J, Silva M T, Azevedo L C and Faria L C 2015 "Tail-Chopped Lightning Impulses Time Parameters Estimated According to Standard IEC 60060-1:2010", *IEEE Trans. Inst. Meas.*, vol. 64, nº. 6, pp. 1369-1372.

[4] Barbosa C R H, Hällström J, Silva M T, Azevedo L C and Ferreira L S 2015 "Alternative Method for Processing Tail-Chopped Lightning Impulses Waveforms", *IMEKO 2015 Proceedings*, Prague,Czech Republic.

DC resistance comparison between a current comparator bridge and the quantum Hall system at Inmetro

M C da Silva[1], H R Carvalho[2] and R T B Vasconcellos[1]

[1] Electrical Standardization Metrology Laboratory, Inmetro, Duque de Caxias, Brazil
[2] Quantum Electrical Metrology Laboratory, Inmetro, Duque de Caxias, Brazil

E-mail: mcsilva@inmetro.gov.br

Abstract. This paper presents a comparison results between the Quantum Hall System (QHS) under development at the Quantum Electrical Metrology Laboratory (Lameq) and the current comparator calibration system, traceable to the Bureau International des Poids et Mesures (BIPM), at the Electrical Standardization Metrology Laboratory (Lampe), both part of the Electrical Metrology Division, at Inmetro. Comparisons were performed with 1 Ω, 10 Ω, 100 Ω, 1 kΩ and 10 kΩ resistors. The results obtained over two years of work are presented here, showing that the comparison contributed to improve the calibration systems of both Lampe and Lameq.

1. Introduction

Recently Lampe acquired a new Current Comparator Bridge (CCB) [1] allowing an improvement of the calibration procedure and reduction of the uncertainties for dc resistors calibration.

The measurement of Lampe standard dc resistors in the range of 1 Ω to 10 kΩ are performed through a CCB. At this resistance range, the new CCB allows measurements at any ratios between 1:10 and 10:1. The traceability at this resistance range is performed by two 1 Ω (TH1 and TH2) and one 10 kΩ (R1) reference standard resistors which are periodically calibrated by BIPM.

The measurement in two ratios, 10:1 and 1:10, (i.e., direct and reverse measurements) reduces the CCB measurement errors, when the arithmetic mean of the two results are applied. The uncertainties specified by the CCB manufacturer for the range from 1 Ω to 10 kΩ is 0.04×10^{-6}, and for the ratio 10 kΩ : 10 kΩ is 0.1×10^{-6}.

At Lameq QHS, the measurements are performed through a Cryogenic Current Comparator (CCC). Two 100 Ω resistors are calibrated directly with the quantum Hall device. Then 1 Ω, 10 Ω, 100 Ω, 1 kΩ and 10 kΩ resistors are calibrated from these 100 Ω resistors.

After the validation of CCB at Lampe, in 2014, it was possible to perform several calibrations, which allowed reviewing the uncertainty budgets and the history of the standards resistors [2]. Thus, it was possible to increase the reliability of the measurements and reduce the calibration uncertainty budget. With the reduced uncertainty we were able to perform reliable comparisons between the results of Lampe and Lameq, allowing an improvement in the calibration system of both laboratories.

2. Methodology

Initially the calibration history of the standard resistors at BIPM was re-evaluated. These calibrations results were then applied to analyze the standards behavior tendencies, thus, determining their temporal stability and estimated corrected values. This correction is made on a monthly basis through a table of values (TabVal), both for the reference standard resistors calibrated at BIPM, and for the working standards calibrated at Lampe [3].

The application of low values of power on resistors during a calibration reduces possible errors due to the Joule effect and, consequently, the propagation errors in the range from 1 Ω to 10 kΩ. So it is possible to obtain the value of 1 Ω resistor from the 10 kΩ resistor with a smaller uncertainty than that of the manufacturer specification.

The QHS can perform measurements, using power from 0.1 mW to 1 mW. Although there are differences between the currents applied on the resistors on CCB and QHS, the Joule effect does not cause significant change in the result of this comparison, due to the small power values selected for both systems.

3. Results

The following results are divided into nominal values of 1 Ω, 10 Ω, 100 Ω, 1 kΩ and 10 kΩ. During the comparisons, different resistance standards were measured, during a short time interval, for each nominal value. The comparisons between Lampe calibration systems and QHS have been started in 2013, even before validation of new CCB.

To simplify the analysis, the results of these measurements were normalized by the results obtained with the QHS, and the relative difference graphically represented by the equation (1).

$$\Delta(CCB - QHS) = \frac{\text{CCB value} - \text{QHS value}}{\text{QHS value}} \tag{1}$$

The QHS is the reference of this study, and its uncertainty is less than 0.10 μΩ/Ω. Thus, to simplify the graphic presentation, the only uncertainty represented graphically is Lampe proposed (reduced) CMC uncertainty for each range.

This paper shows how the review of the calculations and procedures, the acquisition of a new CCB, and the comparisons with the QHS led to the reduction of the difference Δ(CCB − QHS) between the two systems.

3.1. Resistance Standards of 1 Ω and 10 Ω

The graph in figure 1 shows the comparison between the results obtained with the CCB system and QHS for the nominal resistance values of 1 Ω and 10 Ω.

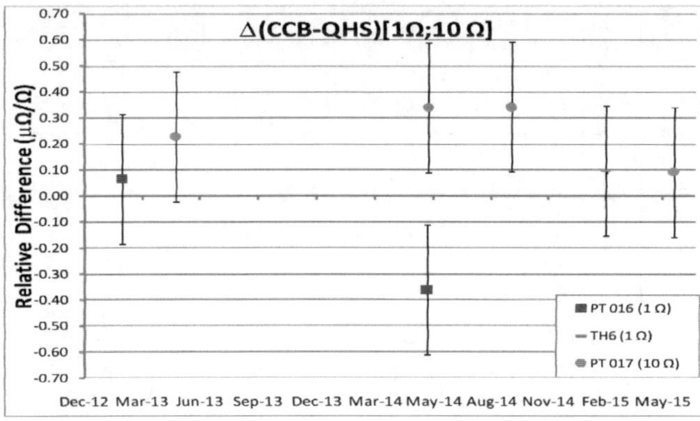

Figure 1. Comparison with nominal values of 1 Ω and 10 Ω.

The unsatisfactory results found between 2013 and the first half of 2014 led Lameq to make some improvements in initial settings procedures of CCC, during resistor calibration, in this measurement range. After these modifications, the new results obtained between the second half of 2014 and early 2015, with $\Delta(CCB - QHS) < 0.2\,\mu\Omega/\Omega$, led Lampe to propose a reduction of the CMC uncertainty budget, at this range, from 0.65 μΩ/Ω to 0.25 μΩ/Ω.

3.2. Resistance Standards of 100 Ω

This is one of the most important comparisons because the standard resistors are directly compared with the quantum Hall device during the measurements in Lameq. These results are shown in figure 2.

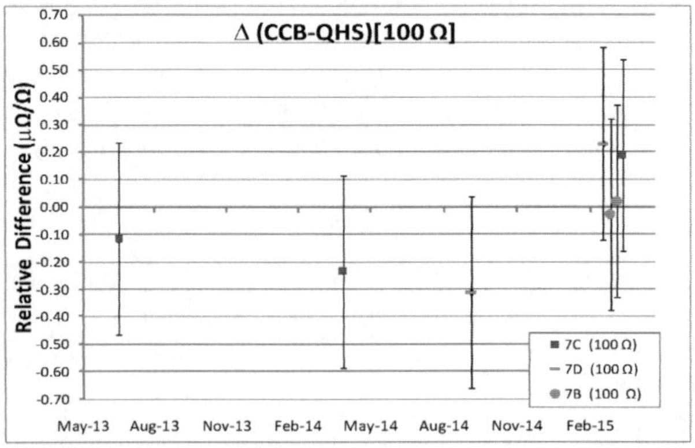

Figure 2. Comparison with nominal values of 100 Ω.

In this range, some adjustments made in Lampe calibration procedure led to a significant reduction of $\Delta(CCB - QHS)$, especially if we consider 2015 measurements. In view of these results, Lampe proposed a reduction of the CMC uncertainty budget, at this range, from 0.61 μΩ/Ω to 0.35 μΩ/Ω.

3.3. Resistance Standards of 1 kΩ and 10 kΩ

In this range Lampe CCB operates with its lowest uncertainty. The working standard resistors of 1 kΩ and 10 kΩ can be measured directly by comparison with the standard resistor R1 (10 kΩ) which presents better stability and lower temperature dependence.

Before the acquisition of the new CCB, it was only possible to calibrate 1 kΩ or 10 kΩ resistors from a 100 Ω standard resistor, with significantly higher uncertainty. Figure 3 shows the measurement results obtained before (2013) and after (2014-15) the implementation of the new CCB and the review of Lampe calibration procedures.

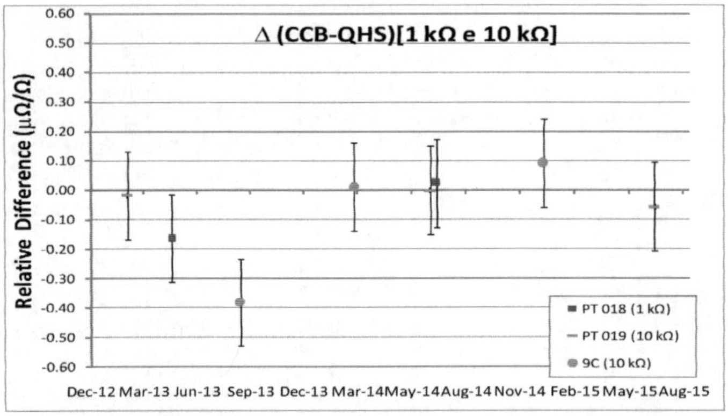

Figure 3. Comparison with nominal values of 1 kΩ and 10 kΩ.

The results obtained after 2014 present $\Delta(CCB - QHS) \leq 0.1\ \mu\Omega/\Omega$, thus leading Lampe to propose a reduction of the CMC uncertainty budget, at the 1 kΩ range, from 0.60 μΩ/Ω to 0.15 μΩ/Ω; and at the 10 kΩ range, from 1.0 μΩ/Ω to 0.15 μΩ/Ω.

3.4. Summary of the results found

Table 1 shows the relative difference $\Delta(CCB - QHS)$ and the Normalized Error Ratio (NER) of the last comparison (2014-15) between QHS and CCB, for each standard resistor. The NER is calculated according to equation (2). For values of $NER < 1$, we can consider the comparison results are coherent with the estimated uncertainty budget.

$$NER = \frac{|\text{CCB value} - \text{QHS value}|}{\sqrt{U_{CCB}^2 + U_{QHS}^2}} \quad (2)$$

Where:
U_{CCB} = Expanded absolute uncertainty of Lampe;
U_{QHS} = Expanded absolute uncertainty of Lameq.

At table 1, it is possible to observe that the values of $\Delta(CCB - QHS)$ are in between 0.02 μΩ/Ω (range of 1 kΩ) and 0.23 μΩ/Ω (range of 100 Ω), all values much below the proposed CMC uncertainty for the respective range.

Table 1. Summary of the results found.

Nominal Value	Standard	Month/Year	Relative Difference (μΩ/Ω)	EN
1 Ω	TH6	feb/2015	0.096	0.38
10 Ω	PT 017	jun/2015	0.090	0.35
	7C	feb/2015	0.187	0.53
100 Ω	7D	feb/2015	0.229	0.65
	7B	apr/2015	0.020	0.06
1 kΩ	PT 018	jun/2014	0.024	0.15
10 kΩ	PT 019	jul/2015	-0.057	0.36
	9C	jan/2015	0.091	0.56

The reliability of the comparison results is also confirmed by the NER value, between 0.06 and 0.65, i.e., smaller than 1 for all ranges.

4. Conclusion

The comparisons made between Lampe and Lameq contributed to improve calibration procedures for both laboratories, allowing not only to reduce uncertainties as well as to increase the reliability of the measurement results.

In the next few months, after a programmed bilateral comparison with BIPM, Lameq QHS should be internationally recognized as a primary electrical resistance laboratory. From this moment on, it will not be necessary to send Lampe reference standard resistors to be calibrated at the BIPM. Therefore Lampe standards will not be subject to drifts due to transportation, and it will be possible to perform dc resistance calibrations with even smaller uncertainties at Inmetro.

Acknowledgments
The authors would like to thank the teams of both Lampe and Lameq that have provided the necessary interaction to carry out this comparison.

References

[1] Measurement International 2005 *Automatic Resistance Bridge : Model 6010D - Operator Manual*, Ontario

[2] Carvalho H R, Briones R E M and Silva J R B 2013 *J. Phys.* : Conf. Ser. 575 012014

[3] Silva M C da, Petkovic S G and Schechter H 2011 Avaliação da estabilidade em resistores padrão *IX SEMETRO* Natal

Journal of Physics: Conference Series **733** (2016) 012075 doi:10.1088/1742-6596/733/1/012075

Metrological traceability for AC High-Voltage in Inmetro up to 40 kV

P C O Vitorio; V R de Lima; O Borges Filho; L A A de Souza and O W G Asencios

Divisão de Metrologia Elétrica – (Diele), Diretoria de Metrologia Científica - (Dimci), National Institute of Metrology, Quality and Technology – Inmetro, Av. Nossa Senhora das Graças, nr. 50, Zip Code 25250-020, Xerém, Duque de Caxias, RJ, Brazil

pcoliveira@inmetro.gov.br

Abstract. This paper refers to a project carried out in Inmetro aiming to provide internal metrological traceability for 60 Hz AC High-Voltage up to 40 kV. It presents details about the method used, its equations and obtained results. A capacitance and tanδ bridge, with a built-in current comparator, was used in combination with two standard capacitors to calibrate a standard potential transformer (PT), both in ratio and phase angle. The results obtained by Inmetro showed good agreement with PTB ones, for the same PT. The maximum estimated uncertainty was 0,0049 % for ratio error and 104 μrad for phase angle error.

1. Introduction

One goal of a National Institute of Metrology (INM) is to provide internal traceability for its standards based on primary references of quantities that comprise its metrological infrastructure.

For voltage quantity in 60 Hz alternating current (AC), Inmetro already has internal traceability up to 1000 V. Recently, a project was carried out in its High Voltage Metrology Laboratory (Lamat), which resulted in the implementation of a method to providing internal traceability for AC High-Voltage (HVAC) up to 40 kV. This paper presents information about this project and the results obtained by Inmetro compared with those obtained by the Physikalisch-Technische Bundesanstalt (PTB).

2. Method used

The measurement principle of the method for traceability in HVAC implemented in Inmetro is shown in figure 1 [1]. A commercial capacitance and tanδ bridge (C and Tanδ bridge), with a built-in current comparator, in combination with two standard capacitors are used to calibrate a standard potential transformer (PT), both in ratio and phase angle.

2.1. Formulation of the method

2.1.1. Equation to obtain the ratio of PT

From figure 1, under bridge balanced conditions ($I_S \cdot N_S = I_X \cdot N_X$), being C_X the value measured by the bridge for C_{HV} and neglecting the voltage drops in the resistances of N_S and N_X windings, we have:

$$j\omega \cdot C_{LV} \cdot U_2 \cdot N_S = j\omega \cdot C_X \cdot U_1 \cdot N_X \quad (1)$$

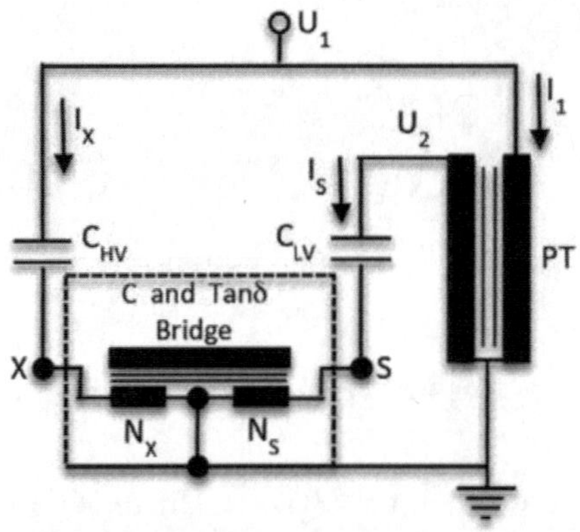

Figure 1. Circuit for PT calibration with a C and Tanδ bridge, C_{LV} = standard low voltage capacitor and C_{HV} = standard HV capacitor.

With $U_1/U_2 = a_N$, being "a_N" the ratio previously attributed to the PT under calibration, from equation (1), we have:

$$j\omega \cdot \left(C_{LV}/a_N\right) \cdot U_1 \cdot N_S = j\omega \cdot C_X \cdot U_1 \cdot N_X \tag{2}$$

If we consider $U_1/U_2 = a$, being "a" the actual ratio of the PT under calibration, the bridge will measure the actual capacitance C_{HV} ($C_X = C_{HV}$). Therefore, from equation (1), we have:

$$j\omega \cdot \left(C_{LV}/a\right) \cdot U_1 \cdot N_S = j\omega \cdot C_{HV} \cdot U_1 \cdot N_X \tag{3}$$

With equations (2) and (3), the expression for the PT ratio obtained in the calibration process becomes:

$$a = \left(C_X/C_{HV}\right) \cdot a_N \tag{4}$$

Where:
C_X = Capacitance measured by the bridge;
C_{HV} = Known capacitance of the HV capacitor;
a_N = Ratio previously attributed to the PT;
a = Actual PT ratio obtained in the calibration.

Considering the original circuit found in the operating manual of a commercial C and Tanδ bridge, without PT, the capacitor C_X to be measured and the known standard capacitor named C_S are connected directly to the voltage source: U_1 in figure 1. In this case, under balanced bridge conditions, we have:

$$j\omega \cdot C_S \cdot U_1 \cdot N_S = j\omega \cdot C_X \cdot U_1 \cdot N_X \tag{5}$$

The bridge requires the characteristic parameters of C_S (capacitance and tangδ values) as very accurate input data. By comparing equations (2) and (5), it is concluded that the best possible C_S value to be set in the bridge, when using the circuit of figure 1, is given by:

$$C_S = \left(C_{LV}/a_N\right) \tag{6}$$

With equations (4) and (6), we also obtain "a" as function of the relationship between capacitances:

$$a = \left(C_X/C_S\right) \cdot \left(C_{LV}/C_{HV}\right) \tag{7}$$

2.1.2. Equation for the phase angle of PT

The phase angle (ϕ) of the PT is determined from the diagram shown in figure 2, which represents the voltages and currents of figure 1, resulting in:

$$\varphi = \delta - \delta X + \delta S \quad (8)$$

Where:
δ = Phase angle between I_X and I_S, measured by the bridge
δ_X = Known dissipation factor of C_{HV} (δ_{HV})
δ_S = Known dissipation factor of C_{LV} (δ_{LV})

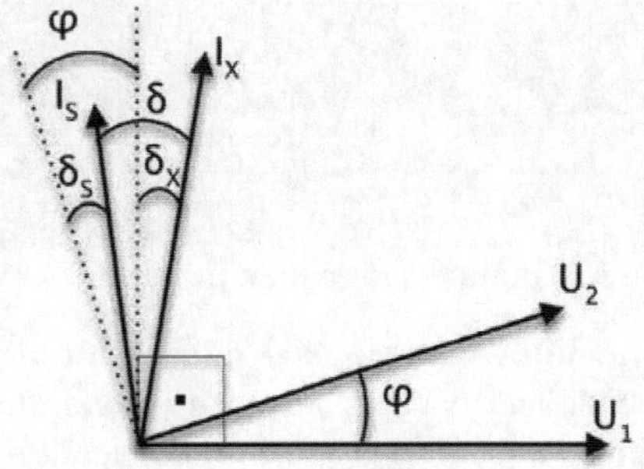

Figure 2. Diagram for determination of the phase angle of the PT under calibration.

3. Equipment used

The C and Tanδ bridge used was a 7010B model manufactured by MI – Measurements International. C_{HV} was a 100 pF/200 kV standard SF_6 insulated capacitor. For C_{LV}, four thermally stabilized [4] units of 1000 pF/750 V N_2 insulated standard capacitors were available, allowing capacitances values of 1000 pF, 2000 pF, 3000 pF and 4000 pF. For the low voltage measurement of C_{HV} and C_{LV} and their respective dissipation factors (δ_{HV} and δ_{LV}), a high accurate capacitance meter AH2700A model, manufactured by Andeen-Hagerling, was used.

The standard PT calibrated during the implementation of the method was a NVRD40 PT model, manufactured by Epro. Nine PT taps were calibrated with rated voltages from 3kV to 40kV. The reasons for choosing this PT were: a) Up to its maximum voltage (40 kV), the variation of C_{HV} due to its voltage coefficient is less than 1 ppm [2]; b) Its ratio values, combined with C_{HV} and C_{LV} available at Inmetro, enable measuring ranges at the bridge with low uncertainties; c) This PT was recently calibrated at PTB (2013), enabling comparison for validation purposes under equivalent conditions.

The C and Tanδ bridge was calibrated in Inmetro considering its best operating parameters that lead to minimum contribution to the uncertainty in the PT calibration results, based on the PT, C_{HV} and C_{LV} known characteristics. The circuit arrangement in the HV hall of Lamat is shown in figure 3.

4. Corrections and uncertainties

For metrological purposes, it must be considered that the temperature variation (Δ_T) and the applied voltage level (U) affect a capacitance (C).

Studies and experiments were performed in Inmetro for C_{HV} and C_{LV} capacitors in order to consider the effects of temperature and voltage either for correction purposes or for including them in the uncertainty of the results, together with the contribution due to the uncertainty of the capacitance bridge itself [3].

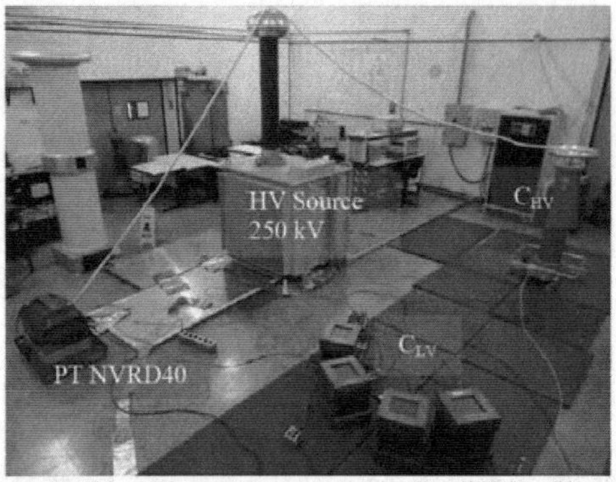

Figure 3. Circuit arrangement in the HV hall of Lamat.

For the PT phase angle, corrections were applied in equation (8) to consider the effects of stray capacitances and resistances between C_{HV} and C_{LV} capacitors and their respective inputs C_X and C_S in the bridge, as well as the correction of the bridge itself for dissipation factor measurement.

The uncertainty for phase angle result was estimated considering the uncertainties of the corrections and those related to the bridge: resolution, repeatability and its calibration for dissipation factor [3].

5. Results

The results obtained by Inmetro for the PT calibration are shown in tables 1 and 2, together with information about the results obtained by PTB and those given by the manufacturer.

The maximum expanded uncertainty estimated by Inmetro in the PT calibration process was 0,0049 % for ratio error and 104 μrad for phase angle error. The PTB expanded uncertainty was 0,0040 % for ratio error and 44 μrad for phase angle error. The manufacturer gives 0,0050 % for ratio error and 145 μrad for phase angle error.

6. Comments and conclusions

A method for internal metrological traceability of 60 Hz HVAC up to 40 kV, based on a circuit that uses a C and Tanδ bridge with two standard capacitors and a standard potential transformer, was successfully implemented in Inmetro.

By this method, the obtained results for PT ratio and phase angle are linked to the electrical resistance quantity (ohms), obtained from the quantum Hall effect in Inmetro.

The maximum estimated uncertainties obtained by Inmetro were 0,0049 % for ratio error and 104 μrad for phase angle error.

For the PT ratio, the results obtained by Inmetro, both error and uncertainty values, were in good agreement with those obtained by PTB.

Table 1. Discrepancies between Inmetro and PTB for ratio error (Δa) [$x10^{-6}$] and phase angle (Δϕ) [μrad].

Tap	3 kV		15 kV		40 kV	
(%)	Δa	Δφ	Δa	Δφ	Δa	Δφ
40	00	04	00	-13	-02	-28
60	09	02	-01	-14	00	-26
80	09	-08	-02	-14	-02	-27
100	06	-08	-02	-15	-03	-29
120	07	-07	-03	-15	-03	-37

Table 2. Inmetro uncertainties for PT error in ratio $U_r(a)$ [$x10^{-6}$] and phase angle $U_r(\phi)$ [μrad].

Tap (%)	3 kV		15 kV		40 kV	
	$U_r(a)$	$U_r(\varphi)$	$U_r(a)$	$U_r(\varphi)$	$U_r(a)$	$U_r(\varphi)$
40	25	104	46	104	36	104
60	26	104	46	104	36	104
80	26	104	46	104	37	104
100	28	104	47	104	38	104
120	30	104	49	104	41	104

References

[1] Anderson W E 1988 NBS (U.S. Special Publication) NBS Measurement services: A calibration service for voltage transformers and HV capacitors pp 250-33

[2] Latzel H G, Schon K 1987 IEEE/TIM Precise Capacitance Measurements of HV compressed Gas Capacitors Vol IM-36 Nr 2

[3] Lima V R et al 2014 CPEM Implementation of a HV Primary Standard Method using a Capacitance Bridge (Rio de Janeiro, Brazil)

[4] Kyriazis G A and Melcher J 2008 CEPEM Thermally-controlled gas-dielectric standard capacitors constructed at Inmetro (Broomfield, USA) pp 560-561

Acknowledgments

The authors would like to thank Dr. Renata T. de Barros e Vasconcellos, head of the Inmetro's Electrical Standardization Metrology Laboratory, for her assistance with technical discussions and laboratory work on calibrations of the C_{LV} and the C and Tanδ bridge, and the technical staff of Lamat for the support.

Using a Programmable JVS for evaluation of Zener voltage standards stability and secondary uncertainty under controlled temperature variation

Marcelo R. de Andrade and Regis P. Landim

Instituto Nacional de Metrologia, Qualidade e Tecnologia – Inmetro, Brazil

rplandim@inmetro.gov.br

Abstract. A study was carried out with two 732B model Zener-diode based electronic voltage standards, in order to analyze the influence of external temperature variation in their outputs. The temperature was controlled and the external variations were monitored. Inmetro Programmable Josephson Voltage Standard (PJVS) and Secondary systems were used for this purpose. Studies have shown that, despite the internal temperature control, there is a strong correlation between the external temperature and the internal thermistor value, for both Zeners. Inmetro Zener calibration uncertainty budget was revisited and the achieved uncertainties are ±0.09 μV (at 1.018 V) and ±0.4 μV (at 10 V), k=2.

1. Introduction

Zener-diode based solid state electronic voltage standards (Zeners) are used worldwide by National Metrology Institutes (NMIs) [1]. The Instituto Nacional de Metrologia, Qualidade e Tecnologia (Inmetro) uses the Fluke 732A and 732B models, in order to disseminate the Brazil legal volt and to maintain internationally consistent and traceable voltage standards tied to the SI units. Since 1998 Inmetro has been using a Conventional Josephson Voltage Standard (CJVS) as a primary reference standard of voltage. In 2012 Inmetro started using its Programmable Josephson Voltage Standard (PJVS) for this task, too. The Zeners calibrated using a JVS are used as a reference standard (in the Secondary system) to calibrate Inmetro client's Zeners (from the Brazilian Network Calibration), which are used to provide traceability to all other laboratories and research institutes in Brazil [2].

Zener standards, in spite of their short-term good stability, are affected by (a) deterministic effects (drift and external variations, like pressure, humidity and, mainly, temperature) and (b) random variations (white noise and $1/f$ noise). In fact, $1/f$ noise in Zeners are the ultimate limit to Zener voltage measurement uncertainties; hence, calibration using Zeners as reference standards are limited to one part in 10^8 [3]. The present work shows some results related to the Zener output voltage variation under the presence of controlled external temperature, targeting lower uncertainties calibration in the Secondary system.

2. Experimental Procedure

We mounted an experimental setup, composed by an air bath (temperature controlled), a thermometer, a humidity meter and a barometer to monitor those quantities inside the air bath (hence, "external" in this paper refers to the quantities inside it and surrounding the Zeners), as well as two ohmmeters to measure the Zeners' thermistors. Two Fluke 732B Zeners (named Z4 and Z7, connected to the thermistor meters) and the temperature, humidity and pressure sensors were placed inside the air bath. A LABView software was developed to automatically monitor all the measured quantities. In order to keep the Zeners always powered only on batteries (reducing the power line noise effects), each Zener was connected to its own external battery pack (outside the air bath and whose voltages were monitored daily and replaced when needed). Hence, there was no need to open the air bath during the whole measurement cycles.

An intralaboratory comparison (at controlled external temperature) was performed using two systems (at each measurement cycle) from the Quantum Electrical Metrology Laboratory (Lameq/Inmetro): (a) the PJVS, used to calibrate both Zeners (during around six hours); and (b) the Secondary calibration system, which used Z4 to calibrate Z7 (during around one hour). Both systems use the differential comparison method. We used the Measurements International 9300 air bath, which provides controlled temperature with ± 0.05 °C stability within 15 °C to 40 °C. The Zeners were connected to the lower *emf* (electromotive force) channels of a scanner, reducing the measurement uncertainties. Table 1 shows the measurement cycles and their respective temperature values. The temperature steps were made non-sequentially, in order to make it easier to analyze separately the temperature drift from the temporal drift effects. Due to the normative laboratory temperature calibration (22.5 °C ±1 °C), the 23 °C temperature was chosen to be measured at the beginning, middle and end of the cycles, in order to help the Zeners' temporal drift analysis.

Table 1. Measurement cycles (from May 14, 2015 through June 1, 2015).

Cycle	Temperature	Cycle	Temperature	Cycle	Temperature
1	23 °C	**6**	21 °C	**11**	21 °C
2	23 °C	**7**	25 °C	**12**	24 °C
3	23 °C	**8**	23 °C	**13**	22 °C
4	20 °C	**9**	22 °C	**14**	26 °C
5	24 °C	**10**	25 °C	**15**	23 °C

3. Results

First, we discarded all the transition measurements (the ones whose thermistors were still changing above 30 Ω). Then, we discarded some of the PJVS measurements, keeping the five ones which are closest (in time) to the ones used in the Secondary system analysis. The remaining measurements are considered the "useful raw data". For each system (either PJVS or Secondary), and for each temperature, the final value is an average from all "useful raw data". The analysis was made considering two approaches: (a) the influence of the external temperature on the internal Zener's thermistor; and (b) the influence of the external temperature on the Zener's output voltage.

3.1. Influence of the external temperature on the internal Zener's thermistor

Right after a step change in the external temperature, the tested Zeners' thermistor experienced a variation towards a new value, after some time (as expected). For instance, after a step change from 20 °C to 24 °C, the thermistors started changing around 10 min later, decreasing their values. After 30 minutes, the external temperature stabilized around 24 °C, but the thermistors kept changing, stabilizing only around 50 minutes after the beginning of the temperature step. The thermistors' resistance stability remains within ± 25 Ω as long as the external temperature remains stable within ± 0.05 °C. Similar situation occurred for all the 15 cycles (table 1). Considering that the temperature at Rio de Janeiro can reach 41 °C (even more) during summer, causing Inmetro clients' Zeners experiencing external temperature variations up to 15 °C (for laboratory room temperature stabilization), we have decided to wait, at least, 8 hours before beginning a new Zener client calibration. This is enough time for the complete Zener stabilization.

Journal of Physics: Conference Series **733** (2016) 012076 doi:10.1088/1742-6596/733/1/012076

Using a Least-Squares Sum (LSS) estimation method, we have got a very good linear model between the external temperature and the internal thermistors' resistance. The thermistor angular coefficients found were -60 Ω/°C (for Z4) and -53 Ω/°C (for Z7). The maximum error between those models and the "useful raw data" was 9 Ω (in 39 kΩ). Using those models and the internal Thermal Coefficient $TC_{Tint} \approx 2$ kΩ/°C [4], we have got a good Zener internal temperature estimator:

$$Tint_{Z4} = -0.03\,Text + 44.48 \tag{1}$$

$$Tint_{Z7} = -0.03\,Text + 44.49 \tag{2}$$

From equation (1), equation (2) and the thermistor angular coefficients found, it is clear that an external temperature increase is followed by a proportional Zener internal temperature decrease, as well as Zener thermistor resistance decrease. However, thanks to its internal temperature control, this variation is small: the Zener internal temperature drops around 0.03 °C for each 1 °C increase in the external temperature (only a factor of 3 % of the external temperature variation reflects in the internal temperature variation), for the tested Zeners. Such Zener temperature stability reflects in its output voltages stability, as well.

3.2. Influence of the external temperature on Zener's output voltages

In order to verify the influence of the external temperature variation on the Zener's output voltages, we first estimated the temporal drift during the measured dates (table 1), from the 22 °C and 23 °C of the "useful raw data" (since the normative temperature range for regular calibration in the laboratory is 22.5 °C ±1 °C). Then we subtracted (accordingly to the time variation) the temporal drift from the "useful raw data". From this new data, for each Zener and each voltage tap (either 1.018 V or 10 V) and using LSS estimation, we have got (table 2):

Table 2. Voltage Temperature Coefficients (*TCv*), in (µV/V)/°C.

Tap (V)	Z4	Z7	Manufacturer
1.018	0.02	-0.01	0.04
10	-0.02	-0.02	0.1

3.3. Inmetro secondary Zener calibration uncertainty

Since 2010, Inmetro CMC for Zener calibration has been ±0.2 µV (at 1.018 V) and ±1 µV (at 10 V) [5]. A new study considering the 22.5 °C ±1 °C temperature range was recently made and the achieved uncertainties were ±0.1 µV (at 1.018 V) and ±0.4 µV (at 10 V), k=2 [5]. In this paper, the maximum temperature variation (at each targeted temperature) was ± 0.05 °C. Table 3 shows the results for this paper at the 1.018 V and 10 V taps (the sensitivity coefficients are one, so they were not presented). U = ±0.373 µV was rounded to U = ±0.4 µV. At 1.018 V, U = ±0.089 µV, rounded to U = ±0.09 µV. All the numbers (for 10 V and 1.018 V) were rounded according to $1/f$ noise limits [3].

According to table 2 (before rounding) the maximum output voltage variation due to 0.05 °C change is 0.01 µV; for 2.00 °C variation, the change is 0.36 µV (within the proposed uncertainty). Also, the uncertainty improvement due to the air bath use was not significant. That means $1/f$ noise limits avoid any better improvement and the use of a more expensive apparatus is useless. Temperature regulation of ± 2.0 °C is enough for our purposes.

Table 4 shows the Secondary system errors (compared to the PJVS system), for each temperature. The errors are smaller than the proposed uncertainties and the lowest errors were achieved at 23 °C, 24 °C and 25 °C. From table 2, one can see that a change from 20 °C to 26 °C may change up to 1.2 µV the output voltage (besides the temporal drift effect). However, since both Zener temperature changes are similar, the errors remain small.

Table 3. 1.018 V and 10 V uncertainty budgets.

Number of measurements (n)	10					
Uncertainty Budget	1.018 V			10 V		
	U(xi)	**Prob Distr**	**Std. Unc.(μV)**	*U(xi)*	**Prob Distr**	**Std. Unc. (μV)**
Reference Certificate, *u(x1)* - μV	0.040	norm	0.020	0.250	norm	0.121
Reference drift, *u(x2)* - μV	0.034	rect.	0.019	0.087	rect.	0.050
Reference Thermal Coef., *u(x3)* - μV	0.008	rect.	0.004	0.087	rect.	0.050
DVM Gain/Linearity error, *u(x4)* - μV	0.001	rect.	0.001	0.001	rect.	0.001
DUT Thermal Coef., *u(x5)* - μV	0.005	rect.	0.003	0.069	rect.	0.040
Thermal *emf*, *u(x6)* - μV	0.056	rect.	0.032	0.056	rect.	0.032
Measurement dispersion, *u(x7)* - μV	0.044	norm	0.014	0.337	norm	0.106
Combined Uncertainty, u_C (μV)			0.045			0.183
Effective degrees of freedom (ν_{eff})			∞			79
Coverage factor *k*			2.003			2.032
Expanded Uncertainty, *U* (μV)			**0.089**			**0.373**

Table 4. The Secondary system errors.

Text (°C)	20	21	22	23	24	25	26
$E_{S(1.018)}$ (μV)	-0.04	-0.02	-0.03	0.01	0.01	0.01	0.03
$E_{S(10)}$ (μV)	-0.3	-0.3	-0.3	-0.2	0.1	0.2	-0.3

Figure 1 shows the consistency between the proposed uncertainty and the measured values, at 1.018 V (a) and at 10 V (b).

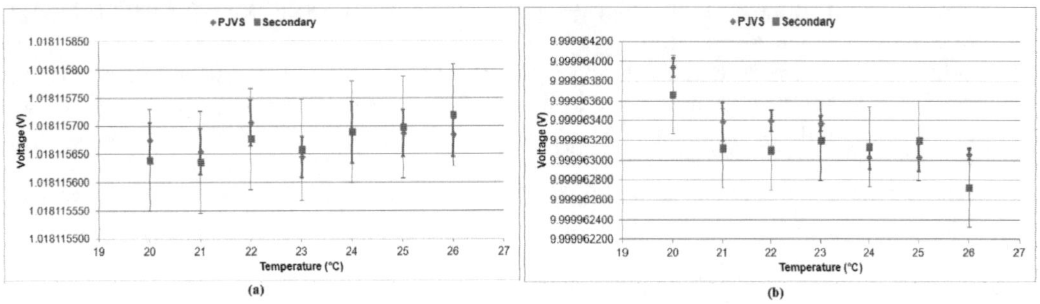

Figure 1. Measurements with the Secondary system and its uncertainty proposal as well as the PJVS with its estimated uncertainty (k=2).

The reference Zener (Z4) value used to calibrate Z7 in the Secondary system was kept unchanged (for each temperature) since day one, according to our current calibration procedure. Updating Z4 value everyday would allow even lower errors, since the reference Zener temporal drift affects the object Zener values. However, it is not worthwhile, since it would be needed to calibrate the reference Zener against a JVS system every day.

Starting from the temporal drift estimation, it was possible to estimate the maximum time needed to update the reference Zener value, in order to keep the proposed uncertainty unchanged. For the 1.018 V tap, it is needed, at least, around 200 days until its uncertainty gets higher than $U = \pm 0.09$ µV. On the other hand, for the 10 V tap, 40 days are enough to get $U = \pm 0.5$ µV; 60 days are enough to get $U = \pm 0.6$ µV; and so on. Of course, those values depend on the Zeners involved (since drift and noise are particular to each Zener/tap).

4. Conclusion

We used Inmetro PJVS and Secondary systems to investigate two 732B Zeners' behavior under controlled external temperature variation. We have found that only a factor of 3 % of the external temperature variation reflects in the internal temperature variation. Due to the $1/f$ noise limits, there is no significant advantage in using an air bath for Zener calibration in a Secondary system. At 10 V (figure 1), the PJVS could detect a higher Z7 temperature drift (above 0.5 µV) at the lowest temperature (20 °C); at 21 °C, 22 °C and 23 °C, no such drift was seen; however, between 23 °C and 24 °C we can notice a 0.4 µV shift; no significant temperature drift was detected with the Secondary system, between 21 °C and 25 °C. At 1.018 V, there was no significant temperature drift (either PJVS or Secondary). Also, the lowest Secondary system errors occurred at 23 °C, 24 °C and 25 °C (table 3). All things considered, we have decided to adopt (23.0 °C ±2.0 °C) as the new temperature range Zener calibration in the laboratory. In this case, the achieved uncertainties (Secondary system) are $U = \pm 0.4$ µV (at 10 V) and $U = \pm 0.09$ µV (at 1.018 V), k=2.

Pressure and humidity (not controlled) maximum variations observed were 14.70 hPa and 14.33 % ur, respectively. Pressure and humidity variations are not significant to the Zeners' output voltages stability, at constant external temperature.

Acknowledgments
The authors would like to thank Franclin S. Sousa, Vinícius G. Santos and Renato S. Costa for the help in this work.

References

[1] Witt T J 2002 Maintenance and dissemination of voltage standards by Zener-diode-based instruments **149** 305-312.
[2] Landim R P, Tang Y h and Afonso E. 2010 Comparison of the INMETRO and NIST Josephson Voltage Standards (part of the ongoing regional key comparison SIM.EM.BIPM-K10.b.1) *Metrologia* **47** Tech. Suppl. 01003.
[3] Witt T J and Reymann D. 2000 Using power spectra and Allan variances to characterise the noise of Zener-diode voltage standards *IEE Proceedings. Science, Measurement and Technology* **147** 177-182.
[4] FLUKE 1997 *732B/732A DC Reference Standard Instruction Manual* REV.3.
[5] Landim R P 2015 *Study for Primary and Secondary Zener CMC Review* Inmetro Technical Report.

Measurements methodology for evaluation of Digital TV operation in VHF high-band

M Pudwell Chaves de Almeida[1], P Vladimir Gonzalez Castellanos[2], J Alfredo Cal Braz[1], R Pereira David[1], R Saboia Lima de Souza[1], A Pereira da Soledade[1], J Rodrigues Nascimento Junior[1] and F Ferreira Lima [3]

[1] Dmtic/Dimci, National Institute of Metrology, Quality and Technology, Inmetro, Brazil
[2] Graduation Program of Electrical and Telecom. Engineering, UFF, Niterói, RJ, Brazil
[3] SCE/DEAA/CGOU-Ministry of Communications - Brasília, Brazil

rssouza@inmetro.gov.br

Abstract: This paper describes the experimental setup of field measurements carried out for evaluating the operation of the ISDB-T$_B$ (Integrated Services Digital Broadcasting, Terrestrial, Brazilian version) standard digital TV in the VHF-high band. Measurements were performed in urban and suburban areas in a medium-sized Brazilian city. Besides the direct measurements of received power and environmental noise, a measurement procedure involving the injection of Gaussian additive noise was employed to achieve the signal to noise ratio threshold at each measurement site. The analysis includes results of static reception measurements for evaluating the received field strength and the signal to noise ratio thresholds for correct signal decoding.

Keywords: Digital TV, ISDB-T$_B$, field measurements, UHF, VHF High-band

1. INTRODUCTION

The Brazilian regulator, the National Agency of Telecommunications (Anatel), approved the allocation of part of the UHF spectrum band (698 to 806 MHz, TV channels 52-69) to mobile cellular services of fourth generation (4G), known as the Long Term Evolution (LTE) system [1]. To clear this frequency band for this usage, the implementation of digital TV transmission of ISDB-T$_B$ standard in the VHF-high band (174-216 MHz) in Brazil [2] became a priority in the Brazilian TV broadcasting scenario. Considering that the majority of ISDB-T$_B$ transmission already operate in UHF band, an assessment of operation in VHF high-band was required.

This paper presents results of field measurements carried out for the evaluation of ISDB-T$_B$ digital TV operation in the VHF-high band [3] [4]. A mobile unit fully prepared with measurement equipment and digital TV signal recording setup performed the trials in urban and suburban areas in a Brazilian city (Gama, near the country capital Brasília). The measurements of static reception were performed at 63 sites. The analysis includes results of field strength values and signal to noise ratio thresholds for correct signal decoding, and the coverage area with static reception.

1.1 Transmission Setup

The experimental setup was composed by two digital TV transmitters in VHF channels 12 (204-210 MHz) and 13 (210-216 MHz). The transmission of adjacent channels 12 and 13 aimed the interference analysis. The directive antennas with 10 dBi gain were installed at a 40-meter tower (15°59'42''S/48°03'08''W), at 32 m (channel 12) and 41.5 m (channel 13) from the ground. Table 1 presents the configuration of the transmission modes employed in this trial, usually employed in current digital TV broadcasting systems. Three transmission ERP levels were employed: 481 W, 48 W and 5 W.

Table 1. Modulation parameters.

Bandwidth	6 MHz	
Mode	3 (8K)	
Guard Interval	1/16	
Layer	A	B
Number of segments	1	12
Modulation	QPSK	64-QAM
Convolutional code	2/3	¾
Time interleaver (ms)	1000	500
Transm. rate (Mbps)	0,44056	17,8416

1.2 Reception Setup

The mobile unit contained the reception setup, which encompasses three distinct receive chains, as sketched in figure 1.

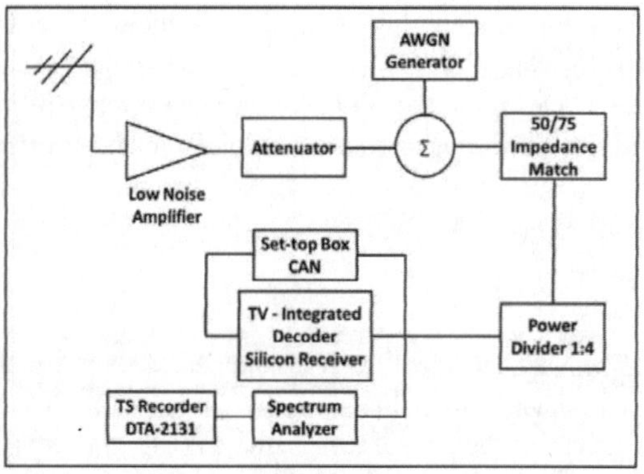

Figure 1. Reception setup.

The digital TV signal analyzer captured the received power level and noise level (in the spectrum analyzer mode), bit error rate (BER), modulation error rate (MER), power delay profile and the transport stream. The two DTV receivers (a can tuner receiver and an integrated silicon tuner receiver) allowed the subjective analysis of the image quality (Layer B).

2. PARAMETERS AND MEASUREMENT METHODOLOGY

2.1. Measurement parameters

The equipment and measurement system provided a set of parameters recorded during testing. Table 3 shows the parameters and the respective equipment and sampling. These parameters, together with the topographic map, aid the analysis of the behavior of the reception signal.

Table 3. Measurement parameters.

Parameter	Equipment	Sampling
Signal power	Spectrum analyzer	3 min
Noise power		3 min
Signal spectrum		1.2 s
BER	ISDB-T signal analyzer	3 min
MER		3 min
Channel impulse response		10 samples
TS, logs, mosaic	DTA-2131	2 min
Lat / Long	GPS	--
Image	TV set	--

Three important parameters were calculated for each measurement site. The field strength E (dBμV/m) employed the received signal power adding the coaxial cable losses and an antenna "K" factor. The signal to noise ratio C/N (dB) was achieved from signal and noise power, which was registered when the signal transmission was turned off. The signal to noise ratio for additive white Gaussian noise (AWGN) injection (C/N_{awgn}) was achieved when a Gaussian noise from an AWGN Generator was injected into the reception setup, forcing the image into the "threshold of visibility" (TOV), according to the guidelines of Report ITU BT.2035-2 [5]. In addition, photos, videos and description of measurement sites provided the environment characterization.

2.2. Measurements methodology

The Report ITU BT.2035 recommends practices for field tests evaluations. The 63 measurement sites were planned to be in a grid with site separation of approximately 2.5 km, aiming at allowing the signal reception in as many environments as possible within the city. They were located inside the service area of the transmitting station of channel 13, as indicated in figure 2.

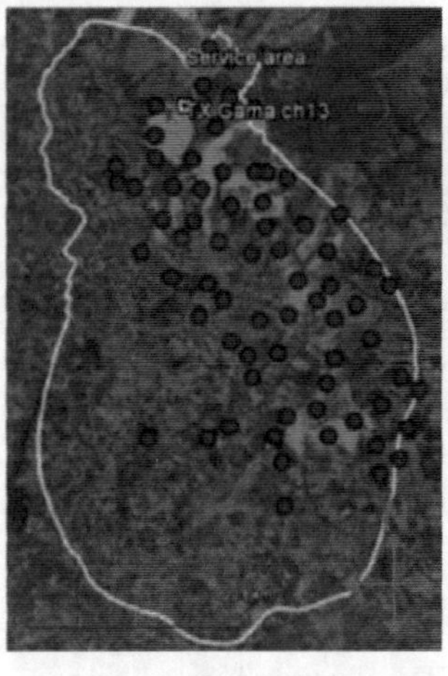

Figure 2. Measurement sites.

The coverage evaluation was performed with the observation of the image at the TV screen, which is the indication whether the digital signal decoding was successful or not. This information is correlated to the field strength values at each measurement site. The service receivability evaluation was done with parameters from the spectrum analyzer of signal power, noise power, error rate, BER and MER, which were registered simultaneously with the signal decoding status.

2.3. Measurement of C/N_{awgn} threshold

This procedure aims the identification of the threshold of C/N at each measurement site with successful signal decoding. Given an optimal received signal level in the range between -40 and -50 dBm, the additive Gaussian noise is injected in the setup until the image reaches the TOV, which means the beginning of image pixelization. The AWGN generator is parallel to the main branch, as depicted in figure 1.

3. RESULTS

The results from static measurements provided information about coverage area, electric field strength threshold, signal to noise ratio threshold (C/N) and signal to noise ration threshold with noise injection (C/N_{awgn}).

The field strength values at each measurement site and the corresponding digital signal decoding status provided the coverage area and the field strength threshold. The coverage area was given by the number of sites with digital decoding. From the set of 63 sites, the digital reception was successful at 56 sites (89%) with the silicon tuner receiver and at 53 sites (84%) with the can tuner receiver. The main reason for no digital decoding at the remaining sites was the attenuation due to obstacles of the terrain profile.

Figure 3 presents the field strength values as a function of the distance and reception condition (green or red). The field strength threshold for good reception is in the range 46 to 48 dBμV/m, depending on the tuner receiver type (silicon or can tuner). These values are 5 to 3 dB lower than the theoretical threshold required for UHF band (51 dBμV/m [6]), and 14 to 12 dB lower than UHF experimental value of 60 dBμV/m [7].

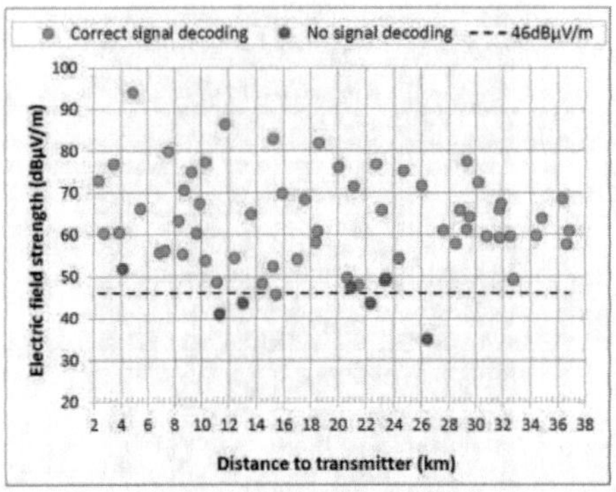

Figure 3. Field strength vs. distance (silicon tuner)

The signal to noise ratio achieved for each measurement site in correlation with the signal decoding status, is presented in figure 4. The results for the three station classes with different transmission levels (Class A, B and C) presented that the signal to noise ratio threshold is in the range from 15 to 18 dB. However, at one site with signal-to-noise ratio 19.7 dB the reception was not successful. In this case, the reception failed due to the influence of multipath, which caused great variability in the received signal.

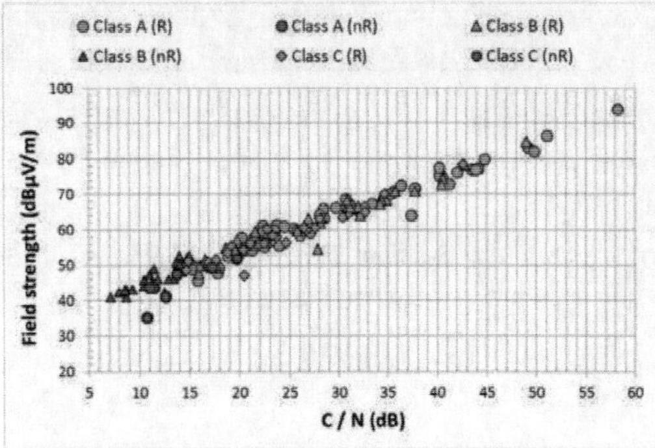

Figure 4. Field strength vs C/N (silicon tuner)

The evaluation of C/N_{awgn} resulted in values in a range of 15 to 19 dB, as shown in figure 5. These results are in accordance to the C/N threshold achieved from measured values, as presented in figure 4, which are in the range of from 15 to 18 dB.

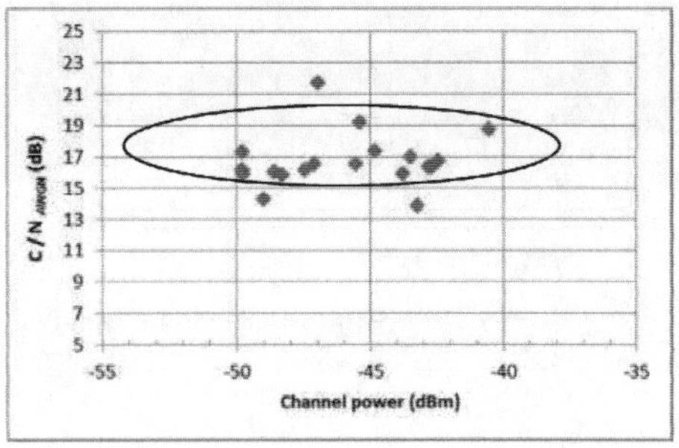

Figure 5. C/N_{awgn} achieved from field tests

4. CONCLUSIONS

This work focuses on the measurement methodology of the field measurements of VHF High-band tests at the city named Gama in the federal district (DF). The main aspects are the measurement procedure to register the environmental noise and the procedure of additive noise injection to the achievement of signal to noise ratio threshold in optimal reception conditions in field.

5. ACKNOWLEDGEMENT

Inmetro and the Brazilian Ministry of Communications thank the SBT, EBC and UNB for the support during the trials.

6. REFERENCES

[1] Anatel Resolution 625/2013

[2] Technical Specifications for Brazilian Digital terrestrial television. http://www.abnt.org.br/

[3] Min.of Communications, Results of field tests in VHF-high band in channels 12 and 13. http://www.comunicacoes.gov.br/documentos/documentos/relatorio-vhfa-assinado.pdf

[4] P.V.G.Castellanos, M.P.C.Almeida, J. Cal-Braz, R.P.David, R.S.LSouza, L.Silva Mello. "Preliminary field tests results of ISDB-TB digital TV transmission in the VHF-high band". Accepted ICEAA-AWPC-2015, Italy Sep.2015

[5] Report ITU-R BT.2035-2, "Guidelines and techniques for the evaluation of digital terrestrial television broadcasting systems including assessment of their coverage areas".

[6] "Norma 01/2010" (Port.276/2010) Technical standard for sound and image broadcasting services deployment using digital technology

[7] Akamine, C. et al., "Field Trial for Brazilian DTV using Space Diversity", IEEE-BMSB 2010

Low cost capacimeter, metrological analysis

M G KRAMAR[1], R U SOUZA[2] and V MACHADO[3]

[1,2,3] Federal University of Technology – Parana (UTFPR), Department of Electronics (DAELN), Curitiba, Brazil

MACHADO, V. email: vmachado@utfpr.edu.br

Abstract. Daily, for electronic professionals, a common need is to measure a capacitance from a capacitor. Often this measurement requires expensive equipments, not portables. This paper describes the development of an electronic circuit capable of measuring capacitance within the range of 100 nF up to 1 mF, providing a reliable and affordable system. Measures had been taken and metrological analyzes were performed on the experimental data. Also, the system theoretical model was evaluated in order to compare the behavior of both: practical and modeled system, investigating the availability of further improvements. A functional circuit with uncertainties compatible to those, provided for the theoretical model was developed. The developed system proved to be accurate, inexpensive and suitable for portable measurements. **Keywords:** capacitance measurement, electrical metrology, capacimeter, calibration.

1. Introduction

Daily, for electronic professionals, a common need is to measure a capacitance from a capacitor. Many technologies are commercial available.

This paper proposes an easy to make and low aggregated cost capacimeter, able to measure in the range of 100 nF up to 1 mF. Software and hardware implementations were carried out in order to achieve low power consumption, allowing portable power source.

2. Measurement Techniques

When it comes to capacitance measurement, there are many techniques applicable. They vary in complexity, in cost and response time.

One widespread measurement technique is called auto balancing bridge [2, 3]. In this circuit there is an operational amplifier (figure 1) whose inverter input is connected to a series association of a sinusoidal source with the element to be measured. Also a negative feedback network with a known value is connected. The positive input is grounded.

This technique is very accurate and its cost is affordable. Commercial equipment prices using this technology are around dozens of hundreds of dollars.

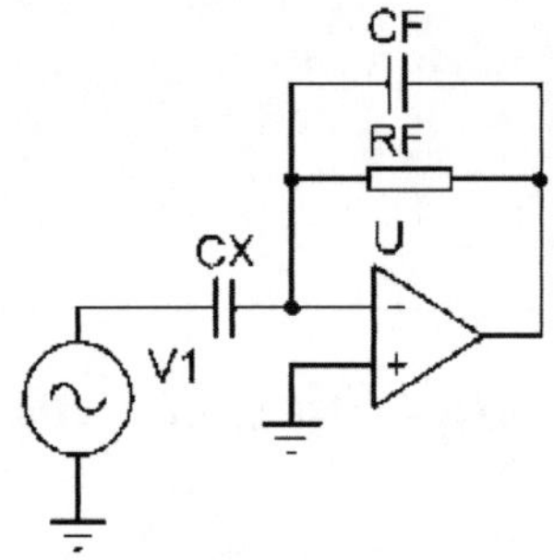

Figure 1. Auto balancing bridge, a common way to measure capacitance.

The Wien Bridge [1] is another procedure with broad use in capacitance measurement. It consists of an electrical bridge in which the left legs are resistors and the right legs are RC associations (series at the top and in parallel at the bottom) as depicted in figure 2.

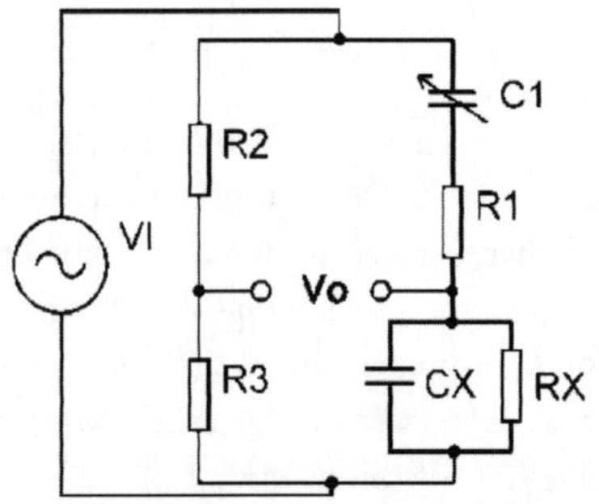

Figure 2. Wien bridge, a precise method to measure capacitance.

The advantage of this kind of circuit is the possibility to choose resistors in the capacitive side (R1 and RX) that diminish effects coming from parasitic resistances of the capacitor constructive characteristics.

Wien Bridge is a technique very accurate, requiring thousands of dollars as investment in equipment carrying this technology.

A third technique, the discharge rate measurement, is a simple and low cost technique, in despite of the fact that it is less accurate than the other two.

It is based in the time needed to a certain voltage be reached in the element whose capacitance is being measured. Another similar version is the connection of a constant current source with the measurand and then measuring the rate of charge.

The employed technique on this paper is a variation of the discharge rate technique. Instead of measuring an analogue voltage level, a digital signal is read.

3. Circuit Design

The developed system consists of an astable oscillator based on an operational amplifier connected to a low-power microcontroller. The microcontroller measures the time between two edges of the produced signal by the op-amp, calculating the capacitance and then displaying it on a liquid crystal display (LCD). The figure 3 shows the circuit block diagram.

To produce the oscillator (figure 4), it was used the op-amp TL082 since it has low cost and good features as: cutoff frequency, input impedance and slew-rate.

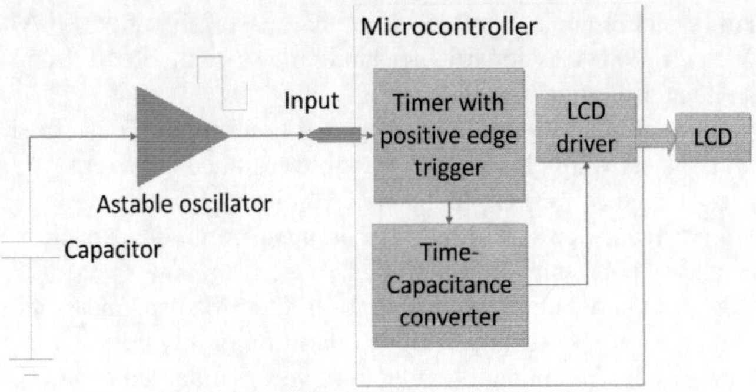

Figure 3. Block diagram of the developed system.

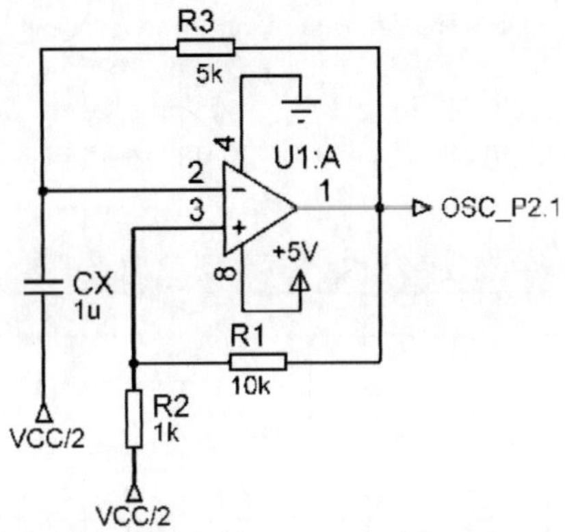

Figure 4. Astable oscillator that generates a signal whose frequency is measurand based.

The period of the square-wave generated by the oscillator is calculated by equation (1) where: T stands for the wave period, Cx is the capacitor to be measured, R1 and R2 are Schmitt trigger resistors and R3 is the negative feedback resistor.

$$\mathrm{T}(\mathrm{C}) = 2R_3 Cx \ln\left(1+2\frac{R_1}{R_2}\right) \quad (1)$$

The microcontroller (MSP430G2553) – by Texas Instruments – was chosen due the low cost and low power consumption.

A simple state machine describes the firmware. When the system is turned on, it stays on low consumption state. If a positive edge occurs, the device's timer triggers the count mode waiting for another positive edge that stops the timer counter and allows the microcontroller calculate the signal period.

After that, the processor leaves the low power state to the print state. When the unknown capacitance (measurand) is calculated it is consequently printed in a LCD 16x2. Finishing the printing state, the system enters in low power mode again and remains in the idle state until another event.

4. Calibration, Results and Discussion

4.1. Experimental Methodology

The produced system (system to be calibrated, calibrate measuring system - CMS) was compared with a Precision LCR Meter E4980A (standard measurement system, SMS) from Agilent Technologies enabling to establish its metrological characteristics.

Three sections of measurements were made, respecting a protocol described as follow: first of all, capacitor's electrical contacts were cleaned and possible residues were removed. The frequency of the SMS was adjusted to a range of 100 Hz to 1 kHz. Admitting RC series model best representing the capacitor, the capacitive reactance is dominant. For most common capacitor used – especially for electrolytic and low quality polyester – this is true.

Secondly, as an instruction manual recommendation, the SMS was measured in open circuit and in short-circuit, in that manner any influence of SMS condition or any parasitic components that it could present would be suppressed. The ambient temperature was maintained at 18° C.

For every section, the capacitance was measured with the SMS. The alligator clips had been put in order to electrical contact was the greatest possible. Thus, the capacitor was discharged and then measured with the CMS. Values were annotated and the process continued until all values were computed.

To measure the stability of the CMS in relation to voltage, every time the voltage was changed, the measurand was removed, the voltage was changed and one more measure was taken.

4.2. Experimental Data

Figures 5, 6 and 7 show the error curves CMS.

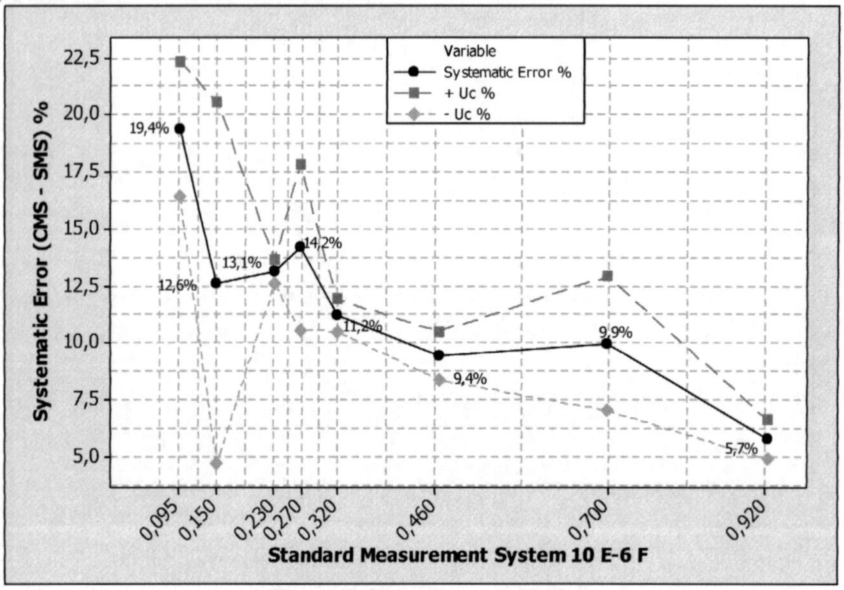

Figure 5. Error Curve 1 CMS.

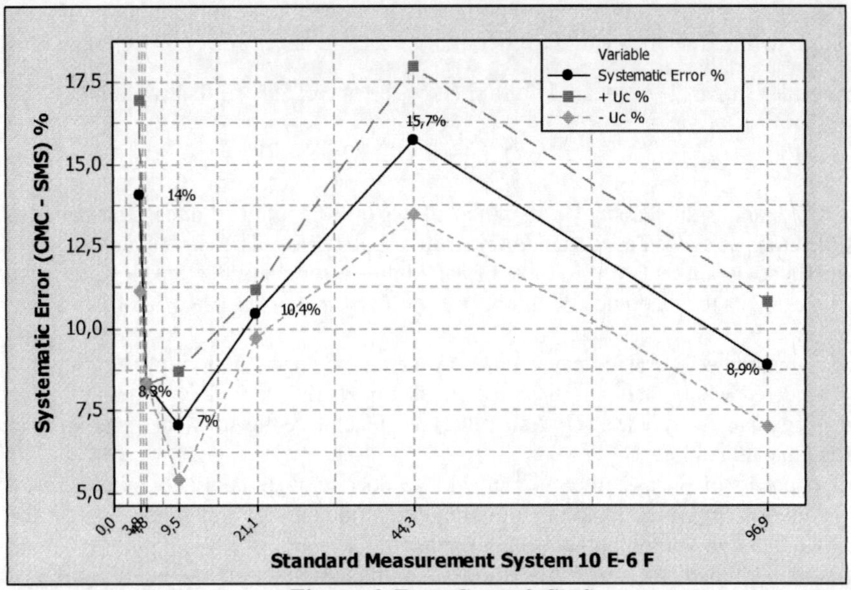

Figure 6. Error Curve 2 CMS.

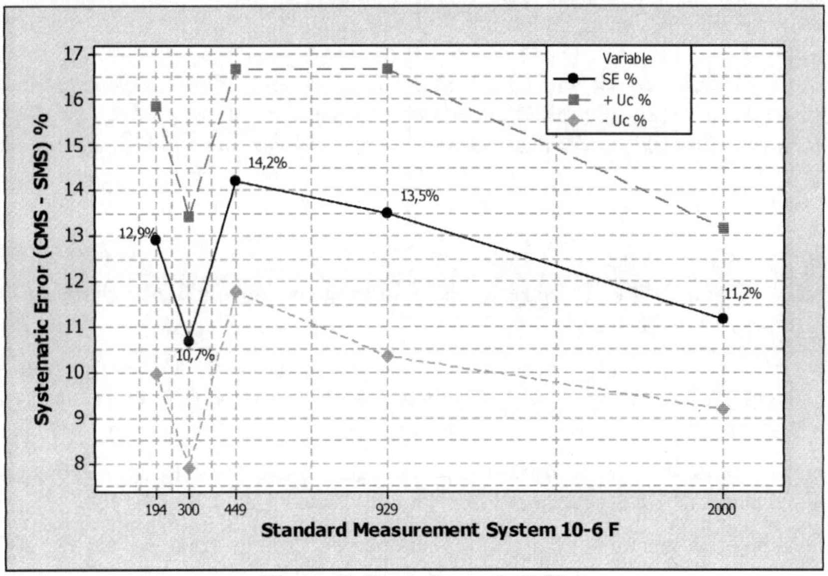

Figure 7. Error Curve 3 CMS.

4.3. Uncertainty analyses of the theoretical model

With the practical combined expanded uncertainty (U_p) calculated it was necessary to calculate the theoretical combined expanded uncertainty (U_t) in order to compare both of them [4].

U_t was calculated based on microcontroller deviation in frequency by not having a crystal as its timing source reaches the 30,000 ppm. Also, there was the uncertainty due to resistors R1-3 that have a deviation from the nominal value about 10 %. Neither the parasitic capacitance in metallic contacts nor amplifier output capacitance, were taken into account.

Equation (2) defines that the theoretical combined standard uncertainty (u_c) squared is equal to the sum of partials derivatives ($\frac{\partial C}{\partial xi}$) squared multiplied by the uncertainty of that portion (u_i).

$$u_c^2 = \sum_{i=1}^{n} (\frac{\partial C}{\partial x_i} u_i)^2 \qquad (2)$$

In figures 5, 6 and 7 the tendency plus practical expanded combined uncertainty and the tendency minus this uncertainty, as well as the tendency itself are presented. The two envelope curves represent the total uncertainty given by CMS. In most of parts in the error curve, the envelope curves stay below the line that represents the theoretical total uncertainty, correspondent to 18,1 % of deviation in relation to the measure.

As already mentioned, parasitic capacitance in metallic contacts and amplifier output capacitance were not considered when analyzing the theoretical model. Either of them is present parallel to the measurand and explains why in the very first values of capacitance the practical uncertainty line exceeds the equivalent theoretical line.

If the components of the system are changed also does its uncertainty. A microcontroller using a crystal as source of frequency (150 ppm in frequency fluctuation) and precision resistors (1 % of fluctuation) would allow the total practical uncertainty be below 1.36%.

Lastly, each portion of the combined uncertainty impact was evaluated. The portion due to the frequency is represented by u_T and u_{R1-3}, that represents the portions due to variations on R1 through R3. The portion that has more impact is the frequency variation.

5. Conclusion

An electronic capacimeter was developed and calibrated. It was possible to measure capacitances within 100 nF to 1 mF with empirically raised uncertainties consistent with those from theoretical analyzes. Also, the theoretical model foresees that more accurate components would lead to better uncertainties. A precision LCR bridge was used to obtain values accurate enough to serve as measurements reference. The developed system is cheap and suits well for portable power sources.

References

[1] Weberster J G and Halit E. The measurement, instrumentation and sensors handbook. 1st ed. Boca Raton: CRC Press; 1999. 2630 p.

[2] Longo J P N et al. Sensor Capacitivo Interdigital para detecção de deposição de parafina em oleodutos [Interdigital capacitive sensor for wax deposition detection on oil pipelines]. Anais do XX Congresso Brasileiro de Automática; 2014; Belo Horizonte, MG, Brazil.

[3] Wrasee A N. Capacitive array sensor for direct imaging of two-phase flows. 15th Brazilian Congress of Thermal Sciences and Engineering; 2014; Belém, PA, Brazil.

[4] Albertazzi A. Fundamentos de Metrologia Científica e Industrial. Barueri: Manole; 2008. 408p.

The quantum vector digital voltmeter of INMETRO

Waldemar G Kuerten Ihlenfeld, Regis P Landim

National Institute of Metrology, Quality and Technology – INMETRO, Av. Nossa Senhora das Graças no. 50 – Xérem, 25250-020 Duque de Caxias/RJ, Brazil.

E-mail: wgkihlenfeld@gmail.com

Abstract. The paper describes the quantum vector digital voltmeter developed at INMETRO, which is based on a programmable Josephson voltage synthesizer. The system employs digital regulation for phase-alignment and frequency synchronization of signals, is fully automated and allows calibration of ac sources and analog-to-digital converters with uncertainties bearing some parts of 10^{-07} up to frequencies of around 500 Hz.

1. Introduction

The Josephson Effect was discovered 54 years ago [1]. Its metrological applications explore its ability of being a nearly perfect quantum converter of frequency into an extremely accurate dc voltage (with typical uncertainties bearing only some 10^{-10} V/V). Josephson systems are nowadays worldwide spread and common in National Metrology Institutes (NMIs). They are used to maintain and disseminate the unit volt (in direct current or dc). Recent advances focus on the development of digital-to-analog converters (DACs) based on the Josephson Effect to provide a quantum reference for the calibration of alternating (ac) signals. Here the developments rely on the quantum generation of ac signals of the same frequency, close resembling the ac signal to be calibrated. These pursuits follow two distinct ways: 1) The generation of ac signals by stepwise-approximated waveforms (like a common DAC) called programmable step-driven Josephson voltage synthesizer - PJVS or 2) pulse driven signal generation, akin to delta-sigma signal generation. Although the first covers low frequencies from dc to 1 kHz up to 10 V peak [2], the second so far generates signals from 1 kHz to 1 MHz (and dc) up to around 1.4 V peak [3]. INMETRO's system is based on a PJVS developed at the National Institute of Standards and Technology (NIST) [4,5]. However, a PJVS alone does not guarantee the traceability of ac signals. INMETRO's PJVS had thus to be integrated as an ac reference into a complex system devised for that purpose. This system is called quantum vector digital voltmeter and differs from other developments done elsewhere in respect with hardware and software. It uses an automated and unique patent pending phase- and frequency- synchronization of signals (to be calibrated against the PJVS), managed by an exquisitely sensitive digital regulation as explained next.

2. The AC Quantum Voltmeter

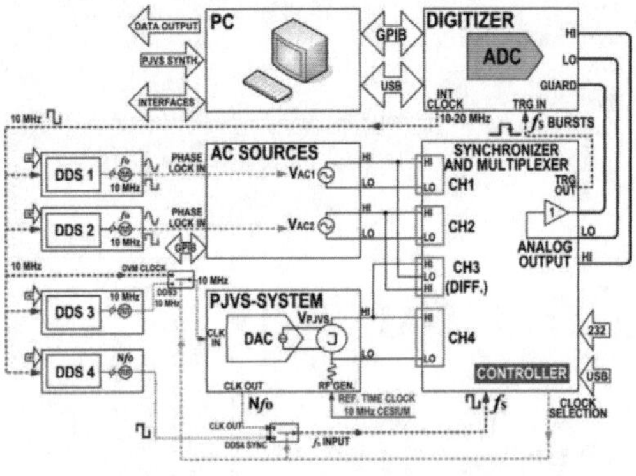

Figure 1. The quantum digital voltmeter (QDVM) uses a PJVS system as a reference, a synchronizer with a multiplexer for comparing signals, ac sources, a 28-bit integrating digitizer and four direct-digital-synthesizers for synchronization purposes (see text).

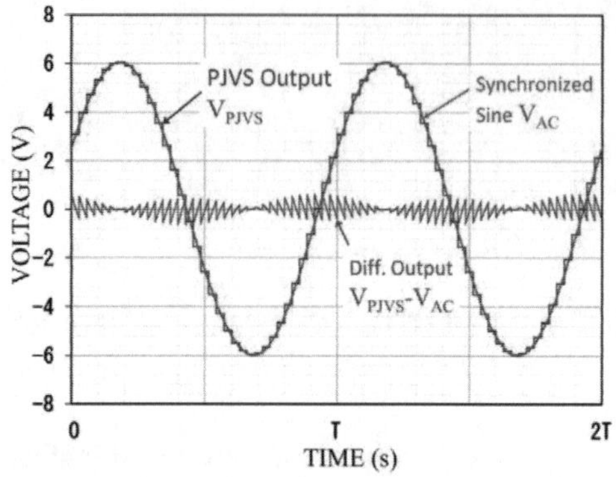

Figure 2. The ac signal is synchronized with the stepwise approximated PJVS waveform by digital regulation. The ADC samples the resulting differential voltage close to its zero crossings to determine the magnitude of V_{AC}.

Figure 1 represents the system developed at INMETRO using a PJVS (at the bottom middle). The PJVS arrays are biased by a current DAC and a microwave signal of nearly 20 GHz, locked to the 10 MHz of a cesium time standard [4].

The devices under test (DUT) are the ac sources V_{AC1} and V_{AC2}, and the ADC is a 28-bit integrating digital converter, which delivers the 10 MHz internal clock for synchronization purposes and signal generation of the PJVS. The ac sources V_{AC1} and V_{AC2} are phase-locked by direct digital synthesizers (DDS), i.e., DDS 1 and DDS 2, whereas DDS 3 allows the synchronization of the PJVS with other sources. DDS 4 generates the sampling frequency f_s to sample the ac signals V_{PJVS}, V_{AC1} and V_{AC2}.

The synchronizer possesses four channels and allows direct and differential measurements [6, 7] of the ac sources with the PJVS to be made. It manages also the synchronization of the system by enabling the selection of clocks (the dotted busses in figure 1) and the synchronous data-acquisition by the ADC. Here a coherent signal generation (of the ac sources and PJVS) and data acquisition (by the ADC) takes place with the 10 MHz ADC clock reference. The ADC data are sampled with N steps (PJVS-steps) or samples per period over a pre-defined M (integer) number of periods. Samples are taken on the flat portion of the PJVS plateaus, allowing enough time for the signals to settle in the ADC circuitry after properly aligning the aperture window on each plateau. This alignment is done by DDS 4 or by using the CLK OUT output of the PJVS system, as described in [7]. The synchronization of the ac sources V_{AC1} and V_{AC2} is accomplished by repeated sampling these signals, determining their phase differences (when compared with the PJVS signal) and by fine trimming DDS 1 and DDS 2 in a feedback loop over a finite time span. Precise and calculable frequency variations yield precise phase-shifts within some nano-radian resolution. A deeper treatment of this patent pending algorithm is described in [7]. This results in perfectly aligned signals, as figure 2 shows.

3. Measurement Capabilities

The synchronizer and multiplexer allow the system to make direct as well as differential voltage measurements. These are depicted schematically in figure 3. Direct measurements are used to first calibrate the ADC and its gain at a particular signal frequency, sampling frequency and aperture time.

Figure 3 A and E shows the multiplexing of either the PJVS and V_{AC1} or V_{AC2} for calibrating V_{AC1} or V_{AC2} with a prior calibration of the ADC against the known PJVS' plateaus.

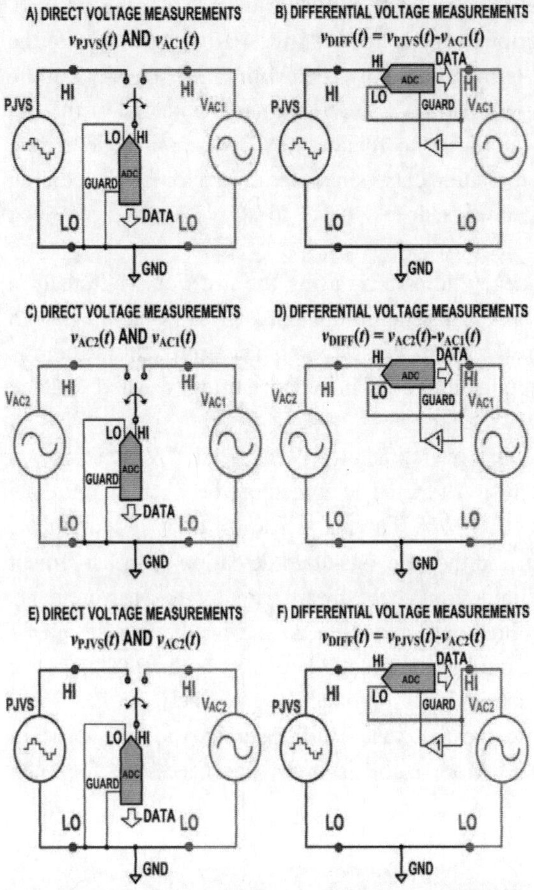

Figure 3. The quantum voltmeter allows direct voltage measurements A) C) and E), as well as the most precise differential measurements B) D) and F) for determining the magnitudes of V_{AC1} or V_{AC2}. Comparisons of two continuous ac voltages is also possible as shown in E) and F).

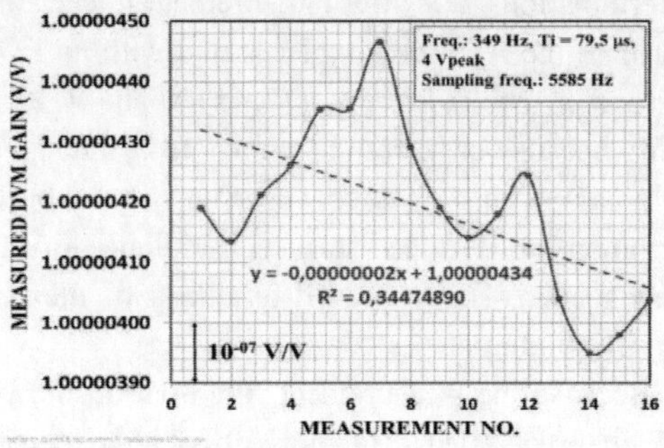

Figure 4. Measured gains of the ADC over 16 measurements (slightly over a 30 minutes measurement time) against the PJVS. A gain drift of 2E-8 V/measurement is noticeable from a linear data fitting as shown. Variations of some parts of 10^{-7} during measurements are common for ADCs.

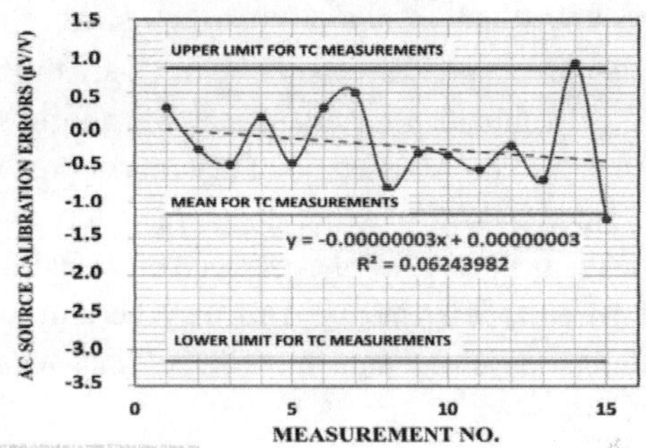

Figure 5. Calibration errors of an ac source operating at 349 Hz and 4 V peak. PJVS differential calibrations indicate that the source error is around ±1E-6 V/V indicating a linear drift of 3E-8 V/V. The mean of 2500 thermal converter measurements lays close to -1E-6 V/V but its uncertainty is around ±2E-6 V/V from the mean (including instabilities of the source).

ADC's errors are computed from a voltage ratio measurement, i.e., the fast Fourier transform (FFT) of the tabulated (or programmed) PJVS plateaus is divided by the FFT of the ADC measured plateau values, resulting in the determination of the ADC gain at the fundamental frequency. Figure 4 shows 16 ADC gain determinations. Variations of ADC's gain as much as some parts of 10^{-7} are common and may vary (during measurements) mainly due to internal temperature fluctuations in the ADC circuitry and ADC´s reference voltage. After sampling the ac sources, the FFT of their measured data is multiplied by the ADC gain at that particular fundamental frequency leading to the correct determination of the magnitude of either V_{AC1} or V_{AC2}, and their harmonic content.

A direct comparison between two ac sources at the same frequency is also possible by substituting the PJVS by V_{AC2}, which can then be compared with V_{AC1} as shown in figure 3 C and 3 D.

Differential measurements as depicted in figure 3 B, D and F demand the signals to be perfectly phase-synchronized (phase aligned) to minimize the amplitude of the differential voltages. The smaller the differential voltage is, the smaller is the effect of the ADC's errors on the accuracy of voltage measurements. Differential voltage measurements result in much more accurate amplitude determinations, allowing measurement uncertainties of some parts in 10^{-7} or even 10^{-8} to be attained. For that, the ADC or sampler is tied with its LO-terminal connected with the HI-output of one of the ac sources (see figure 3 B, D, and F). The ADC HI-terminal is always tied with the HI-terminal of the PJVS because of its fast changing steps. The guard-terminal is driven by the same potential of the HI-terminal of an ac source via a unity-gain buffer amplifier to reduce ADC common-mode errors. Common-mode effects may substantially impair measurements, since some unavoidable coupling among guard-, LO- and HI-terminals of the ADC always do exist due to stray capacitances (also within its internal electronics).

Because the measurements are done in the frequency domain by using the fast Fourier-transform (FFT) on sampled data, it is possible to determine the harmonic content of V_{AC1} and V_{AC2} with quantum (or fundamental) accuracy, including its real and imaginary parts. Therefore, this system is thus called a vector voltmeter. This opens up new applications as in impedance bridges, where voltage ratios are used to determine impedances.

Figure 5 shows calibration errors of an ac source operating at 349 Hz (and 4 V peak) when compared with the PJVS and a primary thermal converter (TC). The agreement between calibrations with PJVS and TC are on the mean smaller than $\pm 10^{-6}$ V/V. Thermal ac-dc transfer measurements show much higher dispersion, so that only the mean and the $\pm 1\ \sigma$ (standard deviation of measurement and TC calibration uncertainty) upper and lower limits around the mean are shown. Measurements with a TC encompassed 2500 determinations over three days and were done after the measurements with the PJVS to avoid the circulation of ground-loop currents between both systems. The measurements suggest the presence of a small systematic deviation of $+5\cdot 10^{-7}$ V/V between PJVS and TC-based measurements. This may be attributed to the fact that TC measurements were done in different days and due to drifts of the ac source. More stable sources are necessary for such investigations.

4. Conclusions and Outlook

The capabilities of the new quantum voltmeter [8] allow a sensible reduction of measurement uncertainties of ac quantities to limits never hitherto attained. Future developments will focus on widening its operation towards the audio frequency range and on investigations of other recondite effects in the ADC.

Research efforts are presently devoted to extend the frequency range of the system depicted in figure 1 towards the audio and MHz frequency range, by using commensurate frequencies and Undersampling [9]. Meanwhile, new ac sources of highest stability and low harmonic content are under development, which will be integrated into the quantum voltmeter to allow even lower measurement uncertainties to be attained.

Acknowledgment

This work is being conducted at Inmetro under research contract no. 52600.057476.

References

[1] Josephson B D 1962 Possible new effects in superconductive tunneling *Phys. Lett.* **1**(7) p. 251-253.

[2] Lee J et al 2013 An ac quantum voltmeter based on a 10V programmable Josephson array Metrologia **50** p. 612–622.

[3] Benz S P and Waltman S B 2014 Pulse-bias electronics and techniques for a Josephson arbitrary waveform synthesizer *IEEE Trans. Appl. Supercond.* **24**(6).

[4] Burroughs C J et al 2011 NIST 10 V programmable Josephson voltage standard system *IEEE Trans. Instrum. Meas.* **60**(7) p. 2482-2488.

[5] Burroughs C J et al 2009 A 10 volt turnkey programmable Josephson voltage standard for dc and stepwised-approximated waveforms *Measure* **4**(3) p. 70-75.

[6] Rüfenacht A, Burroughs C J, Dresselhaus P D, and Benz S P 2013 Differential sampling measurement of a 7 V RMS sine wave with a programmable Josephson voltage standard *IEEE Trans. Instrum. Meas.* **62**(6) p. 1587-1593.

[7] Kuerten Ihlenfeld W G and Landim R P 2015 An automated Josephson-based AC-voltage calibration system *IEEE Trans. Instrum. Meas.* **64**(6) p. 1779-1784.

[8] Kuerten Ihlenfeld W G and Landim R P 2015 The quantum vector voltmeter of Inmetro XI Semetro Conf. Digest no. 187 ISBN 978-85-86768-10-1.

[9] Kuerten Ihlenfeld W G and Landim R P 2016 Investigations on extending the frequency range of PJVS based AC voltage calibrations by coherent subsampling *CPEM2016 Conf. Digest*, Ottawa, Canada.

New thermal neutron calibration channel at LNMRI/IRD

A Astuto[1,2], K C S Patrão[3], E S Fonseca[3], W W Pereira[2,3] and R T Lopes[1]

[1]Universidade Federal do Rio de Janeiro – COPPE/UFRJ. Av. Horácio Macedo, 2030, Bloco G/Sala 206 – Centro de Tecnologia – Cidade Universitária, Ilha do Fundão, CEP: 21941-914, Rio de Janeiro, RJ, Brasil.
[2]Fundação Técnico Educacional Souza Marques - FTESM. Av. Ernani Cardoso, 335, Cascadura, CEP: 21310-310, Rio de Janeiro, RJ, Brasil.
[3]Laboratório Nacional de Metrologia das Radiações Ionizantes LNMRI/IRD. Av. Salvador Allende, s/n, Barra da Tijuca, CEP: 22783-127, Rio de Janeiro, RJ, Brasil.

E-mail: achillesbr@gmail.com

Abstract: A new standard thermal neutron flux unit was designed in the National Ionizing Radiation Metrology Laboratory (LNMRI) for calibration of neutron detectors. Fluence is achieved by moderation of four ^{241}Am-Be sources with 0.6 TBq each, in a facility built with graphite and paraffin blocks. The study was divided into two stages. First, simulations were performed using MCNPX code in different geometric arrangements, seeking the best performance in terms of fluence and their uncertainties. Last, the system was assembled based on the results obtained on the simulations. The simulation results indicate quasi-homogeneous fluence in the central chamber and H*(10) at 50 cm from the front face with the polyethylene filter.

Keywords: Neutron Metrology; Neutron Fluence; Monte Carlo Simulation; Calibration.

1. Introduction

Due to the growth of soil logging activity in prospecting oil and natural gas, where over 60% of all neutrons sources in the country are concentrated [1], the amount of neutron personal area monitors and survey meters in Brazil have increased significantly. It generates a growing demand for calibration and mechanisms to ensure control of their metrological parameters.

The purpose of this project was to develop an irradiation system with thermal neutrons to calibrate personal dosimeters and survey meters, using radioisotopic neutron sources and moderators materials. It derives from a system previously built with paraffin with graphite blocks (PG) and pure graphite blocks [2]. The results achieved were not satisfactory due to problems in the PG composition, especially regarding the spectra at points of interest.

Although the International Organization for Standardization (ISO) recommends the use of a moderated-reactor or an accelerator-produced neutron to produce thermal neutron fields [3], the use of radioisotopes in a suitable thermal device is desirable, mainly due to its stable fluence and easy operation. In addition, facing technological advances, where new equipment and processes have been producing more accurate measurements, several calibration centres around the world are remaking or rebuilding their thermal fields [4,5].

The current facility remained the external dimensional length of the previous one, built on a cube with the same materials used before, with dimensions of 1.2 x 1.2 x 1.2 m on a steel platform which is about 50 cm above the ground. A central chamber 10 x 10 x 10 cm size in the pile centre, connected to the outside by a central channel (10 x 10 x 55 cm). Four equally spaced ^{241}Am-Be neutrons sources are used in order to obtain a central field, with homogeneous thermal neutron fluence for calibration purposes. The great difference to the previous one is the positioning of moderator material, because while the old were used PG blocks and graphite blocks among the source and the central chamber, the current one uses only graphite in the device core and PG around that graphite core to form the cube, as can be seen in figure 1.

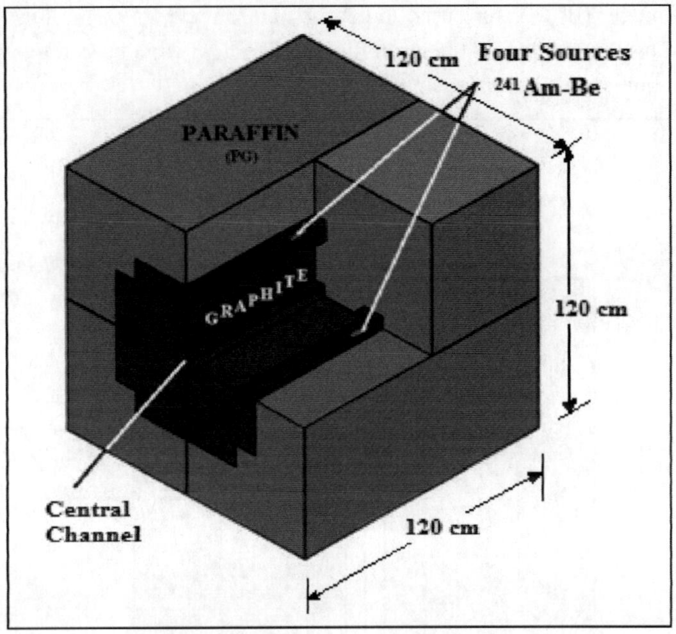

Figure 1 - Schematic 3D view of the new thermal neutron facility.

2. Methodology

This study was divided into two steps. First one, simulations using the MCNPX code [6] on different geometric arrangements of moderator materials and neutron sources were performed. On second step, the pile system was assembled based on results obtained by the new simulation.

2.1 Simulations

Simulations were made to study how the thermal neutrons fluence could be changed by displacing the source distance to the center of the chamber as well as the cadmium rate (CDR). The CDR is defined as the ratio between total neutron fluence and fluence crossing a cadmium cover. Cadmium acts as a filter by absorbing thermal neutrons (neutrons below 0.5 eV).

To setup the creation of an ambient dose equivalent (H*(10)) low-band variation at a distance of 50 cm from the front face, a polyethylene filter was simulated. When this filter was placed at the outlet of the central channel, an 'homogeneous' H*(10) band over about 35 cm was established, allowing dosimeter calibration that should be carried out on the 30 cm x 30 cm x 15 cm ISO phantom [3]. Various thicknesses were simulated to find one that would provide the desired effect.

3. Experimental Procedure and Measurements

The PG simulations as in the first design were affected by difficulties in obtaining the material density and the correct mass fraction, since the mass of the heavier and lighter blocks varied up to 1 kg. These blocks were radiographed and the images showed the presence of air bubbles and sand impurities inside. The elemental chemical analysis, CNHX, was used to indicate the mass fraction of carbon, hydrogen and nitrogen. This fraction was 89.6% for carbon and 7.4% for hydrogen. The analysis showed the presence of about 3% of impurities in the PG mass fraction, and then 3% of air and sand (SiO_2) were included in the simulated PG composition. The average density of graphite blocks was 1.328(14) g/cm^2.

Six polyethylene spheres with diameters ranging from 5.08 to 30.5 cm of the LNMRI Bonner Multisphere Spectrometer (BMS) were used to get the neutron spectra of the four ^{241}Am-Be sources (S1, S2, S3 and S4) used in the thermal neutron flux facility. The same procedure will be used for the spectra analysis on some points around the pile (figure 2).

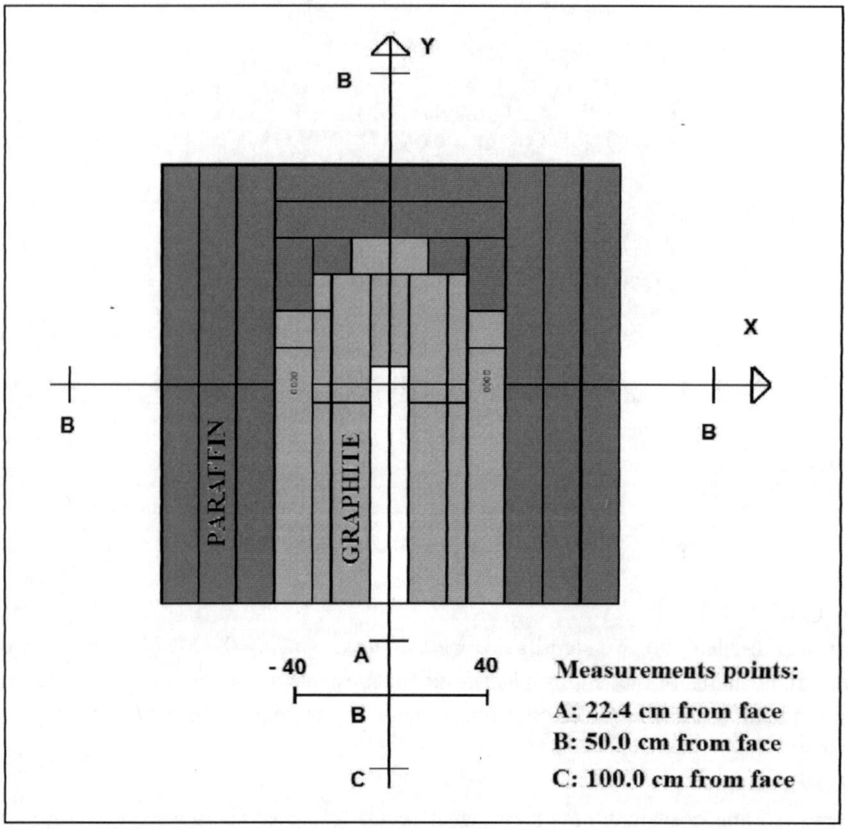

Figure 2 - Measurement points scheme in the facility room, at Z = 0 slice.

To test the thermal neutrons beam homogeneity, the fluence was simulated at 19 points within the central chamber and points around the installation (preferably at a distance of 50 cm from the faces) to determine the net fluence for calibration. In the front face, fluence variation over a distance of 50 cm was investigated. The determination of the ambient dose equivalent rate, H*(10), on points of interest, was performed through the definition of point detectors F5 (F5 tally) for neutrons, available on MCNPX program. The results were modified with the use of ED/DF dose function cards, with the introduction of conversion coefficients for ambient dose equivalent H*(10)/Φ, in $pSv.cm^2$, given by ICRP 74 [7].

One of the four sources used (S1) was calibrated on a Primary and Absolute Neutron Standardization System traceable to BIPM by CCRI-III K9 international key comparison relative to ^{241}Am-Be emission rates [8]. The other three sources (S2, S3 and S4) emission rates were intercompared with the source S1 using the BSS at a distance of 1 m and then normalized.

The polyethylene filter was simulated, designed and built with 29 discs with diameters ranging from 5 to 34 cm and a thickness of 4 mm each. A commercial monitor WENDI, FHT 762 model, manufactured by ThermoElectron, was used to perform measurements of H*(10).

4. Results

Table 1 shows the emission rate of the four sources used in the pile, measured with BSS and the contribution of them to total fluence. It can be seen that the major difference between the sources emission is about 10%, a fact that was weighted during the positioning and the system simulations.

Table 1 - Source emission and weight

LN code	Source	Emission (n/s)	Weight	Weight (n)
A-639	S1 (*)	3.952(59)E+07	1.00	0.254
A-441	S2	3.868(58)E+07	0.9788	0.249
A-397	S3	3.627(54)E+07	0.9177	0.233
A-597	S4	4.097(62)E+07	1.037	0.264
	Total	1.554(23)E+08	3.933	1.00

The results obtained for fluence simulation in the central chamber, showed less than 2% variation over CDR calculations among the 19 points, indicating a probable homogeneous fluence inside the central chamber. The average fluence rate calculated in the chamber center was $1.934(23)x10^5 n/cm^2.s$. The external points values are listed in table 2.

Table 2 - Fluence in external points around the pile

Reference position	Fluence rate ($n.cm^2.s^{-1}$)	Percentage of total fluence (%) < 0.5eV	0.5eV-100keV	>100keV
(0,0,0)	$1.934(23)x10^5$	7.3	6.9	85.8
A(-y)[1]	1948.8(6)	67.0	22.7	10.2
B(-y)[1]	715.4(4)	67.3	21.7	11.0
C(-y)[1]	242.5(6)	65.7	22.3	12.0
B(y)[2]	24.26(2)	44.6	16.4	39.0
B(-x)[2]	57.13(4)	42.6	14.2	43.2
B(x)[2]	55.85(5)	42.5	14.6	42.9

[1](-y) = frontal face values at points A, B and C;
[2](y), (-x), (x) = back and lateral face values at point B.

The graph of figure 3 shows the results of H*(10) variations over a range of 36 cm at a distance of 50 cm from the front face (point B) with MCNPX simulations and WENDI measurements.

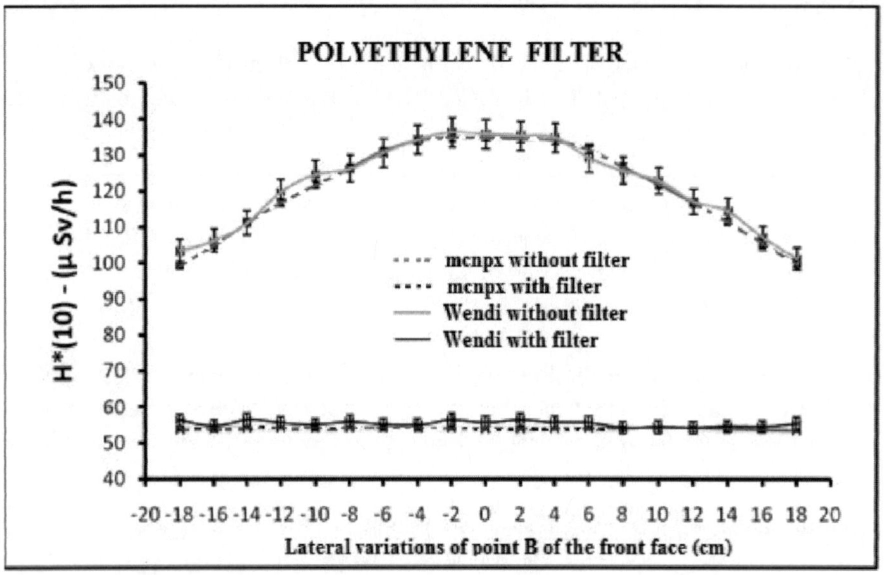

Figure 3 - Measurements along the dimension -50 cm (B), on the front face of the installation.

5. Conclusions

In order to confirm the thermal neutron field, fluence rate and H*(10) at points around the pile and in the central chamber (inside the channel), measurements will be performed using the BMS at first, and then survey meters and their respective response functions. In the central chamber it will be necessary to use another measuring method, such as the gold foils activation with measurement traceability. These steps will be performed in a future stage.

The CDR in the new arrangement is lower than the previous arrangement, however, due to sources proximity, greater fluence is achieved.

The Figure 3 indicates that results obtained with polyethylene filter reached the proposed objectives.

6. References

[1] Martins M M, Maurício C L P, Cunha P G, Almeida C E V, Fonseca E S, 1995. In anais do III Encontro Nacional de Aplicações Nucleares, **v. 2**, pp. 1173-77, Águas de Lindóia.

[2] Astuto A, Salgado A P, Leite S P, Patrão K C, Fonseca E S, Pereira W W and Lopes R T, 2014, Radiation Protection Dosimetry; **161**, 1-4; 185-189.

[3] International Organization for standardization, 2000, **ISO 8529-1.**

[4] Uchida Y, Saegusa J, Kajimoto Y, Tanimura Y, Shimizu S and Yoshizawa M, 2005, JAERI-Tech **2005-012**, Japan Atomic Energy Research Institute.

[5] Lacoste V, Gressier V, Muller H, Lebreton L, 2004, Radiation Protection Dosimetry **110** (1-4), 135-139.

[6] Briesmeister. J F, 2003, *Los Alamos National Laboratory*. **Version 5**.US .

[7] International Commission On Radiological Protection, 1996. ICRP publication **74.**

[8] Roberts, N.J. Jones et al. 2011 Metrologia **48** 35, Tech. Suppl. 06018.

Experimental and Monte Carlo evaluation of an ionization chamber in a ^{60}Co beam

A P Perini[1], L P Neves[1], W S Santos[2], L V E Caldas[2]

[1]Instituto de Física, Universidade Federal de Uberlândia (INFIS/UFU), Caixa Postal 593, 38400-902, Uberlândia, MG, Brasil

[2]Instituto de Pesquisas Energéticas e Nucleares Comissão Nacional de Energia Nuclear (IPEN/CNEN-SP), 05508-000, São Paulo, SP, Brasil

E-mail: anapaula.perini@ufu.br

Abstract. Recently a special parallel-plate ionization chamber was developed and characterized at the Instituto de Pesquisas Energéticas e Nucleares. The operational tests presented results within the recommended limits. In order to determine the influence of some components of the ionization chamber on its response, Monte Carlo simulations were carried out. The experimental and simulation results pointed out that the dosimeter evaluated in the present work has favorable properties to be applied to ^{60}Co dosimetry at calibration laboratories.

1. Introduction

Radiotherapy demands high precision and accuracy in the absorbed dose delivered to the tumor and a good protection to the surrounding healthy tissues. There are three main risks for inaccuracy in the dose delivered to patients: dosimetry, treatment planning and patient treatment [1]. The dosimetry area involves the following issues: commissioning of treatment systems and sources; dose calibration and dose calculation of treatment systems and sources; and problems related to equipments.

At the Calibration Laboratory of the IPEN (LCI) some dosimeters were developed to be used in diagnostic radiology [2], radiotherapy [3] and quality control of X-ray facilities [4]. Recently, an ionization chamber prototype was developed to be used in diagnostic radiology dosimetry [5]. As many hospitals and calibration laboratories worldwide still use ^{60}Co sources for radiation therapy treatments and for calibration of dosimeters, in this work this prototype was also evaluated and characterized to verify its applicability for the dosimetry of ^{60}Co beams.

The undertaken experiments followed the recommendations of the IEC 60731 standard [6]. As the ionization chamber, evaluated in this work, presents differences in relation to the dosimeters usually employed for ^{60}Co dosimetry (cylindrical type), Monte Carlo simulations were adopted to study the chamber configuration (design and materials).

2. Materials and Methods

2.1. Experimental Evaluation

The ionization chamber studied in this work is presented in figure 1. It was manufactured using the following materials: PMMA coated with graphite for the walls and collecting electrode,

Table 1. Ionization chamber technical specifications

Characteristics	Specifications
External height (mm)	16.97±0.02
Wall thickness (mm)	2.00±0.02
Electrode diameter (mm)	42.00±0.02
Sensitive volume (cm^3)	6.3±0.1

PMMA for the insulators, and co-axial cables. Details of the ionization chamber are given in table 1.

The ionization chamber was tested in the ^{60}Co teletherapy unit, Gammatron II S80, used to calibrate clinical dosimeters (radiotherapy). The irradiation conditions for all measurements in the ^{60}Co unit were a field size of 10×10 cm^2 and a source-detector distance of 100 cm. The measurements were taken utilizing an electrometer, model UNIDOS E, Physikalisch-Technische Werkstätten (PTW), Germany. All readings were corrected for standard environmental conditions of pressure and temperature.

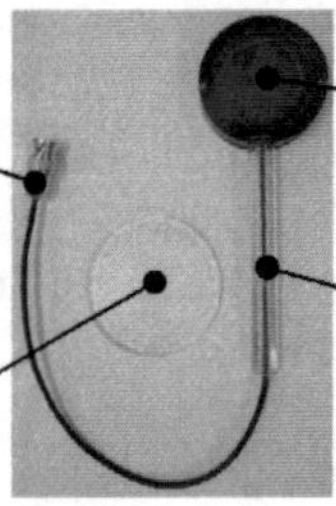

Figure 1. Photo of the ionization chamber tested in this work with its PMMA build-up cap

2.2. Monte Carlo evaluation

As this dosimeter was designed at the LCI, all dimensions and composite materials are well known. This information is very important in order to correctly simulate the dosimeter. The simulations were undertaken utilizing the PENELOPE/penEasy Monte Carlo code [7, 8]. The simulated ionization chamber is shown in figure 2.

As input parameter to the PENELOPE/penEasy code, all dimensions and materials were used. The configuration adopted in the present paper is given in table 2.

The ^{60}Co spectrum utilized in the simulations was provided by Tedgren *et al.* [9], which was obtained for the ^{60}Co therapy machine, model Gammatron I. The radioactive source of this machine is similar to the one available at the LCI. This spectrum was already utilized in a work of Neves *et al.* [3], and it is very suitable to represent our ^{60}Co source.

In order to better understand the influences of the components of the chamber response, the sensitive volume was studied as three different parts in relation to the collecting electrode: above, below and around its side. This scheme is also shown in figure 2.

The experimental uncertainties, calculated for all measurements, are expanded uncertainties, obtained by the combination of type A and B uncertainties, using a coverage factor of 2. The type A uncertainties were adopted for Monte Carlo Simulation, also with a coverage factor of 2.

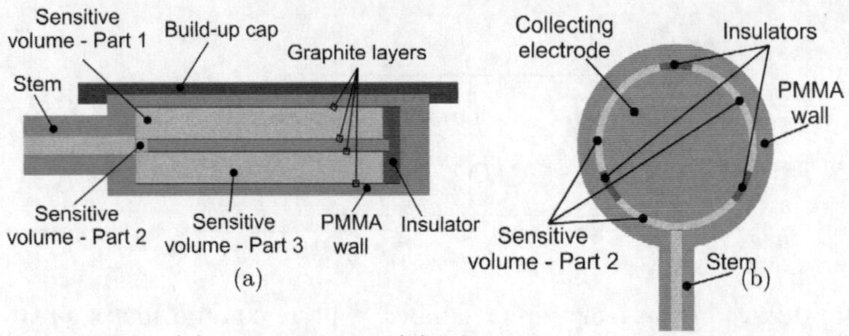

Figure 2. Axial view (a) and top view (b) of the geometry used as input for the Monte Carlo simulations. It is also possible to observe the sensitive volume separated in three distinct parts. The stem was reduced in this figure, for better visualization of all components of the ionization chamber.

Table 2. Simulation parameters used on the PENELOPE/penEasy code [7]

Parameters	Values
Cutoff energy for photons	1 keV
Cutoff energy for e^+ and e^-	1 keV
Average angular deflection (C_1)	0.05
Maximum average fractional energy loss between consecutive hard elastic events (C_2)	0.05
Cutoff energy loss for hard inelastic collisions (W_{CC})	0.1 keV
Cutoff energy loss for hard Bremsstrahlung (W_{CR})	1 keV
External electron step-length control (S_{max})	10^{30} cm
Number of histories	10^{10}

3. Results and Discussions

3.1. Saturation, ion collection efficiency and polarity effect

The saturation test was the first evaluation undertaken with the ionization chamber in order to determine the most appropriate voltage that shall be applied to it. The saturation curve is shown in figure 3. It can be seen that for all applied voltage values, no significant changes in the collected charge were observed. To reduce any possible leakage current, the chosen applied voltage to be used in all tests was set as +100 V.

From the results obtained in the saturation curve, two other tests were conducted: ion collection efficiency and polarity effect. The ion collection efficiency was determined by equation (1) [10].

$$K_S = \frac{(V_1/V_2)^2 - 1}{(V_1/V_2)^2 - M_1/M_2} \tag{1}$$

Where M_x is the collected charge at a V_x voltage, and V_1=V_2 = 2, for V_1 = ±200 V and V_2 = ±100 V. The ion collection efficiency was better than 99.99%.

The polarity effect was obtained comparing the collected charges at similar voltages of opposite signs. The maximum polarity effect observed was 0.4%, therefore within the maximum limit of 1% recommended by the IEC 60731 standard [6].

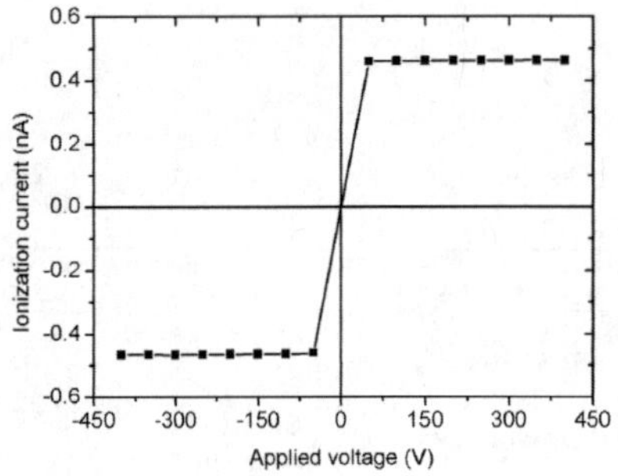

Figure 3. Saturation curve of the ionization chamber. The maximum uncertainty was 0.1%, and is therefore not visible in the figure.

3.2. Short- and medium-term stabilities

The stability test was undertaken for a period of 12 months, as part of the quality control program of this dosimeter. During the short-term stability test, ten consecutive readings were taken. The maximum variation obtained was 0.17%, which is in accordance with the limit of 0.25% [6].

The medium-term stability was obtained by taking the mean value of ten measurements of the short-term stability tests during a period of 12 months under the same reproducible conditions. The maximum variation coefficient (0.3%) is well within the recommended limit of 0.5% [6].

3.3. Leakage current

The leakage current was determined following the instructions of the IEC 60731 standard [6]. Within 5 s after the end of a 10 min irradiation, the leakage current shall have decreased to ±0.5% of the ionization current produced in the measuring volume during the irradiation. For the ionization chamber developed and characterized in this work, this behavior was observed.

3.4. Monte Carlo evaluation

To evaluate the new chamber design for the dosimetry of ^{60}Co beams, some components of the ionization chamber were chosen: collecting electrode, insulators and stem, in order to determine their influence on the chamber response.

The effect of each studied component was obtained as the ratio of the dose to the gas in the ionization chamber (atmospheric air) without the studied component to that with the whole chamber. The influences are listed in table 3. From the results it is possible to observe a small influence of the collecting electrode.

Furthermore, the energy deposited in Parts 1 and 3 of the sensitive volume presented a difference of just 0.06%, which is well within the uncertainties (0.2%). This result indicates that the chamber design of the collecting electrode position, in the middle of the sensitive volume, and material (PMMA coated with graphite) does not alter significantly the energy deposited. This may even improve the dosimeter response, because the electric field becomes more homogeneous in the sensitive volume.

The insulators are very small, and no significant contribution was expected. As the stem presents a considerable size in relation to the sensitive volume of the chamber, its influence is higher. It is important to notice, however, that in this study the whole chamber, including the stem, was in the primary beam.

Table 3. Influence of the chamber components on its response, at the three distinct parts of the sensitive volume.

Studied Component	Influence (%)		
	Part 1	Part 2	Part 3
Collecting electrode	2.5	2.7	2.2
Insulators	0.1	0.7	0.2
Stem	6.3	4.9	7.0

4. Conclusions

The ionization chamber characterized in this work presented results within the limits recommended by international standards. Furthermore, due to the knowledge of all dimensions, materials and geometrical arrangement, the Monte Carlo analyses were undertaken to study this ion chamber configuration for ^{60}Co beams. These results point out a small effect of the studied components on the chamber response, and it presented several advantages: low-cost, easy assembling, robustness, good performance in the characterization tests and a configuration with low influence on the chamber response.

Acknowledgments

The authors received support from the Brazilian agencies: Fundação de Amparo à Pesquisa do Estado de Minas Gerais (FAPEMIG, Grant no. APQ-03049-15), CAPES (Grant Pro-Estratégia no. 1999/2012), CNPq (Grants no. 304789/2011-9 and 157593/2015-0) and INCT for Radiation Metrology in Medicine. The authors would like to thank Dr. Å. Carlson Tedgren (Linköping University, Sweden) for providing the energy spectra of the ^{60}Co.

References

[1] IAEA TECDOC 1040, 1998. Design and implementation of a radiotherapy programme: Clinical, medical physics, radiation protection and safety aspects.

[2] Afonso L C, Vivolo V and Caldas L V E 2010 Characteristics of a tandem system of ionization chambers in X-ray beams, mammography level. *Appl. Radiat. Isot.* **68** 610–3

[3] Neves L P, Perini A P, Fernández-Varea J M and Caldas L V E 2013 Application of a Pencil Ionization Chamber (0.34 cm^3 Volume) for ^{60}Co Beams: Experimental and Monte Carlo Results *IEEE Trans. Nucl. Sci.* **60** 746–50

[4] Maia A F and Caldas L V E 2006 A simple method for evaluation of HVL variation in CT equipment. *Phys. Med. Biol.* **51** 1595–601

[5] Perini A P, Neves, L P and Caldas L V E 2013 Investigation of the applicability of a special parallel-plate ionization chamber for x-ray beam dosimetry *Radiat. Phys. Chem.* **95** 94–97

[6] IEC 60731 2011. International Electrotechnical Commission: Medical electrical equipment Dosimeters with ionization chambers as used in radiotherapy.

[7] Salvat F, Fernández-Varea J M and Sempau J 2008 PENELOPE – A Code System for Monte Carlo Simulation of Electron and Photon Transport, OECD/NEA, Issy-les-Moulineaux.

[8] Sempau J, Badal A, Brualla L, 2011 A PENELOPE – based system for the automated Monte Carlo simulation of clinacs and voxelized geometries – application to far-from-axis fields *Med. Phys.* **38** 5887–95

[9] Tedgren A C, Luelmo S and Grindborg J E 2010 Characterization of a ^{60}Co unit at a secondary standard dosimetry laboratory: Monte Carlo simulations compared to measurements and results from the literature *Med. Phys.* **37** 2777–86

[10] IAEA TRS 398 2001. Absorbed dose determination in external beam radiotherapy: An international code of practice for dosimetry based on standards of absorbed dose to water.

Activity determination of a ^{201}Tl solution by 4πβ-γ and sum-peak coincidence methods

A Ruzzarin[1], M A L da Silva[2], A Iwahara[3], R L da Silva[3], O L T Filho[3], R Poledna[3], R T Lopes[1]

[1] Laboratório de Instrumentação Nuclear/Programa de Engenharia Nuclear/Instituto Alberto Luiz Coimbra de Pós-Graduação em Pesquisa de Engenharia /Universidade Federal do Rio de Janeiro (LIN/PEN/COPPE/UFRJ), Rio de Janeiro, Brasil – CEP 21945-972
[2] Comissão Nacional de Energia Nuclear (CNEN), Rua General Severiano 90, Botafogo, Rio de Janeiro, Brasil – CEP 22294-900
[3] Laboratório Nacional de Metrologia das Radiações Ionizantes/Instituto de Radioproteção e Dosimetria/Comissão Nacional de Energia Nuclear (LNMRI/IRD/CNEN), Av. Salvador Allende s/no, Barra da Tijuca, Rio de Janeiro, Brasil – CEP 22783-127

E-mail: aruzzarin@nuclear.ufrj.br

Abstract. ^{201}Tl is used in nuclear medicine in cardiac imaging for evaluating the injury level in cardiac muscle at rest and exercise. In this work the activity concentration of a ^{201}Tl radioactive solution has been absolutely determined using the 4πβ-γ coincidence and sum-peak coincidence methods. The presence of ^{202}Tl radioactive impurity that imposes some difficult in the activity measurements was taken into account in the measurements. In the sum-peak method a planar germanium detector was used. The half-lives were evaluated by the reference source method and the results obtained were (3.033 ± 0.004) d and (12.320 ± 0.163) d, respectively, for ^{201}Tl and ^{202}Tl.

1. Introduction

The determination of the activity or disintegration rate of radionuclides and the nuclear decay parameters such as half-lives and photon emission intensities, mainly for those used in nuclear medicine, are necessary to allow the correct calibration of the measuring instruments and consequently, the optimization of the internal dose delivered to the patients.

The principal method used for activity standardization of radionuclides by most National Metrology Institutes worldwide is the primary 4πβ-γ coincidence counting method. Another method currently used is the sum-peak coincidence, mainly for those radionuclides which decay by electron capture with short half-lives. The sum-peak method is a primary measurement technique that combines gamma-ray spectrometry and sum coincidence counting, requiring only one photon detector with associated electronics. The activity is determined from the count rates of the full absorption peaks of a single X- and gamma-rays, the sum-peak count rate of photons in true coincidence and the total count rate of the spectrum.

The half-life was determined by the reference source method which uses the simultaneous measurements of gamma spectrometry of the sample and the reference in a germanium detector. This feature allows optimizing the analysis of time and eliminating the influence of instrumental and environmental interferences. Unlike other methods such as utilizing the ionization chamber, it does not depend on the device stability, and in the particular case, not depends on the purity of the sample.

The half-life is an important nuclear parameter in identifying and quantifying a radionuclide, so there is a demand for research to determine more accurate half-lives of radionuclides used in calibration of equipment used to measure gamma and X-rays emitters in nuclear medicine, radiation protection and dosimetry, environmental monitoring, nuclear security programs, reactors and treatment of radioactive waste.

This work describes the standardization of a ^{201}Tl solution by two primary methods, the quantification of ^{202}Tl impurity by gamma spectrometry and the determination of half-lives of ^{201}Tl and ^{202}Tl by the reference source method.

2. Methodology

2.1. 4πβ-γ coincidence counting method

There is an abundant literature on the application of 4πβ-γ coincidence method for standardization of radioactive sources. It is the primary method that has great versatility as may be used for those radionuclides which decay by alpha-gamma and beta-gamma emission, electron capture and pure beta emitters [1, 2]. The 4πβ-γ coincidence system used in this work consists of a gas flow 4π proportional counter coupled to a 10.2 cm ×10.2 cm crystal of NaI(Tl). The counting gas of the 4π counter is a mixture of 90% argon and 10% methane operating at 0.1 MPa pressure. Characteristics X-rays and electron Auger originating from 201Tl are counted in the proportional counter and the gamma counts were confined to a window covering the 167 keV line in the scintillator crystal. In order to avoid the interference effects of the sum-coincidences of X-rays originating from electron capture events (69 to 83 keV) in the gamma counting, a cadmium absorber of 2 mm thick was placed between the source and the scintillator of NaI (Tl). To obtain the activity it is applied the method of efficiency extrapolation. In this method the plot of $(N_\beta N_\gamma)/Nc$ as a function of $(1-N_c/N_\gamma/ N_c/N_\gamma)$ gives the source activity N_0 when $N_c/N_\gamma = \varepsilon X,A \rightarrow 1$. Here εX,A is the count efficiency of proportional counter for X-rays and electron Auger and N_β, N_γ and N_c are β, γ and coincidence count rates, respectively. The variation of efficiency is carried out using aluminized mylar absorbers of various thicknesses on the counts of the proportional counter originating from the electron capture events. In this work we use the expression developed by Funck [3]

$$\frac{N_\beta N_\gamma}{N_c} = N_0\left(1 + c_1 + c_2\frac{1-\varepsilon_{X,A}}{\varepsilon_{X,A}}\right) \qquad (1)$$

where c_1 and c_2 are constants depending on decay scheme data.

2.2. Sum-peak coincidence method

It is a primary method of measuring the activity being determined from the total count rate of photon spectrum, the sum-peak count and emission of X- and gamma-rays, and the relationship between them given by equation 2 [4]

$$N_0 = N_T + \frac{N_X N_\gamma}{N_{X\gamma}} - \frac{N_{XX} N_\gamma^2}{N_{X\gamma}^2} \qquad (2)$$

where N_0 is the activity, N_T is the spectrum total count, N_X is X-ray count, N_γ is the gamma count, $N_{X\gamma}$ is the sum count of X- and gamma-rays and N_{XX} is the sum count of X-rays.

2.3. Reference source method for half-life measurements

The reference source method is based on the ratio of the counts of the radionuclide to be measured and the standard, according to equation 3 [5].

$$R_t = \frac{C_t}{P_t} \tag{3}$$

Where R_t is the ratio between the count of radionuclide to be measured and the reference in time t, C_t is the count of the radionuclide to be measured in time t and P_t is the count of the reference at the same time t. Based on the radioactive decay law, the equation 3 can be rewritten as

$$R_t = \frac{C_0 e^{-\lambda_C t}}{P_0 e^{-\lambda_P t}} = \frac{C_0}{P_0} e^{-(\lambda_C - \lambda_P)t} = R_0 e^{-\left(\frac{\ln 2}{T_C} - \frac{\ln 2}{T_P}\right)t} \tag{4}$$

where C_0 and P_0 are the gamma ray count of the radionuclide to be measured and the reference in time t=0, respectively, λ_C and λ_P are decay constants of the radionuclide to be measured and the reference, respectively, T_C is the half-life of radionuclide to be measured and T_P is the reference half-life. The equation can be rearranged b

$$ln \frac{R_t}{R_0} = -\left(\frac{\ln 2}{T_C} - \frac{\ln 2}{T_P}\right) t \tag{5}$$

and

$$T_C = \frac{\ln 2}{\frac{\ln 2}{T_P} - \left(\frac{\ln R_t}{t} - \frac{\ln R_0}{t}\right)} \tag{6}$$

where $\ln(R_t/t) - \ln(R_0/t)$ is the slope a of the plot $\ln(R_T)$ as function of time t. The final equation can be rearranged as

$$\mathbf{T_C = \frac{1}{\frac{1}{T_P} - \frac{a_1}{\ln 2}}} \tag{7}$$

3. Results and Discussion

3.1. Activity determination

Five sources for the determination of the activity by $4\pi\beta$-γ coincidence counting method were prepared from a solution of ^{201}Tl provided by IPEN/SP. One or two drops of the solution (10 to 20 mg) were deposited on thin VYNS films (copolymer of vinyl chloride and vinyl acetate), approximately 15 μg/cm^2 thick, coated with a gold layer of 20 μg/cm^2 thick on both sides of the film. The extrapolated value for determining the activity must be multiplied by a correction factor due to non-detection of low energy X-rays and Auger electrons. This factor depends upon the nuclear decay scheme parameters such as conversion coefficients, abundance ratios, capture probability and fluorescence yields and for ^{201}Tl is estimated to be 0.9908 for the counting system conditions used in this work (Equation 1). Besides, the radiochemical impurities of ^{200}Tl and ^{202}Tl, detected by gamma spectrometry, should be subtracted. The percentage of impurity was 1.15% and 0.01% of ^{202}Tl and ^{200}Tl, respectively, relative to ^{201}Tl in this exercise in the reference date. From the weighted average of five activity results obtained, the value of (1898.46 ± 13.67) kBq/g at the reference date of May 01, 2015, 12h00 was evaluated. A typical extrapolation curve of one source is showed in Figure 1.

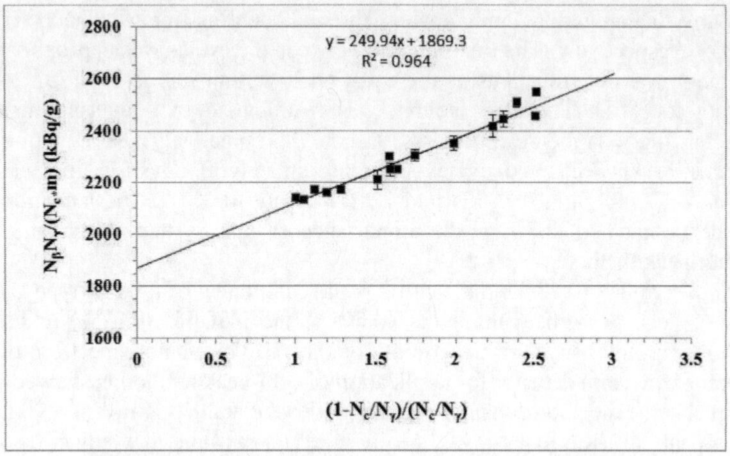

Figure 1. Typical extrapolation curve of experimental data for coincidence measurements of one source of ^{201}Tl. The extrapolated value for $(1-N_c/N_\gamma)/(N_c/N_\gamma)$ gives the activity concentration of the source.

The uncertainty components for activity determination of ^{201}Tl by the 4πβ-γ coincidence method is presented in Table 1.

Table 1. Uncertainty components for activity determination of ^{201}Tl by coincidence method.

Uncertainty component due to	Relative standard uncertainty (%)
Counting statistic (included on fitting procedure	0.90
Fitting procedure	0.60
Weighing	0.05
Dead time	0.13
Resolving time	0.09
Delay mismatch between gamma and beta pulses	0.34
Background	0.02
Decay (due to uncertainty on ^{201}Tl half-life)	0.11
Combined standard uncertainty	0.72
Expanded uncertainty (k=2)	1.4

For applying the sum-peak coincidence method for activity determination, one point source of ^{201}Tl on polystyrene film support was submitted to two measurement cycles with the source placed on the top of the high purity planar germanium detector. The real counting time was 3600 seconds with dead time of 56 and 35%, respectively. The time interval between the cycles was approximately 3 days. The activity concentration was determined using the equation 2, section 2.2, giving 1751.24 kBq/g for ^{201}Tl and 16.708 kBq/g for ^{202}Tl. The same source was resubmitted to 2 measurement cycles in the germanium detector, but with the dead time of 19 and 7%, respectively, real counting time of 3600 s and the time interval between two cycles of approximately 3 days. In this case, the activity concentration was 1887.99 kBq/g for ^{201}Tl and 22.218 kBq/g for ^{202}Tl. The latter result is consistent with the coincidence method showing the importance of the system dead time in the activity calculation by sum-peak method.

The statistical uncertainty (0.28%) in the count is the component of type A; type B components are weighing (0.05%), decay scheme parameters (0.50%), the half-life (0.31%) and the background radiation (0.22%) given total standard uncertainty of 0.69%. The photon spectrum of a201Tl source obtained by planar germanium detector for application of sum-peak method is showed in Fig. 2. In this spectrum one can observe the photopeaks of characteristics Kα and Kβ X-rays (68.90 to 80.7 keV), the coincidence sum-peak of (K_α+K_β X-rays, the 167 keV γ-ray and the coincidence sum-peak of (K_α+167) keV γ–ray and (K_β +167) keV γ–ray.

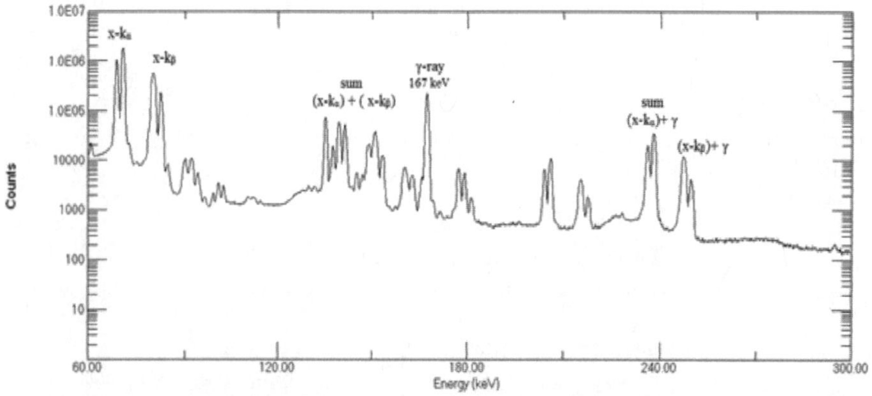

Figure 2. Photon spectrum of a ^{201}Tl source.

The gamma spectrometry with a coaxial germanium detector of 50% efficiency, aluminum window of 1.00 mm thick and energy resolution of 0.90 keV and 1.90 keV for 122.06 keV of ^{57}Co and 1332.5 keV of ^{60}Co, respectively, was used to validate the results so far obtained. An efficiency curve as a function of energy (ε versus E) was determined using standard point sources of various photon-emitting radionuclides with energies ranging from 40 to 2500 keV. A second order hyperbola $Ln(\varepsilon)=A+B/(LnE)+C/(LnE)^2$ was fitted to the experimental data using the Levenberg-Marquardt algorithm based on the curve fitting least square method. The fitting parameters calculated were A = -15.966 ± 0.221, B = 97.590 ± 2.595 ± and C = -239.385 ± 7.547, degree of freedom 74 and Chi square 13.895. The critical Chi square is 95.081 for α = 0.05. The photopeak efficiency for 167 keV gamma ray of ^{201}Tl, 440 keV for ^{202}Tl and 368 keV for ^{200}Tl are calculated by interpolating ε versus E curve for these energies. The interpolated efficiencies from the fitted curve are presented in Table 2.

Table 2. Efficiencies obtained by interpolation of experimental efficiency versus energy calibration curve.

Energy (keV)	ε	Standard relative uncertainty of ε (%)
167 (^{201}Tl)	0.002389	018
368 (^{202}Tl)	0.001825	0.25
440 (^{200}Tl)	0.001671	0.23

The activity is determined by using the equation below:

$$A = \frac{C}{\epsilon * I_\gamma} \tag{8}$$

where A is the activity (Bq), C is the count rate of photopeak of energy E (count/s) and I_γ is the gamma intensity of energy E. Figure 3 shows the ε versus E calibration curve of germanium detector.

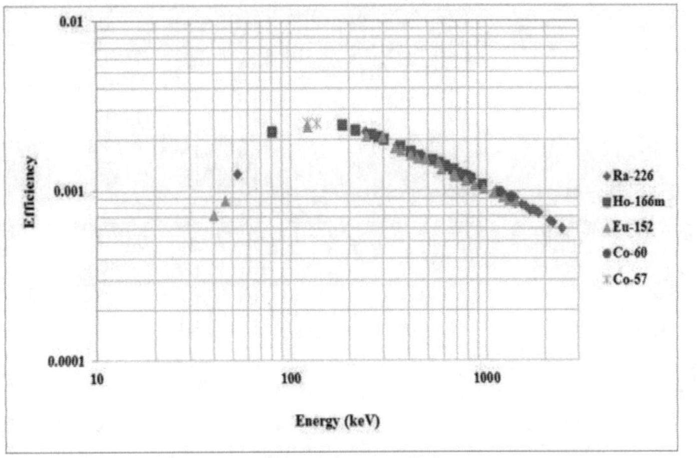

Figure 3. Efficiency versus energy calibration curve of germanium detector.

Three point sources of ^{201}Tl prepared in the polystyrene film holder were measured at 10 cm from the top of the germanium detector. The count times were 74000 s (4.80% of dead time), 78000 s (5.17% of dead time) and 86000 s (1.91% of dead time). The results are shown in Table 3. Table 4 presents the results of radiochemical impurities of the ^{200}Tl and ^{202}Tl.

Table 3. Activity concentration results of ^{201}Tl obtained from de efficiency curve method.

Source	Activity concentration (kBq/g)	Standard uncertainty (kBq/g)
01	1883.64	12.82 (0.68%)
02	1883.88	12.74 (0.68%)
03	1882.58	12.50 (0.66%)
Weighted mean	1883.36	7.32 (0.39%)

Table 4: Radiochemical impurities detected in the ^{201}Tl solution.

Source	Activity concentration (kBq/g)		Standard uncertainty (kBq/g)	
	^{202}Tl	^{200}Tl	^{202}Tl	^{200}Tl
01	21.69	0.100	0.156 (0.72%)	0.002 (1.9%)
02	21.56	0.100	0.142 (0.66%)	0.002 (1.8%)
03	21.17	0.083	0.112 (0.53%)	0.004 (4.6%)
Weighted mean	21.40	0.098	0.229 (1.1%)	0.005 (5.5%)

3.2. Half-life determination

The half-life of ^{201}Tl was determined using the methodology of source reference described in section 2.3. The reference sources used were 137C and 54Mn. A mixed point source of ^{201} Tl + ^{137}Cs + ^{54}Mn in polystyrene film support was subjected to two counting cycles in the germanium detector. The spectral inspection revealed the presence of gamma emitting impurities of ^{202}Tl (440 keV) as can be seen in Figure 4.

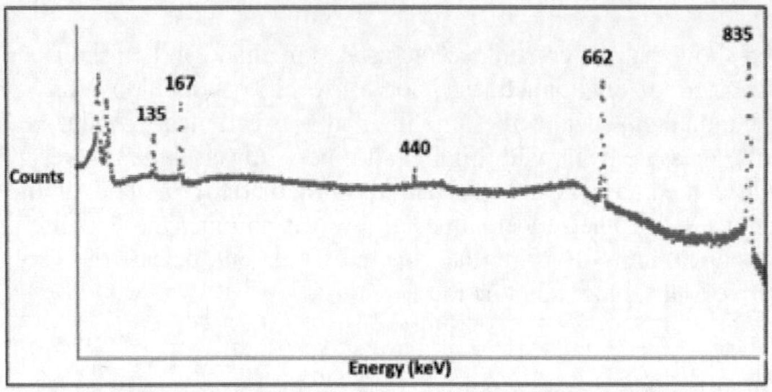

Figure 4. Gamma spectrum of ^{201}Tl+^{137}Cs+^{54}Mn point source by germanium detector.

In the first cycle 40 spectra were obtained with dead time ranging from 6.8 to 6.6%. Measurements of the second cycle performed a week later, originated 50 spectra obtained with dead time ranging from 5.4 to 5.3%. The total counting time, including the two cycles, corresponded to approximately 4 half-lives. Table 4 shows the values of the half-lives determined for each one of the reasons, emphasizing each gamma line and each reference. The values found show that the use of the gamma line of 135.34 keV (gamma intensity $P\gamma$ of 2.56%), for both reference, showed a large difference from the value determined by the gamma line of 167 keV ($P\gamma$ =10.2), the latter being consistent with the values found in the literature. Using the same spectra used to determine the half-life of ^{201}Tl, the half-life of the radiochemical impurity ^{202}Tl was calculated yielding a value of (12.320±0.163) d. Table 5 shows the results of half-lives of ^{201}Tl evaluated according equation 7, section 2.3.

Table 5. Half-lives of ^{201}Tl evaluated by source reference method.

Gamma energy (keV)	Reference	Half-life (d)
135.34	^{137}Cs 661.66 keV	2.928 ± 0.009 (0.32%)
	^{54}Mn 834.84 keV	2.929 ± 0.009 (0.32%)
167.41	^{137}Cs 661.66 keV	3.033 ± 0.009 (0.14%)
	^{54}Mn 834.84 keV	3.033 ± 0.004 (0.14%)

For comparison, the half-lives of ^{201}Tl and ^{202}Tl found in the literature [6] are respectively, 3.043±0.003 and 12:23±0.02 d.

4. Conclusions

The results obtained show that the coincidence sum-peak method can be used as an alternative to the primary standardization of radionuclides. Some specific conditions are required for the implementation of the method, such as the time interval between measurement cycles, the activity of the solution since the measurements with high dead times showed large discrepant results regarding the $4\pi\beta$-γ coincidence method. The results also showed that the method of the source reference associated with gamma spectrometry techniques allows determining the half-life of the radionuclide either as the radiochemical impurity using the same measurement, despite the very low concentration of the impurity relative to the main radionuclide.

References

[1] Kawada Y 1972 *Extended application and improvement of the $4\pi\beta$-γ coincidence method in the standardization of radionuclide* Researches of the Electrotechnical Laboratory No. 730
[2] Watjen U *et al* 2014 *Appl. Radiat. Isotopes* **87** pp 1-488
[3] Funck E *et al* 1983 *Int. J. Nucl. Med. Biol.* **10** pp 137-140
[4] Brinkman G A *et al* 1977 *Int. J. Appl. Radiat. Isot.* **28** pp 271-275
[5] Morel J *et al* 1992 *Bulletin BNM* **88** pp 19-25
[6] Schötzig U and Schrader H 1989 *Halbwertszeiten und Photonen-Emissionswahrscheinlichkeiten von häufig verwendetenRadionukliden* PTB-Ra-**16/3**

Application of patent BR102013018500-0 in well type ionization chambers

C H S Sousa [1,2] and J G P Peixoto[1,2]

[1] Universidade do Estado do Rio de Janeiro – UERJ
[2] Instituto de Radioproteção e Dosimetria – IRD

E-mail: chenriques2@ird.gov.br

Abstract: The definition of the radioactive sample position in a well type ionization chamber is the largest source of uncertainty in the measurement of quantity activity. The determination of this parameter in two activimeters helped to improve their accuracies, from 2.62 and 2.59% to 3.87 and 1.74%, with and without the use of the positioning device, concluding, that with their use has reached an uncertainty of U = 0.2276 and 0.2677 % (k = 2) 95.45%.

1. Introduction

The determination of the radioactivity source activity, needs to be better studied, if the spatial dependence been considered sample radiation in the detector. The instrumentation in charge of these measurements is the activimeter, it should be aided by systems compatible in determining the greatness activity.

The detector spatial dependence is derived from the radioactive sample placement in and established both by use of different geometries and materials [1], and by the different volumes and densities [2], resulting directly in its sensitivity, being necessary to use correction factors for efficiency to avoid biased processes, that could cause sub or super dosage for a clinical or therapeutic procedures, producing unwanted impacts to biological organisms [3,4].

The sample position within the ionization chamber is one of the main factors affecting the measurements accuracy of the greatness activity [5], and this positioning variation can be caused by sources different geometries and volumes [6]. The correlation shown between the two input quantities directly uncertainties related influence to its accuracy [7], so the spatial dependence is determining the uncertainties obtained for the results sensitivity and accuracy [8].

The activimeters are set by manufacturers using standard procedures, with ^{137}Cs, ^{133}Ba and ^{57}Co tracked sources, with volumes and different characteristics to those used daily in nuclear medicine services. These instruments are traded with an accessory used for the radioactive samples insertion in the detector, with different regions to different positions, depending on the manufacturer, providing a blurring of the ideal measurement point.

In order to present a response to the amount of radiation administered to patients, and considering the accuracy fundamental for measurements, we tested the device developed in patent BR102013018500-0, to assess the associated uncertainties activity.

2. Experimental arrangement

The device [9], shown in figure 1, was produced with low scattering material for determining the depth coefficient in activimeters. A conical base is fixed by a screws system positioned in the detector recess and it is capped in the center by bubble level and a dial indicator. There is a rail system, which enables the graduated sliding rod, jogging longitudinally and transversely, fixed by screws. The sliding rod has a handle on its upper end, and in the lower one, a holder sources where radioactive samples are deposited.

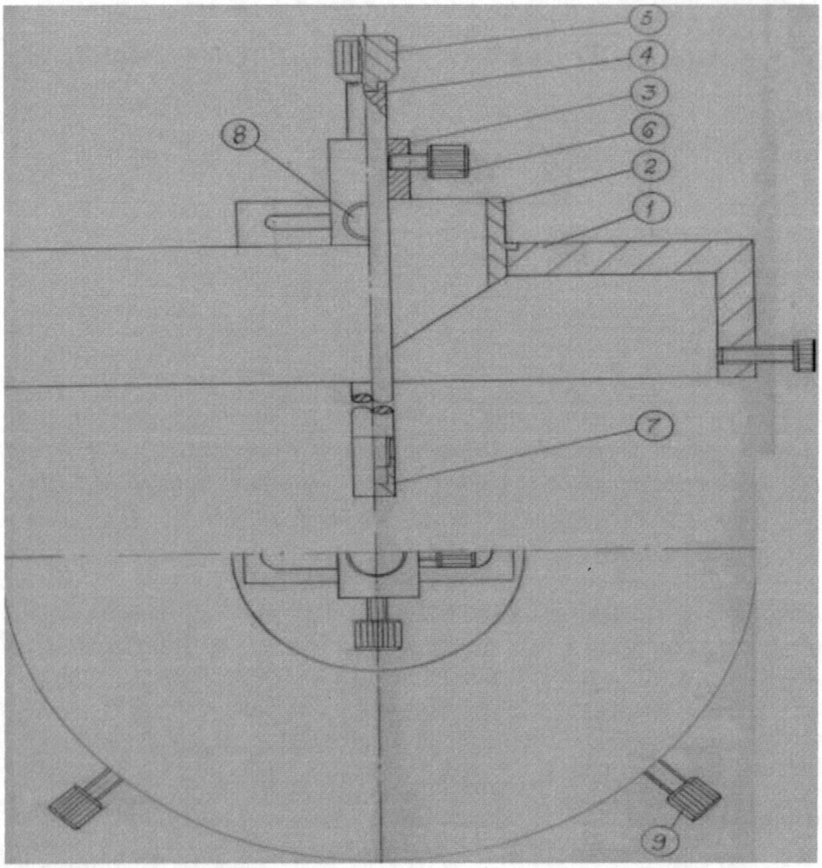

Figure 1: Project Design, presenting the view of the longitudinal and cross section of the device, consisting of nine parts, as follows: (1) Mounting base for coupling to the activimeter, that supports the rail system and sliding-stem, (2) Conical rail in shape which is positioned a sliding rod for transverse and longitudinal axes movement, (3) Base of the tracks where sliding rod moves, (4) graduated sliding rod (5) to handle moving the sliding rod (6) Screw fixer ofl slide rod (7) Holder sources (8) Screw fixer for tracks base (9) Screws to coupling base.

Was used a technetium-99m source (^{99m}Tc), the monoenergetic gamma emitter radionuclide widely used in nuclear medicine procedures. The radioactive volume of the sample was obtained by dripping sodium pertechnetate ($Na^{99m}TcO_4^-$), with hypodermic needle with 25.0 mm length by 0.7 mm diameter aided. The measured activity in a reference activimeter was considering the radioactive decay.

The device was used in two ionization chamber (IC), with sensitive volume of 3.12 cm3 and 3.16 cm3 respectively for detectors 1 and 2. Both IC were sealed and filled with argon gas submitted to 1,2 Pa E5 of pressure.

Input quantities of type A and B were defined according to the Guide to the expression of uncertainty in measurement [10], and later determined the combined and expanded uncertainties as described by Peixoto [11,12], adopting a confidence level of 95.45% [13].

3. Experimental procedure

The device (figure 2), leveling with the bubble level device. Deposit the radioactive source in the carrier sources and determine its volume and activity at the time of measurement.

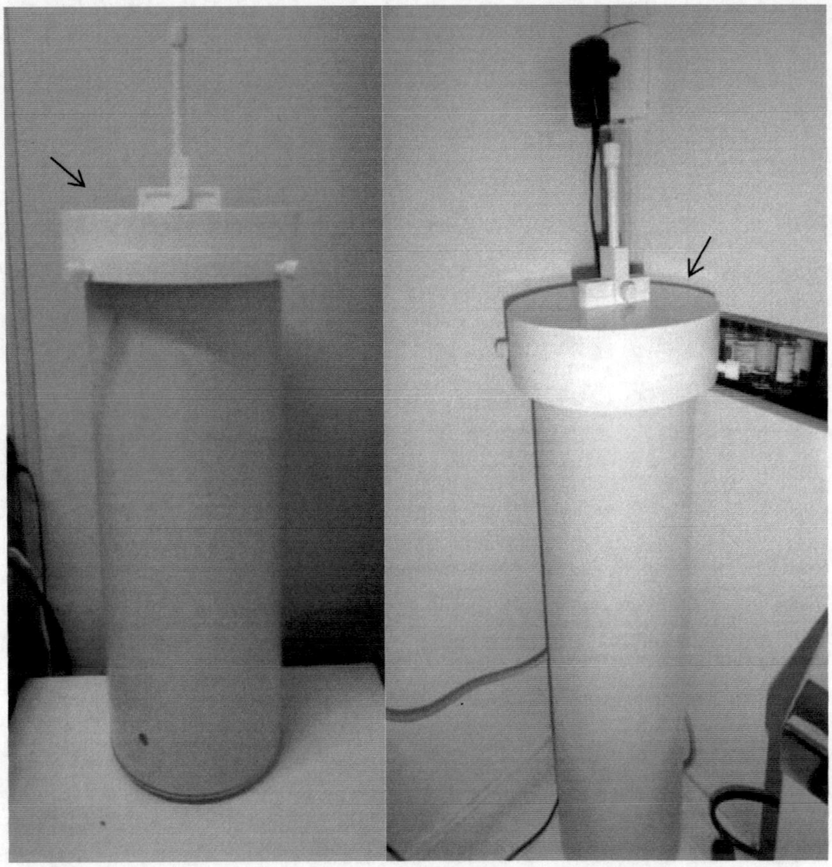

Figure 2: The black arrows show the device mounted on top of the both IC, demonstrating the fit and system functionality. The convenience and ease of handling device, allows the procedure to be performed in the activimeters origin sites.

Calculate the initial activity based on data obtained directly from the radiopharmaceutical supplier, and subsequently validate it with measurement referenced activimeter.

Fix the holder sources on the sliding rod and position it like position 1, recess of the detector background, moving it longitudinally upward, at intervals of one centimeter, performing three measurements every fifteen seconds.

4. Results

The leveling docking showed an uncertainty of 0.07 mm/m. The sliding rod movement in both, longitudinal and the transverse axis, showed ease operation and precision in the indicative degree view, with an uncertainty of 0.01 mm. The sample had a nominal volume of 0.35 μl [10] and the initial activity (1500 ± 114.3) E5 Bq.

The accuracy results before use of the device was (2.62 ± 0.10) % and (2.59 ± 0.14) %, and accuracy of (0.23 ± 0.19) % and (0.31 ± 0.18) %, respectively, for detectors 1 and 2.Was determined the greatest sensitivity range in both detectors and set the optimum depth (O_d) as the point of maximum sensitivity for both detectors (figure 3).

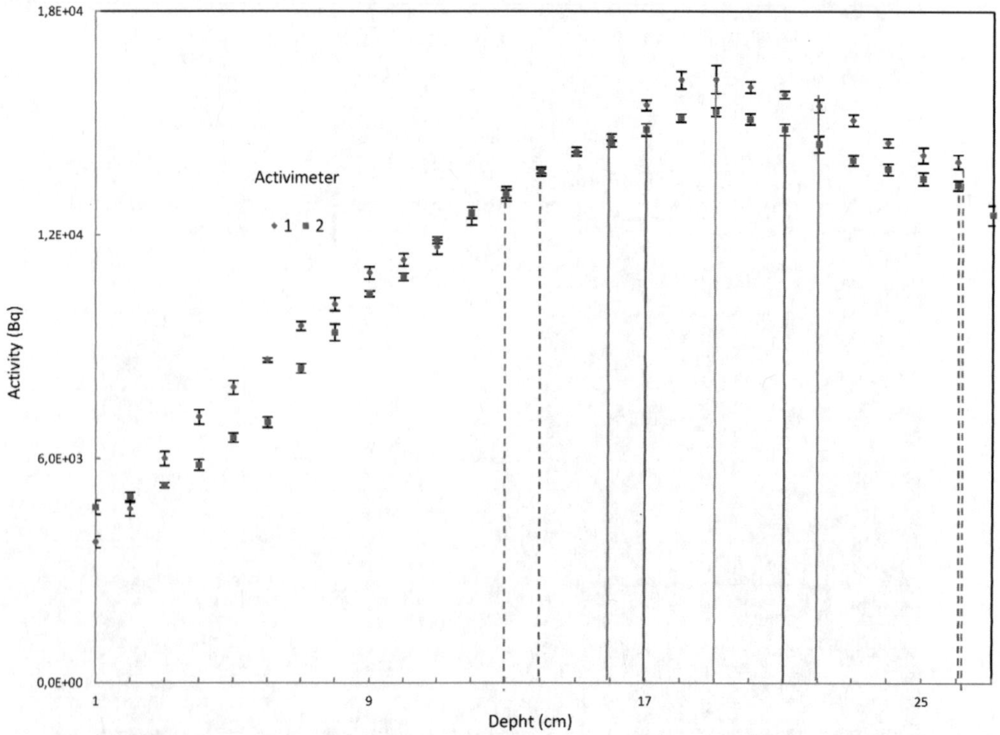

Figure 3: The space between the blue and red lines show the source range variation around the conventional true value (CTV). The dotted lines represent the upper and lower limits equivalent to 10% of the CTV. The green line represents the detector greatest sensitivity point.

The activity result was called equivalent depth (E_d), a function of the activity in relation to depth, allowing minimum and maximum depth limits (L_d) determination (1).

$$L_d = E_d \pm 10\% \quad (1)$$

where Ed is the equivalence depth point, Ld is the maximum and minimum limits based on the measurement result (Figure 3). After determined Ld it is possible to determine the correction factors by depth and uncertainties (table 1).

Table 1: The uncertainty estimation considered uncertainty Type A as obtained by statistical calculationand Type B all of others, as manufacturer information.

Uncertainty (%)	N	Activimeter 1		Activimeter 2	
		A	B	A	B
Accuracy	10	0.097		0.114	
Precision			0.019		0.018
Rod resolution			0.003		0.003
Background			0.001		0.001
Volume			0.020		0.020
Bubble evel			0.001		0.001
	csu =	0.1009		0.1172	
	U =	0.2276		0.2677	
	k =	2		2	

5. Conclusions

The point of greatest sensitivity in both detectors was determined with reproducibility of 96% and 97%, respectively. A linear decrease of the measured values towards the detector recess, may be the key to determine correction factors.

The information obtained in the device test has knowledge of the spatial dependence of IC used as activimeters, allowing to provide correction factors for different geometries and evaluation of associated uncertainties to the measurement process of greatness activity.

The results showed that the insertion of input variables, such as radioactive source decay and volume gradient must be studied to determine the associated uncertainty.

The device reached its goal, as a low-cost design, good portability and precise results for determining the greater sensitivity point in activimeters enabling the estimate of the associated uncertainty.

6. References

[1] Martins EW and Albuquerque P. Avaliação dos fatores de influência da geometria dos frascos utilizados para calibração de ativímetros no IPEN. **VI Congresso Brasileiro de Metrologia. Natal; 2011. p. 4–5.**

[2] Hare DL, Hendee WR, Whitney WP and Chaney EL. Accuracy of well ionization chamber isotope calibrators. **J Nucl Med. 1974;15(12):1138–41.**

[3] Icrp. Annals of the ICRP 60. **Compendium of Dose Coefficients based on ICRP Publication 60. 2012.**

[4] IAEA. Quality Assurance for Radioactive Measurements in Nuclear Medicine. **IAEA Library Cataloguing in Publication Data. Viena; 2006.**

[5] Calhoun JM, Golas DB and Harris SG. Effects of varying geometry on dose calibrator response: cobalt-57 and technetium-99m. **J Nucl Med. 1987;28(9):1478–83.**

[6] Iwahara A, Tauhata L, Oliveira AE De, Nícoli IG, Alabarse FG and Xavier AM. Teste de proficiência para medições de radioatividade na medicina nuclear. **Radiol Bras. 2009;42(4):255–9.**

[7] Sousa CHS, Teixeira GJ and Peixoto JGP. Estimated of the associated uncertaintes with the linearity test of activimeters. J Phys Conf Ser [Internet]. 2015;575:012028. Available from: **http://stacks.iop.org/1742-6596/575/i=1/a=012028?key=crossref.a5dee393760d3297187de5c72dccac8**

[8] Schrader H and Švec a. Comparison of ionization chamber efficiencies for activity measurements. **Appl Radiat Isot. 2004;60(2-4):369–78**.

[9] C. H. S. de Sousa, Peixoto JGP, Cardoso R de S, Bossio F and Cummins JB. Deposito patente BR102013018500-0.pdf. Brasil: Instituto Nacional de Propriedade Industrial INPI; **BR102013018500-0, 2013. p. 22**.

[10] Inmetro. GUM 2008 - **Guia para a expressão de incerteza de medição. 1a edição. Rio de Janeiro; 2012. 141 p**.

[11] Amiot MN, Chisté V, Fitzgerald R, Juget F, Michotte C, Pearce A, et al. Uncertainty evaluation in activity measurements using ionization chambers. **Metrologia. 2015;52(3):S108–22**.

[12] Cardoso R S and Guilherme J G P. Avaliação de incertezas do teste de desempenho de um padrão ´primário para raios x de baixa energia.pdf. volume 7, n° 1. Rio de Janeriro: **Revista Brasileira de Pesquisa e Desenvolvimento; 2005 Mar;5**.

[13] Inmetro. Vocabulário internacional de termos fundamentais e gerais de metrologia. [Internet]. 2007. **Available from: http://www.inmetro.gov.br/infotec/publicacoes/vim.pdf**

Determination of correlated uncertainties of sestamibi- ^{99m}Tc marking

C H S Sousa [1,2], A Gama[3], T G N Camilo[3], C T Mesquita[3] and J G P Peixoto[1,2]

[1] Universidade do Estado do Rio de Janeiro – UERJ
[2] Instituto de Radioproteção e Dosimentria – IRD/CNEN
[3] Hospital Universitário Antonio Pedro - HUAP/UFF

E-mail: chenriques2@ird.gov.br

Abstract: The input quantities determination involved in radiopharmaceutical marking used in heart scans allowed to estimate the combined and associated standard uncertainty with the process. The U value demonstrated that any parameter of the quality control process can be compared and correlated to obtain a real value and validation method, indicating or not, the adequacy of institutional practices and reinforcing the importance of the uncertainties associated to the results in medicine.

1. Introduction

Nuclear medicine is a medical procedure that uses labeled molecules with radionuclides, commonly called radiopharmaceuticals, for diagnostic and therapeutic procedures with different objectives. The products quality is important for the diagnostic images accuracy and needs to be controlled [1]. Approximately 95% of all nuclear medicine procedures diagnostic use sodium pertecnetate ($^{99m}TcO_4^-$) obtained from molybdenum-99 (^{99}Mo) generator elution as marker molecules.

The nuclear medicine diagnostic success is a complex process, which relies on the quality procedures, such as radiopharmaceuticals labeling efficiency and accuracy response in measurement and imaging instrumentation. According with technical recommendations [2], all instrumentation used in medicine must pass for periodically performance tests where results will note the operating conditions tools, resulting in unnecessary exposure to ionizing radiation reduction, thereby reducing the radiobiological interactions [3]. The acceptance of the results should be compared to standards values that quantify the variation percentage in relation to the reference values. [4].

It is very often the presentation of measurement results with no metrological parameters such as traceability, or the associated uncertainties determination. The lack of these qualities can compromise any conclusions based on the obtained results [5]. The quantity value determination is never exact, and is therefore necessary to address the uncertainty on the value obtained, which is a result associated parameter and characterizes the values dispersion, making it possible to display separately from independent input quantities and confronting them with the experimental results.

This study aimed to determine the associated uncertainties of the radiopharmaceutical sestamibi-99mTc marking procedure [6] used in cardiac scintigraphy, correlating it with the quality control of measurement and imaging instrumentation results.

2. Materials and methods

2.1 Materials

Was used a one detector SPECT scintillation camera and a 3.16 cm^3 sensitive volume ionization chamber activimeter. The pharmaceutical drug, sestamibi and technetium-99m used in your markup, were provided by the Instituto de Pesquisas Energéticas e Nuclearers (IPEN).

A flat cobalt-57 source and a vial barium-133 source were used to instrumentation tests, both traceable to Institute of Standards and National. Technology (NIST) with nominal activity 1,3E-2 and 1.8E-5 Ci, respectively. The refrigerator temperature was checked by a digital thermometer with 0.01°C sensitivity.

The radionuclide purity test used a 6.0 mm lead thickness cylindrical container and the remaining tests were used filter paper, chromatographic bowls, specific solvents and universal pH paper.

2.2 Methods

The eluate radiochemical and radionuclide qualities were tested according to standard recommendations on RDC 38 (1), using as samples, aliquots of first generator elution with a 1,0E0 Ci nominal activity.

The radionuclide purity is the ratio expressed in percentage of the radionuclide activity in relation to the total radiopharmaceutical activity. It was verified by the attenuation method, which were measured the activity of the samples and background radiation, adjusting the activimeter for ^{99m}Tc channel, and then for ^{99}Mo channel.

The measurements were performed with samples inside the lead cylinder used as 99Mo half value layer. [6]. Thus, it is possible to determine the solution purity (molybdenum break through), and compare it with the recommended limits of 0.15 µCi/mCi (1).

$$MTB = \frac{A_l\,(^{99}Mo)}{A_l\,(^{99m}Tc)} \tag{1}$$

The determination process for chemical purity, which determined the molecule mass percentage of the interest compound on its chemical state shown in relation to the total preparation mass [6] used the semi-quantitative colorimetric method, visually evaluating the results of aluminum concentration.

The activity percentage of target radionuclide in relation to the total activity of the radiopharmaceutical preparation, called radiochemical purity [6] used the thin layer chromatographic method, with stationary and mobile phase.

After the chromatographic run, the activity was determined in activimeter, and the results expressed as a function of the reference factor (2).

$$R_f = \frac{distance\ travelded\ by\ the\ radioactive}{distance\ travelded\ by\ the\ solvent} \tag{2}$$

The potential hydrogen (pH) was determined by visual examination by color of the universal paper. The activimeter precision test for quality results was determined calculating the percentage difference between individual measurements activities [2,4], A_i, and the measurements average, $\bar{A}$, considering a maximum percentage of ± 5% with 95% of confidence level and correlation between the measured values (3).

$$P = \frac{A_i - \overline{A}}{\overline{A}}\,.\,100\% \tag{3}$$

The activimeter accuracy test measure the percentage between the average activity measurements, A_i, and the activity of standard reference source, A_v, correcting the radioactive source decay for the measurement day. The results considered a maximum of ± 10% percent, with a confidence of 90% [2,4], showing the degree correlation between the measured value and the true conventional value (4).

$$A = \frac{1 - A_i}{A_v} . 100\% \tag{4}$$

Was used as quantitative parameters for gamma camera image uniformity evaluating, integral and differential uniformity test, which is considered as interest area, the central field of view, equal to 75% of full image area [8]. The variation of counts accumulated between two pixels of the image, determined the integral percentage (5),

$$U_i = \frac{pixel_{max} - pixel_{min}}{pixel_{max} + pixel_{min}} . 100\% \tag{5}$$

and the maximum variation between any 5 pixels in any axis determined the differential, given by equation (6).

$$U_d = \frac{pixel_{high} - pixel_{low}}{pixel_{high} - pixel_{low}} . 100\% \tag{6}$$

The type A uncertainties, considered the standard deviation percentage of the accuracy test series measurements, and type B uncertainties, all other contributions that were not considered repeatability [9]. To the linear correlation between the input quantities x_i and y_i was used the Pearson coefficient (r), where x_i corresponds to independent readings performed measurement, x their arithmetic mean, y_i independent readings sestamibi the labeling efficiency and y the arithmetic mean (7).

$$r_{xy} = \frac{\left(\left(\sum_{i=1}^{n} x_i y_i\right) - \left(\left(\sum_{i=1}^{n} x_i\right)\left(\sum_{i=1}^{n} y_i\right) \Big/ n\right)\right)}{\sqrt{\sum_{i=1}^{n}\left(x_i - \overline{x}\right)^2 \sum_{i=1}^{n}\left(y_i - \overline{y}\right)^2}} \tag{7}$$

Once r is defined, the correlation was analyzed (Table 2), and the covariance associated was determined with the x_i and y_i input quantities, which are then type A (8).

$$S(\overline{x}, \overline{y}) = \left(\frac{1}{n(n-1)}\right)\sum_{i=1}^{n}(x_i - \overline{x})(y_i - \overline{y}) \tag{8}$$

3. Results

The refrigerator temperature presented a 0.6 °C variation, the measure instrumentation don´t show the sensitivity to determine any variation on stabilized electrical network. The automatic zero was checked and adjusted before the measurement series.

The percentage ratios of the radionuclide qualities, radiochemical, labeling efficiency and pH were calculated considering measurements series as normal distributions, and measuring instruments and image quality control testing as rectangular distributions. The combined and expanded standard uncertainties conditions were determined for *U* (table 1).

The correlation coefficient between independent input quantities accuracy and labeling efficiency showed a weakly positive correlation equal to 0.00002 and covariance equal to 0.450 (table 2).

Table 1: Uncertainty type A obtained by statistical calculation and type B providedes and not calculated.

Input quantities	Type A (%)	n	Type B (%)
volume			0,0306
radionuclide purity			0,0004
chemical purity			0,0004
uniformity GC			0,1466
standard source			1,7321
accuracy	1,0210	10	1,7321
precision			0,0387
background radiation			0,0024
temperature			0,0006
pH			0,0041
u =	3,0528		
U =	6,1163		
k =	2,00		

Table 2: The correlation coefficient between independent input quantities accuracy and labeling efficiency

Parameter	% results
accuracy	2,015 ± 1,00
efiiciency	2,35 ± 0,84
r	0,00002
$s(\bar{x}, \bar{y})$	0,455

5. Conclusion

The associated uncertainty determination to the sestamibi-^{99m}Tc marking process was equal to 6.11% which shows a dependency on extrinsic factors, which if not controlled, can prejudice the outcome of the diagnostic procedure.

The correlation coefficient between the two independent quantities, accuracy and labeling efficiency, showed a weakly positive correlation, showing little linear relationship, however, all parameters must be controlled so that the results remain within the limits of tolerance for the same conditions ***U***.

This paper concludes by suggesting a correlation between the sestamibi99mTc marking process and the images obtained from them, considering the physician perception as a paradigm to be faced by metrology and medicine.

5. References

[1] ANVISA, 2008. RESOLUÇÃO DA DIRETORIA COLEGIADA - **RDC No 38, DE 4 DE JUNHO DE 2008. Brasilia; 2008**.

[2] IAEA, 1991. Quality control of nuclear medicine instruments 1991. **Quality control of nuclear medicine instruments 1991. Viena; 1991**.

[3] ICRP, 2012. Annals of the ICRP 60. **Compendium of Dose Coefficients based on ICRP Publication 60. 2012**.

[4] CNEN, 2013. **REQUISITOS DE SEGURANÇA E PROTEÇÃO RADIOLÓGICA. CNEN - 3.05 Brasil: Comissão Nacional de Energia Nuclear; 2013 p. 26**.

[5] Sousa CHS, Teixeira GJ and Peixoto JGP, 2015. **Estimated of the associated uncertaintes with the linearity test of activimeters. J Phys Conf Ser [Internet]. 2015;575:012028**.

[6] Nishizaki K, Kameda Y, Kawata T, Mizuguchi K, Imai T and Taniguchi S, 2001. Functional and perfusional assessment with electrocardiograph-gated single photon emission computed

tomography after minimally invasive direct coronary artery bypass grafting. **Ann Thorac Cardiovasc Surg. 2001;7(2):99–102**.

[7] ANVISA, 2009 . **Resolução RDC no 64, de 18 de dezembro de 2009. Brasilia, Brasil: Agência Nacional de Vigilância Sanitária; 2009**.

[8] Mora-Ramirez E, 2010. Intrinsic Flood-Field Uniformity Evaluation. **12th Congress of the International Radiation Protection Association. Buenos Aires: International Atomic Energy Agency; 2010. p. 1–10**.

[9] Inmetro, 2008. **GUM 2008 - Guia para a expressão de incerteza de medição. 1a edição. Rio de Janeiro; 2012. 141 p**.

OSL response bleaching of BeO samples, using fluorescent light and blue LEDs

D P Groppo and L V E Caldas

Instituto de Pesquisas Energéticas e Nucleares, Comissão Nacional de Energia Nuclear, Av Prof Lineu Prestes 2242, São Paulo, 05508-000, Brazil

E-mail: daniela.piai.groppo@gmail.com and lcaldas@ipen.br

Abstract. The optically stimulated luminescence (OSL) is widely used as a dosimetric technique for many applications. In this work, the OSL response bleaching of BeO samples was studied. The samples were irradiated using a beta radiation source (^{90}Sr+^{90}Y); the bleaching treatments (fluorescent light and blue LEDs) were performed, and the results were compared. Various optical treatment time intervals were tested until reaching the complete bleaching of the OSL response. The best combination of the time interval and bleaching type was analyzed.

1. Introduction

The optically stimulated luminescence (OSL) technique is widely used for personal dosimetry, geological and archeological dating, and in radiation diagnostic imaging [1]. Beryllium oxide (BeO) has been used as an OSL dosimeter presenting advantages as high sensitivity to ionization radiation, linear dose response and effective atomic number (Zeff = 7.2) similar to human soft tissue (Zeff ~ 7.6) [2].

The blue stimulation light for the OSL technique from BeO was reported in some papers emphasizing its use in personal dosimetry [3-5].

Most of the OSL dosimeters present a variation in readout due to light exposure; therefore, the dosimeters have to be stored in the dark before the dosimetry evaluation.

In this work fluorescent light was used in bleaching treatments. Different optical treatment time intervals were tested until reaching the complete bleaching of the BeO samples response. A study with a bleaching system with blue LEDs was also performed, and the results were compared.

2. Materials and methods

2.1. BeO samples

The BeO samples (figure 1) were in form of discs (diameter: 4.0 mm, thickness: 0.8 mm and weight: 27.9 mg ± 0.45 mg). The samples are pressed pellets with Si, K, Al, Fe and Mn impurities as reported by Groppo and Caldas [6].

2.2. Irradiations

The irradiations of BeO samples were carried out using beta radiation. It was a $^{90}Sr+^{90}Y$ source available in the Risø TL/OSL-DA-20 reader (figure 2), which was also used to perform the OSL readings. The absorbed dose rate was 0.1 Gy/s, according to the manufacturer's manual. The nominal activity was 1.48 GBq (calibration certificate, 10/06/2010). The appropriate corrections were applied for the measurement date. All samples were irradiated with 1 Gy of absorbed dose.

Figure 1. BeO disc samples.

2.3. OSL measurements

The OSL measurements were taken using a Risø TL/OSL-DA-20 reader (figure 2). The OSL measurements were carried out using the continuous wave mode (CW-OSL); the stimulation light type was composed by blue LEDs; an UV transmitting broad-band glass filter Hoya U-340 was used in front of the photomultiplier tube. Due to the high sensibility of the BeO samples, a collimator with 2 mm diameter was used. Each OSL measurement was carried out during 100 s of LED stimulation with 90% of power, and for each sample a calibration factor was obtained.

Figure 2. RISÖ TL/OSL (TL/OSL-DA-20) reader system.

2.4. Bleaching treatment

The established and conventional bleaching treatment of BeO samples was thermal. The samples were heated at 750 ºC during 15 minutes in a muffle furnace with microwave heating (MFLO1000 by Provecto Analítica). This procedure was enough to empty their traps. Moreover, after the OSL measurements it was possible to bleach the response in a homemade light box with blue LEDs (figure 3) (420-500 nm; 2.5 W; 7000 lux) [7].

Figure 3. Homemade light box with blue LEDs.

Another treatment was also tested. It consisted of the use of a fluorescent lamp system (figure 4): a white wood box (66x21x17) cm^3 containing two fluorescent lamps (Sylvânia, model F16W/78) with 20 W [8]. The lamp-sample distance was 9 cm and the illuminance was 26000 lux.

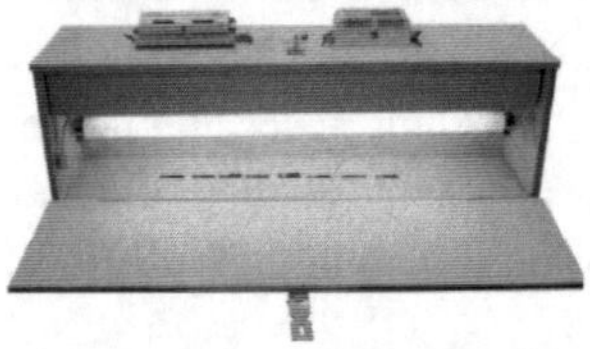

Figure 4. Homemade light box with fluorescent lamps [8].

3. Results

To determine an appropriate optical treatment for reutilization of the BeO samples, initially the samples were thermally treated at 750 °C during 15 minutes in a MFLO1000 furnace, and irradiated with 1 Gy of absorbed dose (beta radiation source). The OSL measurements were taken and then the samples were optically treated during different time periods. In the fluorescent lamp treatment case, the time variation was between 15 min and 10 days. In the bleaching treatment with blue LEDs, the time interval varied between 5 min and 2 h. After each set of irradiation, optical bleaching treatment and OSL measurement, the samples were thermally bleached, as before. The OSL response was normalized for the reading obtained of samples without any treatment.

A typical OSL decay curve obtained by irradiating the sample at a dose of 1 Gy ($^{90}Sr+^{90}Y$) is presented in figure 5.

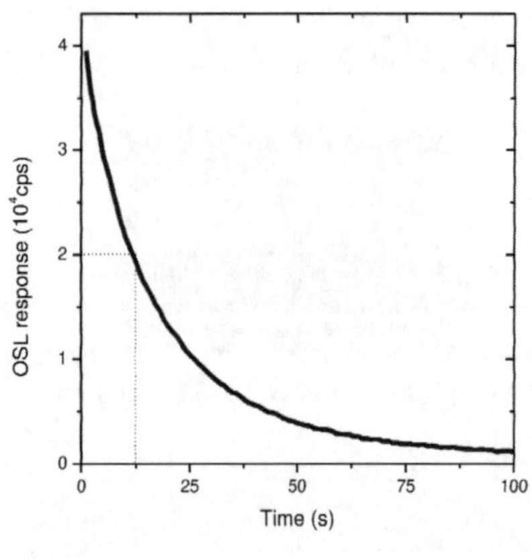

Figure 5. OSL decay curve of BeO samples irradiated with 1 Gy of absorbed dose ($^{90}Sr+^{90}Y$ source).

The fluorescent light treated samples at different interval times were evaluated using the OSL technique, and some of the curves obtained are presented in figure 6.

Figure 7 presents the OSL relative response for each fluorescent treatment performed. It can be noted that until 168 h (7 days) of fluorescent light exposure, the OSL response was bleached and then the OSL response showed constancy with the same value as the OSL response of non-irradiated samples (0R).

The results obtained by optical treatment with the blue LEDs are shown in figure 8, and this procedure was already adopted in previous studies [6,7], with good results.

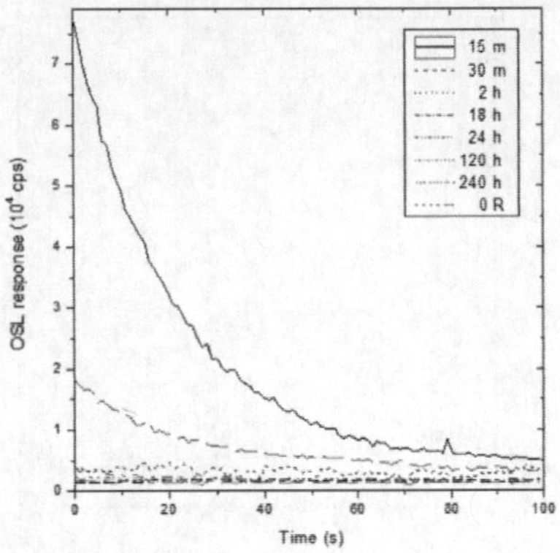

Figure 6. OSL decay curves of BeO samples irradiated with 1 Gy of absorbed dose ($^{90}Sr+^{90}Y$ source), exposed to different time interval of the fluorescent system. 0R represents OSL response of non-irradiated samples.

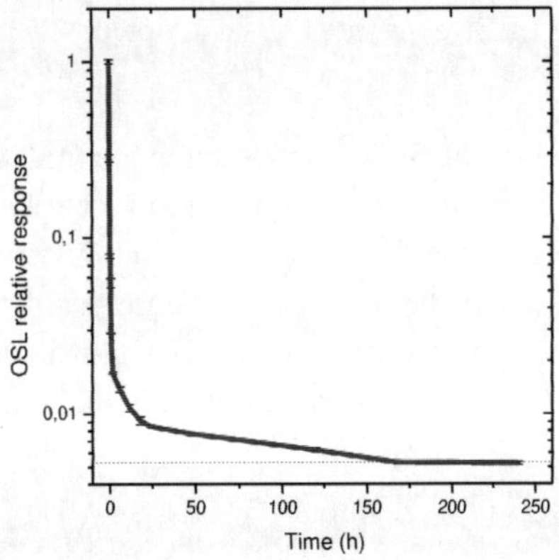

Figure 7. OSL relative response of BeO samples irradiated with 1 Gy of absorbed dose ($^{90}Sr+^{90}Y$ source), exposed to different time intervals of the fluorescent system. The dashed line represents the OSL response of non- irradiated samples (0R).

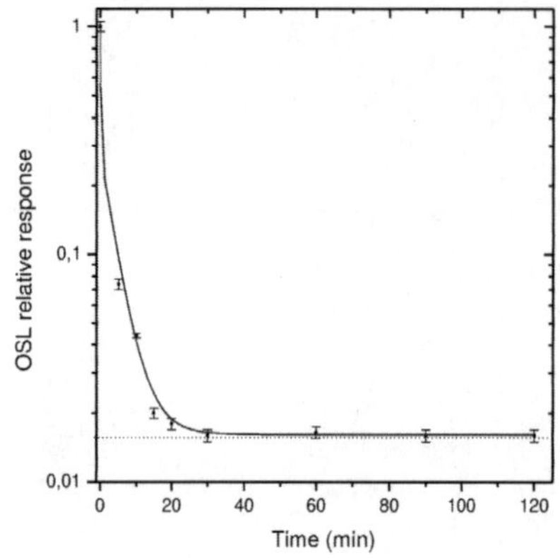

Figure 8. OSL relative response of BeO samples optically treated by blue LEDs.

In figure 8 it is possible to observe that from 30 min of exposure to blue LEDs the OSL response remained constant. It means that the traps involved in the OSL phenomenon were emptied, therefore the treatment with 30 min of blue LEDs exposure can be adopted.

4. Conclusion

For the purpose of response bleaching of BeO samples using the OSL technique, the system with blue LEDs was proved to be more effective in relation to the fluorescent light treatment. In 30 min of blue LED treatment it is possible to bleach the OSL response, much faster than the 7 days required by the fluorescent light. This work shows also the influence of fluorescent light in emptying traps, above all the rapid decrease in the first minutes of exposure; this fact contributed to the adoption of a careful storage of dosimeters after irradiation, mainly under fluorescent light used in most of the laboratories.

Acknowledgments

The authors are grateful for partial financial support from the Brazilian agencies CAPES, FAPESP (Project 2010/16437-0), CNPq (304789/2011-9) and MCT (Project INCT for Radiation Metrology in Medicine, 573659/2008-7).

References

[1] Akselrod M S, Bøtter-Jensen L and Mckeever S W S 2007 *Radiat. Meas.* **41** S78
[2] Bøtter-Jensen L, McKeever S W S and Wintle A G 2003 *Optically Stimulated Luminescence Dosimetry.* Amsterdam: Elsevier
[3] Sommer M and Henniger J 2006 Radiat. *Prot. Dosim.* **119** 394
[4] Sommer M, Freudenberg R and Henniger J 2007 *Radiat. Meas.* **42** 617
[5] Sommer M, Jahn A and Henniger J 2008 *Radiat. Meas.* **43** 353
[6] Groppo D P and Caldas L V E 2014 *Radiat. Meas.* **71** 81
[7] Groppo D P 2013 *MSc Dissertation* Instituto de Pesquisas Energéticas e Nucleares, University of São Paulo (In Portuguese)
[8] Gronchi C C 2009 *PhD Thesis* Instituto de Pesquisas Energéticas e Nucleares, University of São Paulo (In Portuguese)

Journal of Physics: Conference Series **733** (2016) 012086 doi:10.1088/1742-6596/733/1/012086

Mammography calibration qualities establishment in a Mo-Mo clinical system

E L Corrêa, L R dos Santos, V Vivolo and M P A Potiens

Nuclear and Energy Research Institute, University of São Paulo, Av. Prof. Lineu Prestes, 2242, São Paulo, SP, Brazil

Abstract: In this study the mammography calibration qualities were established in a clinical mammography system. The objective is to provide the IPEN instruments calibration laboratory with both mammography calibration methods (using a clinical and an industrial system). The results showed a good behavior of mammography equipment, in terms of kVp, PPV and exposure time. The additional filtration of molybdenum is adequate, air-kerma rates were determined and spectra were obtained.

1. Introduction

Second most frequent in the world, breast cancer is most common among women, accounting for 22 % new cases each year. Prognosis is relatively good if diagnosed and treated properly [1].

Mammography procedure allows the premature breast cancer detection, due to its capability to show injuries in its initial stage [2]. Currently, mammography is the only modality approved for screening application [3]. But, in order to obtain premature and reliable diagnosis, assurance that the mammography system is calibrated and working properly is necessary, and makes a good quality control of these equipment very important.

In Brazil, few laboratories have mammography qualities established in their systems. One of these laboratories is the Laboratório de Calibração de Instrumentos (LCI), at IPEN, which has calibrated about 40 mammography ionizing chambers during 2013/ 2014, and this number means about 80% of this sort of ionizing chambers in Brazil. Mammography chambers calibration are performed in an industrial X-ray system, with a tungsten anode and additional filtration of aluminum and molybdenum [2].

This study presents the results obtained during mammography qualities establishment in a clinical X-ray system, which has already been used in a homemade mammography chamber tests [4]. The goal is to provide information for comparative studies between calibrations performed in an industrial system, which presents low uncertainties and high accuracy, but different energy spectra and measurement geometry, and in clinical equipment, with same irradiation conditions found in hospitals but higher uncertainties.

2. Materials and methods

Mammography qualities were established based on the international standard IEC 61267[5], and the International Atomic Energy Agency (IAEA) code of practice, Technical Report Series No. 457 [6].

2.1. Mammography clinical equipment

Calibration qualities were established in a Philips VMI Graph Mammo AF system with a molybdenum target and molybdenum additional filtration. The voltage can be set from 20 kV to 35 kV. During the tests it presented a temporary malfunction in the Automatic Exposure Control (AEC) system. Therefore the determination of the mean glandular dose and the incident kerma was not possible.

Radiation field intensity variation has already been studied as part of quality control tests [7].

2.2. Voltage and exposure time measurements

For the mean and maximum kVp (peak voltage), PPV (Practical Peak Voltage) and exposure time measurement a non-invasive kVp meter PTW, Diavolt model, was used. This instrument has presented a good behavior both in mammography and conventional radiology beams [8, 9].

Five exposures for each one of four mammography calibration voltages (25 kV, 28 kV, 30 kV and 35 kV) were made. In this system is not possible to set the exposure time directly, but the values of current and current.time were set in order to obtain one second of exposure time. Tube current was set to 80 mA and the product current.time was of 80 mAs.

2.3. Verification of the additional filtration and air-kerma rate determination

Dose measurement was made using an ionization chamber Radcal, RC6M model, with 6 cm^3. This chamber was calibrated at the Physikalisch-Technische Bundesanstalt (PTB), Germany. For this reason same PTB mammography nomenclature (MMV N, where 'N' is the voltage applied to X-ray tube) and half-value layers (HVL) were used [10].

HVL was determined for every four voltages used in mammography to verify if the additional filtration of molybdenum is appropriate for this system. Aluminum attenuators thickness uncertainties were below 5 μm, in accordance with TRS-457 [6].

For air-kerma rate ($\dot{k}$) determination equation 1 was used.

$$k = M f_{t,p} N_k k_Q \tag{1}$$

where M is the measurement made with the reference ionization chamber, $f_{t,p}$ is the correction factor for the pressure and temperature, N_k is the calibration coefficient and k_Q is the factor that indicates the beam quality. Both N_k and k_Q were given by PTB after chamber calibration.

2.4. Spectroscopy

It is important to obtain the system spectra in order to know the energy distribution. It is known that different ionization chambers present different behavior when the combination anode-filtration is changed [11].

For the spectroscopy a CdTe spectrometer Amptek, XR-100T model, connected to an Amptek digital pulse processor, PX4 model, was used.

3. Results and discussions

3.1. Voltage and exposure time measurement

Results obtained for kVp, PPV and exposure time are shown in table 1.

Table 1. Mammography system kVp, PPV and exposure time values obtained using Diavolt. Uncertainties for kVp and PPV were smaller than 4.0 %.

Nominal Voltage (kV)	kVp (kV)	PPV (kV)	Exposure Time (ms)
25	25.7 ± 1.0	25.2 ± 1.0	1005.6 ± 45.3
28	28.4 ± 1.1	28.1 ± 1.1	1006.2 ± 52.3
30	30.5 ± 1.2	30.2 ± 1.2	1007.4 ± 40.3
35	35.8 ± 1.4	35.5 ± 1.4	1005.8 ± 50.3

Results present in table 1 show a good response of this equipment in terms of the presented quantities. The maximum variation between the nominal voltage and the values obtained with the Diavolt was of 0.8 kV, for 35 kV, which is in accordance to standard IEC 61267 [5].

For a current of 80 mA and product current.time of 80 mAs an exposure time of one second was expected. In this case maximum variation was 0.7 %, for 30 kV.

3.2. Verification of the additional filtration and air kerma rate determination

The results for these tests are shown in table 2.

Table 2. Beam intensity reduction (after the HVL is placed) and the air-kerma rate obtained for each quality.

Nominal Voltage (kV)	HVL PTB (mmAl)	Intensity reduction (%)	kQ	Air-Kerma rate (mGy/min)
25	0.29	50.7 ± 0.9	1.000	387.0 ± 25,2
28	0.32	50.9 ± 0.8	1.000	551.3 ± 28.1
30	0.33	52.0 ± 0.8	1.000	678.6 ± 32.5
35	0.37	51.1 ± 0.7	1.000	1020.3 ± 58.2

According to TRS-457[5], beam intensity reduction must be of (50.0 ± 1.5) % when the HVL is inserted. This test showed the reduction is in accordance to TRS-457 for all qualities (when uncertainties are considered). Air-kerma rate was determined for all calibration qualities and can now be used as reference mammography ionization chambers calibration.

3.3. System spectroscopy

System spectroscopy was made for all mammography calibration qualities, and result is presented in figure 1.

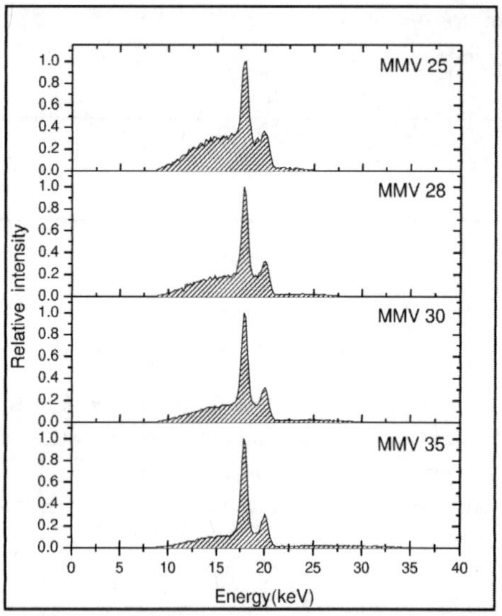

Figure 1. Clinical system spectroscopy made for all mammography calibration qualities.

The spectrometry results show what is expected for Mo-Mo system. Molybdenum energy peaks, due to the anode material, can be noted, and the count reduction in the region after 20 keV, due to the k-edge effect [12].

4. Conclusions

The mammography qualities MMV 25, MMV 28, MMV 30 and MMV 35 were established in the clinical equipment. HVL reductions were within the range for all qualities, and the air-kerma rates were obtained using the calibration coefficient given by PTB. The spectra obtained were according to what is expected from a Mo-Mo mammography system. Despite the problem presented by AEC the results obtained indicate that this mammography system is working properly, and can be used in calibration procedures. More studies will be made in order to develop a calibration method as close as possible to procedures used in medical clinics and hospitals.

Acknowledgements

The authors acknowledge to CNPq, CAPES, FAPESP and FINEP/MCTI project No. 01.10.0650.01 for partial financial support.

References

[1] INSTITUTO NACIONAL DE CÂNCER (INCA), Rio de Janeiro, RJ, Brazil Available: http://www2.inca.gov.br/wps/wcm/connect/tiposdecancer/site/home/mama/cancer_mama+ Accessed in: August, 27th, 2015 (in Portuguese).

[2] Corrêa E L, Vivolo V and Potiens M P A 2012 *Applied Radiation and Isotopes* **70** 1396 .

[3] Kashif A S, Lotz T F, McGarry M D, Pattison A J and Chase J G 2013 *Medical Physics* **40.**

[4] Silva J O, Corrêa E L, Vivolo V, Potiens M P A and Caldas L V E 2013 *IEEE Transactions on Nuclear Science* **60** 708.

[5] IEC 61267 2005 *Medical diagnostic X-ray equipment-Radiation conditions for use in the*

determination of characteristics (Geneva: International Eletrotechnical Comission)

[6] Technical Report Series No. 457 2007 *Dosimetry in diagnostic radiology: an international code of practice* (Vienna: International Atomic Energy Agency)

[7] Corrêa E L, Silva J O, Vivolo V, Potiens M P A, Daros K A C and Medeiros R B 2013 *Radiation Physics and Chemistry* Online **95** 116

[8] Lucena R F, Dias D M, Franciscatto P C, Corrêa E L, Vivolo V and Potiens M P A 2009 *Proceedings of V Congresso Brasileiro de Metrologia* (in Portuguese)

[9] Corrêa E L, Lucena R F, Potiens M P A and Vivolo V *Revista Brasileira de Física Médica* 2011 **4** 23 (in Portuguese)

[10] Physikalisch-Technische Bundesanstalt (PTB). Available: http://www.ptb.de/cms/fileadmin/internet/fachabteilungen/abteilung_6/6.2/6.25/ptb_rad_qual_2015_01_07.pdf
Accessed in: August, 25th, 2015

[11] DeWerd L A, Micka J A, Laird R W, Pearson D W, O'Brian M and Lamperti P 2002 *Medical Physics* **29** 2649

[12] Dowsett D J, Kenny P A and Johnston R E 1998 *The Physics of Diagnostic Imaging* first ed. (Chapman & Hall Medical) Great Britain

Influence of the exposure time in the area monitors at radiodiagnostic

F Leyton[1,2], M J Ferreira[3,4], E M Macedo[3], V C C Navarro[3], I F M Garcia[3,5], L L S Pereira[3], M V T Navarro[3]

[1] Faculty of Health and Odontology, Diego Portales University, Santiago, Chile
[2] Study Center in Radiological Sciences (CECRAD), Tarapacá University, Arica, Chile
[3] Laboratório de Produtos para a Saúde do IFBA (LABPROSAUD-IFBA), Salvador, Bahia, Brazil
[4] Universidade Estadual da Bahia, Salvador, Bahia, Brazil
[5] Bahian School of Medicine and Public Health, Salvador, Bahia, Brazil

E-mail: leyton.fernando@gmail.com

Abstract: Area monitoring ensures radiation exposure at an acceptable level, which must be lower than the legal limit. Experimental measurements were taken in an ionizing radiation calibration laboratory. The specified reference radiation to radiation protection N80 was used. Five area monitors were used. The ranges of dose rate inaccuracy measured in rate mode for times ≤ 2 and ≥ 3s were from 10 to 48 % and from 1 to 15 %, respectively. The inaccuracy ranges measured in integrated mode for times ≤ 2 and ≥ 3s were from 4 to 8 % and from 0 to 22 %, respectively.

1. Introduction

These Medical exposures, among all the practices that involve ionizing radiation, are responsible for the greatest contribution to population exposure [1]. The International Atomic Energy Agency (IAEA) [2] recommend monitoring areas related with medical exposure in addition to supervising public areas. In this regard, the Secondary Standards Dosimetry Laboratory (SSDL) for each country has responsibility to ensure that area monitors are traceable and properly calibrated.

Area monitors are used for the real-time monitoring of exposure and hence to estimate the potential risk to which members of the public and employees are exposed. The IAEA's Technical Reports [3] states that for survey meters, a required accuracy of 20 % in dosimetry measurements should be sufficient. Besides, there is no unified protocol for the calibration procedure of survey meters in SSDLs. The IAEA's Reports [4] shows that exposure times for calibration vary between 5 and 60 s.

For area monitoring at radiodiagnostic, the lowest possible time and current (mA) are generally used to prevent possible harm to the X-ray tube during the completion of the test, with the great majority of area monitoring studies using ionization chambers conducting the test for times less than 2.5 s and in rate mode [5,6]. This means that the survey meter must be capable of measuring exposures lasting seconds or fractions of seconds. Various factors must be considered (range, calibration, energy dependence, etc.) in selecting the appropriate survey meter for area monitoring. There are few studies examining the dependency of the ionization chamber and of solid-state detectors for the measures taken in rate or integrated mode at low exposure times. This paper shows the influence of exposure time on area monitors during dose rate measurement in rate and integrated modes for a tension 80 kV used at radiodiagnostic and interventional facilities.

2. Materials and methods

Experimental measures were taken in LabProsaud, an ionizing radiation calibration laboratory [7]. The instruments of Labprosaud are directly traceable to primary standards maintained at the national metrology institute of Germany, Physikalisch – Technische Bundesanstalt (PTB) [8], with uncertainty within 3 % and a 95 % confidence interval. A General Electric ISOVOLT TITAN E X-ray system with ISOVOLT 160 M2 tube, with output voltage range from 5 to 160 kV, output current range from 0.2 to 10 mA and high frequency was used. The radiation quality required and the calibration arrangement comply with the requirements of reference standards ISO 4037-1, 2 and 3 [9]. The specified reference radiations for the narrow spectrum series dedicated to radiation protection, N80 was used, with the following characteristics: 65 keV, tube potential 80 kV, additional filtration 2.0 mmCu, first HVL 0.58 mmCu and second HVL 0.62 mmCu.

The study assessed five area monitors. The technical characteristics of these systems were taken from the manufacturer manuals. The meters are identified by letters Atomtex AT 1123 (A), Fluke Biomedical 451P (F), PTW 32002 (P), Radcal 9010X5-180 (R) and Radcal 9010X5-1800 (R2). They were positioned at 200 cm from the focal spot of the X-ray system and using reference points and orientations as specified by the manufacturer.

The P ionization chamber is directly traceable to primary standards PTB. The total traceable uncertainties for the P ionization chamber were 6 and 4 % for rate and integrated mode respectively (at a 95 % confidence interval). Hence, calculations of the inaccuracy of measurements were taken as against the values obtained for the PTW chamber. Meters A, F, R and R2 were calibrated at Labprosaud. The measurements were taken at the normal working rate for each meter and corrected for pressure and temperature and for ambient dose equivalent H*(10), which is used as an estimate of ambient dose equivalent at 10 mm depth [10,11].

The measurements of the dose (D) and dose rate (DRm) were measured in integrated and rate modes, respectively. For each detector evaluated, five measurements were taken at each exposure time of 1, 2, 3 and 5 s, respectively, with 0.7 mA for N80 radiation quality.

The calculated dose rate values (DRc) were calculated using the quotient between D measured in integrated mode divided by experiment time. Experiment time was measured by PTW equipment (NOMEX, T11049) directly traceable to primary standards maintained at the PTB, with uncertainty within 1 % with a 95 % confidence interval. The values of experiment time were 1.145, 2.143, 3.146 and 4.953 s for nominal times of 1, 2, 3 and 5s, respectively.

3. Results and discussion

3.1. Rate mode measurements

Figure 1 shows the measurements taken for 80 kV. When exposure time was increased, the DRm became closer to the true value. The highest dose rate inaccuracy was 48 % for the A detector at 1 s with respect to the P meter, and the lowest inaccuracy was -1 % for the R2 meter at 5s.

The ranges of DRm inaccuracy for the area monitors with respect to the true value (PTW) for times $\leq$ 2s and $\geq$ 3s were from 10 to 48 % and from 1 to 15 %, respectively. The R chamber was not able to measure dose rate for times $\leq$ 2s at 80 kV. The average values for dose rate inaccuracy for each area monitors for times ≤ 2 s and ≥ 3s were 36, 32 and 18% for A, F and R2, respectively and 6, 11, 10 and 4% for A, F, R and R2, respectively.

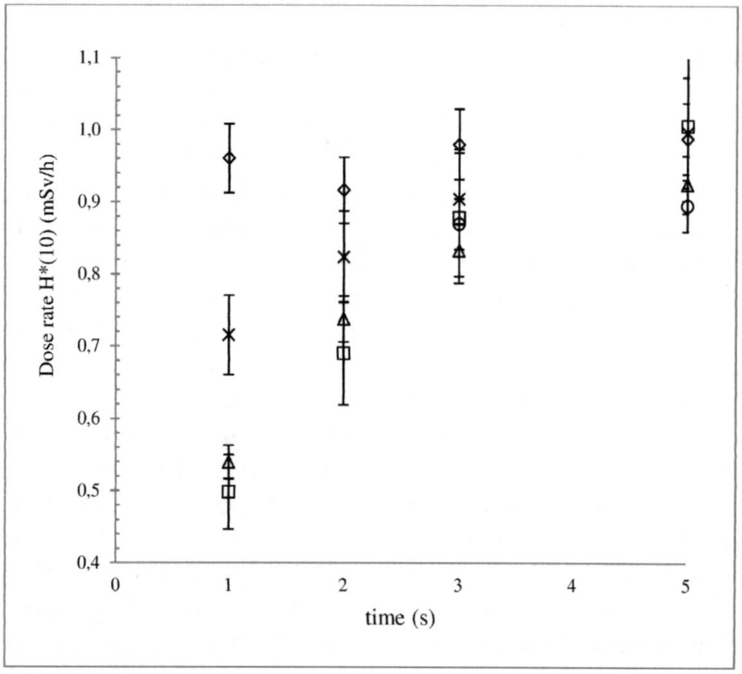

Figure 1: Average dose rate (DRm) and standard deviation (sd) values measured in dose rate mode for different area monitor meters Atomtex (□), Fluke (Δ), PTW (◊), Radcal 180 cc (○) and Radcal 1800 cc (x) at different exposure times with a tension of 80 kV.

At 80 kV, the greatest percentage difference between the readings for the P chamber was 7 % between 2 and 5 s. The greatest percentage differences for dose rate for the A detector, F and R2 ionization chamber were 50, 42 and 28 % between 1 and 5 s, respectively. The percentage differences for each area monitor for the dose rate between 3 and 5 s were 13, 10, 3 and 9 % for the A, F, R and R2, respectively.

3.2. Integrated mode measurements

Figure 2 shows the dose rate calculated taken at 80 kV using integrated mode. The dispersion of DRc values is lower than the dispersion of values obtained at figure 1.

The ranges of in accuracy between the DRc measurements for integrated mode (figure 2) were lower than DRm measurements for rate mode (figure 1). The inaccuracy ranges of the area monitors with respect to the true (PTW) value for times ≤ 2 s and ≥ 3s were from 4 to 8 % and from 0 to 22 %, respectively. On taking integrated mode measurements at 80 kV, the R meter was able to measure the dose for all exposure times used.

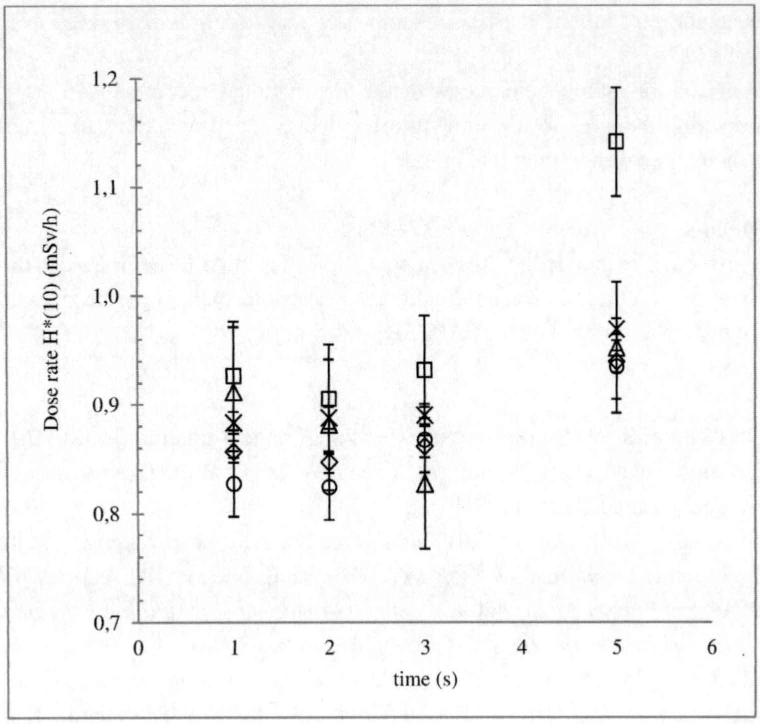

Figure 2: Mean dose rate (DRc) and standard deviation (sd) values calculated in integrated mode for different area monitor meters Atomtex (□), Fluke (Δ), PTW (◊), Radcal 180 cc (○) and Radcal 1800 cc (x) at different exposure times with a tension of 80 kV.

The average values for DRc inaccuracy for times ≤ 2s and ≥ 3s were 8, 5, 3 and 4% for Acal, Fcal, Rcal and R2cal, respectively and 15, 3, 0 and 3% for the Acal, Fcal, Rcal and R2cal, respectively. For integrated mode, only one measurement (6%) has a deviation in accuracy of more than 20 %, while for rate mode (figure 1), 36 % of measurements had a deviation in accuracy of more than 20 %.

The greatest percentage difference between the readings for the Pcal meter was 10 % between 2 and 5s. The percentage differences in DRc for the Acal, Fcal, Rcal and R2cal were 19 and 21 %, 4 and 7 %, 11 and 12 % and 9 and 8% for times of 1 and 2s in comparison with the 5 s readings for each meter, respectively. The percentage differences in DRc between 3 and 5s were 18, 13, 7and 8 % for the Acal, Fcal, Rcal and R2cal monitors, respectively.

4. Conclusions

All the area monitors evaluated in this study show a tendency to underestimate dose rate when measurements are taken in rate mode for times ≤ 2s. On taking integrated mode measurements, the dispersion of dose rate values is lower than for the values obtained in rate mode. Moreover, in integrated mode it was possible to calculate dose rate for all exposure times and different radiation beam qualities.

The values obtained in rate and integrated modes (figures 1,2) suggest that this type of evaluation is necessary to develop knowledge regarding each area monitor and to improve the accuracy and precision of measurements. The results of this study show that it is highly recommendable to work in integrated mode and precautions must be taken if area monitoring is performed in rate mode with exposure times of ≤ 2s.

The results of this study coincide with the international recommendations [12,13], which recommends measuring using exposure times greater than 3 s and advise that measurements of the monitored area should be taken in integrated mode.

4. Acknowledgments

The authors wish to acknowledge the support of the FAPESB/ProParq, through project "Desenvolvimento Tecnológico, inovação e calibração de equipamentos médicos". Garcia wishes to acknowledge the master fellowship of the FAPESB.

5. References

[1] United Nations Scientific Committee on the Effects of Atomic Radiation (UNSCEAR) 2008 *Report to the General Assembly, with scientific annexes. Volume I: Report to the General Assembly, Scientific Annexes A* (New York: United Nation) pp 31 - 36

[2] International Atomic Energy Agency 2011 *Radiation Protection and Safety of Radiation Sources: International Basic Safety Standards, No. GSR Part 3 (Interim)* (Vienna: IAEA) pp 22 - 426

[3] International Atomic Energy Agency 2007 *Dosimetry in diagnostic radiology: an international code of practice. Technical reports series no. 457* (Vienna: IAEA) pp 8 - 10

[4] International Atomic Energy Agency 2011 *Implementation of the international code of practice on dosimetry in diagnostic radiology. Human Health Reports No. 4, Pub 1498* (Vienna: IAEA) pp 24 - 25

[5] Vlachos I, Tsantilas X, Kalyvas N, Delis H, Kandarakis I, Panayiotakis G 2015 Radiat Prot Dosimetry **165**, 382.

[6] Fernando Leyton, Carlos Ubeda, Otto Delgado, Alfonso Espinoza, Jorge Díaz, Carlos Oyarzún, Santiago Mansilla 2007 Rev. chil. radiol. **v.13** n.4, 213.

[7] http://www.labprosaud.ifba.edu.br/

[8] https://www.ptb.de/cms/en.html

[9] https://www.iso.org/obp/ui/#search

[10] International Atomic Energy Agency 2000 *Calibration of Radiation Protection Monitoring Instruments. Safety Reports Series No. 16* (Vienna: IAEA) pp 6 - 76

[11] http://www.iso.org/iso/iso_catalogue/catalogue_tc/catalogue_detail.htm?csnumber=23727

[12] National Council on Radiation Protection and measurements 2005 *Structural Shielding Design for Medical X-Ray Imaging Facilities. NCRP REPORT No. 147* (Bethesda, MD 20814, NCRP) p 149

[13] International Atomic Energy Agency 2014 *Diagnostic Radiology Physics: A Handbook for Teachers and Students* (Vienna, IAEA) pp 548 - 637

Attenuation coefficient determination of printed ABS and PLA samples in diagnostic radiology standard beams

G R Veneziani[1], E L Corrêa[1], M P A Potiens[1], L L Campos[1]

[1] Instituto de Pesquisas Energéticas e Nucleares – IPEN/CNEN – SP
Avenida Lineu Prestes, 2242 – São Paulo - Brazil

Abstract. IAEA code of practice TRS-457 states that standard phantoms should offer the same primary attenuation and scatter production as relevant body section of a representative patient. Material cost, availability and dimensional stability must also be considered. The goal of this study is to determine the attenuation coefficient of printed ABS and PLA samples in standard X-ray beams, verifying if phantoms printed with these materials could be an easier-handle substitute for PMMA, enabling the creation of different designs in an easier and cheaper way. Results show that PMMA presents higher attenuation coefficient, followed by PLA and ABS, which means that thinner PMMA layer creates higher radiation attenuation.

Keywords: Half-value layer, quality assurance, 3D printer, phantom, diagnostic radiology

1. Introduction

According to the International Atomic Energy Agency (IAEA) code of practice Technical Report Series No. 457 (TRS 457), standard phantoms *should be designed and constructed so that they offer the same primary attenuation and scatter production as the relevant body section(s) of a representative patient* [1], and the energy range should also be considered.

Other important characteristics to be considered when building a phantom is the material cost, availability and dimensional stability [1]. A good option in this case is the polymethylmethacrylate (PMMA), which shows all characteristics presented previously.

However, in the past years, 3D printers have been applied in medicine. When technology began to be used for dental implants and prostheses customization this technology has been rapidly expanding and revolutionizing the field of health, being used for several purposes such as organs, implants, anatomical models, customization of prosthetic, to design drugs formats, among other applications [2].

An important use is phantoms construction from magnetic resonance, tomography, and X-ray images converted into printer compatible files. Another application is the construction of anatomical models for use in surgical preparation such as craniofacial, pelvic, spine and brain surgery. This is possible through diagnostic images acquisition and structure reconstruction in software and 3D printing [3, 4].

Two materials are the cheapest and the most commonly used in this kind of 3D printer. ABS (Acrylonitrile Butadiene Styrene), which presents a density from 1060 kg/m³ to 1080 kg/m³ [5], and PLA (Polylactic Acid), which density varies from 1210 kg/m³ to 1430 kg/m³ [6].

The goal of this study is to determine the attenuation coefficient for both ABS and PLA in standard diagnostic radiology X-ray beams, verifying the 3D printed phantom use viability as an easier-handle substitute for PMMA, enabling the creation of different designs in an easier and cheaper way.

2. Materials and methodology

2.1. Materials

The plates used in this study were printed in an UP 3D printer, plus 2 model (figure 1).

Figure 1. UP 3D printer, plus 2 model.

Different thicknesses of ABS and PLA plates (from 2.0 to 10.0 mm) were printed, and seven commercial PMMA plates of 4.7 mm each were also used (figure 2) for results comparison.

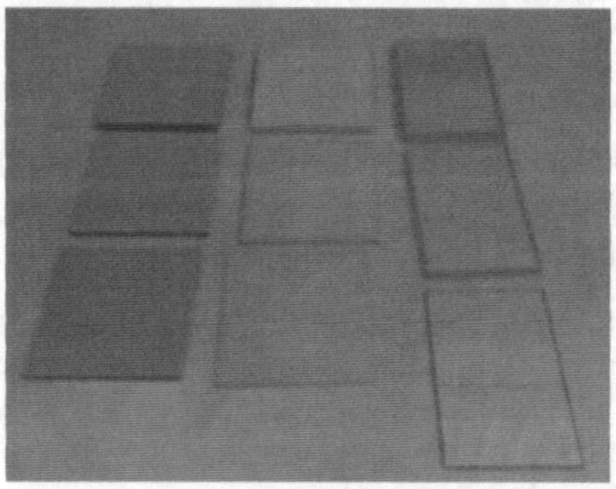

Figure 2. From left to right: PLA, ABS and PMMA plates.

Irradiation tests were performed using an industrial X-ray system Pantak/Seifert used for instruments calibration in different energy ranges [6-8]. The diagnostic radiology qualities main characteristics are presented in table 1.

Table 1. Diagnostic radiology qualities main characteristics

Radiation quality	Tube voltage (kV)	HVL (mmAl)
RQR-3	50	1.78
RQR-5	70	2.58
RQR-8	100	3.97
RQR-10	150	6.57

X-ray beam intensity was measured using a Radcal ionization chamber, RC6 model, connected to a Keithley electrometer, 6517A model.

2.2. Methodology

X-ray beam attenuation measurements were performed using a plate-holder (figure 3), designed using SolidWorks® software and printed specifically for this procedure. This holding device was placed in half distance from the X-ray anode and measurement surface.

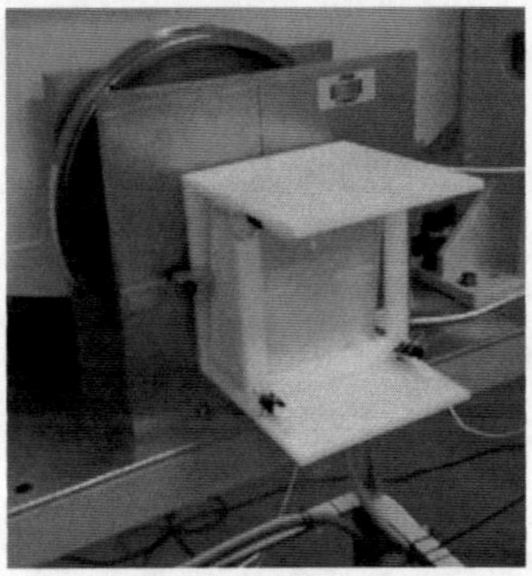

Figure 3. Printed plate-holder used in beam attenuation measurements for different sample plates.

The sample thickness was increased from zero (without filter) to a value high enough to reduce the beam intensity in more than 50 %.

The values obtained in this procedure were analyzed using Origin® software. For each quality an exponential fit was performed in order to obtain the attenuation coefficient (μ), according to equation 1:

$$I = I_0 e^{-\mu x} \tag{1}$$

where I_0 is the initial beam intensity, and I is the beam intensity when some material of thickness x is placed between radiation source and ionization chamber.

3. Results

Attenuation curves of ABS, PLA and PMMA for 50 kV, 70 kV, 100 kV and 150 kV for X-rays beams are shown, respectively, in figures 4, 5, 6 and 7.

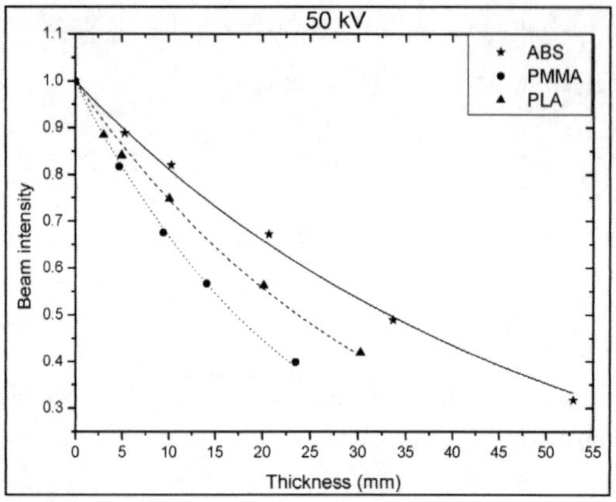

Figure 4. Attenuation curves for 50 kV X-ray beam.

The exponential fit provided attenuation coefficients of 0.020 mm^{-1}, 0.029 mm^{-1} and 0.040 mm^{-1} for ABS, PLA and PMMA, respectively.

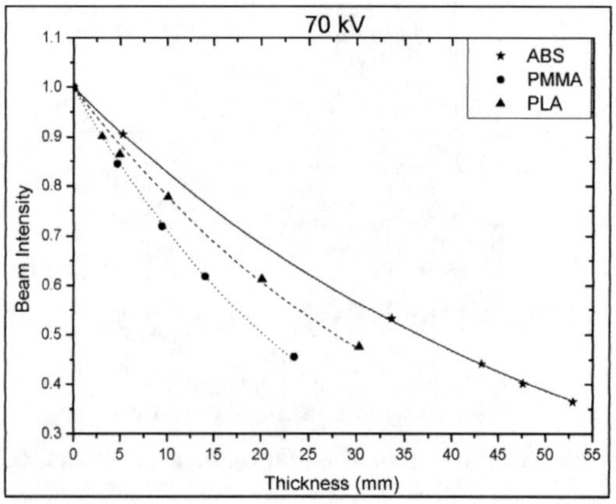

Figure 5. Attenuation curves for 70 kV X-ray beam.

The exponential fit provided attenuation coefficients of 0.018 mm^{-1}, 0.024 mm^{-1} and 0.030 mm^{-1} for ABS, PLA and PMMA, respectively.

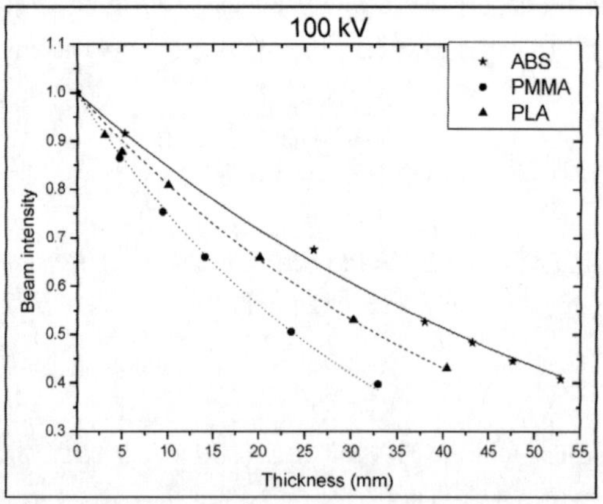

Figure 6. Attenuation curves for 100 kV X-ray beam.

The exponential fit provided attenuation coefficients of 0.016 mm^{-1}, 0.021 mm^{-1} and 0.028 mm^{-1} for ABS, PLA and PMMA, respectively.

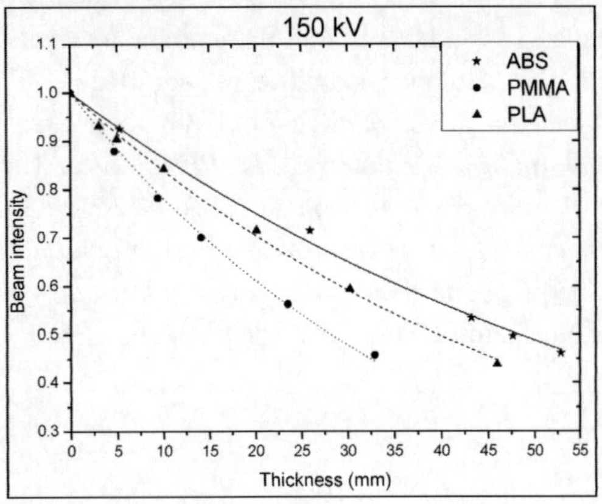

Figure 7. Attenuation curves for 150 kV X-ray beam.

The exponential fit provided attenuation coefficients of 0.014 mm^{-1}, 0.017 mm^{-1} and 0.024 mm^{-1} for ABS, PLA and PMMA, respectively.

The obtained results show a clear difference between materials attenuation coefficients. PMMA presented a higher attenuation coefficient, followed by PLA, from 20 % to 30 % lower, and ABS, from 40 % to 50 % lower, which means that a thinner layer of PMMA creates higher radiation attenuation.

In order to design and print a phantom a thickness correction, for different materials, must be made.

4. Conclusions

Easier access to 3D printers, followed by comparable attenuation coefficient values and ease of design make the use of printed phantoms applied to quality control in diagnostic radiology a reality.

New phantoms could be printed in different dimensions and at relatively low cost considering the large number of different shapes applicable.

Acknowledgments
The authors acknowledge the financial support CNPq, CAPES, FAPESP Project no. 2010/16437-0, FINEP/MCTI project No. 01.10.0650.01.

References

[1] INTERNATIONAL ATOMIC ENERGY AGENCY. Dosimetry in Diagnostic Radiology: An International Code of Practice. IAEA, Vienna, 2007 (Technical Report Series No. 457).

[2] C. LEE VENTOLA. Medical Applications for 3D Printing: Current and Projected Uses. P&T, Vol. 39 No. 10, pp 704 – 711, Outubro 2014.

[3] HESPEL, A.M.; WILHITE, R.; HUDSON, J.: Invited review – Applications for 3D printers in veterinary medicine, Vet Radiology and Ultrasound, Vol55 No. 4, pp 347-358, 2014.

[4] "MATBASE: ABS General Purpose," http://www.matbase.com/material-categories/natural-and-synthetic-polymers/commodity-polymers/material-properties-of-acrylonitrile-butadiene-styrene-general-purpose-gp-abs.html#properties (2015).

[5] "MATBASE: PLA monomere (Polylactic Acid)," http://www.matbase.com/material-categories/natural-and-synthetic-polymers/agro-based-polymers/material-properties-of-polylactic-acid-monomere-pla-m.html#general (2005).

[6] E. L. CORRÊA, V. VIVOLO, M. P. A. POTIENS, "Quality control methodology and implementation of X-radiation standards beams, mammography level, following the standard IEC 61267," *Applied Radiation and Isotopes*, **v. online**, p. 1-4 (2012)

[7] P. C. Franciscatto, *Caracterização das qualidades de radiação x seguindo as recomendações da norma IEC 61267 no laboratório de calibração do IPEN*, Master Thesis - Instituto de Pesquisas Energéticas e Nucleares, Universidade de São Paulo, São Paulo (2009).

[8] D. M. Dias, *Estabelecimento de um novo método de calibração de câmaras de ionização tipo lápis para dosimetria em feixes de tomografia computadorizada*, Master Thesis – Instituto de Pesquisas Energéticas e Nucleares, Universidade de São Paulo, São Paulo (2010).

Development and evaluation of a technique for *in vivo* monitoring of ^{60}Co in human lungs

J Q de Mello[1], E A Lucena[1], A L A Dantas[1] and B M Dantas[1]

Instituto de Radioproteção e Dosimetria-CNEN

E-mail: jquince@bolsista.ird.gov.br, eder@ird.gov.br, adantas@ird.gov.br, bmdantas@ird.gov.br

Abstract: ^{60}Co is a fission product of ^{235}U and represents a risk of internal exposure of workers in nuclear power plants, especially those involved in the maintenance of potentially contaminated parts and equipment. The control of ^{60}Co intake by inhalation can be performed through *in vivo* monitoring. This work describes the evaluation of a technique through the minimum detectable activity and the corresponding minimum detectable effective doses, based on biokinetic and dosimetric models of ^{60}Co in the human body. The results allow to state that the technique is suitable either for monitoring of occupational exposures or evaluation of accidental intake.

1. Introduction

^{60}Co is a synthetic radioactive isotope with a half-life of 5.27 years, produced artificially in nuclear reactors. Measurable quantities are also produced as a by-product of typical nuclear power plant operation and may be detected externally when leaks occur.

^{60}Co is the result of multiple stages of neutron activation of iron components of the reactor structure. It decays to the stable isotope ^{60}Ni, emitting β particle. The activated nickel nucleus emits two γ rays with energies of 1.17 and 1.33 MeV [1]. Such emissions represent a risk of occupational internal exposure to workers involved in the handling and maintenance of parts and equipment potentially contaminated in nuclear power units.

Internal monitoring procedures aimed to identify and quantify intakes by inhalation can be performed through *in-vivo* measurements in laboratories so-called whole-body counters. The *In-Vivo* Monitoring Laboratory of IRD is able to perform this type of measurement using a NaI(Tl)8x4 scintillation detector installed in a heavy shielded room [2].

This work describes the standardization and evaluation of the sensitivity of a technique for the *in-vivo* measurement of ^{60}Co in human lungs.

2. Materials and Methods

The calibration of the detection system is performed using a LLNL thorax phantom [3] containing a certified ^{60}Co lung phantom. The procedure allows obtaining the calibration factor, which correlates the net count rate in specific regions of interest (ROI) and the activity present in the phantom. This parameter is necessary to calculate the activity content in the lungs of an individual measured *in-vivo*.

2.1. Determination of the calibration factor

The LLNL thorax containing a pair of lung phantoms with a total of 33574 Bq of ^{60}Co was placed on the monitoring chair located inside the shielded room of the IRD whole-body counter and the NaI(Tl)8x4 detector was positioned horizontally at a distance of 2 cm between the surface of the phantom and the front face of the crystal, as shown in Figure 1.

A series of five counts of 300 seconds was performed at the standard geometry and the spectra saved for subsequent calculations. Figure 2 is a γ spectrum of ^{60}Co showing the two characteristic peaks of 1.17 and 1.33 MeV. Three regions-of-interest (ROI) were defined including a range of channels of [347-408] for the first peak, [414-470] for the second and a third region of [347-470], including both peaks. An additional count of an inert phantom was also performed for the calculation of the net count rate in these regions of interest.

Figure 1. Calibration setup for the measurement of ^{60}Co in the lungs using NaI(Tl)8x4

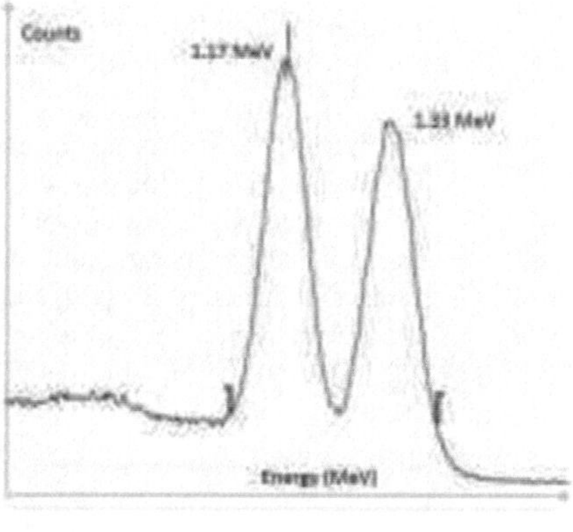

Figure 2. Gama spectrum of ^{60}Co obtained with a NaI(Tl) scintillation Detector

After the series of counts were completed, the calibration factors (CF) in each ROI was calculated. The CF is expressed as the ratio between the average net count rate of the five counts, in cps, and the activity of the phantom, in Bq, as follows:

$$CF_{cps/Bq} = cps/A \quad (1)$$

Where: cps is the net count rate (averaged cps of the phantom subtracted by background count rate) and A (Bq) = ^{60}Co activity content of the lung phantom.

2.2. Evaluation *of sensitivity*

The evaluation of the sensitivity of the method is based on the calculation of the *minimum detectable* activity and the corresponding *minimum detectable intake* and the *minimum detectable effective dose*, using the biokinetic and dosimetric models of ^{60}Co in human body applied to a realistic internal exposure scenario [4]. The MDA of the method is calculated as follows [5]:

$$MDA = \frac{4.65\sqrt{N}}{E \times T} + \frac{3}{E \times T} \quad (2)$$

Where: N = Total counts of the background in 300 seconds; CF = Calibration Factor (cps/Bq) and T = count time.

The Minimumm Detectable Intake (MDI), is a function of the MDA and depends on the exposure scenario and time elapsed between intake and the *in-vivo* measurement. It is calculated as follows:

$$MDI = MDA \,/\, m(t)_{inh} \quad (3)$$

Where: MDA = Minimum Detectable Activity (Bq) and m(t) = Retention fraction for inhalation in the thoracic compartment.

The last parameter to be calculated is the Minimum Detectable Effective Dose. It is based on the MDI and in the dose coefficients associated, e(g), to the corresponding intake scenario adopted in the simulation. It is calculated as follows:

$$MDED_{Bq} = MDI_{inh} \times e(g)_{inh} \quad (4)$$

Where: MDI = Minimum Detectable Intake (Bq) and $e(g)_{inh}$ = Dose coefficients (mSv/Bq).

The values of "m(t)" and "e(g)", used for the calculations of the MDI and the MDED are presented in Table 1. In this work it has been used m(t) values corresponding to 1 day after the intake and the dose coefficient, e(g), corresponding to AMAD of inhaled particles of 1 µm and 5 µm. Such values are available in the Publication 78 of the ICRP [4] and can also be generated for specific exposure scenarios and times through the software AIDE [6].

Table 1. Biokinetic and dosimetric parameters of the ^{60}Co model used for the calculation of the Minimum Detectable Intakes and the Minimum Detectable Effective Doses.

Intake Scenario	Retention fractions in thoracic compartment m(t) (Bq/Bq)			Total Retention Fraction in 1 day m(t) (Bq/Bq)	Dose Coefficient (mSv/Bq)
	Bronchial	Bronchiolar	Alveolar		
1 µm/ Type S	6.07×10^{-3}	9.33×10^{-3}	1.06×10^{-1}	1.21×10^{-1}	2.88×10^{-5}
5 µm / Type S	6.09×10^{-3}	5.34×10^{-3}	5.28×10^{-2}	6.42×10^{-2}	1.66×10^{-5}
1 µm / Type M	5.45×10^{-3}	8.36×10^{-3}	9.48×10^{-2}	1.09×10^{-1}	9.61×10^{-6}
5 µm / Type M	5.46×10^{-3}	4.78×10^{-3}	4.73×10^{-2}	5.75×10^{-2}	7.13×10^{-6}

In order to be considered useful for internal dosimetry purposes, the technique should, at least, be able to detect an activity that would result in an effective dose below 1 mSv per year for the most likely internal exposure scenario [7].

3. Results and discussion

Table 2 presents the count rates, the calibration factors and the MDA for the three ROI. The ROI #3 was elected for subsequente calculations since it has shown to be the most sensitive based on its corresponding MDA.

Table 2. Results of the calibration of the NaI(Tl)8x4 detector for in vivo measurement of 60Co in the lungs, using three regions of interest

Count #	ROI 1	ROI 2	ROI 3
1	167121	142423	313991
2	167309	143071	314674
3	165739	141771	311731
4	166133	142073	312491
5	166269	142602	313139
Mean	166269 +/- 672	142423 +/- 498	313139 +/- 1169
cps	553 +/- 2	474 +/- 2	1042 +/- 4
CF (cps/Bq)	0,01648 +/- 0.00024	0,01412 +/- 0.00021	0,03104 +/- 0.00045
AMD (Bq)	8.1	7.5	5.7

Table 3 presents the values of MDI and MDED derived from the MDA, considering four possible intake scenarios via inhalation in which it has been assumed AMADs of 1 and 5 μm and solubility classes in lung fluid suggested in ICRP Publication 78 [4] as Type S and Type M.

Table 3. Minimum Detectable Intakes and Minimum Detectable Effective Doses assuming four exposure scenarios.

Scenario	MDI (Bq)	MDED (μSv)
1 μm / Type S	46.95	1.35
5 μm / Type S	88.74	1.47
1 μm / Type M	52.48	50.4
5 μm / Type M	99.06	70.6

The most critical internal exposure scenario in terms of radiological risk to the workers would be the intake via inhalation of a particle with AMAD of 1 μm and solubility class Type S. In this case a higher fraction of the inhaled particles would be deposited in the thoracic region - bronchial (BB), bronchiolar (bb) and alveolar–interstitial (AI) – which is the most radiosensitive, resulting in a higher health risk.

4. Conclusions

Based on the criteria for the evaluation of sensitivity adopted in this work it can be stated that the proposed technique is able to detect an activity of ^{60}Co in the lungs derived from an intake via inhalation that would result in a committed effective doses in the order of microsieverts. Such values are far below the Registry Level suggested by the IAEA and stablished by the National Regulatory Board.

The results obtained in this work allow concluding that the proposed technique is sensitive enough for the control of intakes of ^{60}Co via inhalation. The technique is suitable for *in vivo* monitoring of workers in nuclear power stations in special tasks performed during reactor maintenance. It can also be useful for the evaluation of accidental intakes.

5. Acknowledgment

The authors would like to thank the support of FINEP /MCTI (Project 01.10.0650.01)

6. Reference

[1]US EPA [http://www.epa.gov/radiation/radionuclides/cobalt.html#wheredoes Radiation Protection: Cobalt. Accessed 30/07/2015

[2] Oliveira CAN, Lourenço MC, Dantas BM, Lucena EA and Laurer GR 1989 "The IRD/CNEN whole body counter: Background and calibration results" *Radiat. Prot. Dosm.* **29** (3) 203

[3] Spitz, H.B, Glover, S., Liu, N., Smith, B., Hickman, D., Kruchten, D. & Anderson, L 1994 Measurement of the attenuation coefficient for Livermore thoracic phantom lungs fabricated using contemporary materials. Health Phys., v.67, n.1, p.39-46

[4] International Commission On Radiological Protection (ICRP) 1998 Publication 78: Individual Monitoring for Internal Exposure of Workers

[5] Health Physics Society (HPS) 1996 Performance Criteria for Radiobioassay, N13.30

[6] Bertelli, L., Melo, D. R.; Lipsztein, J., Cruz-Suarez, R. 2008 "AIDE – Internal Dosimetry Software". Radiation Protection Dosimetry 130(3), p. 358-367

[7] International Atomic Energy Agency (IAEA) 1999 Assessment of Occupational Exposure Due to Intakes of Radionuclides – Safety Standards Series, no. RS-G-1.2, IAEA

Characterization of a free air ionization chamber for low energy X-rays

N F Silva, M Xavier, V Vivolo and L V E Caldas

Instituto de Pesquisas Energéticas e Nucleares (IPEN/CNEN-SP) Av. Professor Lineu Prestes, 2242 05508-000 São Paulo, SP

nsilva@ipen.br, mxavier@ipen.br, vivolo@ipen.br, lcaldas@ipen.br

Abstract. Free air ionization chambers are used by most primary metrology laboratories as primary standards of the quantities air kerma and exposure in X-ray beams. The free air ionization chamber for low energies of the Calibration Laboratory (LCI) of IPEN showed in a characterization test a problem in the set responsible for the variation of its sensitive volume. After a modification in the support of the micrometers used for the movement of the internal cylinder and the establishment of a new alignment system protocol, the tests were redone. The objective of this work was to present the results obtained in the new condition.

1. Introduction

Most of the primary standard calibration laboratories use the free air ionization chambers as primary standards of the quantities air kerma and exposure to X-ray beams of low and medium energies [1].

The free air ionization chambers do not possess a material in the chamber entry window as in other ionization chambers, thus the primary photons and secondary electrons interact only with the air in the sensitive volume [2]. There are different types of free air ionization chambers; the model most used is the free air ionization chamber with parallel plates (called also conventional model) and the free air ionization chamber with concentric cylinders [3].

The ionization chamber with cylinders, proposed by Attix in 1961, allows the variation of the sensitive volume lenght. The movement of the cylinders is performed with a high precision equipment that allows a precise determination of the air mass and consequently of the exposure and air kerma values. In this case, the independence of the response to the uniformity of the electric field within the sensitive volume eliminates the need for guard rings, required in the case of the parallel plate model. Once the collecting electrode extends throughout the sensitive volume, the collection region is across the internal volume of the cylinder [4].

The Calibration Laboratory (LCI) of the Instituto de Pesquisas Energéticas e Nucleares (IPEN/CNEN-SP) has a free air ionization chamber of the cylindrical type for low energies. A first characterization of this ionization chamber was already performed and reported in a previous study. These tests showed good results for the chamber, except in the volumetric response linearity test [6]. After a modification in the support of the micrometers used for the movement of the internal cylinder devices, the tests were redone using a new alignment protocol described in a previous report [7]. The objective of this work is to present the new results of characterization tests in the new condition.

2. Materials

The free air ionization chamber utilized was of the cylindrical type, Victoreen, model 481.

An X-ray equipment, Pantak/Seifert, model ISO - VOLT 160 HS, operating between 5 kV and 160 kV, was utilized in this work. The standard radiation quality of the mammography beam used in the tests was RQR-2M, with the following characteristics: additional filtration of 0.07mmMo, half value layer of 0.37mmAl, 28kV and 10 mA. Formatting the title

The title is set 17 point Times Bold, flush left, unjustified. The first letter of the title should be capitalized with the rest in lower case. It should not be indented. Leave 28 mm of space above the title and 10 mm after the title.

3. Results and Discussion

3.1. Stability Tests

The results on the repeatability tests were within the limit recommended, by the standard IEC 61674 [8], which is 1.0%; the highest value obtained was 0.7 %. Long-term stability was observed with the results obtained in the repeatability tests over a period of three months. According to the IEC 61674 [8], the maximum variation on the measurements results should be 2.0%; the highest value obtained was only 0.5 %.

3.2. Characterization Tests

The results obtained for the characterization tests include: saturation curve, ion collection efficiency, polarity effect, the linearity response with the X ray tube current and linearity of response to the volume variation.

3.3. Saturation curve

As the free air ionization chamber of cylindrical type requires two measurements to obtain the air kerma, two saturation curves must be obtained: one with the smallest possible length of the sensitive volume (collapsed position) and the other with the greatest possible length of the sensitive volume (expanded position). Figure 1 shows the curves obtained in these cases.

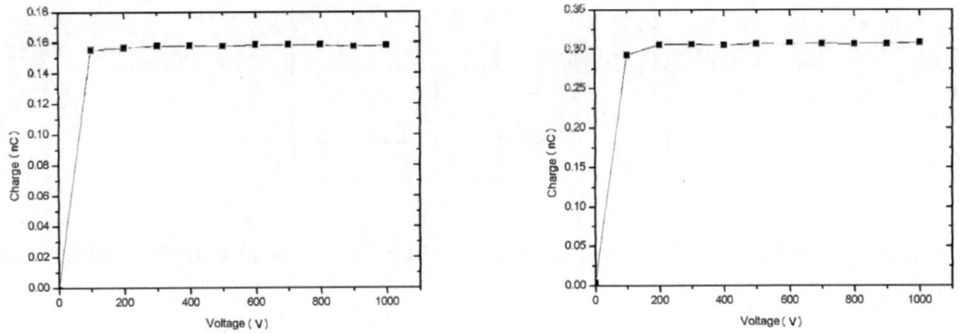

(a) (b)

Figure 1. Saturation curves of the free air ionization chamber in the (a) collapsed position; (b) expanded position.

3.4. Ion collection efficiency

The ion collection efficiency was determined through data obtained from the saturation curve in the extended position for the voltages of 400V, 600V, 800V and 1000V, using the equation (1) [9].

$$k_s = \frac{\left(\frac{V_1}{V_2}\right)^2 - 1}{\left(\frac{V_1}{V_2}\right)^2 - \left(\frac{M_1}{M_2}\right)} \quad (1)$$

where k_s is the correction factor for ion recombination, V_1 is the voltage used in measurement M_1 and $V_2 = V_1/2$ with its measurement M_2. Table 1 shows the values obtained.

The IEC 61674 [8] states that the ion collection efficiency should be greater than or equal to 95% and the IEC 60731 standard [10] recommends a minimum of 99.0%. Observing the table 1, it can be concluded that the results obtained are in accordance with both standards.

Table 1. Ion collection efficiency and the correction factors for ion recombination at different voltages.

Voltage (V)	Efficiency (%)	ks
400	99.67	1.0033
600	99.95	1.0005
800	99.98	1.0002
1000	99.95	1.0005

3.5. *Polarity effect*

For the polarity effect, measurements for the saturation curve were taken by varying the voltage from + 1000V to -1000V, with intervals of 100V. The curve obtained is presented in Figure 2.

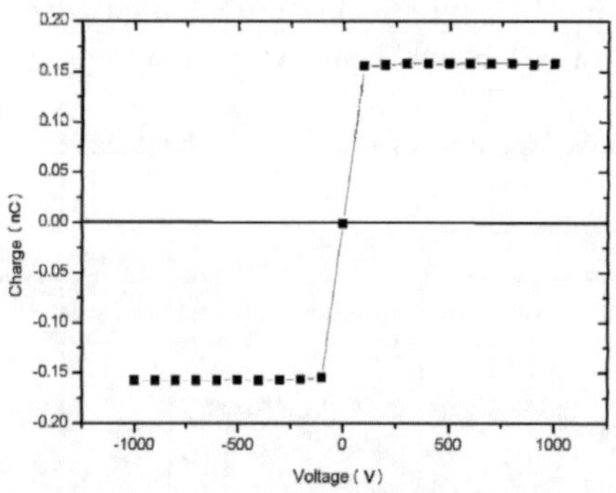

Figure 2. Saturation curve obtained of the free air ionization chamber in the collapsed position for the voltage in both polarities. The maximum uncertainty of the measurements was 0.5%, not visible in the graph.

According to IEC 60731 [10], the variation of the responses obtained by the ratio between the charge collected by the positive polarity and the charge collected by the negative polarity should not be higher than 1.0%. All values were obtained according to this recommendation: the maximum variation was 0.74%.

3.6. *Linearity of response with the current variation in the tube*

In this test the current in the X-ray tube was varied between 5 mA and 35 mA, keeping constant the voltage and the additional filtration of the RQR-2M quality. In figure 3 the resulting curve can be observed. The measured ionization charge was normalized for the result obtained at 5 mA of tube current.

The linear correlation coefficient (R^2) was obtained equal to 0.9999, showing the linearity of the response with the X ray tube current.

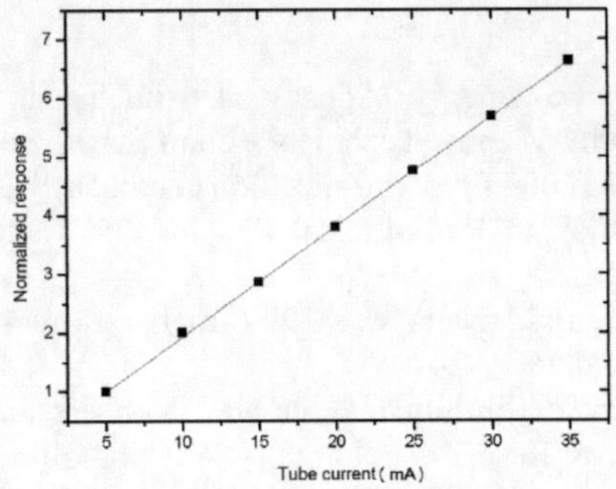

Figure 3. Linearity of response with the X ray tube current variation. The maximum uncertainty of the measurements was 0.4%, not visible in the graph.

3.7. Linearity of response with the volumetric variation
This test was performed by increasing the length of the sensitive volume of 1 cm each time, until 6 cm. This test allows the verification of another volumetric variation to be used for determining the air kerma, since this quantity shall be estimated by means of two measurements from different volumes. Figure 4 shows the results.

The linear correlation coefficient (R^2) was obtained equal to 0.9999, showing the linearity of the response to the variation in the sensitive volume length.

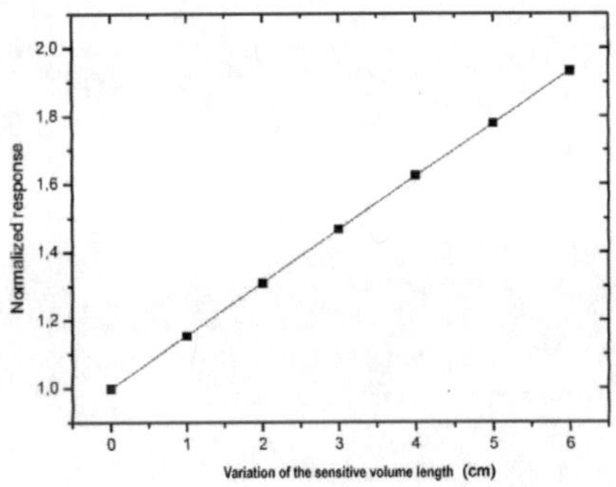

Figure 4. Linearity of response to the variation of the sensitive volume length of the ionization chamber.

4. Conclusions
In this work the characterization test results of a free air ionization chamber are presented using standard mammography X-ray beams of low energy. These results correspond to the expected ones, showing a proper functioning of the ionization chamber. Therefore, the modification of the micrometer set and the new alignment protocol have corrected the problem obtained in the response linearity test with the volumetric variation.

From the results of the ion collection efficiency test, it was also possible to determine the correction factors for ion recombination at voltages of 400V, 600V, 800V and 1000V.

5. Acknowledgments
The authors acknowledge the financial support from the agencies: CNEN, CAPES, CNPq and MCT (Project INCT- Radiation Metrology in Medicine).

References

[1] IAEA, International Atomic Energy Agency, Dosimetry in Diagnostic Radiology: An International Code of Practice, 2007 Vienna (Technical Report Series No.457)

[2] Burns D T and Büermann L 2009 Free-air ionization chambers *Metrologia* **46** 9-23

[3] Attix F H 2004 Introduction to Radiological Physics and Radiation DosimetryWilley-VCH Verlag Wisconsin

[4] Chen W L, Su S H, Su L L and Hwang W S 1999 Improved free-air ionization chamber for the measurement of X-rays *Metrologia* **36** 19-24

[5] Attix F H 1961 Eletronic Equilibrium in Free-air Chambers and a proposed New Chamber Design NRL Report 5646 US Naval Research Laboratory Washington

[6] Lima M H 2014 Characterization of a Free-air Ionization Chamber in Direct X-ray Beams as used in Mammography, MSc. Dissertation Instituto de Pesquisas Energéticas e Nucleares, University of São Paulo (In Portuguese)

[7] Silva N F, Xavier M, Vivolo V and Caldas L V E 2015 Stability results of a free air ionization chamber in standard mammography beams, International Nuclear Atlantic Conference São Paulo

[8] IEC, International Commission on Radiation Units and Measurements 1997 Medical Electrical Equipment – Dosimeters with Ionization Chamber and/or Semi-Conductor Detectors as used in X-ray Diagnosis Imaging, Geneva (IEC 61674)

[9] IAEA, International Atomic Energy Agency, Calibration of Radiation Protection Monitoring Instruments, 2000 Vienna IAEA Safety Report Series No.16

[10] IEC, International Commission on Radiation Units and Measurements 2011 Medical Electrical Equipment – Dosimeters with Ionization Chamber as used in Radiotherapy, Geneva (IEC 60731)

Performance of three pencil-type ionization chambers (10 cm) in computed tomography standard beams

Maysa C de Castro, Marcos Xavier, Linda V E Caldas.

Instituto de Pesquisas Energéticas e Nucleares (IPEN / CNEN - SP). Av. Professor Lineu Prestes, 2242. 05508-000 São Paulo, SP.

E-mail: mcastro@ipen.br, mxavier@ipen.br, lcaldas@ipen.br

Abstract. The use of computed tomography (CT) has increased over the years, thus generating a concern about the doses received by patients undergoing this procedure. Therefore, it is necessary to perform routinely beam dosimetry with the use of a pencil-type ionization chamber. This detector is the most utilized in the procedures of quality control tests on this kind of equipment. The objective of this work was to perform some characterization tests in standard CT beams, as the saturation curve, polarity effect, ion collection efficiency and linearity of response, using three ionization chambers, one commercial and two developed at the IPEN.

1. Introduction

The use of diagnostic imaging making use of ionizing radiation has been shown an increase over the years, especially computed tomography (CT). It allows the obtention of images with better resolution than by other techniques, consequently it is responsible for increasing the radiation dose during the procedure. Therefore, there is an increased concern regarding the dose received by the patients undergoing this kind of imaging procedure.

For the dosimetry of CT beams, the radiation detector has a special geometry, because the tube rotates around the patient table, and it presents a uniform response to the incident radiation beam from all angles [2]. This detector receives the name of pencil-type ionization chamber. These commercial ionization chambers present a length of the sensitive volume of 10 cm, and they are used for quality control testing of the equipment.

The Calibration Laboratory of Instruments (LCI) of the IPEN has developed different kinds of ionization chambers, including a pencil-type [3,4]. However, this ionization chamber presents some differences in its construction in relation to the commercial models. The main differences were in relation to the material used in manufacturing the chamber body and the position of the BNC connector. This homemade configuration provides a low cost in its construction, and the response to the test met the internationally recommended limits, as noted in previous studies [3,4].

This work aims to show results on the characterization tests as saturation curve, polarity effect, ion collection efficiency and response linearity, for each of the pencil ionization chambers in standard CT X radiation beams at the LCI.

2. Materials and Methods

Three pencil-type ionization chambers were tested in this work: one of them was a commercial chamber (Victoreen 660-6) called C1 (sensitive volume of 3.2 cm^3), and the other two chambers were prototypes developed at the LCI called C2 and C3; all of the ionization chambers present a sensitive volume length of 10 cm.

Table 1 shows the specifications of the ionization chambers developed at LCI, and the difference between them are based on the collector electrode material and on the dimensions of their components.

Table 1: Technical specifications of the pencil ionization chambers developed at the LCI.

Characteristics	**Specifications**	
	C2	**C3**
Collecting electrode material	Aluminum	Aluminum with graphited layer
Wall material	PVC with graphited layer	PMMA with graphited layer
Collecting electrode diameter	1.22 mm	3.20 mm
Ionization chamber diameter	6.72 mm	7.40 mm
Wall thickness	0.26 mm	0.26 mm
Sensitive volume length	10 cm	10 cm
Sensitive volume	3.40 cm^3	3.50 cm^3

The tests were performed using the reference standard CT X radiation quality at the LCI: RQT 9. This radiation quality presents the following characteristics: tube voltage of 120 kV, tube current of 10 mA, half value layer of 8.4 mmAl, additional filtration of 3.5 mmAl + 0.35 mmCu, and air kerma rate of 34.0 mGy/min.

The polarity effect for each ionization chamber should be within the internationally recommended limit of 1% [5]. The ion collection efficiency was obtained using the voltages V1= +200 V and V2= +100 V for all three ionization chambers. The results should be within the recommended limit of 95 % [5].

The X-ray system Pantak/Seifert (ISOVOLT model 160HS) of LCI/IPEN was utilized; it operates up to 160 kV.

3. Results

Initially are presented the results obtained for the saturation curves, polarity effects and ion collection efficiencies for the ionization chambers. Figure 1 shows the saturation curves.

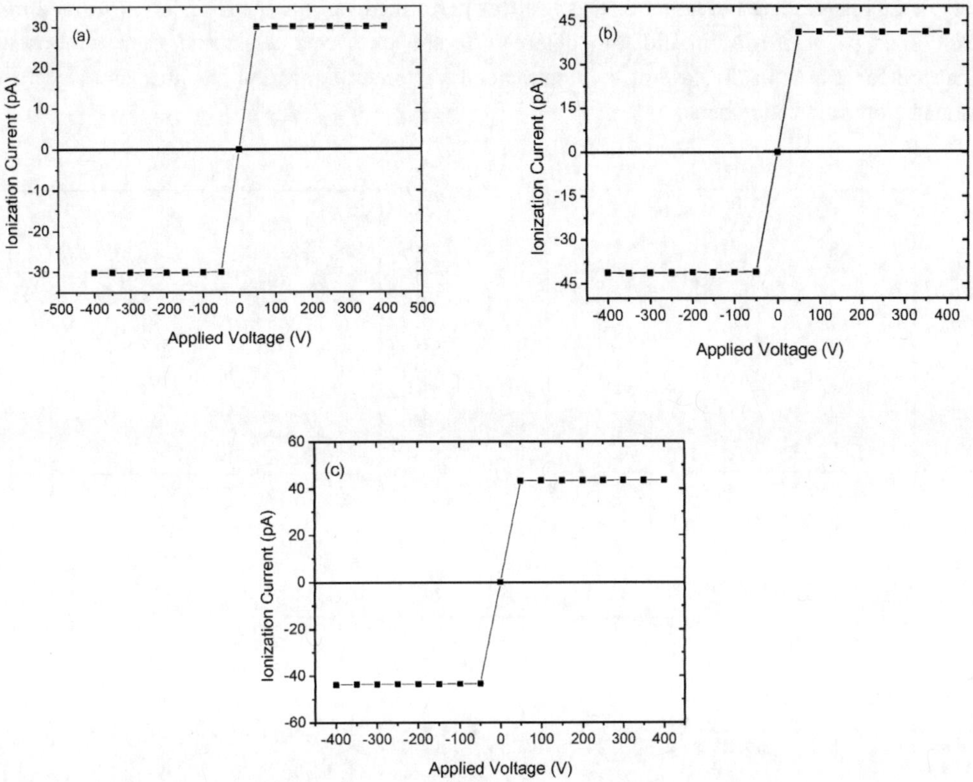

Figure 1: Saturation curves of the ionization chambers for the radiation quality RQT 9: (a) C1; (b) C2; (c) C3. The maximum measurement uncertainty was 0.59%.

It is possible to observe that all three pencil-type ionization chambers present good results for the saturation curves, but they reach saturation at different values, because of their slightly different volumes and constituent materials.

The results for the polarity effects and ion collection efficiencies were obtained through the saturation curves, and they can be seen in table 2.

Table 2: Polarity effects and ion collection efficiencies for the ionization chambers.

Ionization chamber	Polarity effect (%)	Ion collection efficiency (%)
C1	0.32	99.97
C2	0.57	99.99
C3	0.29	99.98

As can be observed, the three ionization chambers with a sensitive volume length of 10 cm presented results within the internationally acceptable limits. The operating voltage used for all characterization tests for all three ionization chambers was chosen as +100 V.

The response linearity obtained for each ionization chamber in the standard CT reference radiation quality of LCI is presented in figure 2. This test was performed varying the tube current from 2 mA to 18 mA in steps of 2 mA, in order to obtain different air kerma rates, and each response was normalized for 2 mA, in the case of the commercial Victoreen ionization chamber and also for the homemade ionization chambers.

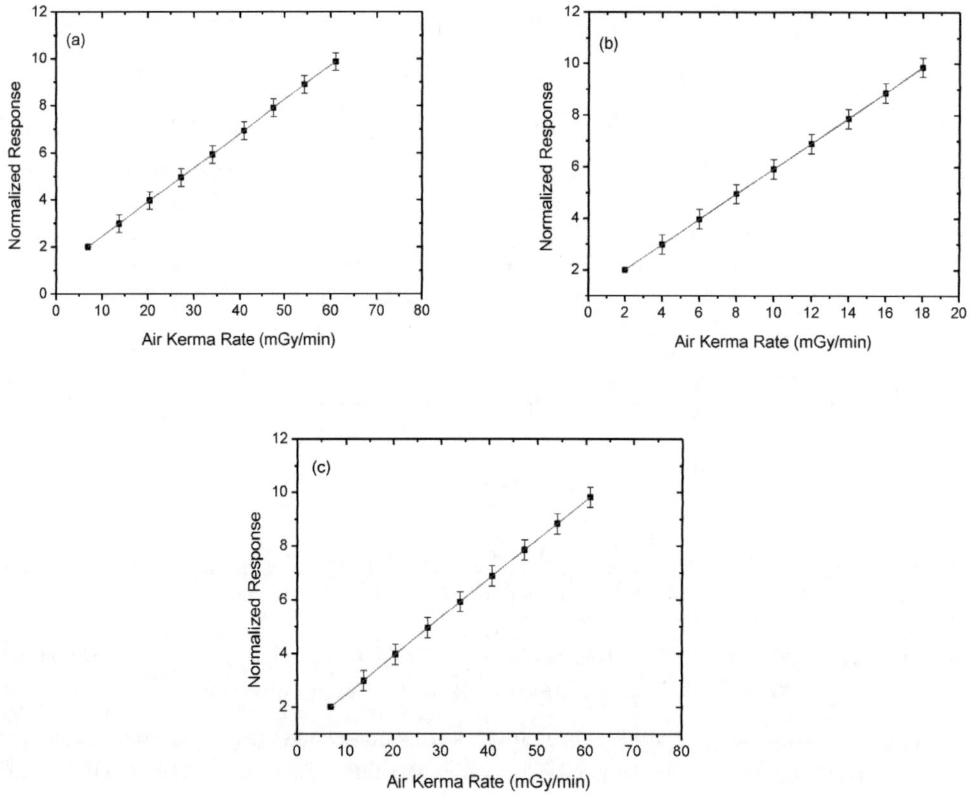

Figure 2: Linearity of response to quality RQT 9 of the ionization chambers: (a) C1; (b) C2; (c) C3. The linear correlation coefficient in all cases is R^2=1.000.

As can be observed, all three ionization chambers with a sensitive volume length of 10 cm presented the linear correlation coefficients of 1.000, showing that the ionization chambers have a linear response, as required by the quality control program.

4. Conclusions

The test results of the commercial ionization chamber (10 cm) were all within the internationally recommended limits. The homemade ionization chambers (10 cm) presented also results within the limits, although of their differences in size and construction material.

References

[1] Boone J M The trouble with CTDI100, 2007 Med. Phys. 34(4) 1364-1371.

[2] Suzuki A and Suzuki M N Use of a pencil-shaped ionization chamber for measurement of exposure resulting from a computed tomography scan, 1978 Med. Phys. 5(6) 536-9.

[3] Perini A P, Neves LP, Fernández-Varea J M, Büermann L and Caldas L V E Evaluation and simulation of a new ionization chamber design for use in computed tomography beams, 2013 IEEE Transactions on Nuclear Science 60(2) 768-773.

[4] Neves L P, Perini A P, Fernández-Varea J M and Caldas L V E Application of a pencil ionization chamber (0.34 cm^3 volume) for 60Co beams: Experimental and Monte Carlo results, 2013 IEEE Transactions on Nuclear Science 60(2) 746-750.

[5] IEC International Electrotechnical Commission, Medical diagnostic X-ray equipment. Radiation conditions for use in the determination of characteristics, 2005 2nd ed. IEC Genève.

The influence of different ^{192}Ir sources geometries to the energy deposition

W S Santos[1], P E Gonalves[2] W Belinato[3], L V E Caldas[1], A P Perini[2], L P Neves[1,2]

[1]Instituto de Pesquisas Energéticas e Nucleares, Comissão Nacional de Energia Nuclear (IPEN-CNEN/SP), São Paulo, SP, Brasil.

[2]Instituto de Física, Universidade Federal de Uberlândia (INFIS/UFU), Caixa Postal 593, 38400-902 Uberlândia, MG, Brasil.

[3]Departamento de Física, Universidade Federal de Sergipe (UFS), São Cristovão, SE, Brasil.

E-mail: lucio.neves@ufu.br

Abstract. In this paper, various simplifications of the HDR source Varian VariSource Classic model, in which ^{192}Ir as a radionuclide is used, were compared. These simplifications were carried out by Monte Carlo simulations, using the MCNPX 2.7.0 code. The different sources were compared through a distribution of energy deposition in a water phantom. Our results indicated that small simplifications will present no influence on the source response, and the removal of the entire capsule surrounding the radionuclide will present a difference of just 0.53% in the final response.

1. Introduction

In 1869, radioactivity was discovered by Henry Becquerel and in 1898 the Curies discovered radium. With these new horizons of science, in 1901 brachytherapy was first applied by a doctor, using a small amount of radium for the treatment of epithelial lesions. At this time, the Curie Institute in Paris and the St Luke's Hospital and Memorial Hospital in New York pioneered the application of techniques in radiotherapy.

Unlike External Beam Radiotherapy (EBRT), where the tumor receives high dose x-rays from an external place, brachytherapy consists in putting a radioactive source in the tumor, or near it, to kill the cancer cells. The source can be in the format of wires or seeds, containing sealed radioactive material. With it, we can reduce the exposure of healthy tissue to high doses of radiation [1]. This study will use sources of ^{192}Ir, considered a HDR (high-dose rate) source. The source to be studied is a Varian VariSource Classic, using the N-particle transport MCNPX 2.7.0 Monte Code [2].

The purpose of the paper is to evaluate the quantity of radiation that is released if we simplify the source's shield in a Monte Carlo simulation. With this, we may determine whether the geometry can be simplified without any changes in the energy deposited to its surroundings.

2. Materials and Methods

In this paper, various simplifications of the HDR source Varian VariSource Classic model were compared. These simplifications were carried out using the the Monte Carlo MCNPX 2.7.0 code [2]. The different sources were compared through the energy deposition in a water phantom.

The evaluated source was the Classic model of HDR Varian VariSource, which uses ^{192}Ir as a radionuclide. Produced and distributed by Varian Oncology Systems, the active part presents a density of 22.7 g/cm^3, 10.0 mm in length and a diameter of 0.34 mm. The source's activity is considered evenly distributed. The active core is wrapped with nitinol with a density of 6.45 g/cm^3, made of titanium and nickel, with a length of 15 mm and a total diameter of 0.59 mm.

The distal tip of the capsule has rounded edges with a radius of curvature of 0.295 mm. The source is fixed in a stainless steel cable with a diameter of 0.59 mm. As the part of the cable near the source is straight, 2 mm of the cable was simulated. The complete geometry of this source is shown in Figure 1.

Figure 1. Design of the source HDR Varian VariSource Classic.

The evaluation consisted in 6 simplifications of this source: (1) replacing the rounded edges for rectangular edges; (2) replacing the superior edge for a rectangular edge; (3) removal of the stainless steel cable; (4) the inferior part of the source's shield was taken; (5) the components of nitinol were taken (6) the source's shield was taken completely. The description of each one of the simplifications is described next.

2.1. Configurations of the source of ^{192}Ir evaluated in this paper

The first simplification consisted in removing the rounded edges of the distal tip of the source and changing it to a rectangular form without any changes in the length and diameter. This configuration is shown in figure 2.

Figure 2. Design of the first simplification of the HDR source Varian VariSource Classic: removal of the rounded edges.

In the second simplification, the distal tip of the original source, the rounded form was replaced with a triangular form, with an opening angle of 40° and no changes in the total length of the source. This configuration is shown in figure 3.

Figure 3. Design of the second simplification of the HDR source Varian VariSource Classic: replacement of the rounded tip for a triangular one.

The third simplification consisted in the removal of the source's cable, but keeping the original diameter and length. This configuration is shown in figure 4.

Figure 4. Design of the third simplification of the HDR source Varian VariSource Classic: removal of the stainless steel cable.

In the fourth simplification, the portion of the capsule that stood between the active part of the source and the portion simulated of the cable in the original source was removed. This thereby reduced the length of the source in relation to the original one by 4 mm, and kept the same diameter of the original source. This configuration is shown in figure 5.

Figure 5. Design of the fourth simplification of the HDR source Varian VariSource Classic: removal of the portion of the capsule that stood between the active part of the source and the cable.

The fifth simplification consisted in the removal of the portion of nitinol between the rounded edges and the active core in the proximal tip and the removal of the portion of nitinol between the active core and the cable in the distal, removing the portion of the cable simulated in the original source. Therefore, the length of the source was reduced by 10.295 mm, but the diameter was kept the same. This configuration is shown in figure 6.

Figure 6. Design of the fifth simplification of the HDR source Varian VariSource Classic: removal the portion of nitinol.

In the sixth simplification, the entire capsule surrounding the radionuclide was removed, and, as a consequence, the source was just the radioactive material. It had 10 mm of length and 0.34 mm of diameter. This configuration is shown in figure 7.

Figure 7. Design of the sixth simplification of the HDR source Varian VariSource Classic: removal of the entire capsule surrounding the radionuclide.

To proceed with the analyses, each seed was placed inside a spherical phantom, composed of water. This geometrical arrangement allowed the evaluation of the angular anisotropy of the irradiation field and the maximum energy deposition.

The anisotropic distribution is in 10° increments, from 0° to 180°. The center of the sphere passes through the center of the ^{192}Ir source. The energy deposition was computed in each of the segments, enclosed by the 10° segments.

With this study, we may determine the influence of each geometrical simplification on the simulated phantom.

3. Results and Discussion

The energy deposition, for each geometrical simplification and angle, are listed in table 1.

As seen from table 1, there were not any differences from the original source (source with no simplifications) for the smaller simplifications, and 0.53% for the source with no capsule surrounding it. This variation may be considered small, in relation to other sources of errors, that may occur during brachytherapy treatments [3].

This shows that making small simplifications that could be time consuming, such as choosing the type of source, may be simplified, with no differences on the final responses.

Table 1. Energy deposition for all the geometrical differences evaluated in this work, for all the wedges. S1 means source 1 (the original source) and S2 to S6 simplifications 2 to 6.

Wedge	Energy deposition ($\times 10^{-3}$MeV)					
	S1	S2	S3	S4	S5	S6
1	1.283	1.283	1.283	1.284	1.284	1.290
2	1.306	1.306	1.306	1.306	1.306	1.306
3	1.326	1.326	1.326	1.326	1.326	1.327
4	1.332	1.332	1.332	1.332	1.332	1.340
5	1.340	1.340	1.339	1.338	1.339	1.347
6	1.340	1.340	1.340	1.339	1.339	1.344
7	1.350	1.350	1.350	1.351	1.350	1.353
8	1.348	1.348	1.347	1.348	1.348	1.358
9	1.345	1.344	1.345	1.344	1.344	1.355
10	1.342	1.342	1.342	1.342	1.342	1.349
11	1.351	1.352	1.352	1.351	1.351	1.353
12	1.344	1.345	1.344	1.344	1.345	1.355
13	1.338	1.338	1.338	1.338	1.338	1.348
14	1.333	1.333	1.333	1.333	1.334	1.347
15	1.322	1.322	1.322	1.322	1.323	1.333
16	1.313	1.313	1.313	1.313	1.314	1.323
17	1.299	1.299	1.299	1.300	1.300	1.309
18	1.286	1.286	1.286	1.287	1.287	1.290
Mean Value	1.328	1.328	1.328	1.328	1.328	1.335
Difference from the original source	-	0%	0%	0.01%	0.01%	0.53%

4. Conclusion

In this study, we evaluated the effect of geometrical simplifications on the simulation of a ^{192}Ir source. Our results indicated that small simplifications will present no influence on the source response, and the removal of the entire capsule surrounding the radionuclide will present a difference of just 0.53% on the final response. This variation may be considered small, in relation to other sources of errors, that may occur during brachytherapy treatments. Therefore, this indicates that one may simplify some parts of the source geometry, with no influence on the final response.

Acknowledgments

The authors received support from the Brazilian agencies: CAPES (Grant Pro-Estratégia no. 1999/2012), CNPq (Grants no. 304789/2011-9 and 157593/2015-0) and Project INCT for Radiation Metrology in Medicine. The authors would like to thank Dr. Å. Carlson Tedgren (Linköping University, Sweden) for providing the energy spectra of the ^{60}Co beam.

References

[1] Bernier J, Hall E J and Giaccia A 2004 Radiation oncology: a century of achievements *Nat. Rev.* **4** 2004 737–47

[2] Pelowitz D B 2011 *MCNPX user's manual, version 2.7.0. Report LA-CP-11e00438* Los Alamos National Laboratory
[3] Elhanafy O A, Das R K, Paliwal B R, Migahed M D, Sakr H A and Elleithy M 2002 *Appl. Clin. Med. Phys.* **3** 1

Determination of true null electrode spacing of an extrapolation chamber for x-ray dosimetry

M T T Figueiredo, F M Bastos, T A Da Silva
Centro de Desenvolvimento da Tecnologia Nuclear, CDTN, Belo Horizonte, Brazil
E-mail: mttf@cdtn.br

Abstract. An accurate determination of the true null distance between electrodes, that defines the sensitive volume of an extrapolation chamber, is critical for the establishment of a primary measurement method for absorbed dose in tissue. In this work, a critical analysis of two methodologies for determining the true null electrode spacing of a 23392 Böhm model PTW extrapolation chamber was done.

1. Introduction

An extrapolation chamber to be used as a primary standard dosimeter for measuring the absorbed dose to tissue requires to be characterized by determining the true null spacing between its electrodes and effective volume. Besides, additional parameters should be determined like: the saturation current, the efficiency of ion collection, the polarity effect, repeatability and reproducibility of the chamber response, etc. A prior knowledge of the absolute distance between the chamber electrodes is required for majority of tests [1].

Electrode plate spacing exists because manufacturers keep a safe distance between them considering that it cannot be too short that the electric field would exceed the dielectric rigidity of the medium; besides, there are doubts regarding the parallelism between the plates and the material roughness. The nominal distance between the electrodes is showed with high resolution and low uncertainty in a micrometer cursor of the chamber; however, the absolute distance between them depends on the knowledge of the true null spacing value, which is not provided by the manufacturer [2].

The true electrode spacing of an extrapolation chamber is often determined by analyzing the experimental relationship between the ionization current and the electrode relative distances. The curve that defines this relationship is a line that intersects the horizontal axis at the correspondent true null spacing. It is quite reasonable to assume that, for constant intensity electric field and homogeneous radiation beam, the ionization current measured by the extrapolation chamber will be proportional to the sensitive volume and, therefore, for the absolute distance between the electrodes, since the overlapping area of the electrodes is constant.

The adjustment of the electric field intensity value, which is a proportionality condition, depends on the prior knowledge of the absolute distance between the electrodes. The relative distance given by the micrometer cursor has been used as a reference for adjusting the intensity of the electric field, which may introduce significant errors for smaller electrode spacing values. Assuming the constancy of electric field strength and uniformity of the radiation beam are necessary and sufficient conditions of proportionality, the use of relative distance, instead of the absolute distance for adjusting the electric field intensity, will produce inevitable loss of proportionality.

Considering that an inherent capacitance is associated with any parallel-plate ionization chamber, there should be a well-defined relationship among the capacitance, the effective colleting area and the electrode spacing of an extrapolation chamber. The knowledge of the capacitance allows the effective electrode spacing of the extrapolation chamber, to be calculated within reasonable uncertainties.

The aim of this work was to analyze the results of measurements of the true null electrode spacing in a 23392 Böhm type PTW extrapolation chamber by two different procedures: one based on ionization current and the other on the capacitance measurement.

2. Materials and Methodology

In order to promote international standardization and metrological coherency, the International Organization for Standardization, ISO, established four series of x- reference radiations for calibrating and typing-test of dosimeters [3]. ISO narrow spectrum series (N) with maximum energy from 10 up to 30 keV were used in this work. Table 1 shows the x-ray beam parameters that define the reference radiations that were reproduced in a Seifert-Pantak x-ray machine. [4]The procedure for determining the true null spacing of the 23292 Böhm type PTW extrapolation chamber based on ionization measurements was done with the extrapolation chamber positioned in the x-ray beams (Figure 1). The ionization current was measured during 60 s accumulated electrical charge for the electrode distance from 1.00 up to 5.00 mm. Each measured ionization current was corrected to the ambient reference conditions of 20°C and 101.3 kPa; to correct the polarity effect, measurements were done for both voltage polarities applied to the chamber and the mean value was adopted; leakage current was also measured in order to assure that it was negligible in comparison to the measured ionization current. .

Table 1. Characterization parameters of low energy ISO narrow spectrum x-ray beams used in this work.

Reference radiation code	Spectral resolution (%)	Mean energy (keV)	Tube voltage (kV)	Additional filtration (mmAl)	1st half-value layer (mmAl)	2st half-value layer (mmAl)
N 10	28	8	10	0.1	0.047	0.052
N 15	33	12	15	0.5	0.14	0.16
N 20	34	16	20	1.0	0.32	0.37
N 25	33	20	25	2.0	0.66	0.73
N 30	32	24	30	4.0	1.15	1.30

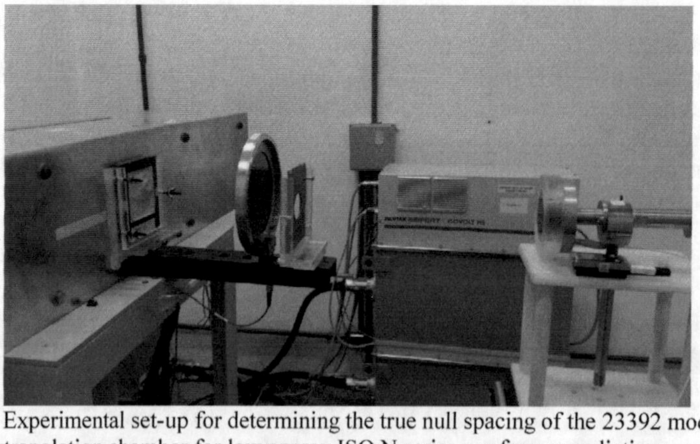

Figure 1. Experimental set-up for determining the true null spacing of the 23392 model PTW extrapolation chamber for low energy ISO N series x-reference radiations.

Capacitance is the ability of a body to store electrical charge; the parallel-plate capacitor is the most common device of energy storage. The capacitance of a parallel plate capacitor, which is defined as the ratio of the charge stored and the voltage applied to the plates, is directly proportional to the area of the conductive plates and inversely to the distance between them, as given by Equation 1.

$$C = \varepsilon_0 \varepsilon_r \frac{A}{d} \tag{1}$$

Where, C is the capacitance, in Farads; A is the area of overlap of the two plates, in square meters; ε_r is the relative static permittivity or dielectric constant of the material between the plate; ε_0 is the vacuum permittivity ($\varepsilon_0 \approx 8.854\times10^{-12}$ F m^{-1}) and d is the separation between the plates, in meters.

An extrapolation chamber can be seen as a parallel plate capacitor, which the distance between its plates can be continuously and accurately changed. The capacitance measurement of the 23392 PTW extrapolation chamber was done in absence of the radiation field using an L-C Bridge that operates by measuring the change in frequency when an unknown capacitance is added to an oscillator tank circuit; they were obtained by keeping the relative distance cursor set to zero.

3. Results and discussions

The results of the variation of the ionization current for nominal electrode distances from 1.0 mm up to 5.0 mm for N10, N15, N20, N25 and N30 ISO series are shown in Figure 2, 3 and 4. The linear fitting was done to obtain the equation and determine the true null spacing for each x-reference radiation [5]. The expanded uncertainties were assessed for the coverage factor equal to 2, according to international standards; the main sources of uncertainties were the repeatability of the measurements, the resolution and calibration of auxiliary equipments (thermometer and barometer) and the extrapolation chamber positioning [6].

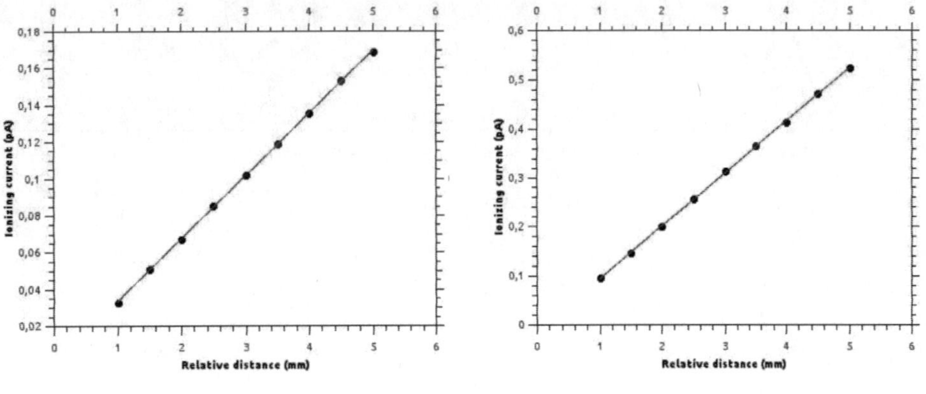

Figure 2. Extrapolation line to determine the true null spacing of the 23392 PTW extrapolation chamber for N 10 (*left*) and N15 (*right*) ISO x-reference radiation

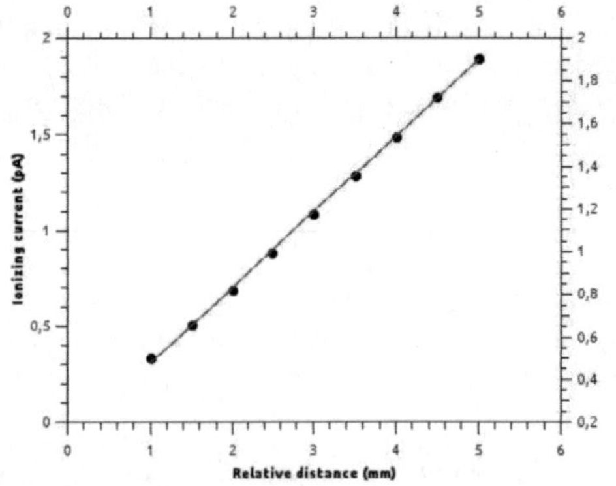

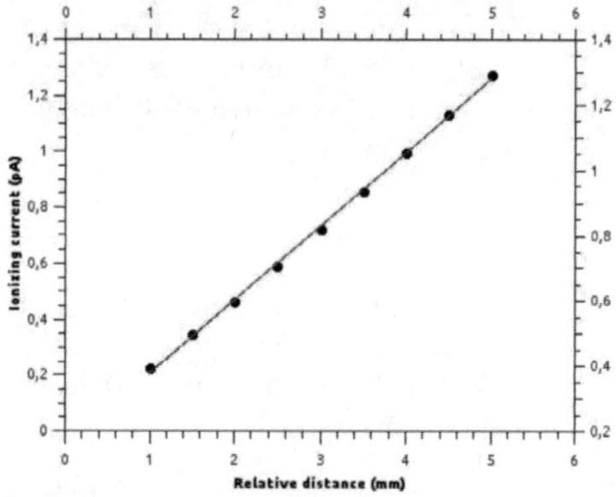

Figure 3. Extrapolation line to determine the true null spacing of the 23392 PTW extrapolation chamber for N 20 (*left*) and N25 (*right*) ISO x-reference radiation.

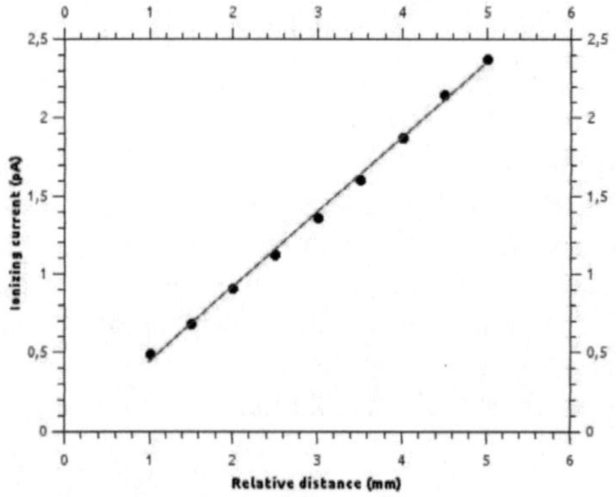

Figure 4. Extrapolation line to determine the true null spacing of the 23392 PTW extrapolation chamber for N 30 ISO x-reference

Table 2. True null spacing of a 23392 PTW extrapolation chamber for ISO x-reference radiations.

Reference radiation code	Linear fit equation	True null spacing (mm)	Expanded uncertainty* (%)
N 10	I = 0.034d - 0.0005	0.015	1.6
N 15	I = 0.108d - 0.013	0.121	1.6
N 20	I = 0.394d - 0.086	0.218	1.6
N 25	I = 0.263d - 0.056	0.213	1.6
N 30	I = 0.394d - 0.086	0.065	1,6

* Coverage factor = 2.

The results of true null spacing of the 23392 PTW extrapolation chamber (Table 1) showed a high sensitivity to the radiation quality; this could be partly explained by the loss of proportionality caused by the nonlinear electric field intensity adjustment.

The result of the direct measurement by means of a LC bridge of the capacitance between the electrodes of the 23392 PTW extrapolation chamber with the micrometer set at zero position is shown in Table 3. The mean value of the capacitance and the standard deviation were obtained from a set of 10 measurements.

Table 3. Capacitance of a 23392 PTW extrapolation chamber measured by an LC bridge with the micrometer cursor set to zero.

Mean capacitance (pF)	Standard deviation (pF)
16.5	0.311

In the absence of the radiation field it is assumed that the effective area of the electrode plate is defined by its diameter of 3.00 mm provided by the manufacturer [2] and it is equal to 7.068 cm^2.

Based on Equation 1, the d value that represents the true null spacing was calculated as 0.380±0.021 mm. Again, uncertainty was assessed based on international standard [5] for the coverage factor equal to 2 and the main sources of uncertainties were the capacimeter resolution and calibration, the repeatability of the measurements and the micrometer resolution.

4. Conclusions

For primary standardization purpose, the accurate determination of true null spacing of an extrapolation chamber is very important because it directly impacts on the calculation of its sensitive volume. The radiation quality dependence of the results obtained by the traditional method can be considered a weakness since, by definition, the true null spacing is just the distance between the electrodes necessary to prevent damage to the measurement procedure and therefore it should not be dependent on the quality of ionizing radiation.

Although the use of an LC bridge for capacitance measurements is a simple and direct method, it is highly dependent on the accuracy of capacimeter that can contribute to high uncertainty.

5. Acknowledgments

Marcus Tadeu Tanuri de Figueiredo is thankful to CNEN for his doctorate fellowship. This work was supported by FAPEMIG (PPM) and CNPQ (PQ); it is part of the project INCT Radiation Metrology in Medicine.

References

[1] Reynaldo S R, Benavente J A and Da Silva T A 2013 Characterization of an extrapolation chamber as a primary standard dosimeter for beta dosimetry *Scientia Plena,* 9 8

[2] PTW-Freiburg, 2002. *Instruction Manual Extrapolation Chamber according to Böhm Type 23392 Freiburg Germany: s.n.*

[3] *ISO International Organization for Standardization* 1996 *ISO 4037. X and gamma reference radiation for calibrating dosemeters and dose rate meters and for determining their response as a function of photon energy* Geneva

[4] Figueiredo M T T 2012 Análise dos procedimentos e critérios de implantação dos feixes de raios X de referência ISO 4037 em baixas energias *Dissertação de mestrado* Belo Horizonte: Centro de Desenvolvimento da Tecnologia Nuclear CNEN

[5] Bastos F M 2015 Caracterização de feixes de raios X de baixa energia com uma câmara de extrapolação *Dissertação de Mestrado* Belo Horizonte Centro de Desenvolvimento da Tecnologia Nuclear CNEN

[6] *Guide to the Expression of Uncertainty in Measurement* 1993 First Edition English (corrected and reprinted, 1995) ISBN 92-67-10188-9

Journal of Physics: Conference Series **733** (2016) 012094 doi:10.1088/1742-6596/733/1/012094

On the response of electronic personal dosimeters in constant potential and pulsed x- ray beams

M.C. Guimarães, C.R.E. Silva, P.M.C. Oliveira[1], T.A. da Silva

Centro de Desenvolvimento da Tecnologia Nuclear, CDTN
Av. Presidente Antônio Carlos 6627, Pampulha
31270-901 Belo Horizonte, MG, Brazil

[1]Faculdade de Medicina – Universidade Federal de Minas Gerais, UFMG
Av. Prof. Alfredo Balena, 190, Funcionários
30130-100 Belo Horizonte, MG, Brazil

E-mail: margaretecristinag@gmail.com, silvata@cdtn.br

Abstract. Electronic personal dosimeters (EPDs) based on solid state detectors have widely been used but some deficiencies in their response in pulsed radiation beams have been reported. Nowadays, there is not an international standard for pulsed x-ray beams for calibration or type testing of dosimeters. Irradiation conditions for testing the response of EPDs in both the constant potential and pulsed x-ray beams were established in CDTN. Three different types of EPDs were tested in different conditions in similar ISO and IEC x-ray qualities. Results stressed the need of performing additional checks before using EPDs in constant potential or pulsed x-rays.

1. Introduction

Radiation protection requires that individual monitoring should be done by reliable dosimeters. The ISO and IEC standards established the performance requirements for the response and calibration of electronic dosimeters in reference constant potential x-ray beams [1]. Electronic personal dosimeters (EPDs) based on solid state detectors have widely been used but some deficiencies in their response in pulsed radiation beams have been reported [2]. Nowadays, there is not an international standard for pulsed x-ray beams for calibration or type testing of dosimeters. In Brazil, reliability of dosimeters is based only in their calibration in a standard ^{137}Cs gamma beam. Performance checks or type testing of radiation protection dosimeters are not legally required but they are informally performed in some Brazilian metrology laboratories for academic studies [3].

The objective of this work was to study the response of some electronic dosimeters in reference constant potential x-ray qualities and in pulsed x-ray beams.

2. Materials and Methodology

The established irradiation conditions for testing dosimeters in both constant potential and pulsed x-ray beams of the Dosimeter Calibration Laboratory, LCD/CDTN, were used. ISO and IEC reference radiations were reproduced in a 320HS constant potential Seifert-Pantak and in an 800 Pulsar Plus VMI medical x-ray machines.

A Xi light UNFORS solid state dosimetric system and a 10X6 RADCAL ionization chamber were used as reference dosimeters to measure the air kerma free in air at the reference calibration point. The performance of both reference dosimeters were already studied in all x-ray beams (Guimarães et al, 2013).

The dosimeters RAD-60 RADOS, PDM-111 ALOKA and EPD-MK2 THERMOELECTRON were calibrated in ^{137}Cs beam and then tested according to their operation ranges in x-rays, in terms of the personal dose equivalent, Hp(10). The true conventional value of Hp(10) was determined based on the air kerma value free in air and conversion coefficients to Hp(10). Each EPD was positioned on the ISO standard slab phantom in the continuous x-ray beams of the Seifert Pantak machine (Figure 1).

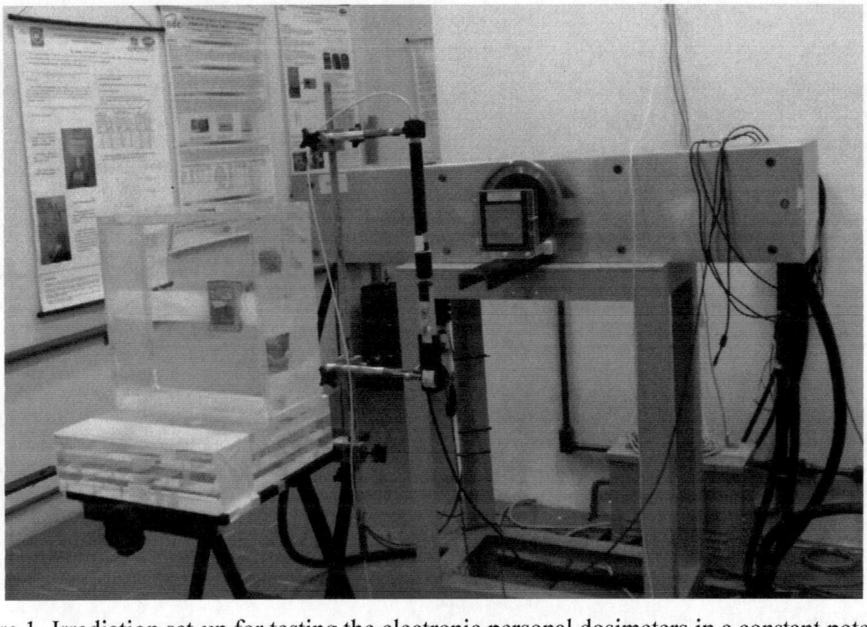

Figure 1- Irradiation set-up for testing the electronic personal dosimeters in a constant potential x-ray machine.

Dosimeters were also tested in pulsed x-ray beams of the VMI medical machine on the ISO slab phantom. The Xi light UNFORS dosimeter was positioned on the surface of the slab phantom almost out of the radiation field to monitor the exposure (Figure 2).

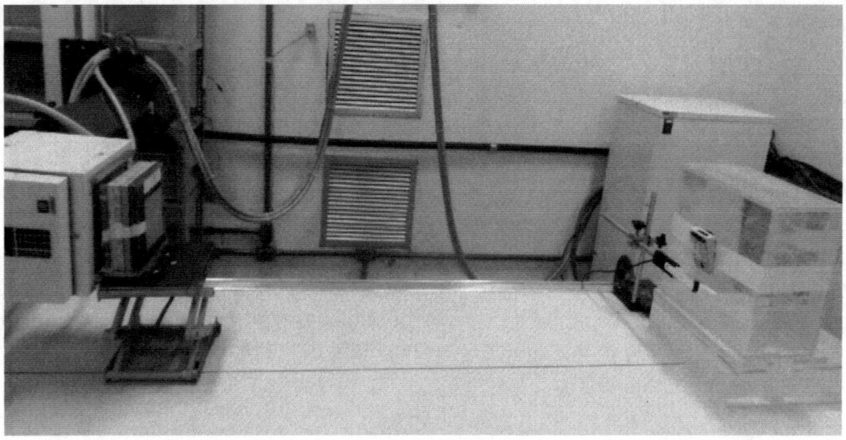

Figure 2 - Irradiation set-up for testing the electronic personal dosimeters in pulsed X-rays of a medical machine.

3. Results and discussions

3.1. Calibration of the EPDs in ^{137}Cs gamma beam

The response of the EPDs was analyzed in terms of relative intrinsic error, i.e., the deviation of the response relatively to the conventional true value of Hp(10) that was determined with the reference dosimeter. The performance of EPDs as far their calibration in ^{137}Cs gamma radiation showed that the relative intrinsic errors were within the acceptable limit of 10% for all dosimeters.

3.2. Performance of EPDs in constant potential X-ray beams

RQA IEC X-ray qualities and N-series ISO reference radiations that were implemented in LCD/CDTN were chosen for testing the EPDs (Table 1).

The response of the EPDs in terms of their intrinsic error relative to Hp(10) showed that even within their range of operation some EPDs did not comply with acceptable limit of ±30% for energy dependence (Figure 3).

Table 1 - IEC and ISO X-ray qualities used for testing the electronic personal dosimeters in constant potential X-ray beams.

Radiation quality code	Tube voltage (kV)	Additional filtration (mm)	Mean energy (keV)
RQA-5	70	21 .0 Al	51.2
RQA-7	90	30.0 Al	62.4
RQA-9	120	40.0 Al	75.8
N-60	60	0.49 Cu+3.18 Al	46.3
N-80	80	1.51 Cu+3.18 Al	62.7
N-100	100	4.00 Cu+3.18 Al	80.6

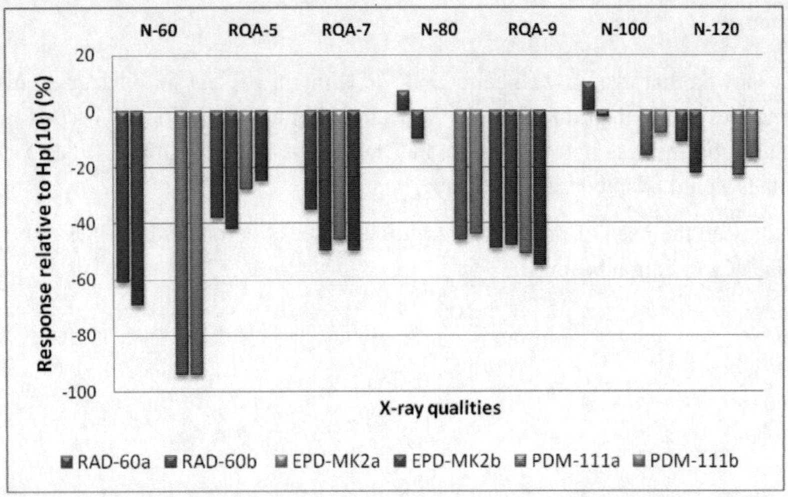

Figure 3 - – Response relative to Hp(10) of EPDs in constant potential X-ray beams.

3.3. Performance of EPDs in pulsed X-ray beams

X-ray beams similar to the RQA-5, 7 and 9 IEC qualities were reproduced in the 800 Pulsar Plus. VMI medical machine. EPDs were exposed to the beam conditions of 150 mA and 50 ms. The response of the EPDs in terms of the intrinsic error relative to Hp(10) is shown in Figure 4.

Results of the EPDs response in those pulsed beams showed that they maintained the trend of underestimation the Hp(10) value. One EPD did not respond at all to radiation and the others showed the same performance as in constant potential beams.

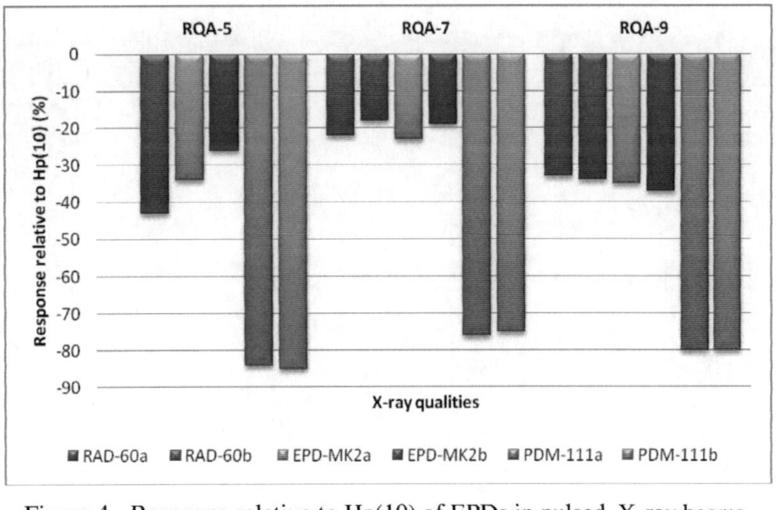

Figure 4 - Response relative to Hp(10) of EPDs in pulsed X-ray beams.

Journal of Physics: Conference Series **733** (2016) 012094 doi:10.1088/1742-6596/733/1/012094

4. Conclusions

The results showed that even a well-calibrated EPDs might present an inadequate response to radiation energies lower than that from ^{137}Cs beam calibration condition. The performance of EPDs in pulsed beams was not conclusive, but the results suggested that studies for different exposure times would be important to be done.

This work stressed the need of performing additional checks before using EPDs that were only calibrated in ^{137}Cs radiation beam.

References

[1] IEC, International Electrotechnical Comission. 2010. IEC 61526. *Radiation protection instrumentation: measurement of personal dose equivalents Hp(10) and Hp(0,07) for X, gamma, neutron and beta radiations - Direct reading personal dose equivalent meters.*

[2] Ankerhold, U.; Hupe, O. Ambrosi, P. *Deficiencies of active electronic radiation protection dosemeters in pulsed fields.* Radiation Protection Dosimetry, vol. 135, No. 3, pp.149–153, 2009

[3] Guimarães M C, Da Silva T A - *Characterization of a medical X-ray machine for testing the response of electronic dosimeters in pulsed radiation fields.* Radiat. Phys. Chem. vol. 104, pp. 321-323, 2014.

Acknowledgments

Margarete C. Guimarães is thankful to CNEN for her fellowship. This work was supported by FINEP/SIBRATEC (METRORADI network), FAPEMIG (PPM) and CNPQ (PQ); it is part of the project INCT Radiation Metrology in Medicine.

Metrological activity determination of ^{133}Ba by sum-peak absolute method

R L da Silva[1], M C M de Almeida [2], J U Delgado[1], R Poledna[1], A Santos[1], E V de Veras[1], J Rangel[1], O L Trindade[1]

[1] Instituto de Radioproteção e Dosimetria (IRD/CNEN), avenida Salvador Allende, s/n, Barra, RJ, Brazil, CEP 22783-127
[2] Comissão Nacional de Energia Nuclear (CNEN), rua Gen Severiano 82, Botafogo, RJ, Brazil, CEP 22290-901

E-mail: marcandida@yahoo.com.br, candida@cnen.gov.br, ronaldo@ird.gov.br

Abstract. The National Laboratory for Metrology of Ionizing Radiation provides gamma sources of radionuclide and standardized in activity with reduced uncertainties. Relative methods require standards to determine the sample activity while the absolute methods, as sum-peak, not. The activity is obtained directly with good accuracy and low uncertainties. ^{133}Ba is used in research laboratories and on calibration of detectors for analysis in different work areas. Classical absolute methods don`t calibrate ^{133}Ba due to its complex decay scheme. The sum-peak method using gamma spectrometry with germanium detector standardizes ^{133}Ba samples. Uncertainties lower than 1% to activity results were obtained.

1. Introduction

The National Laboratory of Metrology of Ionizing Radiation (LNMRI) under the National Nuclear Energy Commission (CNEN) is the body designated by the National Institute of Metrology and Quality Technology (INMETRO) as National Metrology Laboratory for quantities associated with ionizing radiation. This laboratory participates in inter comparison in order to ensure the traceability of measurements in the worldwide network of metrology, coordinated by the Bureau International des Poids et Mesures (BIPM) and the Inter-American Metrology System (SIM). It offers various systems and calibration methods for radioactive sources in various geometries which can be absolute or relative.

The high-resolution gamma spectrometry by high purity germanium detector (HPGe) is applied to the analysis of emitting photons radionuclides [1].

This is a relative technique, by the use of standards for determining the activity of the samples is compared to a standard of the same nature or by efficiency curve which is generated by using various standard (^{241}Am, ^{152}Eu, ^{166m}Ho).

The ^{133}Ba has 10.5 years of half- life and decays by electron capture 100 % into two main branches , E1 (86.2 %) to the level of 437 keV and E2 (13.7 %) to the level of 383.8 keV for ^{133}Cs as seen in figure 1. It is a radionuclide which is in addition to the x – rays characteristics, amount 14 power lines gamma (γ) and x-ray, forming several coincidences in the search for stability [2]. These emitted energies in its excited state cover a range of great interest to analyze, this makes it a standard for laboratories.

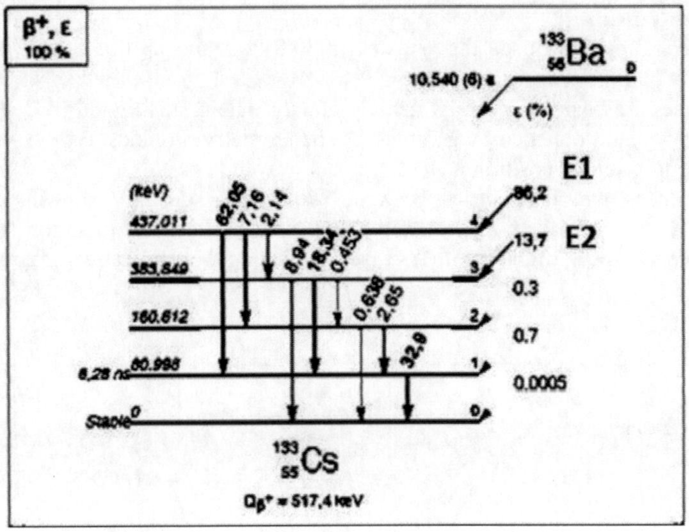

Figure 1. ^{133}Ba scheme decay simplified

The sum-peak method was developed by Brickman [3] and their collaborators presenting a calibration by means of two coincident radiation, X and γ rays. Over several years articles were published in which it is used HPGe [4,5] . This method allows a direct measurement and uses the same principle of coincidence counting by gamma spectrometry with a single detector.

His equation is:

$$[(A_1 \,.\, A_2) \;/\; A_{12}] + T = N \tag{1}$$

where:
A_1 and A_2 represent the areas of photopeaks range of energies observed in coincidence.
A_{12} represents the energies area of photopeaks observed in sum.
T counts of full spectrum.
N is the absolute activity of the source.

The objective of this work is the calibration of the ^{133}Ba using the sum-peak method, showing speed and providing direct calibration values comparable to literature.

2. Methodology

Five sources of ^{133}Ba were made in point form geometry in thin film and the approximate activities were calculated according to the distance source - detector to avoid pile-up effect during measurement.

It used a planar germanium detector coupled to a voltage divider that allows the spectrum energy range of 300 keV to 1200 keV , allowing full spectrum display of ^{133}Ba .

The distance of the sources in relation to the detector are 0 cm (top) and 10 cm. Uncertainties recommendations follow the ISOGUM [6].

3. Results and discussion

In activities determining were used the x-ray, kα= 30.8 keV and kβ= 35.5 keV of ^{133}Ba decay, and their sum-peaks of gamma energies.

The uncertainties are basically type A, namely counts. The relative combined standard uncertainty associated with any result obtained by gamma-ray spectrometry provides: type A - peak area; type B - mass, decay, half-life, source position.

Figure 2 introduces gamma spectrum between 270 keV to 450 keV showing these energies and sum-peaks with x - rays. Peaks not identified come from background radiation (Bg). They are well evident the simple sum-peak of the γ with X- rays, suggesting the application of equation 1.

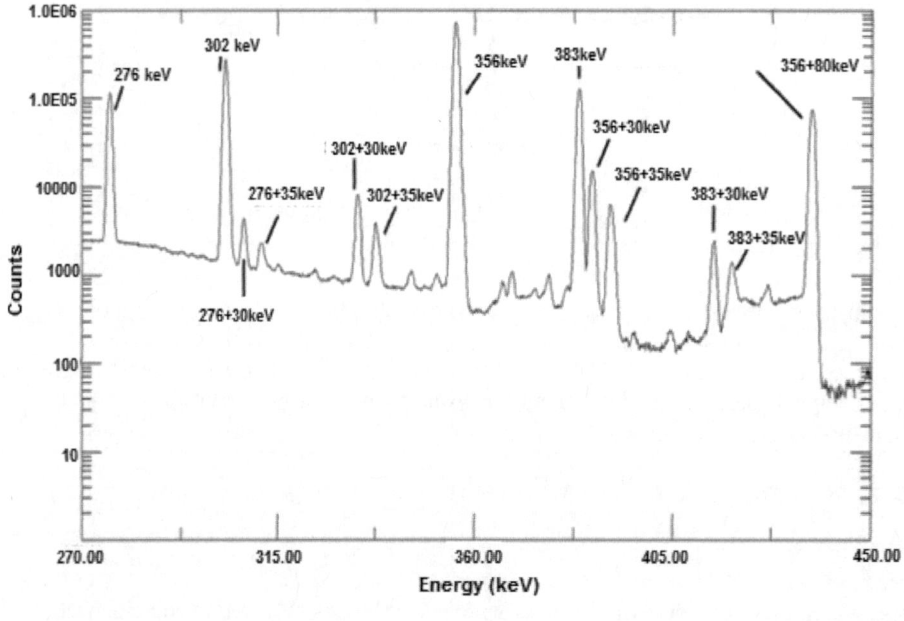

Figure 2. Energy spectrum between 270 keV to 450 keV of ^{133}Ba decay and their sum-peaks with the source on the top of detector

Table 1 shows the results from different sources in the 10 cm position detector by comparative method with the BIPM standard.

Table 1. Results of ^{133}Ba activities by comparative method using BIPM standard (2015/03/25)

Source	**Activity kBq/g**	**U_a**	**U_b**	**U_t (K = 2)**
1	156.46	1.2	0.48	1.30
2	156.79	0.92	0.48	1.04
3	156.96	0.93	0.48	1.05
4	157.37	1.2	0.48	1.30
5	157.66	0.49	0.48	0.69

The table 2 shows the values obtained at the sum-peak method and the table 3 shows the decay in the source that generated the five samples, all calibrations are referenced to the 12:00 h of 2015/03/25.

Table 2. Results of ^{133}Ba activities by sum-peak method (2015/03/25)

Source	**Activity kBq/g**	**U_a**	**U_b**	**U_t (K=2)**
1	156.13	0.51	0.48	0.70
2	156.21	0.51	0.48	0.70
3	156.24	0.51	0.48	0.70
4	156.23	0.51	0.48	0.70
5	156.25	0.51	0.48	0.70

Table 3. Some calibration methods for ^{133}Ba sources (2015/03/25)

Method	Activity kBq/g	U_a	U_b	U_t (K=2)
Sum-peak	156.21	0.51	0.48	0.70
Decay	156.28	0.06	0.66	0.66
Comparative	157.05	0.95	0.48	1.06

They were only analyzed the sum-peaks of the characteristic x-ray and γ energies, and some of these suffer interference and require deconvolution in order to obtain consistent results.

4. Conclusion

The values found among the methods presented here are compatible. The ^{133}Ba analysis can be performed of the absolute sum-peak method with the advantage of being fast, simple and to present low uncertainty results. The continuing search must be performed to ^{133}Ba calibrate by this method due to peaks relating to the sums and , they should be better observed due to the possible generation of other peaks relating to the triple sums , changing the final result.

References

[1] Lépy M C, Pearce A and Sima O 2015 *Bureau International des Poids et Mesures Metrologia* **52** 123.

[2] Nucléide – Lara Bibliothèque d'émissions gamma et alpha 2013 http://laraweb.free.fr/.

[3] Brinkman G A , Aten Jr A H W and Veenboer J Th 1963 *Applied Radiation and Isotopes* **14** 153.

[4] Iwahara A , Poledna R , Silva C J and Tauhata L 2009 *Applied Radiation and Isotopes* **67** 1887.

[5] Silva C J , Iwahara A, Poledna R , Bernardes E M O , Prinzio M A R R and Lopes R T 2008 *Applied Radiation and Isotopes* **66** 231.

[6] ISOGUM 2004 ISO GUIDE to expression of uncertainty in measurement..

Spectra Transmitted by Mortar Barite in x-ray Qualities Applied in Diagnostic Radiology as Shielding

A T Almeida Jr[1, 2], F G S Araújo [2], M S Nogueira [3] and M A P Santos [4]

[1] Brazilian Institute for Safety and Health at Work – FUNDACENTRO, Minas Gerais, Brazil
[2] Federal University of OuroPreto – UFOP / REDEMAT, Minas Gerais, Brazil
[3] Center of Development of Nuclear Technology – CDTN / CNEN, Minas Gerais, Brazil
[4] Regional Center for Nuclear Science – CRCN / CNEN, Pernambuco, Brazil

E-mail: **airton.almeida@fundacentro.gov.br**

Abstract. Concrete which contains water, cement and aggregate, is widely used in building construction such as medical hospitals. The CdZnTe spectrometry system was used to acquire the transmitted spectra in the RQR qualities and the stripping procedure was performed by taking into account both the contributions of efficiency and x-ray escape fraction, experimentally determined. The samples were prepared in rectangular plate format with dimensions of (5 x 5) cm with thicknesses varying from 0.2cm to 2cm and exposed to x-ray beams generated. The HVL and the mean energy in this energy range was determined.

1. Introduction
Adequacy of the shielding, walls as well as equipment must be monitored by means of radiometric assessment. The equipment must be kept under adequate technical conditions and a quality control programme must be implemented. Therefore, aiming at protecting the premises and people against ionising radiation in radioactive facilities, the use of shielding materials is being studied by various researchers with a must-be-implemented objective to determine the curves of attenuation of the ionising radiations. Amongst the materials used as protection barriers, are lead, concrete and iron[1,2]. Relevant studies on the subject of barrier calculation in x-ray facilities were employed to consolidate the methodology of primary and secondary barriers proposed by the National Council on Radiation Protection (NCRP) in 2004, expressed in NCRP Report 147 [3]. The acknowledgement of the characteristics of attenuation of barite is fundamental, from the radioprotection point of view to the viabilisation, to design and execution of projects of ionizing radiation shielding in radioactive facilities. The main aim of this work is to characterise the barite mortars found in different regions of Brazil by measuring the x-ray spectra emerging from the x-ray radiation beams attenuated by the various samples generated in diagnostic radiology qualities, so that the operational quantities equivalent of ambient doses (H*) can be determined for comparison with the effective dose.

2. Experimental
This study aims at characterising samples of purple, white and cream barite occurring in different parts of Brazil, which are used in radioprotection such as x-ray shielding materials.
These materials are used in radioactive facilities as X radiation shielding for they present advantages such as: low cost, high efficiency in shielding and are easy to handle besides being widely available in the local market. Table 1 depicts the materials used in this research, as well as the state of the country where each of them were found. Enhancing the fact that, of all materials studied in this work, the purple barite from the state of Bahia is the hardest to mine.

Table1. Materials used as x radiation shielding.

Shielding Materials	Origin
White barite mortar	State of Paraíba
Cream barite mortar	State of São Paulo
Purple barite mortar	State of Bahia

In order to perform the measurements of x-ray spectra, the incident beams as well as the ones transmitted by samples, the following spectrometry was applied: Detection system made up of a CdTe detector connected to a pre- amplifier made by Amptec model XR -100T- CdTe, as described in the table 2.

Table2. Specifications of the AmptekCdTe detector.

Parameter	**Specification**
Detector	CdTe model XR-100T-CdTe nominal operation voltage: 400 V
Detection characteristics	Resolution (keV): 0.53 to 14.4 keV 0.85 to 122 keV
Geometric Characteristics of the crystal	Area: 3x3 mm^2
Characteristics of the electronic system connected to the detector to collect data	Shaping time: 3μs, Gain: 4.5 RTD: on

The measurements of the spectra were checked with the detector placed at 340 cm from the focal point, and the barite mortar plates or a set of them were positioned at 50 cm from the focal point.
The positioning of the detector was done by using the calibration setup with a beam collimator, and its alignment with the central ray of the radiation beam was done by employing laser convergent. Reference x-ray qualities were used as recommended by IEC 61267[4] (RQR – level of diagnosis guidelines), as presented in the table 3.

Table3. Characteristics of RQR qualities implanted at CRCN/CNEN.

Qualities	**Voltage (kV)**	**1º HVL (mm Al)**	**2º HVL (mm Al)**
RQR4	60	2.90	2.19
RQR6	80	3.05	3.01
RQR9	120	3.90	5.00
RQR10	150	4.40	6.57

After the measurements of the radiation spectra, some corrections are necessary to provide various interactions of radiation with materials that make up the detector. These corrections are due to materials that may be attenuating the radiation and the determination of this energy influences the mean energy. The corrections to primary x-rays transmission, secondary x-rays escape, K-fluorescent x-rays and Compton-scattered x-rays from the attenuation due to materials of the detector are usually carried out using the stripping method [5] that was also applied in this work.

3. Results and Discussion

3.1. Calibration of the spectrometer

The XR-100T-CdTe detector was calibrated for correct measurements of spectra by using standard sources of radiation 241Am, 109Cd and 133Ba. By considering the results of the spectra achieved for each source in this study and the value of their activities, it was possible to calculate the efficiency of the detector CdTe in function of the energy of the photopeak. The results of the values calculated by the Monte Carlo method, as well as those obtained experimentally. Figure 1 shows a good agreement between the average results measured and simulated by Monte Carlo. Three steps observed in the respective figure can be associated to the "escapes-k" of Cadmium (Cd) and Tellurium (Te) between energies of 20 and 30 keV [6]. In the corrected spectra these degrees are partially eliminated.

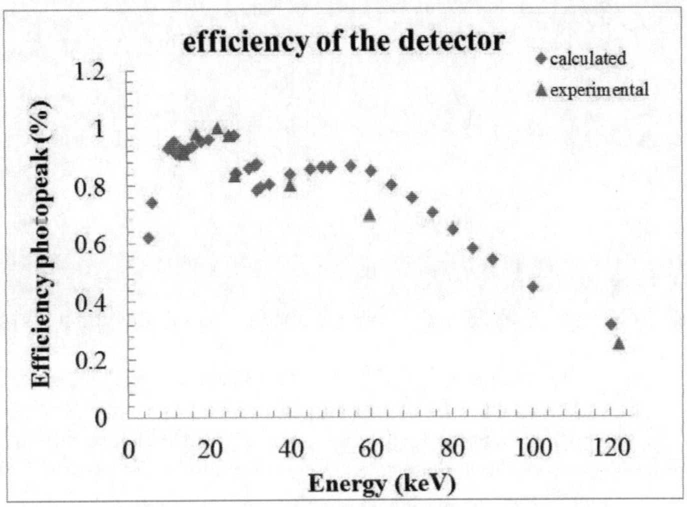

Figure1. Efficiency photopeak obtained through ^{241}Am, ^{133}Ba, ^{109}Cd and ^{152}Eu sources.

3.2. Incident and transmitted spectra transmitted by barite matters

The figure 2 presents the comparisons between the non-corrected spectra (distribution and pulse height) and the corrected ones (photon spectra) which were obtained according to RQR radiation qualities recommended in the IEC guidelines.

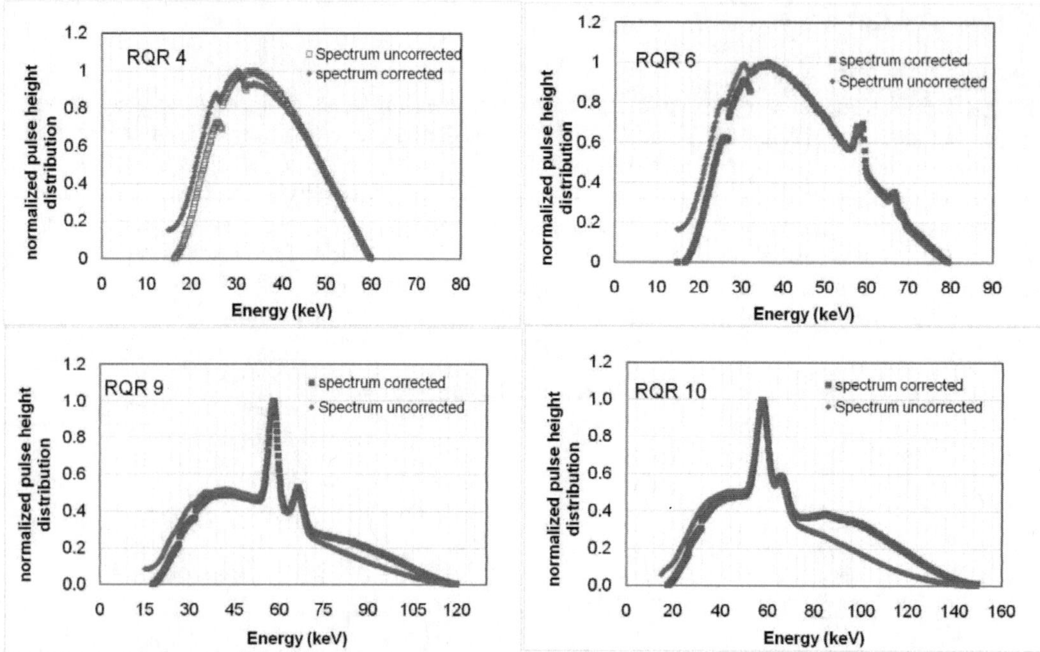

Figure2. Pulse spectra and incident photon spectra with and without correction.

The difference between the measured spectra and corrected spectra is visible, especially in the region of low energy of the measured spectra. Furthermore, this spectra presents depressions or ′′dips′′ at 27keV and 32keV, being more noticeable in the low energy spectra. These dips were also related in the study by Miyajima [6], where the author states they are related to the fraction of escape of the K layer of the Cd and Te, respectively. In the corrected spectrum these dips are partially corrected, thus having their intensity reduced as a result figure 3.

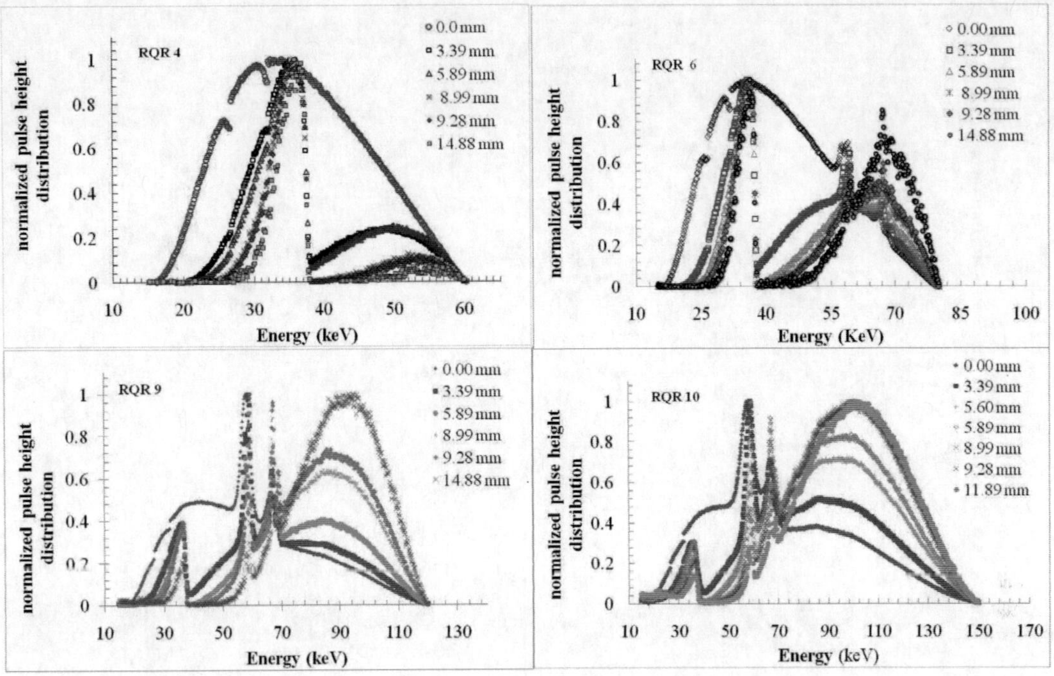

Figure3. Transmitted x-ray spectra in qualities: RQR4, RQR6, RQR9 and RQR10 transmitted by white barite plates.

In the figure 3, the spectra transmitted by white barite plates which were measured using the IEC qualities aforementioned are represented. These spectra were normalised for the same maximum photon area. An analysis of figure 3 shows that the primary spectra are highly modified as it crosses the barite thickness. A very important aspect of these spectra is the peak of absorption of Barium, around 40KeV, which cuts the photons above this energy deeply. In lower energy spectra which are the case of RQR4, this situation is well noticeable.
The modification of the primary spectrum, when crossing the thickness of the barite will have a great impact on the quantities, which depend on the photon spectra, such as H*(d) and the effective dose and what represents an important aspect in metrology, mainly in area monitoring, where it is necessary to estimate the effective dose. Figure 4 presents the spectra transmitted for each type of barite mortar, according with the RQR qualities used. It can be observed that the cream barite presents a better shielding efficiency in this energy range for diagnosis x-ray, followed by the white barite. This difference is better observed in qualities RQR4 and RQR6 for they are of lower energies.

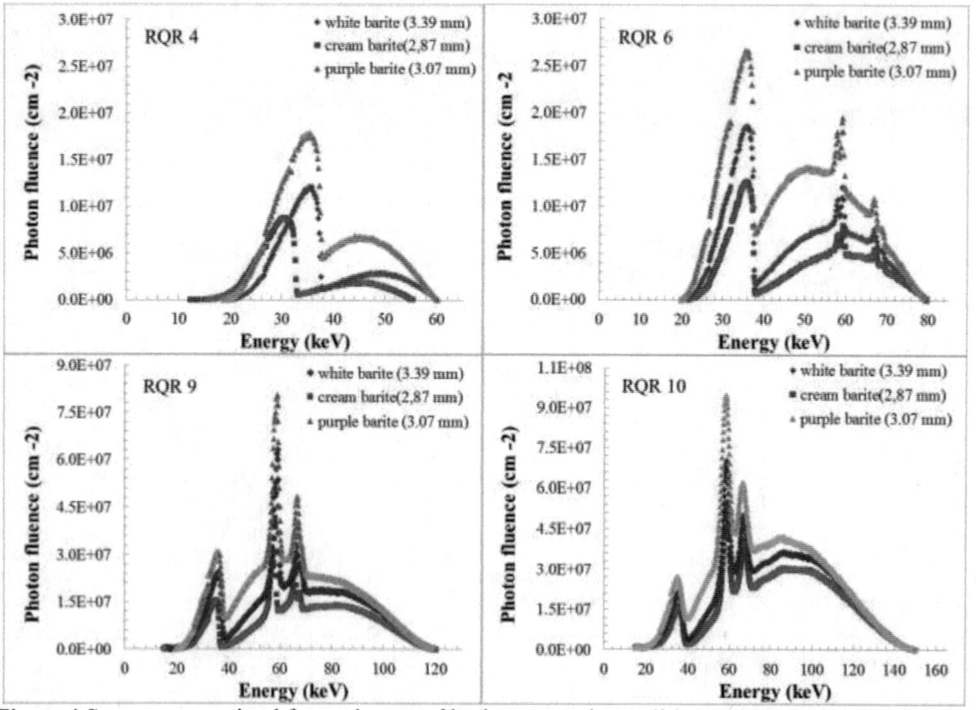

Figure4. Spectra transmitted for each type of barite mortar in qualities RQR4, RQR6, RQR9 and RQR10.

The atomic composition of these materials may vary according to different manufacturers and may result in a variation in the attenuation of the radiation in function of energy and thickness, thus presenting a modification in the behaviour of the transmission spectrum.

4. Conclusion

The characteristics of attenuation of barite influence the transmission curves in the range of energy of diagnostic radiology. Also, it was demonstrated that the atomic composition of the barite influences the transmission curves of the x-rays, hence showing the importance of atomic determination of these materials. The results achieved in this research made it possible to meet the respective dimensioning of these barriers with precision and safety, when designing shielding using the studied materials as protection barriers against x-rays, thus optimizing the cost of radiology protection

5. Acknowledgments

The authors are thankful to FUNDACENTRO, UFOP/REDEMAT, FUNDAÇÃO GORCEIX, CNPq, and CNEN. This work was supported by FAPEMIG, CNPq and Ministry of Science and Technology - MCT/Brazil, through the Brazilian Institute of Science and Technology (INCT) for Radiation Metrology in Medicine.

Referências

[1] DOUGLAS J. SIMPKIN. General solutions to the shielding of medical X and ray by the NCRP report 49 methods. 1987 *Health Physics* **52(4)** 431-436.

[2] ALMEIDA JR AIRTON T 2014 OuroPreto-MG, Universidade Federal de Ouro Preto/ REDEMAT Tese de Doutorado 127p.

[3] NCRP – Structual shielding design for medical X-ray imaging facilities. Bethesda MD 2004 (NCRP Report 147).

[4] IEC – IEC 61267: Medical diagnostic X-ray equipment – Radiation conditions for use in the determination of characteristics 2005Geneva, Switzerland.

[5] MATSUMOTO M YAMAMOTO A HONDA I TANIGUCHI A AND KANAMORI H. Direct measurement of mammography X-ray spectra using a CdZnTe detector 2000 *Med. Phys*. **27(7)** 1490–1502.

[6] MIYAJIMA S IMAGAWA K and MATSUMOTO M CdZnTe detector in diagnostic X-ray spectroscopy 2002 *Medical Physics* **29** 1421–1429

Determination of the ^{121}Te gamma emission probabilities associated with the production process of radiopharmaceutical NaI[^{123}I]

M T F de Araújo[1,2], R Poledna[1], J U Delgado[1], M C M de Almeida[1], R T Lopes[2], R L Silva[1] and A C F Cagido[3]
[1]Instituto de Radioproteção e Dosimetria/Comissão Nacional de Energia Nuclear (IRD/CNEN), Av. Salvador Allende s/nº, Barra da Tijuca, CEP 22783-127, Rio de Janeiro/Brasil
[2]Programa de Engenharia Nuclear, Universidade Federal do Rio de Janeiro (PEN/COPPE, UFRJ), Ilha do Fundão, Caixa Postal 68590, CEP 21945-970, Rio de Janeiro, Brasil
[3]Universidade Federal do Paraná, UFPR, Ciências Geodésicas Caixa Postal:19001 CEP:81531-990 Curitiba -PR-Brasil
E-mail: miriamtaina@hotmail.com

Abstract: The ^{123}I is widely used in radiodiagnostic procedures in nuclear medicine. According to Pharmacopoeia care should be taken during its production process, since radionuclidic impurities may be generated. The ^{121}Te is an impurity that arises during the ^{123}I production and determining their gamma emission probabilities (Pγ) is important in order to obtain more information about its decay. Activities were also obtained by absolute standardization using the sum-peak method and these values were compared to the efficiency curve method.

1. Introduction

The presence of impurities that emit gamma rays can be identified through the technique of gamma spectrometry. However, the determination of nuclear decay related parameters of this radioisotope will provide a greater insight into the method of production and the possible effects of the presence of this radionuclide when administered in the form of radiopharmaceutical to the patient. The determination of the gamma emission probability (Pγ) is fundamental for this control, as well as being an opportunity to confirm the efficiency of the spectroscopy system for determination of radionuclidic impurities enabling the measurement procedure both activity and Pγ. The sum-peak method was adopted here since it allows to obtain very low values for activity of radionuclides emitting gamma radiation, as is the case of ^{121}Te. This method had already been used by (Brinkman et al., 1977; Silva et al., 2006) for calibration of ^{123}I, providing good results with low uncertainty.

2. Methods

2.1. Sum-peak method

The spectrum of ^{121}Te with energy peaks can be observed in Figure 1. Gamma spectrometry system used here consisted of an HPGe coaxial detector, with 20 % relative efficiency, and a good resolution in the range from 3 to 300 keV.

The sum-peak coincidence method is an absolute method used in calibration processes of radionuclides that have emission lines x or gamma. Therefore, the combinations that will result in sum-peak can be of type: γ-γ, x-x or x-γ. Thus, the application of the method requires a well-calibrated spectrometry system, combined with associated electronic units, as well as a program for data acquisition, integrated with a multichannel analyzer. According to the decay scheme, ^{121}Te decays by electron capture to ^{121}Sb. The absolute value of the activity is calculated using the following equation:

$$N_0 = [N_T + N] + \frac{N_X N_\gamma}{N_{X\gamma}} \tag{1}$$

where N_0 is the activity; N_T is the total count; N is the counting corresponding to the extrapolation to zero in multichannel analyzer; Nx is the number of counts in the x photopeak; Nγ is the number of counts to the γ photopeak, Nxγ is the number of counts of coincidence due to x and γ peaks. The estimate of uncertainty is given taking into account uncertainty components of A- and B - type for an expanded uncertainty k = 1 (de Almeida et. Al., 2007).

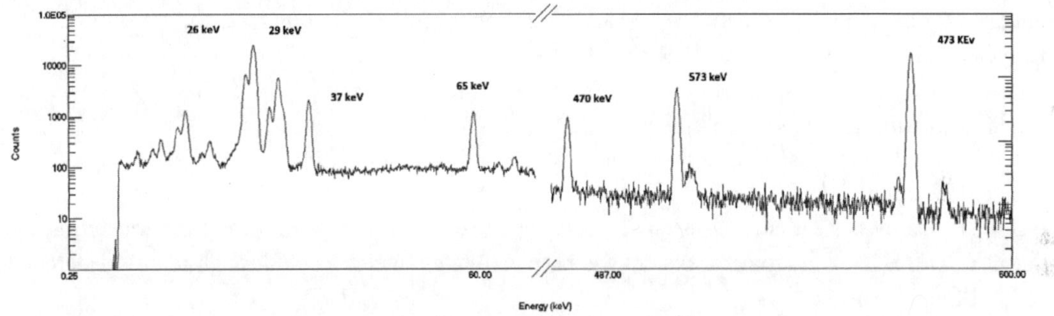

Figure 1 HPGe spectrum obtained by the spectrometry system with the respective energy peaks associated with the radionuclide ^{121}Te.

2.2. *Spectral separation method*

Here was developed a spectral separation method to remove the contribution of ^{125}I that appears in the sample due to the production mode of ^{123}I. This other radionuclidic impurity has a characteristic energy represented by the peak of 35.5 keV, which interferes with quantification of ^{121}Te x-ray due to its proximity and high intensity.

Point sources were prepared from the master solution of ^{123}I, the first step consisted in identifying all energy peaks in the region of 30-40 keV that appear in the spectrum, which is stored in the internal memory of the automated multichannel Analyzer. Then one obtains the gross count values for this region of the spectrum and subtract the corresponding background radiation. Hence, it is bounded on the region of interest in the spectrum that corresponds to the peak of 35.5 keV and, with appropriate command, gets the count for just this peak, generating a peak area correction factor of this region. This factor is obtained by division between the 35.5 keV peak area in the spectrum of the source to be calibrated (mixture) and the corresponding area in another spectrum of a source contained only ^{125}I (pure) in the same conditions of measurements. This area factor obtained in the previous step is considered to subtract the contribution of the ^{125}I contaminant. Due to small discontinuity that appears at the beginning of the detector window is made the extrapolation to zero in multichannel to get the value of N, which will be added to total N so that, finally, is determined the value of N_0, according to equation 1.

3 Experimental Procedure

3.1 Source preparation

The master solution used for these measurements was provided in ampoule by the producer (IEN/CNEN) and then was eluated in the form of iodide (^{123}I) with a sodium hydroxide solution for obtaining sodium iodide (NaI). Before starting the counts, was here set a timeout, around a few half-lives of ^{123}I, in order to reduce the intensity of the sources so that its secondary energy lines do not hide the ^{121}Te characteristic peaks observed in the spectrum. The masses were determined by gravimetric method with differential weighing.

Point sources were prepared with the help of a picnometer, depositing drops of solution of radionuclide in a polystyrene film, with a thickness of 0.05 mm, set in an acrylic ring. The ring has an external diameter of 25 mm, inner diameter of 4 mm and 1 mm thick. Once dried, the sources were covered with the same polystyrene film (Bernardes et al., 2002). HPGe spectrometry system has been calibrated in efficiency through point standards of ^{60}Co, ^{152}Eu, ^{166m}Ho and ^{226}Ra. The efficiency curve obtained for an energy range between 100 and 1000 keV can be seen in Figure 2, and the uncertainties for values of activity varied around 2% (k = 1).

3.2 Radionuclidic impurities assessment

For carrying out of measurements was used a system consisting of a planar HPGe detector with the standard electronic. This detector is known "d4" and the measurements were carried out in two different positions: the first at a distance of 10 cm, "p2", from the detector and the second on the detector "p_0". HPGe spectrometry system has been calibrated in efficiency through point standards of ^{60}Co, ^{152}Eu, ^{166m}Ho and ^{226}Ra.

The efficiency curve obtained for an energy range between 100 and 1000 keV can be seen in Figure 2, and the uncertainties for values of activity varied around 2% (k=1). Significant sum effects were not observed in for the preparation of the efficiency curve, at “p2”, mainly for ^{152}Eu and ^{166m}Ho standards. This allowed the calibration point source, quantification of impurities and determining the gamma emission probability for ^{121}Te. The spectrum acquired in the position "p0" was used by the sum-peak absolute method to get the activity of ^{121}Te.

Radionuclidic impurities of ^{121}Te and ^{125}I, both with half-life more than ^{123}I, identified in the spectrum were quantified in order to assess possible damage to the patient during the period of incorporation. As the main radionuclide decay during the acquisition of multiple spectra it was possible to verify the presence of the peaks associated with impurities. Brazilian law follows the recommendation of the American Pharmacopoeia, which, for the production of ^{123}I, stipulates a limit of 15% for impurities in relation to the main radionuclide (^{123}I). And in this study the presence of ^{125}I was a negative contribution factor, which could be properly evaluated with the aid of the spectral separation technique, ensuring its content to be compared to the percentage indicated by the current recommendations.

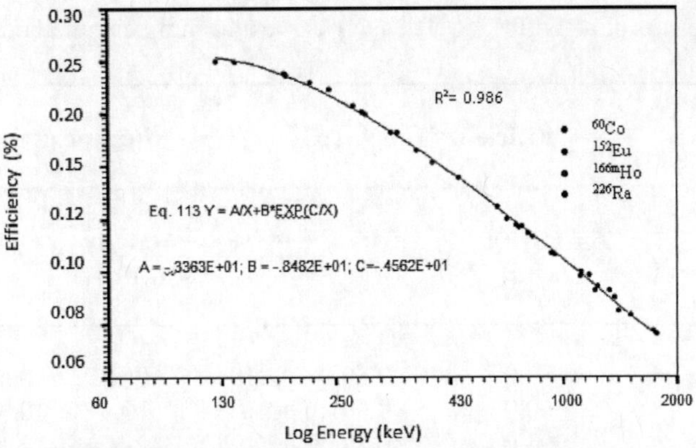

Figure 2. Efficiency curve obtained (linear-log scale) for the spectrometry system using the radionuclidic standards in point source geometry.

3.3 Gamma-ray emission probabilities measurement

In order to associate the main peaks of the spectrum to the radionuclidic is need to obtain the relation energy-channel, for calibrating in energy the spectrometer. After, the total absorption efficiency curve is determined, in function of energy, to calculate the radionuclidic activity from net areas under each interest peak. The expression of activity area is:

$$A = CPS_{(CORRECTED)}/(\varepsilon . P_{\gamma}) \quad (2)$$

where: CPS (CORRECTED) is the count rate of photopeak; $\varepsilon\gamma$ is the photopeak efficiency for specific energy; and $P\gamma$ is the emission probability for specific energy. However, as the source activity was obtained directly by sum-peak method, $P\gamma$ was calculated for two main energies of ^{121}Te, 507 and 573 keV, by means of the following expression, taking into account the corrections as decay, background, position:

$$P_{\gamma} = CPS_{(CORRECTED)}/(A . \varepsilon_{\gamma}) \quad (3)$$

where: CPS (CORRECTED) is the count rate of photopeak; A is the absolute activity measured by sum-peak method; $\varepsilon\gamma$ is the photopeak efficiency for specific energy. The precision to determine $P\gamma$ depends on the precision achieved in efficiency curve with the choice of interest energy of the standards: ^{60}Co, ^{152}Eu, ^{166m}Ho and ^{226}Ra.

The peak area evaluate method taken into account the channel integrations that define the region of each peak, after subtracts background. Spectra were analyzed by Maestro II code. The samples and standards were measured three times at least. All measurements were made with point sources in "p2" position using the same geometry.

4 Results and Discussion

The experimental results for the activity of ^{121}Te are presented in Table 1, accompanied by their respective uncertainties both for efficiency curve method, position p2 as to the sum-peak method, position p0.

Table 1: Results of activities for efficiency curve and sum-peak methods for ^{121}Te

	Efficiency Curve (p2)			Sum-peak method (p0)		
SOURCE	**Activity (kBq)**	$u_A(\%)$	$u_B(\%)$	**Activity (kBq)**	$u_A(\%)$	$u_B(\%)$
172S14	695.0	0.50	2.10	696.4	0.50	0.1
173S14	560.0	0.59	2.10	570.4	0.30	0.1
174S14	600.0	0.39	2.10	599.0	0.15	0.1
175S14	689.0	0.27	2.10	604.7	0.35	0.1
176S14	600.0	0.47	2.10	596.5	0.36	0.1

Positions: p2 = 10 cm (distance between source detector); p0 = 0 cm (top of detector)

The result for the source 175S14 obtained by the efficiency curve method suffered some interference and unable to again measure the value of the activity of the source. The values obtained in both methods are compatible and this enabled the standardization of absolute ^{121}Te.

As can be seen in table 2, it should be noted also the low uncertainties associated with the sum-peak method, indicating values that are more reliable. The identification of impurity for the ^{125}I brought several problems in order to calibrate ^{121}Te absolutely.

Table 2: Uncertainties to ^{121}Te impurity determination by sum-peak method

Uncertainty Component	^{121}**Te**	
type	**A**	**B**
measurement statistic	0.15	
decay	-	< 0.1
live time	-	< 0.1
weighing		0.1
Relative standard uncertainty ***u*** **(%)**	0.18	
Relative expanded uncertainty	0.36	

***U* (%) (*k*=2)**

Table 3 shows the results for the gamma emission probabilities compared to the values of the Physikalisch-Technische Bundesanstalt (PTB) and Laboratoire National Henry Becquerel (Nucleide-LARA/LNHB). In general, the values for the emission probabilities evaluated were consistent and uncertainty values obtained are slightly below the published values.

Table 3: Gamma emission probabilities and their relative standard uncertainties for 507 and 573 keV energies of ^{121}Te by efficiency curve method (k=1)

Energy (keV)	**PTB**	**LARA/LNHB**	**This work**
507	0.177±0.004	0.177±0.006	0.167±0.002
573	0.803±0.017	0.823±0.025	0.804±0.006

Finally, it was possible to develop a methodology for analysis of these impurities that took into account the interference contribution for ^{125}I during the data acquisition. The proper calibration of the photon spectroscopy system, as well as the successful implementation of sum-peak method, made it possible to obtain the values of the emission probabilities with associated uncertainty until 1.2 % for 507 keV and 0.5% for 573 keV. The Pγ value for 507 keV is 5.6 % lower than of reference publications. And for 573 keV the value Pγ is until 2% is compatible with the reference. However, the uncertainties here obtained are lower than of literature data.

Table 4: Percentage of radionuclidic impurities in the ^{123}I sample and associated uncertainties by efficiency curve method

Isotope Impurity	**^{121}Te**		**^{125}I**	
Percentage in ^{123}I sample (%)	0.326		0.394	
Uncertainties (μ) type	A	B	A	B
measurement statistic	0.47		0.88	
efficiency curve	-	0.60*	-	4.0
weight-geometry	-	0.20	-	0.20

half-life	-	0.40	-	0.02
live time	-	< 0.1	-	< 0.1
photon emission probability (Pγ)	-	2.1	-	2.0
relative standard uncertainty ***u* (%)**	2.3		4.6	
Relative standard uncertainty associated with an activity of ^{123}I **	1.1		1.1	
Relative expanded uncertainty ***U* (%) (*k*=2)**	5.1		9.5	

* (Lépy et al., 2015) ** values obtained by ionization chamber method

6 Conclusions

The absolute standardization process of ^{121}Te here developed presented a new approach for the treatment of radionuclidic impurities. Activities obtained by efficiency curve were satisfactory and important enough for the comparison of two methods here adopted. The sum-peak method made it possible to obtain values of activity with uncertainties below 0.5 %. At the same time made it possible to obtain good values for both the activity and the determination of precise nuclear parameters as gamma emission probabilities for the two main radionuclide and its impurities.

7 References

[1]Bernardes. E.M.O., Delgado. J.U., Tauhata, L., da Silva, C.J., Iwahara, A., Poledna,R., Paschoa, A.S., 2002. ^{166}Ho: a multi-γ standard for the calibration of the Ge spectrometers. Appl. Radiat and Isotopes. 56, 157-161.

[2]Brinkman, G. A., Aten, A. H. W., 1963. Absolute standardization with a NaI (Tl) crystal. Int. J. of Appl. Radiat. and Isotopes, 14, 153-157.

[3]Brinkman, G. A., Lindner, L., Veenboer, J. TH., 1977. Sum-peak calibration of ^{123}I. Int. J. of Appl. Radiat. and Isotopes, 28, 271-275.

[4]de Almeida, M.C.M., Iwahara, A.,Poledna, R., da Silva, C.J., Delgado, J.U., 2007.Absolute disintegration rate and 320 keVγ-ray emission probability of ^{51}Cr. Nucl. Inst and Meth. in Phys. Research A, 580, 165-168.

[5]Dias, M. S. and Koskinas, M. F., 1995. Accidental summing corrections in ^{125}I activity determination by the sum-peak method. Appl. Radiat. and Isotopes, 46, 945-948.

[6]Kim, I. J., Park, C. S., Choi, H. D., 2003. Absolute calibration of ^{60}Co by using sum peak method and an HPGe detector. Appl. Radiat. and Isotopes, 58, 227-233.

[7]Laboratoire National Henri Becquerel, LNHB/CEA, 2004. Table of Radionuclides www.nucleide.org/DDEP_WG/123_tables

[8]Lederer, C. M., Shirley, V.S., 1978. Table of Isotopes. Ed. John Wiley & Sons. New York 24.

[9]Lépy, M. C., Pearce, A., Sima, O., 2015. Uncertainties in gamma-ray spectrometry. <etrologia, 52, S123-S145.

[10]Schoetzig, U. und Schrader, H., 1989. Halbwertszeiten und Photon-Emissionswahrsheinlichkeiten von haeufigverwendetenradionukliden. PTB - Ra-16/3 september.

[11]Silva, M. A. L.,Poledna, R., Iwahara, A., Silva, C. J., Delgado, J. U., Lopes, R. T., 2006. Standardization and decay data determination of ^{125}I,^{54}Mn and ^{203}Hg. Appl. Radiat. and Isotopes, 64, 1440-1445.

[12]Wang, Z., Zhang, X., Chang, Y., Liu, D., 2001. The determination of ^{125}I activity using sum-peak method with a well-type HPGe detector based spectrometer. Nucl. Inst. and Meth. in Phys. Research, 459, 475-481.

PHANTOM POSITIONING VARIATION IN THE GAMMA KNIFE® PERFEXION DOSIMETRY

N A Costa[1], M P A Potiens[1], C W C Saraiva[2]
[1] Instituto de Pesquisas Energéticas e Nucleares, IPEN-SP, Comissão Nacional de Energia Nuclear, CNEN, São Paulo, SP, BRAZIL
[2] Hospital do Coração, São Paulo-SP, BRAZIL

E-mail: nathaliaac@ymail.com

Abstract. The use of small volume ionization chamber has become required for the dosimetry of equipments that use small radiation fields such as the Gamma Knife® Perfexion (GKP) unit. In this work, a pinpoint ionization chamber was inserted into the dosimetry phantom and measurements were performed with the phantom in different positions, in order to verify if the change in the phantom positioning affects the dosimetry of the GKP. Four different phantom positions were performed. The variation in the result is within the range allowed for the dosimetry of a GKP equipment.

1. Introduction

The Gamma Knife® Perfexion (GKP) unit, introduced in 2006, is a fully automated unit for the stereotactic radiosurgery and stereotactic radiotherapy of brain lesions including also certain extra-cranial lesions [1].

It has 192 sealed sources of ^{60}Co arranged on eight movable sectors, each sector containing 24 sources. The collimator system consists of a single large tungsten ring with collimator holes of three different sizes of 4, 8 and 16 mm [2].

When a patient treatment is initiated, all eight sectors with their 24 sources quickly move out from their shielded home position at the extreme rear of the radiation cavity and slide over the tungsten ring and past the 8 mm collimators to park in the "off position" in shielded space between the 8 and 4 mm collimators. During this time, no beam is turned on and the treatment does not start to count. Depending on the size of the beam desired from a particular sector, the sources on that sector are aligned with the desired collimator size. The beams from a sector may also remain "blocked" if the sources in that sector remain parked in the off-position [1].

The presence of many critical structures and eloquent areas in the human brain requires careful stereotactic radiosurgery treatment planning as well as dosimetric verification of the plan. The verification between the plan and the delivery is achieved by comparing the planning dose distribution exported from the treatment planning system (TPS) with the measurement data obtained using radiation detectors [4].

In this work, a pinpoint ionization chamber was inserted into the phantom and measurements were performed with the phantom in different positions, in order to verify if the change in the phantom positioning affects the dosimetry of the GKP. Considering that the GKP dosimetry is performed using the largest size collimator (16 mm), the detector volume used in this study is 0.016 cm^3, which justifies the influence analysis of the size detector effect in the dosimetry of the GKP.

2. Materials and methods

The dosimetry of a GKP is performed using a spherical polystyrene phantom having a diameter of 160 mm, which is provided by Elekta (GKP manufacturer). It consists of two hemispheres each one having 130 mm wide and 5 mm deep slot in the central area across the diameter of the hemisphere. The two hemispheres are held together with the help of two small lucite pegs and corresponding holes for the pegs in the peripheral area of the hemispheres. There is a lucite gamma angle plate on the side of the hemisphere that helps the phantom to stay in its ideal position. The slot in the central area of the phantom is used to accommodate a 130 x 160 x 10 mm^3 insert, made of ABS plastic. The insert is rounded off on two ends so that the surface of the insert is flush with the spherical surface of the phantom [3].

This phantom is held in irradiation position by a C-shaped aluminum adaptor, which has stainless steel side-fixation screws. This side-fixation screws have been proved to attenuate the beam in up to 4.6% [2]. Figure 1 shows the setup of the phantom mounted with the ionization chamber inserted in it.

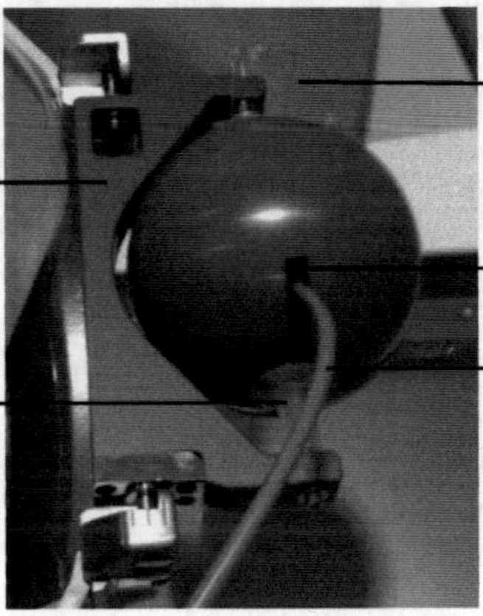

Figure 1. Dosimetry phantom set up

In this study, a pinpoint ionization chamber from PTW-Freiburg with a volume of 0.016 cm^3 was used (Figure 2). The pinpoint ionization chamber was inserted into the phantom having its sensitive volume center in the center of the phantom. A dose-rate calibration of a GKP unit is performed using the largest size collimator (16 mm, shown in Figure 3).

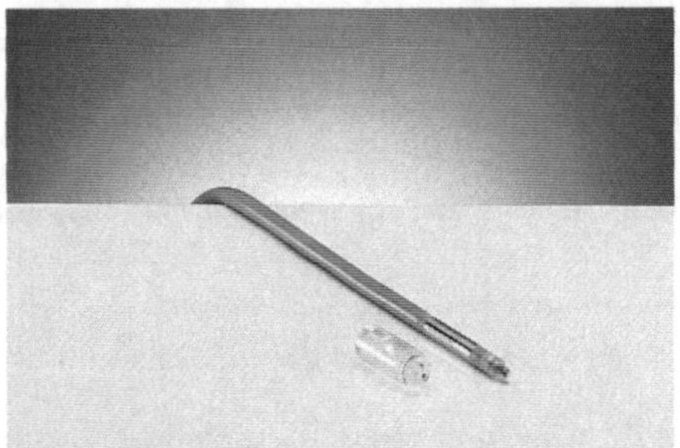

Figure 2. Pinpoint ionization chamber

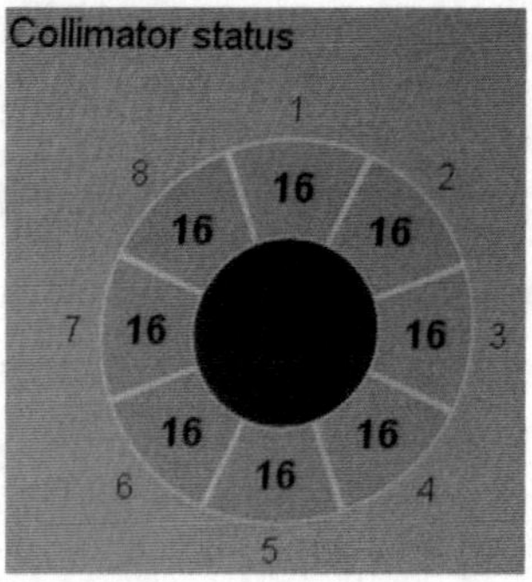

Figure 3. Size collimator used for dosimetry. All sectors opened

Ten readings of one minute each were performed for each position. It must be said that Elekta does not give any requirement related to the position that the phantom must acquire during the GKP dosimetry. The ionization chamber was connected to a PTW Unidos E electrometer.

With the aim of verifying if there is any attenuation that might be considered during the GKP dosimetry, four different phantom positions were performed. All of them are related to the stereotactic XZ plan (PPS plan). Position A: 90° from the PPS plan. Position B: -120° from the PPS plan. Position C: 30° from the PPS plan. Position D: 0° from the PPS plan. For position C, measurements were performed using all collimators opened and also with only sector 4 opened and the other closed. This evaluation was made in order to estimate the scattering contribution on the rod of the ionization chamber. Figure 4 presents the diagram of the possible scattering contribution of beam sector 4 and the Figure 5 shows each position used for the measurements.

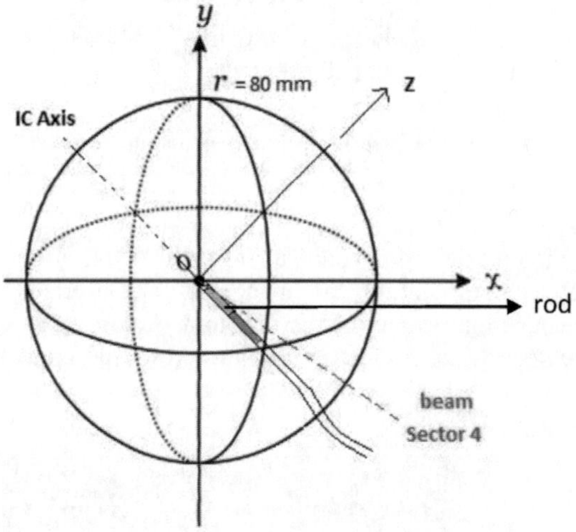

Figure 4. Scattering contribution from beam sector 4

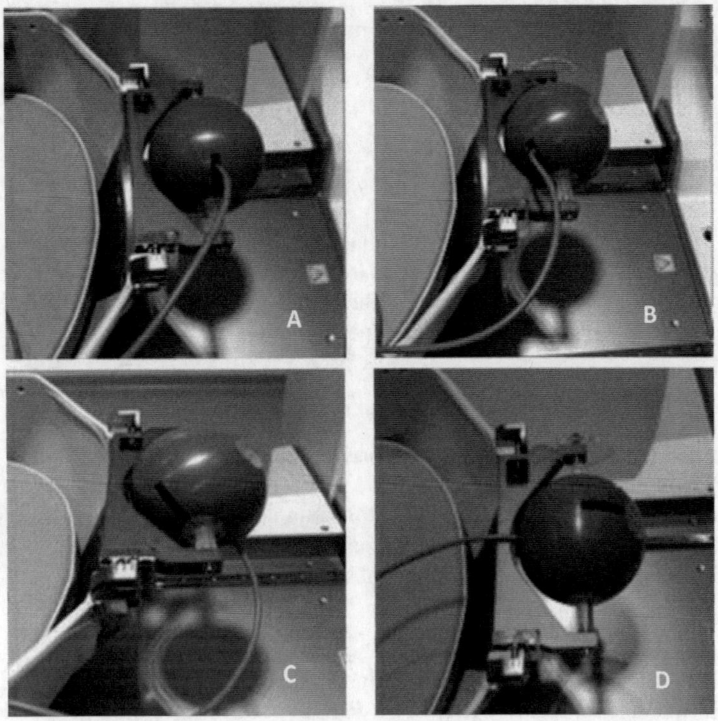

Figure 5. Phantom positions used for the measurements. In A: 90° from the PPS plan; in B: -120° from the PPS plan; in C: 30° from the PPS plan and in D: 0° from the PPS plan.

3. Results

Table 1 shows the results of the readings for the four different phantom positions. To calculate the average charge reading the following equation was used:

$$\bar{L} = \frac{\sum_{i=1}^{10} l_i}{10}$$

Table 1. Results of the collected charge for the four different phantom positions

Positions	Charge (nC)
A	0.9906 ± 0.0038
B	0.9955 ± 0.0029
C	0.9869 ± 0.00013
D	1.000

Table 2 shows the results of the readings for positions A, B and C with only sector 4 of the 16 mm collimator opened.

Table 2. Results of the collected charge for three different phantom positions with only sector 4 opened

Position	Charge (nC)
A	0.1234
B	0.1229
C	0.1154

Considering the average charge values for all positions evaluated, position D showed the highest value, so it is considered the best position to perform the GKP dosimetry.

Position C presented the smallest charge value, which suggests attenuation of the beam at sector 4. The reduction comparing to the other positions was 6.3%.

4. Conclusion

In table 1, the highest charge reading was position D, which means that position D presents the smallest influence of the ionization chamber rod when performing the dosimetry. However, uncertainties of other positions are acceptable for Gamma Knife Perfexion dosimetry. After analyzing the data on the table 2, the impact induced by the rod of the ionization chamber showed a reduction of 6.3% in the reading when compared with positions A and B.

Acknowledgments

The authors acknowledge the partial financial support of Coordenação de Aperfeiçoamento de Pessoal de Nível Superior (CAPES), SIBRATEC FINEP/MCTI project number 01.10.0650.01, coordinated by IRD/CNEN-RJ, Brasil and the partnership with Hospital do Coração – Hcor- São Paulo-SP, Brasil.

References

[1] Bhatnagar J P, Novotny J, Huq M S 2012 Dosimetric characteristics and quality control tests for the collimator sectors of the Leksell Gamma Knife (R) Perfexion (TM) *Med. Phys.* **39** (1) 231-236

[2] Bhatnagar J P, Novotny J, Quader M A, Bednarz G, Huq M S 2009 Unintended attenuation in the Leksell Gamma Knife Perfexion calibration-phantom adaptor and its effect on dose calibration *Med. Phys.* **36** (4) 1208-1211

[3] Bhatnagar J P, Novotny J 2010 Assessment of variation in Elekta plastic spherical-calibration phantom and its impact on the Leksell Gamma Knife calibration *Med. Phys.* **37** (9) 5066-5071

[4] Park J H, Han J H, Lee D H, Suh T K, Chung D G K. 2011 Application of the gamma evaluation method in Gamma Knife film dosimetry *Med.Phys.* **38** (10)

TDCR and CIEMAT/NIST Liquid Scintillation Methods applied to the Radionuclide Metrology

P A L da Cruz[1], C J da Silva[1], A Iwahara[1], J S Loureiro[1], A E De Oliveira[1], L Tauhata[1] and R T Lopes[2]

[1]Instituto de Radioproteção e Dosimetria/Comissão Nacional de Energia Nuclear (IRD/CNEN), Av. Salvador Allende s/nº, Barra da Tijuca, CEP 22783-127, Rio de Janeiro/Brasil

[2]Programa de Engenharia Nuclear, Universidade Federal do Rio de Janeiro (PEN/COPPE, UFRJ), Ilha do Fundão, Caixa Postal 68590, CEP 21945-970, Rio de Janeiro, Brasil

E-mail: palcruz@ird.gov.br

Abstract: This work presents TDCR and CIEMAT/NIST methods of liquid scintillation implemented in National Institutes of Metrology for activity standardization of radionuclides, which decay by beta emission and electron capture. The computer codes used to calculate the detection efficiency take into account: decay schemes, beta decay theory, quenching parameter evaluation, Poisson statistic model and Monte Carlo simulation for photon and particle interactions in the detection system. Measurements were performed for pure emitters ^{3}H, ^{14}C, ^{99}Tc and for $^{68}Ge/^{68}Ga$ which decay by electron capture and positron emission, with uncertainties smaller than 1% (k = 1).

1. Introduction

Although known for a long time, the use of liquid scintillation technique for the radionuclide standardization was only possible from the cooperation between the laboratories of the international metrology network to find theoretical and technological solutions that enable its implementation. Thus, the cooperation between National Institute Standards and Technology (NIST/USA) and the Centro de Investigaciones Energéticas, Medioambientales y Tecnológicas (CIEMAT/Spain) resulted in the development of semi-empirical CIEMAT/NIST method [1].

The cooperation between the Laboratoire National Henri Becquerel/France and Radioisotope Center/Poland (LNHB/RC) was important step towards the rise TDCR (Triple-to Double Coincidence Ratio Method) method for absolute radionuclide standardization [2]. The development of computational codes for the detection efficiency calculation [3] and MAC3 unit (Module d'acquisition de coincidences triples) that uses the extended dead-time technique [4] represented the final step to TDCR method implementation.

Due the consistency in the measurements and minimized uncertainty levels obtained, the CIEMAT/NIST and TDCR methods were adopted by the Bureau International des Poids et Mesures (BIPM) as references for the radionuclide standardization.

The present work uses these methods to perform measurements of beta emitters, ^{3}H and ^{14}C, as well as ^{99}Tc from the key-comparison organized by the BIPM with participation of the national laboratories of the international radionuclide metrology network [5].

Another important step in this work was the application of TDCR method to standardize $^{68}Ge/^{68}Ga$, nuclides in secular equilibrium, which decay by electron capture and positron emission, in cooperation with the LNHB/France. The last two nuclides are important because of their increasing use in nuclear medicine [6].

1.1 Liquid Scintillation Counting

The TDCR and CIEMAT/NIST counting systems consisting of three or two photomultipliers based on statistical Free Parameter model [7] of the scintillation photons distribution and their detection probabilities.

The liquid scintillation process occurs when a radionuclide solution is dissolved in cocktail of the scintillator substances and the kinetic energy of the particles is transferred to the molecules of the medium, with the consequent light emission. The light photons captured in the photomultiplier photocathode produces electrons, which are multiplied by dynodes with the generation of electronic signal of enough amplitude to be processed in the electronic chain.

G. F. KNOLL [8] describes the main process of scintillation light: the fluorescence arising from the transition between of energy levels from the structure of organic molecules of the scintillator substances with aromatic structure, which present alternating double bonds and π electrons with the light emission in order of nanoseconds. BIRKS [9] studied the mechanism of particles interaction with the chemical medium of the scintillator organic substances. In this, the charged particles emitted in scintillator liquid interact with the molecules of the medium, lose kinetic energy and some molecules experiment transformations, such as ionization, free radical formation, excitement or rupture in neutral or ionized fragments.

The quenching from ionization process produced by excited molecules of the solvent manifests itself in those parts of the liquid where local excited molecules concentration is high.

The equation 1presents the Birks's expression for the evaluating quenching from light generated by particles and photons interaction with cocktail molecules that it takes into account the correction due loss light photons intensity like result of the ionization caused by radiation in its trajectory in chemical medium.

$$m(E) = \int_0^E \frac{A\,dE}{1 + kB\,dE/dx} \tag{1}$$

Where, m(E) is the average number photoelectrons; A is a characteristic parameter of the scintillation cocktail; kB is a semi-empirical parameter; and dE/dx is the stopping power coefficient.

The Free Parameter (λ) is the average energy required to produce a photoelectron on photocathode. Thus, if the photons emissions follow the Poisson statistical law, the detection efficiency of the photomultiplier can be determined by the introduction of the Free Parameter concept. The equation 2 and 3 give the detection efficiency for the probability of zero photoelectrons generation when m value is expected.

The detection efficiency for beta emitters takes into account the energy spectrum S(E) obtained by application of the Fermi theory, according to equation 4.

$$P(x/m) = \frac{m^x e^{-m}}{x!} \tag{2}$$

$$\varepsilon = 1 - P(0) = 1 - \frac{(vm)^0 e^{-m}}{0!} = 1 - e^{-vm} \tag{3}$$

$$\varepsilon = \int_0^E S(E)(1 - e^{-vm})dE \tag{4}$$

Journal of Physics: Conference Series **733** (2016) 012099 doi:10.1088/1742-6596/733/1/012099

The computational codes for theoretical calculation of the detection efficiency in TDCR and CIEMAT/NIST methods depending on the Free Parameter and take into account the quenching from the Birks equation, the radionuclide decay schemes, the particles and photon and particle interactions in the scintillation cocktail and system measurement evaluated by use Monte Carlo simulation codes [10].

1.2 CIEMAT/NIST primary method

This counting system consists of two photomultipliers placed in coincidence at 180° to each other. The CIEMAT/NIST method uses a set of standard tracer tritium (^3H) samples for the commercial scintillator characterizing in terms of the quenching parameter by the introduction of the chemical quenching agent in increasing degree. The figure 1 shows the correlation between Free Parameter, quenching and efficiencies for standard tracer and radionuclide in analysis by CIEMAT/NIST method.

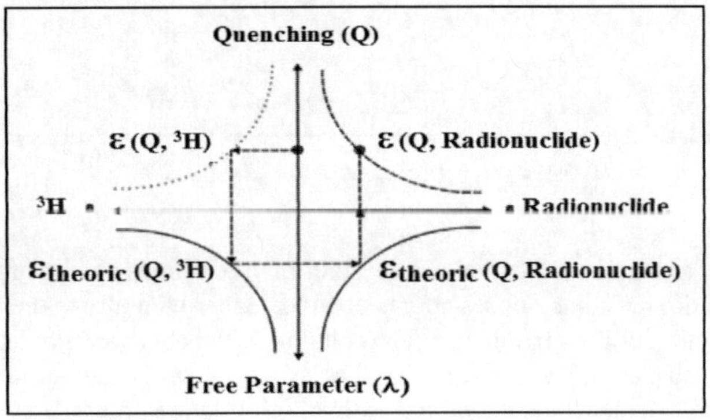

Figure 1. CIEMAT/NIST model for the radionuclide standardization.

The equation 5 gives the detection efficiency for two photomultipliers obtained from Poisson model.

$$\varepsilon = \left(1 - e^{-\frac{vm}{2}}\right)^2 \tag{5}$$

1.3 TDCR method for absolute standardization

The counting system comprises three photomultiplier tubes placed in coincidence at 120° to each other. The figure 2 shows the TDCR systems. The equation 6, 7, 8 and 9 calculate the photomultiplier detection efficiency, detection efficiency for the logical sum of the three double coincidences, efficiency ratio between the triple-to-double coincidences and TDCR ratio for beta emitters.

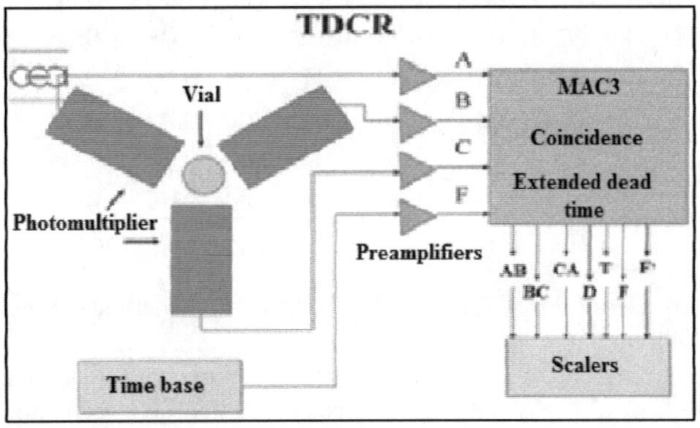

Figure 2. TDCR measurement system.

$$\varepsilon = 1 - e^{-\frac{vm}{3}} \tag{6}$$

$$\varepsilon_D = 3\left(1 - e^{-\frac{vm}{3}}\right)^2 - 2\left(1 - e^{-\frac{vm}{3}}\right)^3 \tag{7}$$

$$\mathrm{TDCR} = \frac{\varepsilon_T}{\varepsilon_D} = \frac{\left(1 - e^{-\frac{\eta}{3}}\right)^3}{3\left(1 - e^{-\frac{\eta}{3}}\right)^2 - 2\left(1 - e^{-\frac{\eta}{3}}\right)^3} \tag{8}$$

Where,

$$\eta = \frac{v}{3}\int_0^E \frac{A\,dE}{1 + kB\,dE/dx}$$

Then,

$$\mathrm{TDCR} = \frac{\int_0^{E_{max}} S(E)\left(1 - e^{\frac{-vm}{3}}\right)^3 dE}{\int_0^{E_{max}} S(E)\left(3\left(1 - e^{\frac{-vm}{3}}\right)^2 - 2\left(1 - e^{\frac{-vm}{3}}\right)^3\right) dE} \tag{9}$$

The absolute standardization of radionuclide solutions combines the interpolating of the ratio between the triple and double coincidences counts obtained experimentally to the theoretical efficiency versus TDCR as a function of Free Parameter. The counting efficiency variation is performed by either, photomultiplier defocusing or use of the filters of increasing optical density around sample glass vials or the addition of chemical quenching agent in the liquid scintillation cocktail.

1.4 Scintillation cocktails

At present, the commercial scintillation cocktails have chemical composition that allows the radioactive solutions dissolution in water, constituting a homogeneous solution. The most common are composed of a DIN solvent (diisopropyl naphthalene), a fluorescent solute such as 2,5 Dipheniloxazole substance and a surfactant. The table 1 presents the density and the stoichiometric composition of the main scintillation cocktails provided by the commercial Perkin Elmer Company.

Table 1. Stoichiometric composition of liquid scintillators by PerkinElmer Company.

Chemical Element Cocktail	C	H	N	O	P	S	Na	Density (g/cm^3)	Z/A	WM
HiSafe3	19.17	28.12	0.02	2.88	0.06	0.00	0.00	0.99	0.5458	306.70
Ultima Gold	16.81	24.54	0.04	1.52	0.11	0.02	0.02	0.98	0.5459	255.76
Ultima Gold XR	18.11	29.80	0.04	2.83	0.11	0.03	0.03	0.99	0.5476	297.98
Ultima Gold AB	18.67	28.49	0.01	2.53	0.01	0.00	0.00	0.98	0.5485	293.47
Ultima Gold LLT	18.57	28.43	0.01	2.56	0.01	0.00	0.00	0.98	0.5486	292.68
Insta-Gel Plus	18.53	30.93	0.01	3.90	0.00	0.00	0.00	0.95	0.549	315.71
Hionic-Fluor	10.83	18.77	0.08	1.97	0.18	0.04	0.04	0.95	0.5449	188.87

2. Experimental Procedure

^3H and ^{14}C solutions were measured by TDCR and CIEMAT/NIST methods in the scintillation cocktails HiSafe3 and Ultima Gold.

The ^{99}Tc solution from the international key-comparison organized by the BIPM was measured by TDCR, Coincidence $4\pi\beta$(PC)–γ(NaI) and Anticoincidence 4π β(LS)–γ(NaI) counting systems.

The $^{68}Ge/^{68}Ga$ solution was submitted to the comparison between LNMRI/Brazil and LNHB/France to measurements in TDCR counting system.

The radionuclide solutions used in the experimental standardization were ^{3}H (IPL 958-62-4) in water; ^{14}C (LMRI) in glucose and water; ^{99}Tc – NPL/UK DA-11337, in NH_4OH 0.1 M and; $^{68}Ge/^{68}Ga$ (Eckert & Ziegler Analytics) in HCl 0.1 N. The standardization ^{99}Tc solution by Coincidence and Anticoincidence counting systems used the tracer standard ^{60}Co.

2.1 Computational codes for liquid scintillation methods

The theoretical calculations of the radionuclide detection efficiency were performed by computational codes CN2003 (11) to CIEMAT/NIST and TDCR07c [12] to TDCR. Some changes were introduced in the TDCR07c code to suit $^{68}Ge/^{68}Ga$ decay scheme [13].

2.2 Coincidence counting system

The 4πβ(PC)-γ(NaI) Coincidence counting system consists of a gas flow 4π proportional counter coupled to a crystal of NaI(Tl). The sources are prepared by dropping known masses of onto VYNS film previously gold coated on both sides.

2.3 Anticoincidence counting system

The 4πβ(LS)-γ(NaI) Anticoincidence counting system with extended dead time is based on specialized modules developed to implement 4πβ–γ coincidence counting without the use of a resolving time. The coincidence rate is obtained indirectly by subtracting the no correlatedγ-rate from the total γ-rate.

3. Results

The theoretical radionuclide efficiencies by CIEMAT/NIST method was calculated using kB parameter value of 0.0075 $cm.MeV^{-1}$, according to Birks' expression. The kB parameter for TDCR method was obtained from the sample measurements by use of filters of increasing optical density around sample glass vials. The choice of the optimum kB was made for the value that produced the slightest variation and lower slope of linear fit curve: ^{3}H (kB = 0.013 $cm.MeV^{-1}$), ^{14}C (kB = 0.008 $cm.MeV^{-1}$), ^{99}Tc (KB = 0.008 $cm.MeV^{-1}$) and $^{68}Ge/^{68}Ga$ (KB = 0.010 $cm.MeV^{-1}$). The assessment took into account the uncertainty components from the atomic and nuclear parameters and the experimental procedure, evaluated according to BIPM [14].

3.1 ^{3}H solution standardization

The table 2 and figure 3 show that the results obtained for the standardization of ^{3}H solution in HiSafe3 and Ultima Gold cocktails performed by TDCR method are consistent from each other, within a standard uncertainty.

Table 2. Uncertainty budget (k = 1) of the ^{3}H solution standardization by TDCR method.

	Type	Hisafe3 (%)	Ultima Gold (%)
Weighing	B	0.05	0.05
Statistic counts	A	0.57	0.58
Source activities	A	0.36	0.66
kB value (grey filters)	A	0.28	0.20
Nuclear and atomic data	B	< 0.01	< 0.01
Combined uncertainty (k = 1)		0.73	0.90

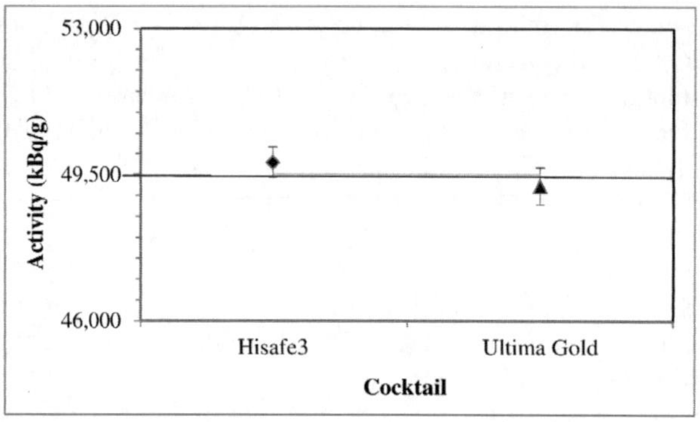

Figure 3. Results of the ^{3}H solution standardization by TDCR method (k = 1).

3.2 ^{14}C solution standardization

The figure 4 and table 3 show that the results of the ^{14}C standardization by TDCR and CIEMAT/NIST methods were consistent from each other for the two cocktails, within a standard uncertainty.

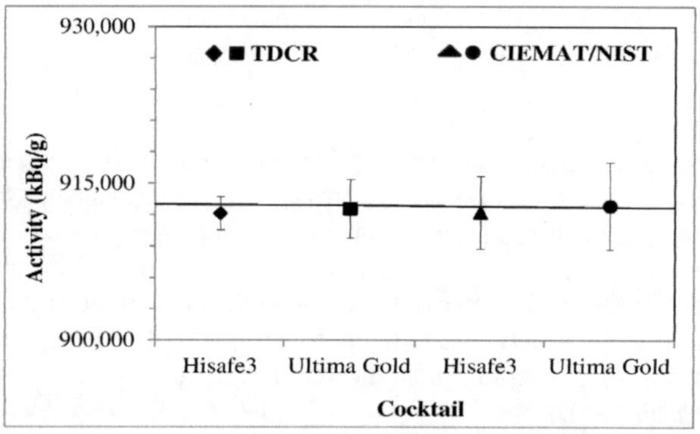

Figure 4. Results of the ^{14}C solution standardization by TDCR and CIEMAT/NIST methods (k = 1).

Table 3. Uncertainty budget (k = 1) of the ^{14}C solution standardization by TDCR method.

Component	Type	Hisafe3 (%)	Ultima Gold (%)
^{3}H activity	B	0.03	0.03
^{3}H weighing	B	< 0.01	< 0.01
^{3}H Quenching (SQPE)	A	0.03	0.04
^{3}H statistic counts	A	0.03	0.03
Nuclear and Atomic data	B	0.07	0.07
Photomulplier asimetry	B	0.25	0.25
kB value	B	0.23	0.23
^{14}C weighing	B	0.06	0.05
^{14}C Quenching (SQPE)	A	0.04	0.03
^{14}C statistic counts	A	0.07	0.08
^{14}C source activities	A	0.10	0.27
Combined uncertainty (k = 1)		0.38	0.45

3.3 ^{99}Tc solution standardization

The figure 5 and table 4 show the results obtained to the standardization of ^{99}Tc-IPL/UK solution, from key-comparison organized by the BIPM in 2012, measured by TDCR and using HiSafe3 cocktail. The results were compared to the obtained by Coincidence 4π β(PC) – γ(NaI) and Anticoincidence 4π β(LS) – γ(NaI) counting systems, using ^{60}Co tracer.

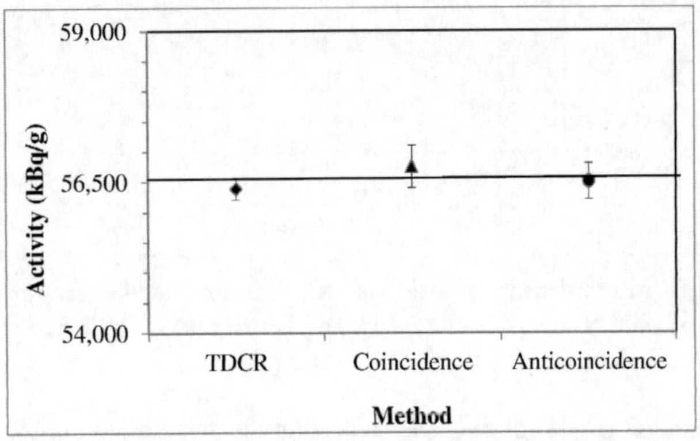

Figure 5. Results of the ^{99}Tc solution standardization by TDCR, Coincidence and Anticoincidence methods (k = 1).

Table 4. Uncertainty budget (k = 1) of the ^{99}Tc solution standardization by TDCR, Coincidence and Anticoincidence methods.

Component	Type	Method		
		TDCR (%)	Anticoincidence (%)	Coincidence (%)
Statistic counts	A	0.21	Included in fitting	Included in fitting
Fitting curve	A		0.38	0.53
Weighing	B	0.05	0.05	0.05
Background	A	< 0.001	0.29	0.2
Live time technique	B	0.01	0.01	
Decay (half-time)	B	< 0.001	< 0.001	< 0.001
Dead time	B			0.04
Resolving time	B			0.04
Gandy effect	B			0.08
^{60}Co activity	B		0.22	0.22
kB value	A	0.05		
Shape form	A	0.21		0.08
Source activities	B	0.08		
Combine uncertainty (k = 1)		0.32	0.53	0.62

3.4 ^{68}Ge/^{68}Ga solution standardization

The comparison between LNMRI/Brazil and LNHB/France of the ^{68}Ge/^{68}Ga solution by TDCR method was performed using the same computer code TDCR07c. The results obtained by both laboratories shown in figure 6 and table 5 were consistent each other, within a standard uncertainty.

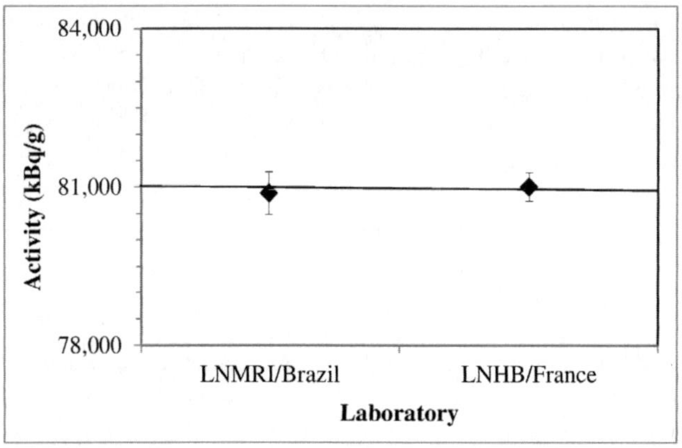

Figure 6. Results of the $^{68}Ge/^{68}Ga$ solution standardization by TDCR method, in the LNMRI/Brazil and LNHB/France radionuclide metrology laboratories (k = 1).

Table 5. Uncertainty budget (k = 1) of the $^{68}Ge/^{68}Ga$ solution standardization by TDCR method of the LNMRI/Brazil and LNHB/France.

Component	Type	LNMRI (%)	LNHB (%)
Weighing	B	0.05	0.05
Statistic counts	A	0.26	0.11
Source activities	A	0.31	0.11
kB value (grey filters)	A	0.06	0.06
Nuclear and atomic data	B	0.28	0.28
Combined uncertainty (k = 1)		0.50	0.33

4. Conclusion

The results obtained by TDCR and CIEMAT/NIST methods based on Free Parameter model were consistent at each other for the standardization of pure beta emitters as ^{3}H and ^{14}C for HiSafe3 and Ultima Gold cocktails, within a standard uncertainty. The standardization of the others radionuclides were performed only in HiSafe3.

The standardization of ^{99}Tc-IPL/UK solution, from key-comparison organized by the BIPM in 2012 was performed by TDCR and using HiSafe3 cocktail. This result was compared with those obtained by Coincidence and Anticoincidence counting systems and were consistent each other, within a standard uncertainty.

The good performance obtained by comparison between LNMRI/Brazil and LNHB/France for the $^{68}Ge/^{68}Ga$ solution standardization performed by TDCR show that this method can be applied to the standardization of radionuclides that present complex decay scheme. In the end of BIPM key-comparison, for $^{68}Ge/^{68}Ga$ in progress will be evaluated the consistency of the results obtained by TDCR counting system compared to the others absolute methods.

References

[1] Malonda AG and Coursey BM 1987 Standardization of Pure Beta Emitters by Liquid Scintilllation Counting. Applied Radiation and Isotopes **39**: 165.

[2] Broda R and Pochwalski K 1992 The enhanced triple to double coincidence ratio (ETDCR) method for standardization of radionuclide by liquid scintillation counting. Nuclear Instruments and Methods **A 312**: 85.
[3] Broda R, Cassette P, Maletka K and Pochwalski K 2000 A simple computing program for application of the TDCR method to standardization of pure-beta emitters. Applied Radiation and Isotopes **52**: 673.
[4] Bouchard J and Cassette P 2000 MAC3: an electronic module for the processing of pulses delivered by a three photomultiplier liquid scintillation counting system. Applied Radiation and Isotopes **52**: 669.
[5] da Cruz PAL, da Silva CJ, Moreira DS, Iwahara A, Tauhata L, Lopes RT 2014 Primary activity standardization of ^{99}Tc by three different methods. Applied Radiation and Isotopes **87**: 175 – 178.
[6] Applied Radiation and Isotopes ^{68}Ga (2013) Special Issue **76**: 1.
[7] Malonda AG 1999 Free Parameter Models in Liquid Scintillation Counting, Editorial CIEMAT/Spain.
[8] Knoll G 1989 Radiation Detection and Measurement. Second Edition, John Wiley & Sons **216**.
[9] Birks JB 1951 Scintillations from Organics Cristals: Especific Fluorescence and Relative Response of Different Radiations. Proc. Phys. Soc. London Sect. **A64**, 874.
[10] OECD/NEA 2008 Penelope: A Code System for Monte Carlo Simulation of Electron and Photon Transport. Nuclear Agency Energy/Organization for Economic Co-operation and Development.
[11] Gunther E 2003 Program CN2003: A program to calculate the LC efficiency of a nuclide vs. efficiency the tracer H-3 (CIEMAT/NIST). PTB/Germany.
[12] Cassette P 2007 TDCR07c: Program TDCR and tracer methods in LSC. LNHB-CEA/LIST/France.
[13] Zimmerman B E, Cessna JT and Fitzgerald R 2008 Standardization of ^{68}Ge/^{68}Ga using three liquid scintillation counting based methods. Journal of Research of the National Institute of Standards and Technology **113**: 265.
[14] JCGM 2008 Evaluation of measurement data: Guide to the Expression of Uncertainty in Measurement. JCGM (BIPM, IEC, IFCC, ILAC, ISO, IUPAC, IUPAP and OIML).

Journal of Physics: Conference Series **733** (2016) 012100 doi:10.1088/1742-6596/733/1/012100

Comparison of surface contamination monitors for *in vivo* measurement of ^{131}I in the thyroid

S M Oliveira, A L A Dantas and B M Dantas

Instituto de Radioproteção e Dosimetria – IRD, Rio de Janeiro, Brasil.

E-mail: salomao.marques@ymail.com

Abstract. The routine handling of radiopharmaceuticals in nuclear medicine represents a significant risk of internal exposure to the staff. The IAEA recommends the implementation of monitoring plans for all workers subject to a risk of exposures above 1 mSv per year. However, in Brazil, such recommendation is practically unfeasible due to the lack of a sufficient number of qualified internal dosimetry services over the country. This work presents an alternative based on a simple and inexpensive methodology aimed to perform *in vivo* monitoring of ^{131}I in the thyroid using portable surface contamination probes. Results show that all models evaluated in this work present enough sensitivity for the evaluation of accidental intakes.

1. Introduction

According to the information available at the Brazilian Nuclear Energy Commission (CNEN) [1], there are approximately 430 authorized nuclear medicine clinics in regular operation in Brazil, resulting in a significant number of occupationally exposed workers who handle routinely a wide variety of radionuclides in the form of unsealed sources. Such professional activity represents a risk associated to the possibility of intake via inhalation and ingestion and the subsequent internal exposure. Currently, the most frequently used radionuclides in the field of nuclear medicine are ^{99m}Tc, ^{131}I, ^{123}I, ^{201}Tl and ^{18}F. ^{131}I is widely used for diagnosis and therapy and is considered the most critical radionuclide in terms of internal exposure of the workers because of its high dose coefficient. The use of other radionuclides is being studied worldwide and can be a subject of radiological protection concern in a near future.

In spite of a consensus that external exposure is predominant, significant intakes may occur simultaneously in nuclear medicine practice. However, depending of the exposures scenario, and based on international criteria of evaluation, the permanent risk of intakes of radionuclides requires the implementation of a routine monitoring plan, aiming to control and limit internal doses [2]. It is a responsibility of the Radiation Safety Officer (RSO) of the facility to manage the Radiation Protection Programme and, based on monitoring results, to implement the necessary measures to keep exposure levels as low as possible. It should also be highlighted that the application of radiation protection principles in a routine pattern relies on a shared responsibility among the RSO and the occupationally exposed workers [3].

The evaluation of internal occupational exposures requires the use of specific methodologies which allow identification and quantification of intakes as well as the estimation of committed effective doses of the workers [4]. Currently, in Brazil, qualified laboratories are not available in all regions of the country to offer internal dosimetry services. Thus, it would represent a high cost to the medical institutions if the CNEN would apply the need for the implementation of internal monitoring programmes according to national radiation protection requirements [5].

This work describes a simple and convenient alternative aiming to implement a low-cost internal monitoring programme applied to workers who handle unsealed sources of ^{131}I used mainly for therapy purposes in nuclear medicine facilities. Such professionals are frequently exposed to height activities of ^{131}I and consequently, present higher risks of intake. The methodology uses portable surface-contamination monitors, a type of equipment available in all nuclear medicine facilities since it is of obligatory use according to the CNEN requirements.

2. Materials

2.1. Neck-thyroid phantom

The neck-thyroid phantom (figure 1) used for the calibration of the detectors for the measurement of ^{131}I in the thyroid was developed at the In Vivo Monitoring Laboratory of IRD. The neck phantom is made of polyurethane-base tissue equivalent material. A filter paper simulating a human thyroid is contaminated with a known amount of ^{133}Ba liquid source. After being sealed with a plastic film, the filter paper is contained in an acrylic part and inserted in the neck phantom. It is conventional practice to use sources of ^{133}Ba as a surrogate for measuring ^{131}I since the energy and yield of photons emitted by ^{133}Ba are very similar to photon emissions from ^{131}I (table 1). Furthermore, the half-life of ^{133}Ba is long (10.5 years) compared with ^{131}I (8 days), so that a certified calibration standard containing ^{133}Ba can be used for many years.

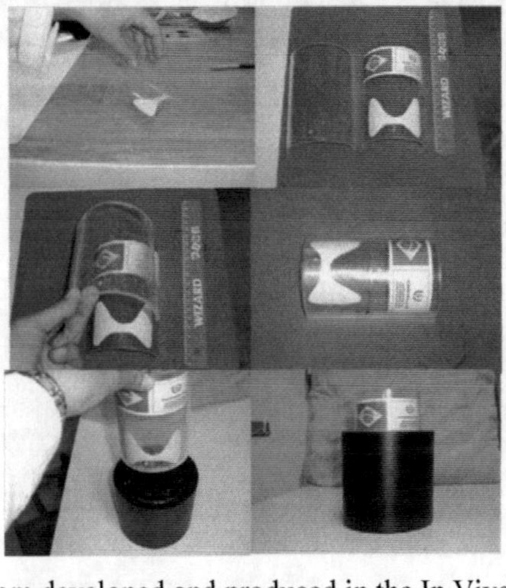

Figure 1. Neck-thyroid phantom developed and produced in the In Vivo Monitoring Laboratory of IRD [6].

Table 1. Energy photons and emission intensity for ^{133}Ba e ^{131}I

Energies γ ^{133}Ba (keV)	**Emission Intensity γ**
276.39	0.071
302.85	0.183
356.01	0.620
383.85	0.089
Energies γ ^{131}I (keV)	**Emission Intensity γ**
284.30	0.061
364.48	0.817

2.2. Portable surface contamination monitors

The table 2 presents a list of sixteen equipment evaluated in this work that were been previously calibrated at the *Laboratório Nacional de Metrologia das Radiações Ionizantes* (LNMRI-IRD) and at the *Laboratório de Ciências Radiológicas* (LCR-UERJ).

Table 2. List of Portable monitors evaluated.

Trade Mark	Model	Quantity	Type
Eberline	E-120	1	Geiger-Muller
Prólogo	PSN-7013	1	Geiger-Muller
Berthold	LB-124 Scint	1	Scintillator
IEN	MIR-7026	2	Geiger-Muller
MRA	GP-500	1	Geiger-Muller
Polimaster	PM 1400	1	Geiger-Muller
Thermo Scient.	RadEye B20	1	Geiger-Muller
Tech. Associates	PUG-7A	1	Geiger-Muller
Dosimeter	3007A	1	Geiger-Muller
Ludlum	Model 3	3	Geiger-Muller
Thermo Scient.	Identifinder	1	Scintillator
SE International	Inspector Exp	1	Geiger-Muller
Saphimo	Mini Trace	1	Geiger-Muller

3. Methods

3.1. Calibration of the monitors

The calibration procedure was conducted as follows:

Step 1: The ^{133}Ba activity content of the phantom was corrected taking into account the time elapsed between fabrication and calibration dates and the half-life of ^{133}Ba. The ^{131}I equivalent activity was then calculated according to equation (1)

$$\text{At Eq }^{131}\text{I}_{Bq} = \text{A }(^{133}\text{Ba}) \text{ x } \frac{\Sigma\,(\gamma\,^{133}\text{Ba})}{\Sigma\,(\gamma\,^{131}\text{I})} \qquad (1)$$

Where:

At Eq ^{131}I = Equivalent activity of ^{131}I ;
A (^{133}Ba) = Activity of ^{133}Ba (Bq) present in the phantom;
Σ (γ ^{133}Ba) = Sum of emission intensities γ of ^{133}Ba; and
Σ (γ ^{131}I) = Sum of emission intensities γ of ^{131}I.

Step 2: The measurement setup was established as shown in Figure 2, and the phantom count rate was recorded over five sequential measurements at standard geometries of 0 and 3 cm between the detector front face and the phantom surface.

Figure 2. Calibration setup applied in four detector models using a Neck-thyroid phantom.

Step 3: A blank phantom was measured in five sequential counts in the same setup for background account.

Step 4: The calibration factors were calculated for the standard geometry according to equation (2):

$$CF_{cpm/Bq} = \frac{\text{cpm}}{A_{\text{calib}}} \tag{2}$$

Where:
cpm = net count rate (total cpm of phantom subtracted by background count rate); and
A_{calib} = ^{131}I equivalent activity content of the phantom.

3.2. Evaluation of sensitivity
The evaluation of the sensitivity of the method for its application in routine internal monitoring is based on the calculation of three parameters: (i) Minimum Detectable Activity (MDA); (ii) Minimum Detectable Intake (MDI) and; (iii) Minimum Detectable Effective Dose (MDED).

The MDA of the method is calculated as follows [7]:

$$MDA_{Bq} = \frac{4.65 \text{ x } \sigma_{bg}}{CF} \tag{3}$$

Where: σ_{bg} = Standard deviation of the BG (counts) in one minute; and
CF = Calibration Factor (cpm/Bq).

The MDI is a function of the MDA and depends on the exposure scenario and time, in days, elapsed between intake and *in-vivo* measurement. In this work MDI and MDED where calculated for retention fractions "m(t)" values of 1 and 7 days. The Minimum Detectable Intake (MDI) is calculated as follows:

$$\mathrm{MDI}_{Bq} = \mathrm{MDA} / \mathrm{m(t)}_{inh\ or\ ing} \tag{4}$$

Where: MDA = Minimum Detectable Activity (Bq); and
m(t) = Retention fraction in the compartment of interest for inhalation or ingestion (Bq/Bq).

The last parameter to be calculated is the Minimum Detectable Effective Dose, based on the MDI, considering the dose coefficients associated to the corresponding intake scenario adopted in the simulation. It is calculated as follows:

$$\mathrm{MDED}_{mSv} = \mathrm{MDI}_{inh\ or\ ing} \times \mathrm{e(g)}_{inh\ or\ ing} \tag{5}$$

Where: MDI = Minimum Detectable Intake (Bq); and
$e(g)_{inh\ or\ ing}$ = Dose coefficient (mSv/Bq).

The values of "m(t)" and "e(g)", presented in table 3, are available in the Publication 78 of the ICRP [9] and may also be generated for specific exposure scenarios and times after intake through the software AIDE [10]. Figure 3 illustrates an example of the AIDE software main screen

Table 3. Retention fractions and dose coefficients generated with the software AIDE.

m(t) (Bq/Bq)				**e(g) (mSv/Bq)**	
1 day		**7 days**		Inh	Ing
Inh	Ing	Inh.	Ing		
0.229	0.252	0.229	0.252	1.98×10^{-5}	2.17×10^{-5}

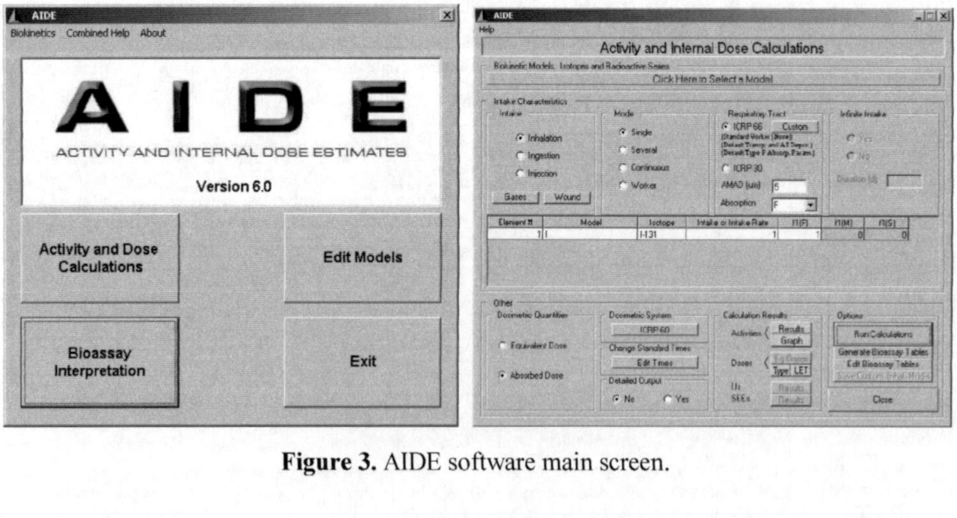

Figure 3. AIDE software main screen.

In order to be considered useful for internal monitoring purposes, the technique should, at least, be able to detect an activity that would result in an effective dose below 1 mSv per year for the most likely internal exposure scenario [8]. Such evaluation is based on the annual projection of MDED considering the sensitivity of the method and frequency of monitoring adopted. The evaluation is carried out by adopting a generic monitoring routine, being realistic with handling procedures routine of ^{131}I in nuclear medicine facilities where radioiodine therapy is performed. The routine is set to a worker that handles ^{131}I once a week for 11 months throughout the year, resulting in a total of 48 monitoring periods. The evaluation of applicability is obtained by the following equation:

$$\mathrm{PCED}_{mSv} = \mathrm{MDED}_{inh\ or\ ing}\ \mathrm{x\ N} \tag{6}$$

Where: PCED = Projected Committed Effective Dose for 1 year; MDED = Minimum Detectable Effective Dose; and N = number of monitoring periods throughout 1 year.

4. Results

Two thyroid phantoms, with distinct ^{133}Ba activities, were used in this work. Phantom #1 and #2 contain, respectively, 15211 Bq (10/06/2011) and 4531 Bq (17/09/2013). The calculation of ^{131}I equivalent activities, as described above, resulted in the following values in the date of the calibration of the monitors: Phantom #1: 16684 Bq and Phantom #2: 4970 Bq.

Tables 4 and 5 present the results of the counts, calibration factors, minimum detectable activities, minimum detectable intakes for each detector and phantom used in each calibration geometry of 0 and 3 cm respectively. Detector count rates were recorded in cpm for harmonization purposes since some models present output in cpm and others in cps (values in cps were multiplied by 60). The values of MDI correspond to an intake by inhalation considering the time of 1 day elapsed between intake and measurement.

Table 4. Results of the calibration performed at the geometry of 0 cm.

Detector	Phantom used	Counts (cpm)	CF (cpm/Bq)	MDA (Bq)	MDI (Bq)
PM 1400	#1	226	0.0173	783	3421
MIR-7026	#1	238	0.0183	4097	17890
RadEye B20	#1	339	0.0261	1677	7323
PUG-7A	#1	433	0.0334	3611	15769
3007A	#1	333	0.0257	3140	13711
Model 3 #1	#1	433	0.0334	1196	5224
Model 3 #2	#1	333	0.0258	3975	17359
Model 3 #3	#1	329	0.0255	1557	6798
Identifinder	#1	53048	4.1366	347	1516
Inspector Exp	#1	306	0.0261	1964	8575
Mini Trace	#1	428	0.0365	1329	5802

Figure 4 shows the results of MDA the geometry of 3 cm, as a parameter for sensitivity comparison. It is possible to observe a significant difference of sensitivity between the LB-124 Scint and Identifinder compared with other detectors evaluated. This is related to the fact that these two models are scintillation detectors, made of ZnS:Ag (zinc sulphide doped with silver) and NaI (sodium iodide doped with thallium), respectively. Because of its intrinsic characteristics, scintillators detectors are significantly more sensitive to photons *Geiger Miller* detectors.

Table 5. Results of calibration performed at the geometry of 3 cm.

Detector	Phantom used	Counts (cpm)	CF (cpm/Bq)	MDA (Bq)	MDI (Bq)
E-120	#1	100	0.0069	6708	29294
GP-500	#2	73	0.0153	3782	27347
PSN-7013	#2	48	0.0100	3844	16788
MIR-7026 #1	#2	33	0.0069	6094	18476
LB-124 Scint	#2	4020	0.8439	232	1014
PM 1400	#1	102	0.0078	1755	7664
MIR-7026 #2	#1	125	0.0096	7218	31519
RadEye B20	#1	174	0.0133	3237	14136
PUG-7A	#1	262	0.0202	5942	25948
3007A	#1	200	0.0154	5097	22258
Model 3 #1	#1	116	0.0089	4396	19197
Model 3 #2	#1	160	0.0124	7430	32446
Model 3 #3	#1	146	0.0113	3615	14120
Identifinder	#1	25811	2.0127	714	3117
Inspector Exp	#1	151	0.0128	3992	17434
Mini Trace	#1	171	0.0146	3322	14505

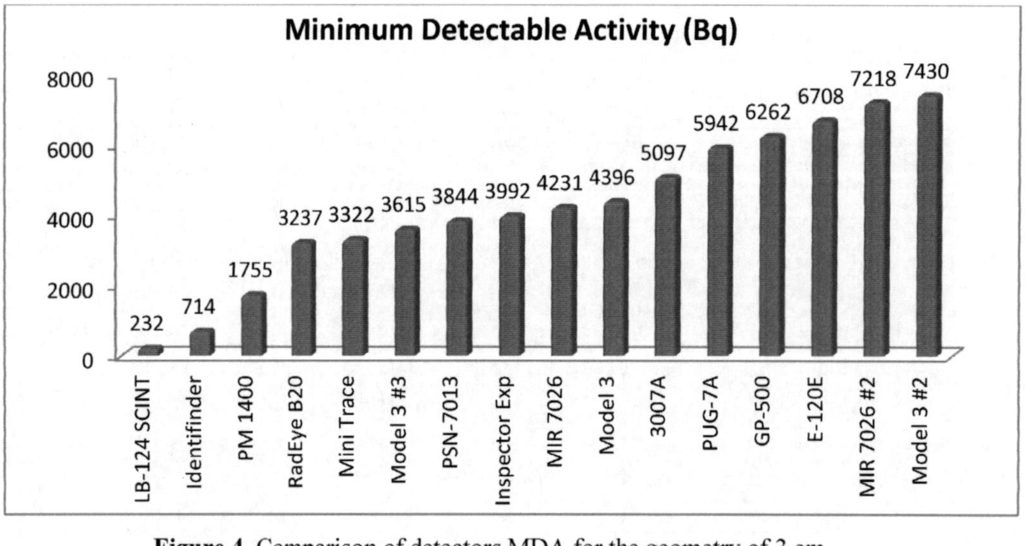

Figure 4. Comparison of detectors MDA for the geometry of 3 cm.

In internal dosimetry, the evaluation of sensitivity of detection system is performed based on MDA value, with the following limits calculated on the basis of this initial parameter and the possible exposure scenarios.

Tables 6 and 7 present the values of MDED calculated according to the methodology described previously. These values correspond to the geometries of 0 cm and 3 cm, considering times of 1 and 7 days elapsed between intake and measurement.

Table 6. Minimum Detectable Effective Doses for 1 and 7 days after intake. Measurements performed at the geometry of 0 cm.

Detector	MDED (mSv) – 0 cm			
	1 day		7 days	
	Inh	Ing	Inh	Ing
PM 1400	0.068	0.067	0.112	0.110
MIR 7026	0.354	0.353	0.584	0.577
RadEye B20	0.145	0.144	0.239	0.236
PUG-7A	0.312	0.311	0.514	0.509
3007A	0.271	0.270	0.447	0.442
Model 3 #1	0.103	0.103	0.170	0.169
Model 3 #2	0.344	0.342	0.566	0.560
Model 3 #3	0.135	0.134	0.222	0.219
Identifinder	0.030	0.030	0.049	0.049
Inspector Exp	0.170	0.169	0.280	0.277
Mini Trace	0.115	0.114	0.189	0.187

Table 7. Minimum Detectable Effective Doses for 1 and 7 days after intake. Measurements performed at the geometry of 3 cm.

Detector	MDED (mSv) – 3 cm			
	1 day		7 days	
	Inh	Ing	Inh	Ing
E-120	0.580	0.578	0.956	0.945
GP-500	0.541	0.549	0.892	0.882
PSN-7013	0.332	0.331	0.548	0.542
MIR 7026 #1	0.366	0.364	0.603	0.596
LB-124 SCINT	**0.020**	**0.020**	**0.033**	**0.033**
PM 1400	0.152	0.151	0.250	0.247
MIR 7026 #2	0.624	0.622	1.028	1.017
RadEye B20	0.280	0.279	0.461	0.456
PUG-7A	0.514	0.512	0.846	0.837
3007A	0.441	0.439	0.726	0.718
Model 3 #1	0.380	0.379	0.626	0.619
Model 3 #2	0.642	0.640	1.058	1.047
Model 3 #3	0.313	0.311	0.515	0.509
Identifinder	0.062	0.061	0.102	0.101
Inspector Exp	0.345	0.344	0.569	0.563
Mini Trace	0.287	0.286	0.473	0.468

In tables 6 and 7, it can be observed that there are not much significant difference between values of MDED calculated assuming inhalation or ingestion. This fact is related to the biokinetic and dosimetric models of ^{131}I and reduces the uncertainty on internal dose estimations.

The figure 5 shows the results of MDED in decreasing order of sensitivity for geometry of 3 cm, which all detectors were evaluated.

Table 8 shows the values of projected committed effective dose for 1 year, calculated by applying the methodology described previously.

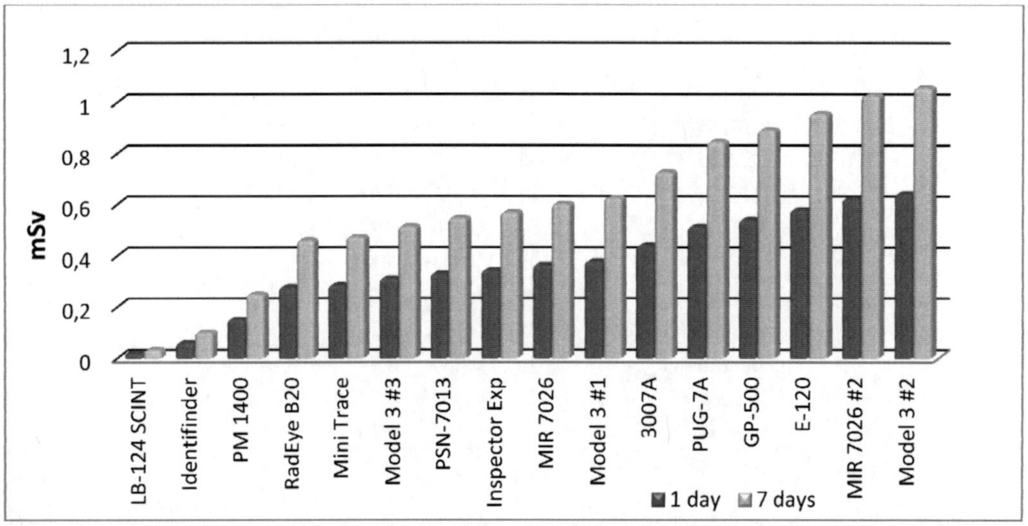

Figure 5. MDED values for 1 and 7 days after the ^{131}I intake, in descending order of sensitivity, for geometry of 3 cm.

Table 8. Projected Committed Effective Dose for the geometry of 0 cm

Detector	Projected Committed Effective Dose (mSv)			
	1 day		7 days	
	Inhalation	Ingestion	Inhalation	Ingestion
PM 1400	3.3	3.2	5.4	5.3
MIR 7026 #2	17.0	17.0	28.1	27.7
RadEye B20	7.0	7.0	11.5	11.3
PUG-7A	15.0	15.0	24.7	24.4
3007A	13.0	13.0	21.5	21.2
Model 3 #1	5.0	4.9	8.2	8.1
Model 3 #2	16.5	16.4	27.2	26.9
Model 3 #3	6.5	6.4	10.7	10.5
Identifinder	1.4	1.4	2.4	2.4
Mini Trace	5.5	5.5	9.1	9.0
Inspector Exp	8.2	8.1	13.4	13.3

Considering the scenario where the *in vivo* measurement is performed 1 day after the intake, Detector Berthold LB-124-SCINT presented the highest sensitivity among all models evaluated in this work (0.96 mSv at geometry 3 cm) and is the only one suitable for application in routine occupational monitoring.

Figure 6 shows a comparison of the detectors performance at geometry 0 cm, including detector LB-124SCINT at geometry 3 cm. The monitoring frequency must be established so that lost dose fractions in each single measurement projected over a period of 1 year does not exceed 1 mSv. In the evaluation performed in this work, the detectors should present MDEDs below 0.02 mSv which is the fraction of 1 mSv in 48 monitoring periods over 1 year.

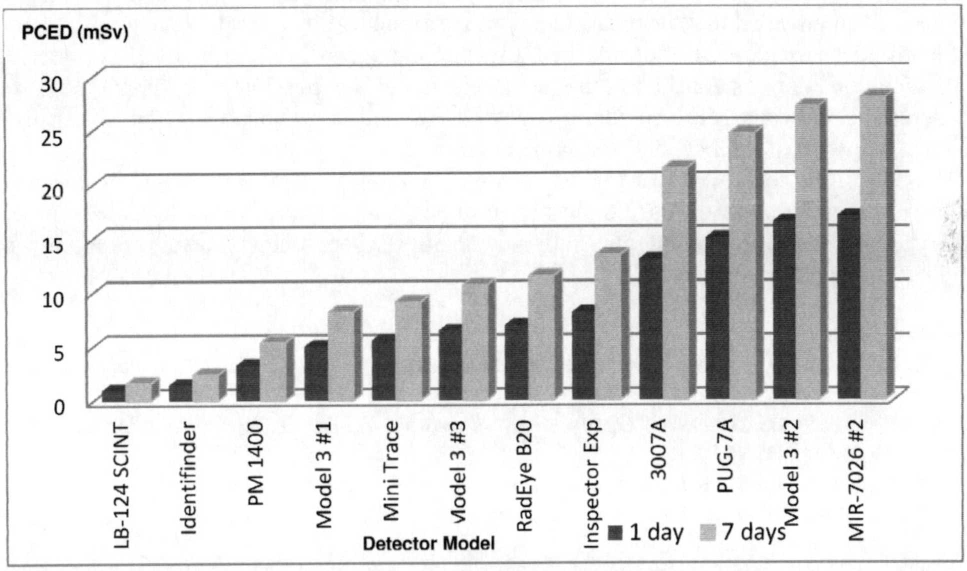

Figure 6. Comparison of detectors performance in terms of Projected Committed Effective Doses (mSv). Scenario: Measurement Geometry 0 cm; Intake via inhalation; In vivo measurement performed 1 and 7 days after the intake.

5. Conclusions

This work demonstrated technical and economic viability of using surface contamination detector for the control of internal exposures of workers in the practice of nuclear medicine. However it has been observed significant differences in sensitivity among the models evaluated. At the standard geometry of 3 cm, the detector Berthold LB-124 SCINT has shown to be the most sensitive, presenting a MDED of approximately 0.02 mSv, for a monitoring scheme supposing an in vivo measurement performed one day after intake by inhalation or ingestion, and approximately 0.03 mSv if the measurement is performed seven days after the intake.

Such performance is related to the higher sensitivity of detectors using ZnS:Ag scintillation crystals, resulting in a higher efficiency for photons if compared to a *Geiger-Müller* detetor.

It was also observed that the Ludlum Model 3 #2, which is the oldest model included in this comparison presented the lowest sensitivity, being able to detect approximately 0.64 and 1.05 mSv for one and seven days after the intake respectively at the geometry of 3 cm.

However, even considering the differences in sensitivity it can be stated that practically all detectors are suitable for application in special in vivo monitoring of the thyroid at the geometries of 0 and 3 cm.

This conclusion relies on the fact that all models presented enough sensitivity at 0 cm, as far as the in vivo measurement is performed up to seven days after the intake. In such monitoring scheme the MDA results in committed effective doses below the registry level of 1 mSv.

On the other hand, at the geometry of 3 cm, Ludlum Model 3 #2 and MRA-MIR 7026 #2 did not present enough sensitivity for measurements performed seven days after the intake, with MDED slightly above 1 mSv.

Based on the results obtained in this work, for occupational routine monitoring, only model Berthold LB-124-SCINT attended the sensitivity criterion, showing enough sensitivity at geometry 3 cm provided that the measurement is performed 1 day after the intake.

Identifinder detector has not reached the minimum sensitivity required, considering the calibration conditions established in this work. However, it is possible to optimize the procedure by increasing the count time so that a lower MDA, and consequently a higher sensitivity, is obtained, allowing its application for routine monitoring.

Thus, it can be concluded that the use of portable monitors is a suitable and inexpensive tool for thyroid monitoring in the case of suspicion of accidental intakes in nuclear medicine clinics, radiopharmaceutical production plants or any facility where occupationally exposed workers handle significant activities of ^{131}I.

References

[1] CNEN 2015 *Instalações Autorizadas*. http://www.cnen.gov.br/seguranca/cons-ent-prof/entidades-aut-cert.asp

[2] IAEA 1999 *Assessment of Occupational Exposure due to Intakes of Radionuclides* (Safety Guide No. RS-G-1.2)

[3] CNEN 2011 *Diretrizes Básicas de Radioproteção* (Norma CNEN-NE-3.01)

[4] IAEA 1996 *Direct methods for measuring radionuclides in the human body* (Safety Series n. 115)

[5] CNEN 2011 *Critérios para cálculo de dose efetiva, a partir da monitoração individual* (Posição Regulatória 3.01/005)

[6] Dantas B M, Cardoso J S, Dantas A L A, Lucena E A, Ramos M A P, Sá M S, Alonso T C, Silva T V, Oliveira C M, Lima F F, Oliveira M L, Lacerda I V B, Fajgel A 2013 Intercomparação Nacional de Medição In Vivo de Iodo-131 na Tireoide Projeto TC IAEA BRA 9055. *Sci. Plen.* 9(8)

[7] HPS 1996 *Performance Criteria for Radiobioassay* (Health Physics Society) N13.30

[8] IAEA 2004 *Methods for Assessing Occupational Radiation Doses due to Intakes of Radionuclides* (Safety Reports Series No. 37)

[9] ICRP 1998 *Individual Monitoring for Internal Exposure of Workers* (ICRP Publication 78) vol. 27/3-4

[10] Bertelli L, Melo D R, Lipsztein J and Cruz-Suarez R 2008 AIDE – Internal Dosimetry Software *Rad. Prot. Dos*. 130(3), p. **358**-367